Interdisciplinary Biotechnological Advances

Series Editors

Jayanta Kumar Patra, Research Institute of Integrative Life Sciences
Dongguk University
Ilsandong, Goyang, Korea (Republic of)

Gitishree Das, Research Institute of Integrative Life Sciences
Dongguk University
Ilsandong, Goyang, Korea (Republic of)

This series is an authoritative resource that explains recent advances and emerging directions in biotechnology, reflecting the forefront of research clearly and reliably, without excessive hype. Each volume is written by authors with excellent reputations and acknowledged expertise in the topic under discussion. The volumes span the entire field from an interdisciplinary perspective, covering everything from biotechnology principles and methods to applications in areas including genetic engineering, transgenic plants and animals, environmental problems, genomics, proteomics, diagnosis of disease, gene therapy, and biomedicine. The significance of these applications for the achievement of UN Sustainable Development Goals is highlighted. The series will be highly relevant for Master's and PhD students in Biotechnology, Nanochemistry, Biochemical Engineering, and Microbiology, medical students, academic and industrial researchers, agricultural scientists, farmers, clinicians, industry personnel, and entrepreneurs.

A.H.M. Nurun Nabi • Hossain Uddin Shekhar
Editors

Biotechnological Advances in Healthomics

Springer

Editors
A.H.M. Nurun Nabi
Department of Biochemistry
and Molecular Biology
University of Dhaka
Dhaka, Bangladesh

Hossain Uddin Shekhar
Department of Biochemistry
and Molecular Biology
University of Dhaka
Dhaka, Bangladesh

ISSN 2730-7069 ISSN 2730-7077 (electronic)
Interdisciplinary Biotechnological Advances
ISBN 978-981-95-3611-5 ISBN 978-981-95-3612-2 (eBook)
https://doi.org/10.1007/978-981-95-3612-2

This Springer imprint is published by the registered company Springer Nature Singapore Pte Ltd.
The registered company address is: 152 Beach Road, #21-01/04 Gateway East, Singapore 189721, Singapore

Foreword

The turn of the twenty-first century has brought significant changes to life sciences, driven by the merging of biotechnology and the “omics” revolution. Genomics, proteomics, metabolomics, transcriptomics, and related fields have transformed our understanding of biology. They have also changed how we approach health, disease, and medicine. This combined field, often called healthomics, represents more than just a set of technologies; it marks a shift from reactive treatment to predictive, preventive, and personalized strategies in healthcare.

The book, *Biotechnological Advances in Healthomics*, comes at a crucial time. As societies face emerging infectious diseases, antimicrobial resistance, chronic noncommunicable diseases, and the challenges of aging populations, the need for broad, systems-level approaches has never been more urgent. Healthomics offers a framework that enables the exploration of genes, proteins, metabolites, and microbes, along with their interactions in human physiology and ecosystems.

This volume features a wide range of scholars and practitioners who share both foundational and cutting-edge expertise. The chapters cover a diverse range of topics. Readers will learn about natural products research driven by metabolomics and microbial genomics and its significant effects on public health, the role of synthetic biology, genome editing in therapeutic development, and how artificial intelligence can help interpret large and complex datasets. Together, these contributions depict the current state of the field and possible future directions.

What sets this book apart is not just the variety of topics but also its focus on practical applications. The insights presented here showcase real ways to turn data into diagnostics, discoveries into therapies, and innovations into better health outcomes. The volume also highlights the global significance of healthomics, pointing out the challenges and opportunities for countries like Bangladesh and other resource-limited areas.

This reinforces the idea that scientific progress should benefit all of humanity rather than just a select few. The editors, Professors A.H.M. Nurun Nabi and Hossain Uddin Shekhar, deserve special recognition for putting together this timely and important work. By uniting diverse voices, they have created a platform that reflects the collaborative nature of modern science and the urgent need for interdisciplinary discussions. Their vision and dedication shine through on every page.

For students, this book will guide them into a dynamic and fast-changing field. For researchers, it provides a focused resource on current methods and new frontiers. For policymakers and healthcare professionals, it makes a strong case for investing in and using omics-based approaches to improve public health. In a time when health challenges are increasingly complex, interconnected, and global, *Biotechnological Advances in Healthomics* is not just timely, but essential. I believe this book will inform and inspire its readers, sparking further research, encouraging new collaborations, and ultimately helping to create a healthier world.

Harvard Medical School
Boston, MA, USA

A. K. M. Khyrul Wara

Preface

In recent decades, biotechnology has changed dramatically due to the rapid growth of "omics" technologies, including genomics, transcriptomics, proteomics, metabolomics, and more. These approaches, known as *healthomics*, allow researchers to explore biological systems in ways that were once unimaginable. The combination of high-throughput data generation with bioinformatics and machine learning is now transforming biomedical research, public health, and clinical practice. This volume, *Biotechnological Advances in Healthomics*, aims to bring together current knowledge, showcase innovations, and outline future research directions.

The main goal of the book is to present a variety of views on how *healthomics* applies to medicine, biotechnology, and global health. Each chapter, written by experts in the field, summarizes the current state of the art while also highlighting new opportunities and challenges in their areas. This collective effort embodies the book's main philosophy: to improve health through data-driven, biotechnological innovations.

The volume covers many themes. The early chapters focus on natural products and metabolomics. They explain how scientists can identify, characterize, and develop secondary metabolites and bioactive compounds using modern omics platforms. Subsequent chapters explore microbial genomics, showcasing how genomic surveillance and metagenomics have improved our ability to monitor infectious diseases, tackle antimicrobial resistance, and understand the complex relationships between microbes and human health. Other chapters discuss how omics integrates with synthetic biology, artificial intelligence, and precision medicine, providing insights into how these combined approaches may transform therapies, diagnostics, and personalized healthcare.

Overall, the chapters highlight several common themes. First, the convergence of omics technologies provides a comprehensive view of biological complexity. Researchers can now reconstruct entire pathways and systems instead of just focusing on individual genes, proteins, or metabolites. Second, improvements in technology are making these tools more accessible. Platforms like next-generation sequencing, CRISPR editing, and AI-powered analytics are now available to more researchers, including those in low- and middle-income countries. This broader access is essential for tackling global health challenges, from infectious disease outbreaks to noncommunicable disorders. Third, the contributions highlight that *healthomics* goes beyond scientific inquiry; its real value lies in informing diagnostics, shaping therapies, and enhancing human well-being.

This book is aimed at a wide audience. For researchers, it serves as a reference point and a starting place for future studies. For students, it offers a straightforward introduction to emerging technologies and their real-world health applications. For clinicians, policymakers, and biotechnologists, it provides a framework for understanding how integrative science can lead to practical solutions. By combining molecular biology, computational advances, and applied biotechnology, this volume seeks to bridge the gap between laboratory discovery and healthcare innovation.

Editing this book has been both challenging and rewarding. Coordinating contributions from authors across various disciplines, institutions, and countries took significant effort, but the end result reflects the collaborative spirit of modern science. Each chapter carries the unique expertise and perspective of its authors, yet all share a common goal: to enhance the role of biotechnology in creating healthier societies.

We are very thankful to the contributors who shared their knowledge and time to make this project possible. Their chapters showcase not only technical expertise but also a real passion for advancing science for the good of humanity. We also acknowledge the institutions and research groups that supported this work, along with the reviewers whose feedback helped improve the content. Our heartfelt thanks go to the publishing team for their professional guidance throughout the editorial process.

Finally, we hope that *Biotechnological Advances in Healthomics* will encourage readers to think broadly, collaborate widely, and innovate boldly. The journey in *healthomics* has just begun. As technologies develop and the lines between fields blur, the chances to improve health through biotechnology will only grow. We invite readers to engage thoughtfully with the ideas in this book, to apply them in their own work, and to join the effort to enhance global health.

Dhaka, Bangladesh

A.H.M. Nurun Nabi
Hossain Uddin Shekhar

Acknowledgments

We sincerely thank all the authors for their valuable contributions to the chapters. The journey from initial outlines to developed chapters took time, and we truly appreciate the dedication and patience everyone showed during the revision process in response to reviewers' comments. We are also thankful to the members of our review panels. Their expertise and thoughtful feedback, often given on short notice, greatly improved this work. Our heartfelt thanks go to the whole team of Springer Nature, especially Kavitha Jayakumar, Mei Hann Lee, Sean Eickhoff and J. RekhaKumar. We also extend our thanks to Jayanta Kumar Patra, Series Editor of Interdisciplinary Biotechnological Advances, for his kind cooperation during the process. Their guidance, support, and help were crucial in making this book a reality. Last but not least, we are profoundly grateful to our families for their constant patience and unfailing support.

Acknowledgments

Contents

Editors and Contributors

About the Editors

A.H.M. Nurun Nabi is currently working as a Professor at the Department of Biochemistry and Molecular Biology, University of Dhaka, Bangladesh. He obtained his PhD degree from Gifu University, Japan, under Monbukagakusho scholarship offered by the Japanese Ministry of Education, Culture, Sports, Science, and Technology (MEXT). He has over 25 years of teaching and research experience. His extensive professional engagement reflects both leadership and versatility. He has served as visiting scholar in Japan and South Africa, acted as student advisor, and contributed to academic quality assurance at the University of Dhaka. Importantly, his role on multiple editorial boards (such as the *Asian-Australasian Journal of Bioscience and Biotechnology*; *International Journal of Biochemistry & Physiology*; *The World Journal of Biological Chemistry*) directly illustrates his editorial competence. He is also an international Advisory Board Member of *Reviews in Agricultural Sciences*; life member of the Bangladesh Society of Biochemistry and Molecular Biology, Graduate Biochemist Association of Bangladesh, Bangladesh Biosafety and Biosecurity Society, and Bangladesh Association of Clinical Biochemists; and member of Asiatic Society of Bangladesh. He has authored more than 100 articles, in national and international peer-reviewed journals. Moreover, his research spans cutting-edge fields such as protein biochemistry, functional genomics, diabetes, cardiovascular diseases, and COVID-19 genetics. In addition to research and editorial strengths, he has shown mentorship by supervising 4 PhD, 2 MPhil, and 36 MS students so far. Currently, his Google Scholar citation is more than 3200 with an h-index of 25 and an i10-index of 45.

Hossain Uddin Shekhar is a Professor in the Department of Biochemistry and Molecular Biology at the University of Dhaka. He earned his PhD in 2003 from Tokyo Metropolitan University and completed his postdoctoral work in 2010 at United Nations University. He has numerous publications in Q1 journals. He is also a member of the editorial boards of *PLOS One*, *Frontiers in Molecular Biosciences*,

BioMed Research International, and Springer Nature. In 2004, he jointly published a paper on gene-deficient yeast with Nobel Laureate Yoshinori Ohsumi. In 2012, he published his first book, *Functional Foods from Bangladesh*. He has also co-edited books for publishers such as Springer Nature, Wiley & Sons, Frontiers in Molecular Biosciences, and Nova Science Publishers. Under his leadership, the Consumers Association of Bangladesh conducted a two-year research project on the institutionalization of healthy street food. He is the president of VOCTA (Voluntary Consumers Training and Awareness Society) and the general secretary of BSBMB (Bangladesh Society for Biochemistry and Molecular Biology). He is also serving as the chief editor of Bioresearch Communications (BRC) and as an editorial board member of DUJBS (*Dhaka University Journal of Biological Sciences*). He is currently serving as the Vice Chancellor of Gopalganj Science and Technology University (GSTU).

Contributors

Rownock Afruza Translational Hepatology Section, National Institute of Diabetes and Digestive and Kidney Diseases, National Institutes of Health, Bethesda, MD, USA

Hosneara Akter Genetics and Genomic Medicine Centre, NeuroGen Healthcare, Dhaka, Bangladesh

Laboratory of Population Genetics, Department of Biochemistry and Molecular Biology, University of Dhaka, Dhaka, Bangladesh

Mohammad Al Mamun Department of Chemistry, Jagannath University, Dhaka, Bangladesh

Mohammad Ali Nuffield Department of Clinical Medicine, University of Oxford, Oxford, UK

Geoffroy Andrieux Institute of Medical Bioinformatics and Systems Medicine, Faculty of Medicine, University of Freiburg, Freiburg, Germany

Anowara Begum Department of Microbiology, University of Dhaka, Dhaka, Bangladesh

Partha Biswas Division of Medical Oncology, Department of Internal Medicine, The Ohio State University College of Medicine, Columbus, OH, USA

Pelotonia Institute for Immuno- Oncology, The Ohio State University Comprehensive Cancer Center, The Arthur G James Cancer Hospital and Richard J Solove Research Institute, Columbus, OH, USA

Alexandra E. Butler Research Department, Royal College of Surgeons in Ireland, Bahrain, Adliya, Bahrain

Sajib Chakraborty Institute of Medical Bioinformatics and Systems Medicine, Faculty of Medicine, University of Freiburg, Freiburg, Germany

Dipankor Chatterjee Department of Molecular Genetics, The Ohio State University, Columbus, OH, USA

Radwan Darwish School of Medicine, Royal College of Surgeons in Ireland, Bahrain, Adliya, Bahrain

Arindita Das Department of Genetic Engineering and Biotechnology, University of Dhaka, Dhaka, Bangladesh

Tonmoy Das Institute of Medical Bioinformatics and Systems Medicine, Faculty of Medicine, University of Freiburg, Freiburg, Germany

Abu Hashem Microbial Biotechnology Division, National Institute of Biotechnology, Dhaka, Bangladesh

Misuzu Hashimoto Laboratory of Biological Chemistry, Faculty of Applied Biological Sciences, Gifu University, Gifu, Japan

Md. Ismail Hosen Department of Biochemistry and Molecular Biology, University of Dhaka, Dhaka, Bangladesh

Md. Raiyan Hosen Department of Biochemistry and Molecular Biology, University of Dhaka, Dhaka, Bangladesh

Anamica Hossain Department of Microbiology, University of Dhaka, Dhaka, Bangladesh

Mohammad Nazir Hossain Department of Genetic Engineering and Marine Biotechnology, Bangladesh Maritime University, Dhaka, Bangladesh

Mohammed Akhter Hossain The Florey, University of Melbourne, Parkville, VIC, Australia

Department of Biochemistry and Pharmacology, University of Melbourne, Parkville, VIC, Australia

Md. Jahirul Islam Molecular Biotechnology Division, National Institute of Biotechnology, Dhaka, Bangladesh

Md. Shahidul Islam Discipline of Biochemistry, School of Agriculture and Science, University of KwaZulu-Natal, Durban, South Africa

Christopher Krolla Institute of Medical Bioinformatics and Systems Medicine, Faculty of Medicine, University of Freiburg, Freiburg, Germany

Ivan Maslov Florey Institute of Neuroscience and Mental Health, University of Melbourne, Parkville, VIC, Australia

Department of Biochemistry and Pharmacology, University of Melbourne, Parkville, VIC, Australia

Abu Saleh Md Moin Research Department, Royal College of Surgeons in Ireland, Bahrain, Adliya, Bahrain

A. H. M. Nurun Nabi Laboratory of Population Genetics, Department of Biochemistry and Molecular Biology, University of Dhaka, Dhaka, Bangladesh

Manjula Nandakumar Research Department, Royal College of Surgeons in Ireland, Bahrain, Adliya, Bahrain

Noor-E-Meem Department of Microbiology, University of Dhaka, Dhaka, Bangladesh

Md Abdur Rahim Human Microbiome Medical Research Center (HM·MRC), School of Medicine, Soonchunhyang University, Asan-si, Republic of South Korea

Neervana Rambaran Discipline of Biochemistry, School of Agriculture and Science, University of KwaZulu-Natal, Durban, South Africa

Isabelle Riches Florey Institute of Neuroscience and Mental Health, University of Melbourne, Parkville, VIC, Australia

Department of Biochemistry and Pharmacology, University of Melbourne, Parkville, VIC, Australia

Hossain Uddin Shekhar Department of Biochemistry and Molecular Biology, University of Dhaka, Dhaka, Bangladesh

Atsuhiro Shimada Biomolecular Chemistry Group, Department of Applied Life Science, Faculty of Applied Biological Sciences, Gifu University, Gifu, Japan

Mohammad Umer Sharif Shohan Center for Human Genetics, University of Oxford, Oxford, UK

Mohammad Kawsar Sharif Siam School of Pharmacy, Brac University, Dhaka, Bangladesh

Gazi Nurun Nahar Sultana Centre for Advanced Research in Sciences (CARS), University of Dhaka, Dhaka, Bangladesh

Munawar Sultana Department of Microbiology, University of Dhaka, Dhaka, Bangladesh

Jannatul Nayeem Tabia Late Phase Haematology, Oxford University Hospital NHS Foundation Trust, Oxford, UK

Andreas Tsouris Institute of Medical Bioinformatics and Systems Medicine, Faculty of Medicine, University of Freiburg, Freiburg, Germany

Mohammed Uddin GenomeArc Inc., Mississauga, ON, Canada

Center for Applied and Translational Genomics (CATG), Mohammed Bin Rashid University of Medicine and Health Sciences, Dubai Health, Dubai, UAE

College of Medicine, Mohammed Bin Rashid University of Medicine and Health Sciences, Dubai, UAE

Diana Vidovic Florey Institute of Neuroscience and Mental Health, University of Melbourne, Parkville, VIC, Australia

Department of Biochemistry and Pharmacology, University of Melbourne, Parkville, VIC, Australia

M. R. Washikur Department of Genetic Engineering and Marine Biotechnology, Bangladesh Maritime University, Dhaka, Bangladesh

Hongkang Wu The Florey, University of Melbourne, Parkville, VIC, Australia

Department of Biochemistry and Pharmacology, University of Melbourne, Parkville, VIC, Australia

Kosei Yamauchi The United Graduate School of Agricultural Science, Gifu University, Gifu, Japan

Graduate School of Natural Science and Technology, Gifu University, Gifu, Japan

Faculty of Applied Biological Sciences, Gifu University, Gifu, Japan

Center for One Medicine Innovative Translational Research (COMIT), Institute for Advanced Study, Gifu University, Gifu, Japan

High-Throughput Omics Data and Bioactive Compounds from Natural Products

Kosei Yamauchi

Abstract Natural products such as medicinal plants, foods, and fungi have been traditionally used for human health. Bioactive ingredients in the natural products have been identified to search for the mechanism, and the unique structures of active compounds have been elucidated so far. Recently, the omics techniques were used to search for secondary metabolites and to conduct a comprehensive analysis of the mechanism of bioactivity. The target proteins could be identified using thermal shift assay with the proteomics analysis using LC–MS^2. We could also comprehensively analyze secondary metabolites by using the database of the fragmentations of LC–MS^2. These innovative methods make it possible to elucidate the ingredients and mechanism effectively and quickly. Recent studies on isolation and identification of compounds from natural products and the search for the mechanism were introduced in this section.

Keywords Metabolomics · Dereplication · Genomics · NMR · LC–MS

Abbreviations

ABS	Access and benefit sharing
ADC	Antibody–drug conjugate
BBNJ	Biodiversity Beyond National Jurisdiction
BGCs	Biosynthetic gene clusters

K. Yamauchi (✉)
The United Graduate School of Agricultural Science, Gifu University, Gifu, Japan

Graduate School of Natural Science and Technology, Gifu University, Gifu, Japan

Faculty of Applied Biological Sciences, Gifu University, Gifu, Japan

Center for One Medicine Innovative Translational Research (COMIT), Institute for Advanced Study, Gifu University, Gifu, Japan
e-mail: yamauchi.kosei.p2@f.gifu-u.ac.jp

A. H. M. N. Nabi, H. Uddin Shekhar (eds.), *Biotechnological Advances in Healthomics*, Interdisciplinary Biotechnological Advances,
https://doi.org/10.1007/978-981-95-3612-2_1

CNS	Central nervous system
CNV	Copy number variation
CRISPR	Clustered regularly interspaced short palindromic repeats
DEGs	Differentially expressed genes
DSI	Digital sequence information
FDA	Food and Drug Administration
HDAC	Histone deacetylase
HPLC	High-performance liquid chromatography
IC_{50}	Half maximal inhibitory concentration
LC-MS	Liquid chromatography–mass spectrometry
MNPs	Marine natural products
NMR	Nuclear magnetic resonance
OSMAC	One strain–many compounds
PKC	Protein Kinase C
TCM	Traditional Chinese medicine
TC-NER	Transcription-coupled nucleotide excision repair
WHO	World Health Organization

1 Introduction

Starting with the extraction of morphine from plants by crystallization in the eighteenth century, research into the application of naturally produced compounds as pharmaceuticals expanded. The invention of chromatographic component fractionation techniques in the nineteenth century, followed by the development of NMR and X-ray crystallography, led to the elucidation of the structures of naturally occurring compounds. A large number of screening tests were conducted in a human-oriented manner to search for plant- and microorganism-derived extracts with biological activity, and the active compounds were clarified by fractionation in pursuit of the activity. After these screening tests were completed, the number of compounds with novel skeletons gradually decreased. However, by incorporating genetic and LC–MS information, new compounds with trace amounts have begun to be discovered.

Secondary metabolic components produced by natural products, especially microorganisms, are diverse, and many active compounds with unique structures have been isolated. Recently, genome mining research has been conducted to efficiently lead to the discovery of new compounds while incorporating genomics to predict secondary metabolic components from DNA information. This is due to the recent rapid development of next-generation sequencers, which enable cheaper and faster analysis of the DNA information of organisms. Furthermore, tools for predicting secondary metabolic components from gene clusters, such as antiSMASH and PRISM, are now available to anyone as open source, making it possible to predict secondary metabolic components from DNA that have not been revealed in the past. Conversely, the development of these technologies has also made it possible to predict the gene clusters involved in the biosynthesis of compounds with unique structures.

In addition, more accurate assembly of DNA and research to eliminate errors has been done (Cornet and Baurain 2022; de Oliveira et al. 2025). From the acquisition of accurate genomic data, it is now possible to accurately predict secondary metabolic components and gene clusters.

On the other hand, the development of LC–MS instrumentation has improved the technology of metabolomics analysis, making it possible to predict secondary metabolic components of plant- and microorganism-derived extracts in the state of mixtures. Until now, structural analysis of compounds required isolation, but with the advent of this metabolomics analysis, it is now possible to infer certain low-throat structures. This has made it possible to predict new compounds from many secondary metabolites. By applying these techniques, it can be said that compounds in extracts can be captured comprehensively, leading to the efficient discovery of new compounds. In this chapter, natural product chemistry has been developed by omics technology, and newly developed technologies in recent years will be introduced with examples of recent research.

2 Metabolomics Analysis

Accumulation of LC–MS–MS fragmentation data enabled the prediction of compounds in extracts by matching against a database. LC–MS data were combined with NMR data to perform multivariate analysis to determine secondary metabolites between *Cassia acutifolia* and *Cassia angustifolia* senna species were revealed (Farag et al. 2015). In particular, the construction of molecular networks that visualize similar compounds by clustering them enables us to visually capture groups of known compounds (Allard et al. 2016). On the flip side, compound groups whose structures are not predicted are more likely to be novel compound groups, thus enabling de-replication, an efficient novel compound search. Salsola species, a salt crop, has been reported to be rich in phenolic compounds, triterpenoids (Boulaaba et al. 2019; Murshid et al. 2022; Rasheed et al. 2013), and various biological activities such as hypoglycemic and antioxidant effects have been revealed (Iannuzzi et al. 2020; Jin et al. 2011; Seo et al. 2020; Tundis et al. 2007). In addition, in vivo studies have reported that salsola reduces pain, inflammation, and edema in experimental animals (Gonfa et al. 2023; Mohammed et al. 2021; El Raey et al. 2016). The chemical metabolic profiles of three Salsola plants from Egypt, *Salsola tetrandra* Forssek, *Salsola vermiculata* L., and *Salsola tetragona* Delile, were performed by UPLCMS/MS and multivariate analysis to predict their anti-inflammatory activity and prediction of contributing compounds were conducted (Zakaria et al. 2025). *Aloe vera* is used worldwide in health drinks and other beverages, milk-issue products, jams and jellies, and other edible products (Basannavar et al. 2014; El-Sayed and El-Sayed 2020; Ramachandra and Rao 2008), and its gel is also used to preserve food (Maan et al. 2021; Salama and Abdel Aziz 2020). Its nutritional components are being studied through metabolomics analysis (Krupa et al. 2025). *A. vera* has been reported to contain lipids such as β-sitosterol, β-carotene, tocopherol, and palmitic acid (López-Cervantes et al. 2018). In addition, proteomic analysis has been conducted in *Nepeta deflersiana*, a traditional Saudi Arabian medicinal plant of the Perilla family (Mannoubi et al. 2025). This plant has been reported to have anti-cancer activity, antioxidant activity, anti-inflammatory activity, and antiviral activity, etc. It is also used in Saudi Arabia for asthma medicine, deceased father role, anti-rheumatism, and so on. These metabolomics technologies have been further developed in recent years, incorporating not only LC–MS but also NMR, genetic information, biological activity, and many other types of information. AI-based predictions are also evolving by integrating these data. The development of these technologies is described below.

2.1 Dereplication Using NMR

In 2008, NMR/LC–MS parallel dynamic spectroscopy (NMR/LC–MS PDS), which combines NMR data with 1H-NMR fingerprinting, was reported for the structural estimation of compounds in extracts for which LC–MS alone was insufficient for metabolomics analysis. MS parallel dynamic spectroscopy (NMR/LC–MS PDS) was reported in 2008 (Dai et al. 2008). This method was used to identify 12 components from *Gossypium herbaceum* L. flower extracts. Grkovic et al. focused on the unique spectral patterns of small molecules from NMR of extract fractions from the Australian sponge iotrochota sp. and identified four new compounds, iotrochotadines A–D, and iotrochotazine A were successfully discovered (Grkovic et al. 2014). Similar methods were used in termite gut-associated Streptomyces species, and six new compounds were discovered: actinoglycosidines A and B, actinopolymorphol D, niveamycin A, B, and C (Romero et al. 2015). In addition, attempts have been made to de-replicate by 13C-NMR database matching; Yang et al. were the first to attempt efficient isolation of new compounds from 13C-NMR databases (Yang et al. 2015). They succeeded in efficiently discovering six flavone dimers, podoverine D, E, F, G, H, and I, by comparing 13C-NMR of fractions of root-derived extracts of *Dysosma versipellis* (Hance) with a database. Furthermore, an attempt was made in 2018 to estimate and de-replicate compounds in *Picea abies* L. extracts by using HMBC and HSQC 2D NMR data (Bakiri et al. 2018). In addition to fortunefuroic acids A, B, C, and D, NMR, EI-MS, and LC-ESIMS data from Keteleeria collected in Yunnan, China, were integrated to discover new fortunefuroic acids K, L, M, N, O, and P (Zhao et al. 2025).

2.2 Biochemometrics

Biochemometrics is a research field that combines chemical informatics with biological data. It has been applied in drug discovery, food science, and other fields, and is also bringing about dramatic changes in natural product chemistry. Chemical data are collected using LC–MS and NMR techniques, while biological data are obtained from genetic information and from biological activity. These are integrated through statistical analysis and machine learning to efficiently and quickly obtain the information researchers want. In other words, this technology makes it possible to infer from extracts compounds that have biological activity and highly novel structures among secondary metabolic components produced by plants and microorganisms, leading to rapid estimation of compounds that can be applied to drug development, cosmetic development, and functional food development. Crockford et al. used statistical heterospectroscopy (SHY) to improve the efficiency of biomarker identification in rat urine samples (Crockford et al. 2005). LC–MS and NMR data were used to estimate the active compounds by applying this method of statistical processing to chemical analysis data. In 2016, statistical processing was performed on biological activities in addition to metabolomics analysis using LC–MS and NMR of fractions to estimate tyrosinase inhibitory compounds from *Morus alba*. It was demonstrated that the actual presumed compound was more active than the positive control, kojic acid (Aligiannis et al. 2016). Wasilewicz et al. combined LC–MS and NMR information to construct chemical data, to which they added biological activity data relevant to dry eye disease (DED) to estimate compounds with antioxidant activity and cytotoxic components in *Buddleja officinalis* Maxim flower extracts. The results suggest that verbascoside and *cis*-acteoside were the compounds with antioxidant activity important for the treatment of DED, while saponins are predicted to be cytotoxic (Wasilewicz et al. 2024). Using the same method, a lanostantriterpene that inhibits steroid sulfatase was also found in a complex extract of the porous fungus *Fomitopsis pinicola* (Grienke et al. 2019). Multivariate data analysis of 1H-NMR spectra of *Orthosiphon stamineus* extracts combined with adenosine A1 binding activity to identify compounds that may bind adenosine A1 in the mixture state (Yuliana et al. 2011).

3 Integration of Genomics and Metabolomics

From genomic information, the secondary metabolic components that are expressed can be predicted. Hence, in addition to metabolomics analysis using LC–MS, attempts to estimate secondary metabolic components by taking genetic information into account have been rapidly developing in recent years. Even if a biosynthesis-related enzyme is encoded in a gene, secondary metabolic components may not be expressed in general environmental conditions. In this case, metabolomics analysis alone cannot discover all the compounds that an organism can make. By predicting the secondary metabolic components that may be expressed based on genetic information, it is expected that new compounds that have not been brought to light will be discovered. It will also be a valuable source of information for predicting which gene clusters are responsible for the biosynthesis of compounds with unique structures. The antiSMASH, published in 2011, is a powerful tool that has enabled the identification of potential gene clusters of secondary metabolic components, which was previously difficult to do. PRISM was then published in 2015. Here, it was reported that PRISM predicted the structure of type II polyketides more accurately than antiSMASH (Skinnider et al. 2015). Then in 2017, PRISM3 was released with its improved features (Skinnider et al. 2017). It has increased its versatility, including the ability to search for new cluster types. antiSMASH 5.0 will be released in 2019, enabling cluster searches that encode the biosynthesis of even more secondary metabolic components. In addition, more detailed predictions can now be made for gene clusters encoding type II polyketide synthases (Blin et al. 2019). Since then, it has been further updated and continually improved to accurately predict a wider range of secondary metabolic components. These new tools have led to the discovery of new compounds one after another. For example, a new peptide has been discovered in the symbiotic fungus of *Pachycephala schlegelii*, a known toxic bird of New Guinea, based on LC–MS information to construct a molecular network. Furthermore, the biosynthesis mechanism of the isolated novel peptides has been elucidated from the genetic information (Seibel et al. 2024).

In 2011, Kerasten et al. proposed a method to link metabolomics information from MS to biosynthetic gene clusters (Kersten et al. 2011). In 2014, NRPquest, which combines mass spectrometry data with genomic information to search for non-ribosomal peptides (NRPs), was published (Mohimani et al. 2014). In addition, Nguyen et al. successfully linked biosynthetic gene clusters to information on secondary metabolic components obtained by MS/MS information, which is expected to lead to further genome mining and discovery of biosynthetic gene clusters (Nguyen et al. 2013). Combining metabolomics with genomic information is also a useful tool for inferring the process of the generation of secondary metabolic components in the evolution of species. Molecules produced by Pseudomonas were subjected to molecular networks using mass spectrometry data to discover new compounds and to examine evolutionary relationships among species (Nguyen et al. 2017). Mass spectrometry revealed great chemical diversity in the secondary metabolic components produced by *Photorhabdus* and *Xenorhabdus*, which have very similar DNA sequences. Genomic and metabolomics methods have led to the elucidation of classes of naturally occurring molecules that were previously unidentifiable (Tobias et al. 2017). In 2021, NRPminer was developed to infer even modification information of non-ribosomal peptides (NRP) based on genetic information and LC–MS metabolomics data (Behsaz et al. 2021). This paves the way for the prediction of the complete structure of NRPs, which has been difficult with conventional genome mining tools. Many of these genome mining attempts have been made in Bacteria, but there have been few reports on fungi. This is because fungi have more dispersed BGCs and are subject to epigenetic regulation than bacteria, making it more difficult to predict secondary metabolic components. In addition, fungi often have silent BGC, which must be analyzed in combination with RNA-seq. However, in 2023, Caesar and others succeeded in statistically associating the resulting compounds with genetic information, even in fungi (Caesar et al. 2023). This research has revealed more than 200 connections between natural compounds and gene cluster families. The accumulation of data from such studies should enable highly accurate compound prediction in fungi. With the development of omics technologies such as genomics and metabolomics, even more detailed structural information can be predicted at the mixture stage, leading to the efficient discovery of new compounds and molecules that can be applied in drug discovery, cosmetics, and health foods in the future.

4 Machine Learning Approaches to Genomics and Metabolomics

Over 60,000 type I polyketide gene clusters have now been discovered, but only 100 compound structures are known to be synthesized, and most metabolites remain unknown (Yan et al. 2024). Therefore, Seq2PKS, a machine learning algorithm for predicting the structure of type I polyketide synthases and biosynthesized polyketides from mass spectral data, was presented (Yan et al. 2024). By using this tool, we predicted the structure of compounds that could not be predicted by antiSMASH or PRISM. Seq2PKS identified polyketides with unknown modifications and predicted multiple molecular structures for each BGC. These were ranked against the mass spectrometry data to improve the accuracy of structure prediction. The group has developed HypoRiPPAtlas, which predicts post-translationally modified peptides (RiPPs) and aggregates virtual natural product structures. It uses seq2ripp, a machine learning tool, to derive the structure of the compound produced from the genetic information, and compares the estimated mass spectrum with the actual mass spectrometry data for more reliable structure estimation (Lee et al. 2023). This is expected to allow us to infer the structure of a large number of secondary metabolic components that have not been revealed from the genetic information.

5 Proteomic Analysis Is Used to Explore the Mechanism of Activity of Natural Products

Omics analysis is used not only to estimate naturally occurring small molecule compounds, but also to investigate the bioactive mechanisms of active compounds. Active compounds found in nature are thought to bind to proteins on the surface or inside cells and express biological activity. Proteomic analysis, which enables comprehensive analysis of binding proteins, is used in the search for target proteins and elucidation of their mode of action. Naturally occurring compounds have multiple target proteins, and their biological activities are expressed through complex interactions with proteins. It used to be extremely difficult to identify multiple target proteins, but with the development of LC–MS, a method to comprehensively reveal target proteins has been invented. The first method to search for target proteins using thermal denaturation of proteins was published in 2013 (Molina et al. 2013). Since then, it has been used as a powerful method to search for target proteins of naturally occurring compounds. Dihydrocaffeic acid, a metabolite of chlorogenic acid, has been reported to have a wide range of biological activities, including anti-inflammatory activity (Botta et al. 2015; Chen et al. 2012; Etzenhouser et al. 2000; Lu et al. 2021; Poquet et al. 2008; Rogozinska et al. 2023). The mechanism that reduces its acute pneumonia was revealed using proteomics (Li et al. 2025). This study identified the target protein of dihydrocaffeic acid as transaldolase 1 (TALDO1) and demonstrated that it regulates the activity of the protein kinase RNA-like ER kinase (PERK), inflammation-inhibitory κB (IκB), and nuclear translocation of nuclear factor-κB (NF-κB), and that it exerts anti-inflammatory activity by regulating these factors. In addition, honokiol, a lignan compound found in the bark of magnolia plants, has a variety of biological activities, including antitumor and antibacterial activity (Cai et al. 2023; Ochoa et al. 2019; Salgado-Benvindo et al. 2023; Taferner et al. 2011; Yang et al. 2022). It is the compound whose target proteins have been identified by proteomics using Cellular Thermal Shift Assay (CETSA) and other methods. Proteomic remains revealed Family 3 Member 2 (SLC3A2) as a target protein of honokiol, which is reported to be involved in its anti-inflammatory activity (Cai et al. 2023). CETSA is a comprehensive and simple method to discover target proteins by identifying proteins whose thermal stability is altered by the interaction. Although the CETSA method is capable of exhaustive target protein searches, it may miss a target protein if the compound is not affected by thermal denaturation upon binding. Therefore, a method is generally used to search for the target proteins using compound probes, analyzing the target proteins by proteomics. Proteomic analysis of the compound probes revealed that honokiol is a potent interactor of dipeptidyl peptidase 3 (DPP3), lyric, 6-phosphogluconate dehydrogenase decarboxylating (PGD), and calpain-1 (CAPN-1), which were characterized as its target proteins because their thermal stability was enhanced by the binding of honokiol. These proteins are involved in biological functions related to cancer cell survival, such as apoptosis, cell proliferation, and cell growth, and are thought to be involved in the anticancer activity of honokiol (Vázquez-Villa et al. 2024).

References

Aligiannis N, Skaltsounis LA, Argyropoulou A et al (2016) Heterocovariance based metabolomics as a powerful tool accelerating bioactive natural product identification. Chem Select 1:2531–2535. https://doi.org/10.1002/slct.201600744

Allard P-M, Roussi F, Péresse T et al (2016) Integration of molecular networking and in-Silico MS/MS fragmentation for natural products Dereplication. Anal Chem 88:3317–3323. https://doi.org/10.1021/acs.analchem.5b04804

Bakiri A, Renault J-H, Nuzillard J-M et al (2018) Reconstruction of HMBC correlation networks: a novel NMR-based contribution to metabolite mixture analysis. J Chem Inf Model 58:262–270. https://doi.org/10.1021/acs.jcim.7b00653

Basannavar S, Sharma RK, Pothuraju R (2014) Effect of Aloe vera (Aloe barbadensis Miller) on survivability, extent of proteolysis and ACE inhibition of potential probiotic cultures in fermented milk. J Sci Food Agric 94:2712–2717. https://doi.org/10.1002/jsfa.6615

Behsaz B, Pevzner PA, Mohimani H et al (2021) Integrating genomics and metabolomics for scalable non-ribosomal peptide discovery. Nat Commun 12. https://doi.org/10.1038/s41467-021-23502-4

Blin K, Medema MH, Steinke K et al (2019) antiSMASH 5.0: updates to the secondary metabolite genome mining pipeline. Nucleic Acids Res 47:W81–W87. https://doi.org/10.1093/nar/gkz310

Botta G, Bizzarri BM, Garozzo A et al (2015) Carbon nanotubes supported tyrosinase in the synthesis of lipophilic hydroxytyrosol and dihydrocaffeoyl catechols with antiviral activity against DNA and RNA viruses. Bioorg Med Chem 23:5345–5351. https://doi.org/10.1016/j.bmc.2015.07.061

Boulaaba M, Medini F, Hajlaoui H et al (2019) Biological activities and phytochemical analysis of phenolic extracts from Salsola kali L. Role of endogenous factors in the selection of the best plant extracts. S Afr J Bot 123:193–199. https://doi.org/10.1016/j.sajb.2019.03.003

Caesar LK, Raja HA, Gupta R et al (2023) Correlative metabologenomics of 110 fungi reveals metabolite-gene cluster pairs. Nat Chem Biol 19:846–854. https://doi.org/10.1038/s41589-023-01276-8

Cai X, Jiang X, Zhao M et al (2023) Identification of the target protein and molecular mechanism of honokiol in anti-inflammatory action. Phytomedicine 109:154617. https://doi.org/10.1016/j.phymed.2022.154617

Chen K-C, Chang S-S, Tsai F-J, Chen CY-C (2012) Han ethnicity-specific type 2 diabetic treatment from traditional Chinese medicine? J Biomol Struct Dyn 31:1219–1235. https://doi.org/10.1080/07391102.2012.732340

Cornet L, Baurain D (2022) Contamination detection in genomic data: more is not enough. Genome Biol 23. https://doi.org/10.1186/s13059-022-02619-9

Crockford DJ, Zirah S, Rainville P et al (2005) Statistical heterospectroscopy, an approach to the integrated analysis of NMR and UPLC-MS data sets: application in metabonomic toxicology studies. Anal Chem 78:363–371. https://doi.org/10.1021/ac051444m

Dai D, He J, Sun R et al (2008) Nuclear magnetic resonance and liquid chromatography–mass spectrometry combined with an incompleted separation strategy for identifying the natural products in crude extract. Anal Chim Acta 632:221–228. https://doi.org/10.1016/j.aca.2008.11.002

De Oliveira RVF, Garrido LM, Padilla G (2025) Decontamination of DNA sequences from a Streptomyces genome for optimal genome mining. Braz J Microbiol 56. https://doi.org/10.1007/s42770-024-01598-2

El Raey M, Wink M, Osman S, El Kashak W (2016) New isorhamnetin derivatives from Salsola imbricata Forssk. Leaves with distinct anti-inflammatory activity. Pharmacogn Mag 12:47. https://doi.org/10.4103/0973-1296.176110

El-Sayed SM, El-Sayed HS (2020) Production of UF-soft cheese using probiotic bacteria and Aloe vera pulp as a good source of nutrients. Ann Agric Sci 65:13–20. https://doi.org/10.1016/j.aoas.2020.05.002

Etzenhouser B, Hansch C, Kapur S, Selassie CD (2000) Mechanism of toxicity of esters of Caffeic and dihydrocaffeic acids. Bioorg Med Chem 9:199–209. https://doi.org/10.1016/s0968-0896(00)00238-8

Farag MA, Porzel A, Mahrous EA et al (2015) Integrated comparative metabolite profiling via MS and NMR techniques for Senna drug quality control analysis. Anal Bioanal Chem 407:1937–1949. https://doi.org/10.1007/s00216-014-8432-1

Gonfa YH, Tessema FB, Bachheti A et al (2023) Anti-inflammatory activity of phytochemicals from medicinal plants and their nanoparticles: a review. Curr Res Biotechnol 6:100152. https://doi.org/10.1016/j.crbiot.2023.100152

Grienke U, Foster PA, Zwirchmayr J et al (2019) 1H NMR-MS-based heterocovariance as a drug discovery tool for fishing bioactive compounds out of a complex mixture of structural analogues. Sci Rep 9. https://doi.org/10.1038/s41598-019-47434-8

Grkovic T, Pouwer RH, Vial M-L et al (2014) NMR fingerprints of the drug-like natural-product space identify Iotrochotazine A: a chemical probe to study Parkinson's disease**. Angew Chem Int Ed Engl 53:6070–6074. https://doi.org/10.1002/anie.201402239

Iannuzzi AM, Moschini R, De Leo M et al (2020) Chemical profile and nutraceutical features of Salsola soda (agretti): anti-inflammatory and antidiabetic potential of its flavonoids. Food Biosci 37:100713. https://doi.org/10.1016/j.fbio.2020.100713

Jin Y-S, Jin L, Chen H-S et al (2011) Chemical and biologically active constituents of Salsola collina. Chem Nat Compd 47:257–260. https://doi.org/10.1007/s10600-011-9896-2

Kersten RD, Moore BS, Fischbach MA et al (2011) A mass spectrometry–guided genome mining approach for natural product peptidogenomics. Nat Chem Biol 7:794–802. https://doi.org/10.1038/nchembio.684

Krupa S, Ruman T, Szuberla W, Nizioł J (2025) Analysis of the spatial distribution of metabolites in Aloe vera leaves by mass spectrometry imaging and UHPLC-UHRMS. Sci Rep 15. https://doi.org/10.1038/s41598-025-88144-8

Lee Y-Y, Liu H, Guler M et al (2023) HypoRiPPAtlas as an Atlas of hypothetical natural products for mass spectrometry database search. Nat Commun 14. https://doi.org/10.1038/s41467-023-39905-4

Li G, Zhang X, Zhang Y et al (2025) Chemo-proteomics reveals dihydrocaffeic acid exhibits anti-inflammation effects via Transaldolase 1 mediated PERK-NF-κB pathway. Cell Commun Signal 23. https://doi.org/10.1186/s12964-024-01958-3

López-Cervantes J, Sánchez-Machado DI, Cruz-Flores P et al (2018) Antioxidant capacity, proximate composition, and lipid constituents of Aloe vera flowers. J Appl Res Med Aromat Plants 10:93–98. https://doi.org/10.1016/j.jarmap.2018.02.004

Lu L, Cho C, Zou K et al (2021) Quinic acid: a potential antibiofilm agent against clinical resistant Pseudomonas aeruginosa. Chin Med 16. https://doi.org/10.1186/s13020-021-00481-8

Maan AA, Reiad Ahmed ZF, Iqbal Khan MK et al (2021) Aloe vera gel, an excellent base material for edible films and coatings. Trends Food Sci Technol 116:329–341. https://doi.org/10.1016/j.tifs.2021.07.035

Mannoubi IE, Aouadi K, Mohamed SA et al (2025) Untargeted metabolomics approach based on UPLC-ESI-QTOF-MS/MS and evaluation of antioxidant and cytotoxic activities of Nepeta deflersiana Schweinf. ex Hedge supported by molecular modeling and ADMET analysis. J Mol Struct 1333:141767. https://doi.org/10.1016/j.molstruc.2025.141767

Mohammed HA, Abd-Elmoniem EM, MSM S et al (2021) Phytochemical analysis, pharmacological and safety evaluations of halophytic plant, Salsola cyclophylla. Molecules 26:2384. https://doi.org/10.3390/molecules26082384

Mohimani H, Liu W-T, Kersten R et al (2014) NRPquest: coupling mass spectrometry and genome mining for nonribosomal peptide discovery. J Nat Prod 77:1902–1909. https://doi.org/10.1021/np500370c

Molina DM, Cao Y, Sreekumar L et al (2013) Monitoring drug target engagement in cells and tissues using the cellular thermal shift assay. Science 341:84–87. https://doi.org/10.1126/science.1233606

Murshid SSA, Abdallah HM, Atoum D et al (2022) Genus Salsola: chemistry, biological activities and future prospective-a review. Plants (Basel, Switz) 11:714. https://doi.org/10.3390/plants11060714

Nguyen DD, Dorrestein PC, De Mot R et al (2013) MS/MS networking guided analysis of molecule and gene cluster families. Proc Natl Acad Sci U S A 110:E2611–E2620. https://doi.org/10.1073/pnas.1303471110

Nguyen DD, Duggan BM, Ghequire MGK et al (2017) Indexing the Pseudomonas specialized metabolome enabled the discovery of poaeamide B and the bananamides. Nat Microbiol 2:16197. https://doi.org/10.1038/nmicrobiol.2016.197

Ochoa C, Solinski AE, Kozlowski MC et al (2019) A bisphenolic honokiol analog outcompetes oral antimicrobial agent cetylpyridinium chloride via a membrane-associated mechanism. ACS Infect Dis 6:74–79. https://doi.org/10.1021/acsinfecdis.9b00190

Poquet L, Clifford MN, Williamson G (2008) Effect of dihydrocaffeic acid on UV irradiation of human keratinocyte HaCaT cells. Arch Biochem Biophys 476:196–204. https://doi.org/10.1016/j.abb.2008.01.019

Ramachandra CT, Rao PS (2008) Processing of aloe vera leaf gel: a review. Am J Agric Biol Sci 3:502–510. https://doi.org/10.3844/ajabssp.2008.502.510

Rasheed DM, El Zalabani SM, Koheil MA et al (2013) Metabolite profiling driven analysis of Salsola species and their anti-acetylcholinesterase potential. Nat Prod Res 27:2320–2327. https://doi.org/10.1080/14786419.2013.832676

Rogozinska M, Lisiecki K, Czarnocki Z, Biesaga M (2023) Antioxidant activity of sulfate metabolites of chlorogenic acid. Appl Sci 13:2192. https://doi.org/10.3390/app13042192

Romero CA, Kurtböke DI, Zhang L et al (2015) NMR fingerprints, an integrated approach to uncover the unique components of the drug-like natural product metabolome of termite gut-associated Streptomyces species. RSC Adv 5:104524–104534. https://doi.org/10.1039/c5ra17553d

Salama HE, Abdel Aziz MS (2020) Optimized alginate and Aloe vera gel edible coating reinforced with nTiO2 for the shelf-life extension of tomatoes. Int J Biol Macromol 165:2693–2701. https://doi.org/10.1016/j.ijbiomac.2020.10.108

Salgado-Benvindo C, Tas A, Van Hemert MJ et al (2023) Honokiol inhibits SARS-CoV-2 replication in cell culture at a post-entry step. Microbiol Spectr 11. https://doi.org/10.1128/spectrum.03273-22

Seibel E, Schirmeister T, Murphy R et al (2024) Bacteria from the Amycolatopsis genus associated with a toxic bird secrete protective secondary metabolites. Nat Commun 15. https://doi.org/10.1038/s41467-024-52316-3

Seo JH, Jin MH, Chang YH (2020) Anti-inflammatory effect of Salsola komarovii extract with dissociated glucocorticoid activity. BMC Complement Med Ther 20. https://doi.org/10.1186/s12906-020-02979-4

Skinnider MA, Dejong CA, Rees PN, et al (2015) Genomes to natural products PRediction Informatics for Secondary Metabolomes (PRISM). Nucleic Acids Res 43:gkv1012. https://doi.org/10.1093/nar/gkv1012

Skinnider MA, Merwin NJ, Johnston CW, Magarvey NA (2017) PRISM 3: expanded prediction of natural product chemical structures from microbial genomes. Nucleic Acids Res 45:W49–W54. https://doi.org/10.1093/nar/gkx320

Taferner B, Huefner A, Ecker GF et al (2011) Modulation of $GABA_A$-receptors by Honokiol and derivatives: subtype selectivity and structure–activity relationship. J Med Chem 54:5349–5361. https://doi.org/10.1021/jm200186n

Tobias NJ, Stinear TP, Ebersberger I et al (2017) Natural product diversity associated with the nematode symbionts Photorhabdus and Xenorhabdus. Nat Microbiol 2:1676–1685. https://doi.org/10.1038/s41564-017-0039-9

Tundis R, Statti GA, Loizzo MR, Menichini F (2007) Inhibitory effects on the digestive enzyme alpha-amylase of three Salsola species (Chenopodiaceae) in vitro. Pharmazie 62:473–475. https://doi.org/10.1691/ph.2007.6.6794

Vázquez-Villa H, Rueda-Zubiaurre A, Fernández D et al (2024) Chemical probes for the identification of the molecular targets of honokiol. Eur J Med Chem 283:117102. https://doi.org/10.1016/j.ejmech.2024.117102

Wasilewicz A, Areesanan A, Kirchweger B et al (2024) Combining the strengths of MS and NMR in biochemometrics: a case study on Buddleja officinalis. J Nat Prod 88:1099–1110. https://doi.org/10.1021/acs.jnatprod.4c00847

Yan D, Mohimani H, Guler M et al (2024) Discovering type I cis-AT polyketides through computational mass spectrometry and genome mining with Seq2PKS. Nat Commun 15:5356. https://doi.org/10.1038/s41467-024-49587-1

Yang Z, Zhou H, Wu Y et al (2015) A novel strategy for screening new natural products by a combination of reversed-phase liquid chromatography fractionation and ^{13}C NMR pattern recognition: the discovery of new anti-cancer flavone dimers from Dysosma versipellis (Hance). RSC Adv 5:77553–77564. https://doi.org/10.1039/c5ra13756j

Yang R, Li S, Zhang T et al (2022) Membrane-targeting neolignan-antimicrobial peptide mimic conjugates to combat methicillin-resistant Staphylococcus aureus (MRSA) infections. J Med Chem 65:16879–16892. https://doi.org/10.1021/acs.jmedchem.2c01674

Yuliana ND, Verpoorte R, Choi YH, Khatib A (2011) Comprehensive extraction method integrated with NMR metabolomics: a new bioactivity screening method for plants, adenosine A1 receptor binding compounds in *Orthosiphon stamineus* Benth. Anal Chem 83:6902–6906. https://doi.org/10.1021/ac201458n

Zakaria A, Kassem FF, Ghareeb DA et al (2025) A comprehensive metabolomic study of three Egyptian Salsola species revealed their potential anti-inflammatory activity. Sci Rep 15. https://doi.org/10.1038/s41598-024-80807-2

Zhao Z-Y, Jiang Z-L, Tong Y-P et al (2025) Phytochemical and biological studies on rare and endangered plants endemic to China. Part XLIV. Integrated NMR/EI-MS/LC-PDA-ESIMS approach for dereplication and targeted isolation of fortunefuroic acids from Keteleeria fortunei across diverse geographical origins. Phytochemistry 235:114453. https://doi.org/10.1016/j.phytochem.2025.114453

Microbial Genomics: Unraveling the Healthomic Code

Munawar Sultana, Anamica Hossain, Noor-E-Meem, and Anowara Begum

Abstract Microbial genomics is a transformative field that reveals the entire genetic content of microorganisms (bacteria, archaea, viruses, and single-cell eukaryotes) in order to understand and predict their biology, pathogenicity, resistance, and interactions with hosts and the environment. Sequencing technologies (Illumina, Oxford Nanopore, and PacBio) have significantly contributed to high-resolution analyses via whole genome sequencing (WGS), metagenomics, and functional genomics. Such tools have drastically increased our capacity to track infectious disease outbreaks, antimicrobial resistance (AMR), and microbiome dynamics. The importance of microbial genomics in health applications. Precision diagnostics, therapeutic protein engineering, vaccine development, and real-time pathogen surveillance during the COVID-19 pandemic are underpinned by microbial genomics in healthcare. It also helps to understand the microbiota gut–brain axis, autoimmune diseases, and metabolic disorders through the study of host–microbe relations using multi-omics technologies. In Bangladesh, genomic analysis has detected resistant pathogens, facilitated outbreak investigations, and guided local stewardship policies. However, the obstacles of poor infrastructure, low bioinformatics capability, high expense, and poor regulatory environments still retard the popularization of NGS. The future growth will need holistic national plans, including genomic literacy, data governance, and infrastructure scaling. Indeed, microbial genomics is a frontier in contemporary science and medicine, promising evidence-based strategies for public health, biotechnology, and personalized medicine—particularly, in low- and middle-income countries ready for transformation, like Bangladesh.

Keywords Next-generation sequencing · Genomics · Antimicrobial resistance · Multi-omics · Metagenomics · CRISPR/Cas9 · Microbiome · Synthetic biology · Artificial intelligence · Healthomic innovations

---

M. Sultana · A. Hossain · Noor-E-Meem · A. Begum (✉)
Department of Microbiology, University of Dhaka, Dhaka, Bangladesh
e-mail: anowara@du.ac.bd

A. H. M. N. Nabi, H. Uddin Shekhar (eds.), *Biotechnological Advances in Healthomics*, Interdisciplinary Biotechnological Advances,
https://doi.org/10.1007/978-981-95-3612-2_2

Abbreviations

AI	Artificial intelligence
AMR	Antimicrobial resistance
BSMMU	Bangabandhu Sheikh Mujib Medical University
CARD	Comprehensive Antibiotic Resistance Database
CASTs	CRISPR-associated transposons
CNS	Central nervous system
CRISPR	Clustered regularly interspaced short palindromic repeats
DETECTR	DNA Endonuclease-Targeted CRISPR Trans Reporter
ENA	European Nucleotide Archive;
FISH	Fluorescence in situ hybridization
FMT	Fecal microbiota transplantation
GDPR	General Data Protection Regulation
GI tract	Gastrointestinal tract
GISAID	Global Initiative on Sharing All Influenza Data;
GLP-1	Glucagon-like peptide-1
HMP	Human Microbiome Project
IBD	Inflammatory bowel disease
icddr,b	International Centre for Diarrhoeal Disease Research, Bangladesh
IEDCR	Institute of Epidemiology Disease Control and Research
MDR	Multidrug-resistant
MetaHIT	Metagenomics of the human intestinal tract
MetaPhlAn	Metagenomic phylogenetic analysis
mNGS	Metagenomic next-generation sequencing
NGS	Next-generation sequencing
NIH	National Institutes of Health
PAL	Phenylalanine ammonia-lyase
PKU	Phenylketonuria
PNA	Peptide nucleic acid
ptNGS	Pathogen-targeted next-generation sequencing
QIIME2	Quantitative Insights Into Microbial Ecology 2
qPCR	Quantitative polymerase chain reaction
ResFinder	Resistance gene finder
SARS-CoV-2	Severe Acute Respiratory Syndrome Coronavirus 2
SHERLOCK	Specific High Sensitivity Enzymatic Reporter UnLOCKing
ST	Sequence type (e.g., ST2, ST361)
TB	Tuberculosis
WGS	Whole genome sequencing
XDR	Extensively drug-resistant

1 Introduction

The comprehensive genetic analysis of microorganisms is now made possible by microbial genomics, which has emerged as an important field in biomedical and public health research. Microbial interactions with the environment and with host organisms, as well as the evolution and physiology of microbes, can now be explored at an unprecedented level, and in fact, entire genomes can be analyzed rather than just subset functions through microbial genomics. This is vital for identifying pathogenic traits, understanding the diversity of microbes, tracking outbreaks, and formulating treatment strategies. Analysis of complex microbial communities, like the human microbiome, as well as the discovery of new microbial species, has been made possible due to advancements in sequencing technologies like NGS and WGS. Now, these essential study tools allow for the precise and prompt analysis of massive genomic datasets. Microbial genomics, on a worldwide scale, is pivotal for public health policy and managing infectious diseases. Countries like the United Kingdom showed real-time genomic surveillance capability and its utility for outbreak response and variant tracking through the Covid-19 Genomics UK (COG-UK) consortium (Saravanan et al. 2022). Also, the United States CDC has incorporated genomic epidemiology into the national antimicrobial resistance (AMR) surveillance system. Together with the rest of the world, Bangladesh is now harnessing the microbial genomics paradigm in nutrition, agriculture, and health within collaborative models (Monir et al. 2023). The domain bridges the gap between laboratory research and health applications, facilitating infectious disease detection, vaccine development, and even tailored medicine. Engaging with the latest advancements and genomic data for innovative medical, agricultural, and public health solutions requires understanding the microbial genomics landscape and its concepts.

This chapter provides a focused introduction to microbial genomics, highlighting its significance in understanding microbial biology, the mechanisms of disease, and the development of antimicrobial resistance. It discusses essential technologies like genome sequencing and bioinformatics, and examines their applications in diagnostics, vaccine design, and precision medicine. Particular emphasis is placed on combining multi-omics strategies with artificial intelligence to take genomic research forward. The chapter also underscores the importance of microbial genomics for public health, especially in resource-limited settings like Bangladesh. In short, the chapter tries to lay out a basic scientific foundation, practical applications, and new directions for microbial genomics in both global and local healthcare contexts.

2 Microbial Genomic Study

Microbial genomics refers to the understanding of the genetic traits, functions, evolutionary patterns, and interactions of microorganisms with their environment or hosts, encompassing the study of bacteria, archaea, viruses, and single-celled eukaryotes. The study of genomics aims to decode the DNA of microorganisms to reveal crucial information on their biology, pathogenicity, ecology, as well as their uses in biotechnology and medicine.

The discipline's commencement is marked by the first free-living microbial genome, called *Haemophilus influenzae,* in 1995. The study benefited from the development of technologies such as the next-generation sequencing (NGS), which increased the speed of sequencing and reduced its cost. Oxford Nanopore and PacBio are long-read sequencing platforms that have recently been able to perform better genome assemblies and explore genomic regions that were inaccessible before, which broadened the study of diverse and complex microbial communities.

The most essential parts of microbial genomics include: *whole genome sequencing (WGS)* facilitates sequencing the DNA of individual microorganisms to determine their full DNA sequence; *comparative genomics* enables cross-strain/spp. genomic comparison to define evolutionary relationships and differences of functions; *metagenomics* helps to analyze the genetic material of entire microbial communities direct from the samples without culturing; *functional genomics* supports studying the expression and regulation of the genes; *population genomics and phylogenomics* assesses the microbial diversity, evolution, and epidemiology within a population. These studies relies on a suite of tools and techniques, including: *sequencing platforms*: Illumina (short-read), Oxford Nanopore and PacBio (long-read), Ion Torrent; *bioinformatics software*: For assembly and annotation (SPAdes and Prokka), comparative genomics (Mauve), metagenomic analyses (QIIME2, MetaPhlAn), and resistance/virulence genes (CARD, ResFinder); *genome editing technologies:* such as cas9 genomics of microbes for biological and therapeutic investigations modifies parts of the genome using CRISPR-Cas systems; *machine learning and AI*: crucial for the analysis of complex genomic and multi-omic datasets.

The impact of microbial genomics has evolved our perception of microbes, unlocking their potential as sophisticated entities not just as pathogens, but in fields including health, the environment, and industry, whilst catalyzing progress in diagnostics, therapeutics, and synthetic biology.

3 Potentials of Microbial Genomics Study in Healthcare

3.1 Microbial Genomics and Disease Understanding

An organism's genome is the primary factor that determines all of its biological traits, including those necessary for reproduction, adaptability, survival, and the development of illness. The genetic material of microbes controls not only the fundamental functions of life but also the systems that underlie host invasion, virulence, immunological evasion, and resistance to antibiotics. Therefore, without a thorough investigation of the microbial genome, a comprehensive understanding of infectious diseases is still lacking.

Microbial genomics has emerged as a transformative approach in modern infectious disease research, allowing scientists to analyze and interpret the complete genetic blueprint of microorganisms. Using techniques such as whole genome sequencing (WGS) and comparative genomic analysis from different strains, researchers can trace evolutionary changes, spot the key pathogenic genes, and figure out the roots of antimicrobial resistance. These findings not only build our understanding of pathogen behavior but also point to critical targets for developing vaccines, diagnostic tools, and better therapeutic strategies.

One striking example comes from studies on *Mycobacterium* TB, which provides a compelling illustration of this. Using WGS, scientists have identified crucial mutations in genes like *kat*G and *rpo*B, which are linked to resistance to isoniazid and rifampicin—two drugs in tuberculosis treatment (Dokht Khosravi et al. 2012). Rapid molecular diagnostic techniques that enable prompt and precise diagnosis of multidrug-resistant tuberculosis have been made possible by these discoveries.

These insights have paved the way for rapid molecular diagnostic methods capable of swiftly detecting multidrug-resistant TB, enabling timely and accurate treatment decisions. The COVID-19 pandemic offered another powerful demonstration of the value of microbial genomics. Faced with the global spread of SARS-CoV-2, the scientific community engaged in an unprecedented level of genomic surveillance. Sequencing efforts quickly revealed key mutations and new variants of concern such as Alpha, Delta, and Omicron, which proved essential in assessing transmissibility, evaluating vaccine performance, and guiding public health policies. Real-time genomic data sharing was made feasible by platforms such as Nextstrain and CoVariants.org, which provided academics and decision-makers throughout the world with an up-to-date picture of viral evolution and propagation. Shiga toxin genes (*stx1, stx2*) in *Escherichia coli* have been identified by genomic analysis as being essential to the severe disease brought on by enterohemorrhagic strains (Zhang et al. 2021). According to Du et al. (2014), the Clostridium difficile toxin genes tcdA and tcdB are essential for identifying hypervirulent epidemic strains and have been connected to severe colitis. These investigations have demonstrated how infections can change their virulence by acquiring, losing, or undergoing point mutations in their genes, which are frequently made possible by horizontal gene transfer. The role of microbial genomics in monitoring antimicrobial resistance (AMR) is crucial. By uncovering the origins and mechanisms of resistance genes, genomics enables scientists to predict the emergence of new resistant strains and to understand their global spread. For example, genomic studies have documented the rise of extensively drug-resistant (XDR) and multidrug-resistant (MDR) organisms, such as methicillin-resistant *Staphylococcus aureus* carrying the *mecA* gene) (Anjum et al. 2024) and carbapenem-resistant *Klebsiella pneumoniae* (harboring *bla*KPC) (On et al. 2025). Hospital infection control procedures, antimicrobial stewardship guidelines, and worldwide monitoring systems have all been influenced by these findings.

Ultimately, microbial genomics reveals the remarkable adaptability of microorganisms, which shows how genetic shifts, whether through mobile genetic elements, point mutations, or the gain and loss of genes, can turn benign species into dangerous "superbugs." By dissecting these evolutionary patterns, scientists gain the knowledge needed to design targeted interventions, including genome-based vaccines, novel antimicrobial agents, and predictive diagnostic tools that anticipate future threats.

3.2 Microbiome Studies and Health Correlations in the Genomic Era

The fact that the human body contains more microbial cells than human cells has long been known, and this statistic emphasizes the significant impact that microbes have on their host. Together, these microbial communities, referred to as the human microbiome, colonize the skin, gastrointestinal (GI) tract, oral cavity, and respiratory tract, among other anatomical locations. These microorganisms are far from being passive inhabitants; they are crucial for controlling the physiology, metabolism, immunological development, and even behavior of their hosts.

A wide range of illnesses, including autoimmune, neurological, and metabolic diseases, have been progressively connected to dysbiosis—disruptions to the natural structure of the microbiome. Conventional microbiology, which involves isolating, cultivating, and characterizing individual microbial members, is time-consuming and resource-intensive, and it is unable to fully represent the complexity of microbial ecosystems, particularly those including unculturable species.

The study of the microbiome has undergone a major shift with the arrival of high-throughput DNA sequencing, particularly metagenomic approaches. Unlike traditional culture-based methods, metagenomics makes it possible to investigate entire microbial communities directly in their natural environments, providing a detailed view of both their diversity and functional capabilities. Among the core techniques in this field is 16S rRNA gene sequencing, which allows researchers to classify bacteria down to the genus or even species level, while also estimating their relative abundance. This method has been instrumental in characterizing the typical human gut microbiome, now estimated to consist of around 160 bacterial species per individual (Wang et al. 2015). For even deeper insights, shotgun metagenomic sequencing analyzes every microbial gene within a sample, revealing important biological functions such as vitamin synthesis, carbohydrate processing, and the breakdown of xenobiotic compounds (Bradley and Haran 2024).

By linking the genetic information with clinical data, researchers are gaining a clearer picture of the connections between the microbiome and human health. For example, *Ruminococcus gnavus,* a species present in over 90% of healthy individuals, has been associated with chronic inflammatory disorders, including Crohn's disease, when it occurs in elevated (Nooij et al. 2025). Likewise, microbiome changes have been linked to conditions such as colon cancer, irritable bowel syndrome, and inflammatory bowel disease (IBD) (Khorsand et al. 2022). In metabolic disorders like type 2 diabetes, genomic studies have reported a shift in genes related to carbohydrate metabolism alongside a decline in butyrate-producing bacterial populations (Dash and Al Bataineh 2020). Microbial genomic analyses have revealed clear variations in metabolic pathways, including those linked to neurotransmitters like GABA and serotonin precursors, in neurodevelopmental and neurodegenerative disorders, such as autism spectrum disorders and Parkinson's disease. These findings lend credence to the growing idea of a gut–brain axis (Metcalfe-Roach et al. 2024). Equally important is the microbiome's impact on the immune system. Certain taxonomic and microbial gene functions have been linked by genomic analysis to the emergence of autoimmune and allergy disorders. For example, dysregulation of taxa such as *Odoribacter, Lachnospiraceae Clostridium,* and *Mogibacteriaceae* has been implicated in autoimmune disease models, indicating possible mechanistic links between gut microbes and immune dysfunction (Volkova and Ruggles 2021).

Additionally, metagenomics analyses of colorectal cancer patient groups have uncovered microbial biomarkers linked to choline metabolism, pointing toward potential applications in early diagnosis (Thomas et al. 2019). The field of precision microbiome therapy is now moving into a more advanced stage, as microbial genomic information is increasingly integrated with host genetic data, transcriptomic profiles, and metabolomic signatures. This integration makes it possible to design truly individualized treatment strategies—ranging from fecal microbiota transplantation (FMT) and tailored dietary interventions to the use of targeted probiotics and prebiotics.

Pioneering large-scale projects such as the Human Microbiome Project (HMP) and the MetaHIT consortium have been instrumental in driving this progress. By assembling extensive catalogs of microbial genes and compiling comprehensive reference genomes, these initiatives have provided an essential foundation for the current wave of innovation in disease modeling, diagnostic tool development, and microbiome-derived therapeutics.

Ultimately, the convergence of microbial genomics with microbiome science not only deepens our understanding of how microbial communities contribute to both homeostasis and disease but also equips researchers with the tools to create precise, targeted interventions aimed at maintaining or restoring human health.

4 Recombinant Technologies for Therapeutic Protein Production

For decades, scientists have relied on heterologous microbial systems to make recombinant proteins, and this approach has become one of the cornerstones of biotechnology. These systems have earned their place in both industrial and medical settings due to their novel advantages, including high production rates, scalability, cost efficiency, and effective use of resources. One of the earliest major breakthroughs in this field was the production of recombinant human insulin, which marked the beginning of leveraging microbial systems for life-saving therapeutic treatments.

The application of microbial genomics has greatly expanded the relative scope and accuracy of recombinant protein technologies. With the entire genome of a host organism as well as the genetic sequence of the target protein in hand, researchers have a strong basis for rational strain design, which helps in overcoming widespread problems like protein misfolding, host protease degradation, or lack of sufficient post-translational modifications by adjusting the enhanceable elements like promoters, regulatory sequences, and codon usage. Recombinant technology has become a systematic and programmable method that can be finely adjusted to meet specific biotechnological goals due to well-mapped microbial genomes. Improvements in technology, such as transcriptomics, whole genome sequencing, and CRISPR-based genome editing, now allow for unprecedented control over microbial hosts. These methods enable the addition, removal, or modification of genes with exceptional accuracy, boosting protein production by allowing precise changes to host pathways. Additionally, synthetic biology allows for the design of custom microbial chassis, equipped with optimized metabolic pathways and engineered circuits for producing specific proteins or metabolites.

Recent applications highlight how genomic engineering is transforming the production of therapeutic proteins. For instance, *Bacillus subtilis* strain BsS RS06551 was CRISPR-engineered to produce butyrate, a short-chain fatty acid with anti-inflammatory and metabolic benefits. By deleting self-lysis genes and introducing a butyrate biosynthesis pathway, researchers observed significant improvements in lipid profiles and glucose tolerance in animal models of obesity (Rahmati et al. 2025). Similarly, an engineered *Escherichia coli* strain (EcN-based TYS8500) designed for phenylketonuria (PKU) incorporated multiple copies of phenylalanine ammonia-lyase (PAL) and enhanced phenylalanine uptake genes, resulting in a marked reduction of serum phenylalanine in PKU (Ma et al. 2022).

Another important example is a CRISPR-edited *E. coli* Nissle strain (EcN-GLP-1) that secretes glucagon-like peptide-1 (GLP-1). This peptide could be a treatment for Parkinson's disease. In mouse models, the probiotic strain improved motor function and lowered neuroinflammation. This shows that live biotherapeutics may be used to treat neurological diseases (Wu et al. 2023). The benefits of live biotherapeutics for neurological conditions are evident in the improved motor function and reduced neuroinflammation in infected mice. Through genetic modification, *Saccharomyces boulardii* was able to release the GLP-1 receptor agonist Exendin-4 (Rahmati et al. 2025). This live yeast treatment, which was administered straight into the gastrointestinal tract, decreased gut inflammation, appetite, and weight gain. It also showed a new method for managing metabolic disorders using in situ peptide administration. Nowadays, altered *Pseudomonas fluorescens* and improved *E. coli* strains that promote improved protein folding and secretion are utilized to generate the CRM197 protein, a non-toxic version of diphtheria toxin that is widely employed as a vaccine carrier, on an industrial scale (Briday et al. 2025).

Apart from manufacturing platforms, genome editing technologies have also made great progress. In *Streptomyces coelicolor*, high-throughput CRISPR interference (CRISPRi) has made it possible to repress several genes at once, improving the pathways involved in the manufacture of antibiotics (Alper and Beisel 2018). Similarly, CRISPR/Cas9 techniques have been successfully applied to *Lactobacillus* species, allowing researchers to enhance traits such as glucose metabolism, stress tolerance, and other key probiotic functions (Mu et al. 2022).

These examples collectively highlight how the convergence of synthetic biology, microbial genomics, and recombinant technologies is ushering in a new era of therapeutic research. As genome-editing methods continue to advance, with emerging tools like CRISPR-associated transposons (CASTs) and TIGR-Tas, the field is poised for even greater innovation. Beyond facilitating the production of complex therapeutic proteins, these cutting-edge approaches are enabling the development of live biotherapeutic interventions tailored specifically for neurological, genetic, and metabolic disorders (Ruden 2025).

5 Advanced Diagnostic Tools

A prompt, thorough, and correct diagnosis is the first step towards effective treatment. While conventional diagnostic methods such as culture, microscopy, and serology have long served as the backbone of clinical microbiology, their limitations in speed, sensitivity, and diagnostic range have become increasingly evident, especially when addressing complex polymicrobial infections, emerging pathogens, or antimicrobial resistance (AMR). The advancement of microbial genomics has transformed diagnostics by enabling rapid, culture-independent, and highly accurate pathogen detection. Technologies such as whole genome sequencing (WGS), shotgun metagenomics, metatranscriptomics, and nanopore-based real-time sequencing now facilitate the generation of rich datasets that can be computationally analyzed for clinical application. With the aid of bioinformatics tools like BLAST, Centrifuge, ResFinder, CARD, QIIME2, and DADA2, clinicians and researchers can accurately identify pathogens down to the strain level, annotate AMR genes, predict resistance profiles, conduct phylogenetic and epidemiological analyses, and profile host-associated microbial communities. Among these innovations, metagenomic next-generation sequencing (mNGS) has emerged as a powerful tool for direct infection diagnosis from clinical samples without prior culturing. A 2025 cohort study reported that mNGS identified 86% of lung infection cases compared to 67% by conventional methods, and led to therapy adjustments in over 40% of patients, underscoring its growing clinical relevance (Chen et al. 2025). When cerebrospinal fluid (CSF) samples are used, mNGSs demonstrated a diagnostic accuracy of 92.9% for central nervous system (CNS) illnesses, identifying pathogens in 14.4% of over 4800 patient cases across a seven-year period, significantly outperforming traditional serological and culture-based methods (Benoit et al. 2024). In a multicenter study conducted in China comparing probe-targeted NGS (ptNGS) and mNGS for detecting meningitis and encephalitis pathogens in CSF, ptNGS showed superior sensitivity and specificity, particularly for known infections (Chen et al. 2024). This highlights the ongoing evolution of genetic diagnostics and suggests that targeted sequencing approaches may be more effective than broad-spectrum methods in certain clinical contexts. Additionally, a benchmarking study in the UK involving 1000 diarrhea cases revealed that metatranscriptomic sequencing is more sensitive for detecting RNA viruses, *Cryptosporidium*, and *Giardia*, while metagenomic approaches are more effective for identifying bacterial infections. This dual-omics strategy enhances both the precision and breadth of gastrointestinal infection diagnostics (Cunningham-Oakes et al. 2025).

Metagenomic next-generation sequencing (mNGS) of synovial and peritoneal fluid has demonstrated significantly increased sensitivity in detecting joint infections and peritoneal dialysis-associated peritonitis, particularly in patients already receiving antibiotics. This improvement has enabled earlier and more effective therapeutic interventions, often reducing the diagnostic turnaround time to just 1 day. In parallel, a novel diagnostic tool utilizing fluorescence in situ hybridization (FISH) based on peptide nucleic acid (PNA) has been developed to identify seven common bacterial infections within 3 h, eliminating the need for sequencing or culture. This rapid method leverages the variability of 16S rRNA for probe design and is especially valuable in settings with limited resources or urgent clinical demands (Oliveira et al. 2024). A deep learning model that was trained on more than 16,000 Gram-stained blood smear pictures was able to identify bacterial and fungal infections with 96.2% accuracy. With the addition of microbial genomic data, this AI system marks a substantial breakthrough in real-time, image-based diagnostics, especially in the management of sepsis, where prompt judgment is essential (Sroka-Oleksiak et al. 2025). A paradigm change in contemporary medicine is represented by the use of microbial genomes in diagnostic procedures. These technologies enable faster, more accurate, and more thorough pathogen detection by overcoming the drawbacks of conventional diagnostic techniques. This is crucial for outbreak control, tailored medicine, and global health readiness. The routine use of genomic platforms in clinical diagnostics is set to become commonplace as they develop further and become more widely available.

6 Omics Technologies for Comprehensive Analysis

The integration of multi-omics technologies, including genomics, transcriptomics, proteomics, metabolomics, and metagenomics, has revolutionized microbial genomics by enabling a comprehensive, systems-level understanding of microbial physiology, interactions, and disease associations. These advanced tools allow researchers to explore gene function, regulatory mechanisms, metabolic pathways, and community dynamics far beyond the capabilities of simple sequencing. Recent studies underscore their expanding impact: in *Mycobacterium tuberculosis,* proteogenomics uncovered hundreds of new microproteins linked to virulence and antibiotic response, revealing potential therapeutic targets (de Souza et al. 2024). The metabolomics of *Klebsiella oxytoca* treated to ciprofloxacin revealed changes in the CoA and glutathione pathways that were connected to oxidative stress (Ahmed et al. 2025). Metabolic genes influencing antibiotic resistance, particularly in purine biosynthesis and respiration, were discovered by a genomic screen of 15,000 *E. coli* mutants (Lubrano et al. 2025).

In order to identify new biomarkers and processes causing gut inflammation, multi-omics in inflammatory bowel disease (IBD) have combined human and microbial data. Location-specific tumor–microbe–metabolite relationships were identified in colorectal cancer by combining host and microbiome omics (Wang et al. 2025). New biomarker routes were made possible by machine learning applied to microbiome and metabolome data, which discriminated early-onset colorectal cancer with approximately 98% accuracy (Jayakrishnan et al. 2024). Acetylated polyamines were shown to be targets in metabolomics of bloodstream infections; antibiotic efficiency was restored by inhibiting the enzyme SpeG. Diet and gut microbial ecology are linked, as evidenced by the significant microbial transfer and diversity found in 20,000 human gut metagenomes and 2500 dietary samples.

These examples show how omics technologies are making it possible to gain profound functional understandings of microbial systems, which are becoming increasingly important for precision medicine, treatments, and diagnosis.

7 Emerging Sectors in Microbial Genomics: Integrating Precision Medicine, Predictive Microbiology, and AI

Owing to developments in data science, systems biology, gene editing, and sequencing technologies, microbial genomics is quickly expanding beyond its historical foundations in clinical and academic research. These developments are spawning revolutionary fields in public health, biotechnology, food safety, and healthcare.

One of the most significant breakthroughs in recent years has been the use of artificial intelligence (AI) and machine learning to understand microbial processes, host relationships, and ecosystem interactions by integrating diverse multi-omics datasets, such as genomics, transcriptomics, proteomics, and metabolomics. This approach speeds up discoveries in different microbial habitats and also improves the accuracy of data annotation.

A particularly fast-growing branch of this work is precision medicine, which leverages both host and microbial genomic information to tailor medical treatments to an individual's specific genetic background, environmental exposures, and lifestyle. Genomics-based interventions are now increasingly used for complex conditions such as inflammatory bowel disease (IBD), obesity, and certain neurological disorders. These interventions often employ next-generation probiotics with targeted metabolic or immunomodulatory functions, as well as specialized microbial therapeutics like *Lactobacillus* or *E. coli* Nissle (Praveschotinunt et al. 2019). With advances in genomic analysis and big data, the precision medicine market is forecasted to expand dramatically, potentially reaching USD 470.53 billion by 2034, growing at an estimated compound annual growth rate (CAGR) of 16.5% between 2025 and 2034. Several national public health bodies—including those in the United States, United Kingdom, and Australia—are already embedding microbial genomics into standard outbreak monitoring, antimicrobial resistance (AMR) surveillance, and biosecurity systems (CDC Australia, 2025–2027).

In parallel, an emerging intersection between predictive microbiology and precision medicine is taking shape. By using computational tools to model microbial behavior in both food systems and human hosts, predictive models can improve pathogen detection and risk assessment. When combined with genomic insights, these models make it possible to design more effective preventive healthcare measures and targeted food safety strategies. This integrated approach is proving valuable in tackling persistent health challenges such as chronic diseases, AMR, and foodborne infections. Looking ahead, the combined force of AI, synthetic biology, and international collaboration is paving the way for a future where microbial genomics drives safer, more intelligent food systems alongside personalized, predictive, and preventive healthcare.

8 Comprehensive Impact of Microbial Genomics Research on the Global Healthcare Sector

Microbial genomics has revolutionized global healthcare by enabling quick and focused responses to infectious diseases. It has improved diagnostic capabilities, guided vaccine development strategies, and reinforced efforts to manage antimicrobial resistance. The COVID-19 pandemic clearly shows this effect. By employing real-time whole genome sequencing, researchers were able to track the genomic epidemiology of SARS-CoV-2, identify key mutations, trace patterns of transmission, and support public health measures on a population scale. This real-time genomic epidemiology now aids in tracking other viral threats like Ebola, Zika, and monkeypox, enabling timely outbreak and containment efforts (Kraemer et al. 2022). The pioneering work in the field of genomics, specifically in the nursing sequencing of viral RNAs and the understanding of host–pathogen interactions, enabled the development of mRNA-based COVID-19 vaccines. Just as in the case of SARS-CoV-2, foundational genomic work allows the triggering of the immune system to specific oncogenic epitopes and controlling other immune evasion mechanisms greatly aids in vaccine construction, as seen in mRNA-based vaccines (Go et al. 2021). High-throughput quantitative polymerase chain reaction (qPCR), and other metagene-based next-generation sequencing (mNGS), as well as SHERLOCK and DETECTR CRISPR-based diagnostics/assays, enable timely and precise target genomic, pathogen, and organism identity capture and greatly transform diagnostics in the field. These platforms are being utilized more and more to customize therapies in real-time since they can identify new infections even when traditional culture techniques don't work (Chiu and Miller 2019). In order to combat AMR, microbial genomics is essential. Healthcare systems can precisely direct antibiotic stewardship initiatives by sequencing resistant bacteria and detecting resistance genes (for example, through the CARD or ResFinder databases). For example, real-time tracking of resistant organisms is made possible by genomic surveillance in hospitals, which improves clinical outcomes and informs infection control programs (Sundermann et al. 2025).

Globally, exchanging genomic data via networks such as GISAID and the Global Microbial Identifier effort promotes equitable pandemic preparedness by allowing under-resourced nations to take part in cooperative research and early warning systems. This fosters a coordinated response to microbial threats and lessens health inequities worldwide (Gardy and Loman 2018). In conclusion, by enabling pathogen detection, speeding up the development of vaccines and drugs, directing customized treatments, and improving worldwide disease surveillance, microbial genomics is revolutionizing healthcare. Its ongoing incorporation into public health infrastructure is crucial to preventing pandemics in the future and promoting universal access to quality treatment.

9 Comprehensive Report on Microbial Genomics Research and Its Impact on Bangladesh's Healthcare Sector

Microbial genomics research in Bangladesh is playing an increasingly important role in improving healthcare outcomes, from strengthening pathogen surveillance and guiding antibiotic stewardship to advancing diagnostic methods and supporting data-driven public health strategies. For example, researchers at the University of Chittagong have performed whole genome sequencing (WGS) on *Klebsiella pneumoniae* strains obtained from clinical samples in Chattogram, identifying core resistance genes, and traced phylogenetic lineages across adjacent hospitals. In another research, researchers studied 82 isolates of *Acinetobacter baumannii* across multiple Bangladeshi hospitals. This genomic epidemiology showed 28 antibiotic resistance genes and common sequence types, such as ST2 and ST575. It further showed high genomic diversity, and the insights obtained could be used to design effective infection control strategies in hospitals. *Staphylococcus aureus* (MRSA) isolates from Dhaka were analyzed, and several resistance mechanisms were found (such as mecA, efflux pumps, and pathogenicity clusters). The dominant clonal complex was found to be ST361, which is important information for infection control policy and antibiotic stewardship. Approximately 34% of MDR-TB transmission is attributable to recent spread rather than acquired resistance, according to a nationwide molecular epidemiology study of multidrug-resistant *Mycobacterium tuberculosis* (n = 544 isolates collected 2011–2017), which also estimated transmission and resistance patterns and classified predominant lineages (Beijing and T1). In 2024, a new genotype (4.3.1.2.B1) with bla_CTX-M-15 on plasmid pCROB1 was implicated in an outbreak of ceftriaxone-resistant *Salmonella enterica* serovar Typhi in Thailand. These revelations led to the investigation of alternate empirical therapies that are successful in combating resistant strains. Nearly 2000 SARS-CoV-2 genomes have been sequenced by the national respiratory virus genomic surveillance consortium, which also includes icddr,b, and IEDCR. The partnership has extended surveillance to influenza and RSV, providing near-real-time information for vaccination policy and epidemic response. High amounts of antibiotic resistance genes were discovered in a WGS investigation of *E. coli* from Dhaka chicken retail markets, underscoring the dangers of foodborne AMR transmission in urban settings. Resistance genes were found when *Leuconostoc citreum* from artisanal buffalo milk curd was sequenced, supporting the idea that food-associated microorganisms should be monitored genomically to safeguard public health. Bangladesh's core microbial genomics infrastructure, including cholera and diarrheal pathogen genomes, is managed by icddr,b, under the direction of specialists like Shah M. Faruque. Its efforts support pathogen sequencing, national outbreak response, and the conversion of genetic data into public health initiatives. Thus, Bangladesh is likewise using genetic research and responding significantly to the findings.

10 Future Scopes of Microbial Genomics Application to Develop Bangladesh's Medical Sector

Microbial genomics in Bangladesh has the potential to revolutionize the nation's healthcare system in the future. The scopes in Table 1 demonstrate how microbial genomics can spur innovation and improvement in healthcare, given the growing availability of sequencing tools, improved bioinformatics capabilities, and growing public health initiatives:

Table 1 Comparative overview of microbial genomics applications and future development opportunities in Bangladesh

Aspect	Bangladesh	Global	Scope for development in Bangladesh
Pathogen surveillance	Limited, event-based surveillance (e.g., dengue, cholera, TB) using PCR and partial genome typing.	Real-time WGS surveillance (e.g., SARS-CoV-2, AMR pathogens) integrated into national public health.	Expand genome sequencing capabilities and build a national pathogen genomics network (e.g., IEDCR-based).
Antimicrobial resistance (AMR)	Focused on phenotypic resistance; genomic AMR studies limited to research projects.	Genomic AMR mapping integrated into clinical decision-making and stewardship programs.	Implement genomic AMR surveillance and create a national resistance gene database.
Diagnostic tools	Use of conventional microbiological techniques with limited molecular platforms in public hospitals.	CRISPR, NGS, and AI-powered diagnostics used for point-of-care and real-time detection.	Scale up rapid genomic diagnostics (e.g., nanopore, CRISPR-Cas systems) in tertiary and district hospitals.
Microbiome research	Very few studies; gut microbiome profiling limited to nutritional/child health studies.	Advanced human microbiome projects across populations and disease models (e.g., HMP, MetaHIT).	Develop a national microbiome biobank and apply it in NCDs, IBD, autism, and undernutrition interventions.
Vaccine development	Relies on external vaccines; local research focuses on serotyping and immunogenicity studies.	Genomics-guided mRNA and recombinant vaccine platforms (e.g., COVID-19, malaria, TB).	Use pathogen genomics to inform regional vaccine design, especially for endemic infections.
Bioinformatics capacity	Limited to a few institutions (e.g., icddr,b, BSMMU, BUET); skill gap in data analytics.	Robust bioinformatics infrastructure and training integrated into health and research systems.	Invest in national bioinformatics training programs and cloud-based genomics platforms.
Data sharing and integration	Scattered databases; limited access and collaboration.	Centralized genomic repositories (e.g., GISAID, NCBI, ENA) with open-access policies.	Create a centralized genomic data hub with FAIR data principles and international interoperability.
Policy and regulation	No national guidelines on genomic data use, ethics, or privacy.	Policies in place for genomic ethics, clinical application, and biobanking (e.g., GDPR, NIH).	Formulate national genomic ethics and data governance frameworks.

(continued)

Table 1 (continued)

Aspect	Bangladesh	Global	Scope for development in Bangladesh
Precision medicine	In nascent stages; no integration with clinical practice.	Widely implemented in cancer, rare diseases, and infectious disease treatment using host/pathogen genomics.	Pilot programs using host–microbe genomic integration for TB, diabetes, and gut-related disorders.
Funding and collaboration	Dependent on donor-driven projects and external grants.	Large-scale national and international funding (e.g., NIH, EU horizon, UKRI).	Increase government and private-sector investment in microbial genomics and global collaborative grants.

11 Limitations and Challenges in Microbial Genomics

11.1 Infrastructure and Capacity Challenges

High barrier of equipment and laboratory infrastructure: Numerous individual hospitals and research centers in low-resource environments cannot afford substantial capital, specialized lab space, sample storage, and instrument maintenance required to establish sequencing facilities (e.g., WGS, NGS, long-read platforms) (Sundermann et al. 2025).

Shortage of skilled workforce: Genomics and bioinformatics require specialized personnel ranging from technicians to data scientists. Many healthcare systems lack professionals with the necessary training to conduct sample prep, sequencing, and downstream analysis (Vashisht et al. 2023).

Weak data infrastructure: Low-and middle-income countries (LMICs) frequently lack or have unstable internet, cloud infrastructure, high-performance computing, genomic data storage, and secure sharing platforms (Vashisht et al. 2023; Sundermann et al. 2025).

11.2 Technical and Methodological Limitations

Fragmented standardization: There's no globally recognized standard for sequencing bioinformatics pipelines, laboratory procedures, or sequencing platforms. This results in inconsistent quality, reproducibility issues, and difficulties comparing results across labs (Waddington et al. 2022; Sundermann et al. 2025).

Sensitivity and specificity concerns: Low-abundance pathogens may have a chance to be undetected, or complex mixed infections may be difficult for sequencing-based detection to handle, host–DNA contamination may occur, or only fragments of interest may be detected. Without deep sequencing, quantitative evaluation may not be accurate (Waddington et al. 2022; Vashisht et al. 2023).

Coverage gaps and assembly challenges: Incomplete genome assemblies and mis-annotations may arise from repetitive regions, large genomes, or microbial species with few references. Chimeric contigs are frequently produced by metagenomic samples, which makes taxonomic resolution more difficult (Waddington et al. 2022).

12 Data Management and Interpretative Complexity

Data volume overload: Sequencing generates enormous datasets (hundreds of gigabases to petabytes), creating major storage, compute, and analysis burdens-often exceeding institutional capacity (Govender et al. 2021).

Complex analytical tasks: Interpreting microbial versus commensal species, detecting resistance or virulence genes, deriving phylogenetic/epidemiologic insights require advanced, validated bioinformatics tools and curated reference databases (Waddington et al. 2022; Filardo et al. 2024).

13 Cost, Economy, and Sustainability

High per-test cost: Although sequencing costs have dropped, metagenomics remains expensive when including sample prep, reagents, library kits, and computational processing. Economic models show it needs a much higher diagnostic yield to justify costs in routine use (Pulivarti et al. 2023).

Difficulty assessing cost-effectiveness: Limited data exist on long-term economic benefits of integrating WGS into routine diagnostics or surveillance, making investment decisions uncertain.

14 Ethical, Legal, and Policy Constraints

Genomic privacy and re-identification risk: Microbial or host-associated genomic data may accidentally reveal the identity of a patient or their family members, which is a big privacy issue without proper protections or consent frameworks (Pulivarti et al. 2023).

Insufficient governance and regulation: Many countries, including Bangladesh, lack clear policies on data ownership, consent, genomic ethics, or guidelines for clinical and public use of microbial genomic data.

15 Context-Specific Barriers in LMIC and Bangladesh

Limited integration in routine health systems: Genomics-based AMR surveillance or outbreak tracking is mostly done as part of a project, not on a regular basis. Culture or PCR is still used in many healthcare settings (Vashisht et al. 2023).

Supply chain and logistics issues: Consumables, reagents, and platforms often need importation under cold-chain conditions with long delays or waste, which impacts continuity.

Low genomic literacy among healthcare professionals and public: Limited understanding of genomics' value hampers adoption; ethical communication and clinical translation require broader training and awareness (Sundermann et al. 2025).

16 Conclusion

Microbial genomics has broadened the understanding of microorganisms and their connections to human health, industries, and ecosystems. Through the genetic unraveling of bacteria, viruses, archaea, and other microbes, we have gained revolutionary understanding of the pathogenic mechanisms, antibiotic resistance, dynamics of the microbiome, and the interactions of the host and microbe. Such understanding has led to advancements in healthomic innovations, including precision diagnostics, targeted therapeutics, recombinant bioproduction, and surveillance of diseases in real time. Omics technologies and CRISPR-based gene editing, as well as machine learning and synthetic biology, have furthered the interest in systems biology and personalized medicine. The range of possibilities in healthcare is being expanded by microbial genomics, which makes it possible to track tuberculosis mutations and design microbiome-based treatments for neurodegenerative diseases. The emergence of genomic epidemiology for pathogens like *Salmonella typhi, Klebsiella pneumoniae,* and SARS-CoV-2 also signals the national interest, which is readily engaging with microbial genomics in Bangladesh.

However, realizing the full potential of microbial genomics, especially in low- and middle-income countries (LMICs), requires overcoming significant limitations. These include infrastructural deficits, shortages of trained personnel, lack of standardized workflows, high implementation costs, complex data analysis pipelines, and limited regulatory frameworks. Additionally, ethical concerns about genomic privacy and equitable data sharing must be addressed through robust governance. To truly harness microbial genomics for global and national health improvement, there must be sustained investment in infrastructure, workforce development, bioinformatics training, and policy reform. In countries like Bangladesh, embedding genomic tools into routine healthcare, expanding AMR surveillance, establishing national microbiome biobanks, and promoting inter-institutional collaboration will be vital steps forward.

Ultimately, microbial genomics stands as both a scientific frontier and a public health imperative, which offers the tools to decode life at its smallest scale while shaping solutions for some of humanity's largest challenges.

References

Ahmed S et al (2025) Metabolic response of *Klebsiella oxytoca* to ciprofloxacin exposure: a metabolomics approach. Metabolomics 21(1). https://doi.org/10.1007/s11306-024-02206-y

Alper HS, Beisel CL (2018) Advances in CRISPR technologies for microbial strain engineering. Biotechnol J. Wiley-VCH Verlag. https://doi.org/10.1002/biot.201800460

Anjum A et al (2024) Deciphering the genomic character of the multidrug-resistant Staphylococcus aureus from Dhaka, Bangladesh. AIMS Microbiol 10(4):833–858. https://doi.org/10.3934/microbiol.2024036

Benoit P et al (2024) Seven-year performance of a clinical metagenomic next-generation sequencing test for diagnosis of central nervous system infections. Nat Med 30(12):3522–3533. https://doi.org/10.1038/s41591-024-03275-1

Briday M, Carvalho N, Oganesyan N, Chang M-J, Lees A, Brier S, Chenal A (2025) Comparative analysis of the structural dynamics of diphtheria toxin and CRM197 carrier proteins used in the development of conjugate vaccines. Int J Pharm 675:125535. https://doi.org/10.1016/j.ijpharm.2025.125535

Bradley E, Haran J (2024) The human gut microbiome and aging. Gut Microbes. Taylor and Francis Ltd. https://doi.org/10.1080/19490976.2024.2359677

Chen S et al (2025) Application of metagenomic next-generation sequencing in pathogen detection of lung infections. Front Cell Infect Microbiol 15. https://doi.org/10.3389/fcimb.2025.1513603

Chen W et al (2024) Evaluation of metagenomic and pathogen-targeted next-generation sequencing for diagnosis of meningitis and encephalitis in adults: a multicenter prospective observational cohort study in China. J Infect 88(5). https://doi.org/10.1016/j.jinf.2024.106143

Chiu CY, Miller SA (2019) Clinical metagenomics. Nat Rev Genet. Nature Publishing Group:341–355. https://doi.org/10.1038/s41576-019-0113-7

Cunningham-Oakes E et al (2025) Enhancing infectious intestinal disease diagnosis through metagenomic and metatranscriptomic sequencing of 1000 human diarrhoeal samples. Genome Med 17(1). https://doi.org/10.1186/s13073-025-01478-w

Dash NR, Al Bataineh MT (2020) Metagenomic analysis of the gut microbiome reveals enrichment of menaquinones (Vitamin K2) pathway in diabetes mellitus. Diabetes Metab J 44. https://doi.org/10.4093/dmj.2019.0202

Dokht Khosravi A, Goodarzi H, Alavi SM (2012) Detection of genomic mutations in katG, inhA and rpoB genes of Mycobacterium tuberculosis isolates using polymerase chain reaction and multiplex allele-specific polymerase chain reaction. Braz J Infect Dis 16:57. Available at: www.elsevier.com/locate/bjid

Du P et al (2014) Sequence variation in tcdA and tcdB of Clostridium difficile: ST37 with truncated tcda is a potential epidemic strain in China. J Clin Microbiol 52(9):3264–3270. https://doi.org/10.1128/JCM.03487-13

Filardo S, Di Pietro M, Sessa R (2024) Current progresses and challenges for microbiome research in human health: a perspective. Front Cell Infect Microbiol 14. https://doi.org/10.3389/fcimb.2024.1377012

Gardy JL, Loman NJ (2018) Towards a genomics-informed, real-time, global pathogen surveillance system. Nat Rev Genet. Nature Publishing Group:9–20. https://doi.org/10.1038/nrg.2017.88

Go CD et al (2021) A proximity-dependent biotinylation map of a human cell. Nature 595(7865):120–124. https://doi.org/10.1038/s41586-021-03592-2

Govender KN et al (2021) Metagenomic sequencing as a pathogen-agnostic clinical diagnostic tool for infectious diseases: a systematic review and meta-analysis of diagnostic test accuracy studies. J Clin Microbiol. American Society for Microbiology. https://doi.org/10.1128/JCM.02916-20

Jayakrishnan TT et al (2024) Multi-omics machine learning to study host-microbiome interactions in early-onset colorectal cancer. NPJ Precis Oncol 8(1). https://doi.org/10.1038/s41698-024-00647-1

Khorsand B et al (2022) Overrepresentation of Enterobacteriaceae and Escherichia coli is the major gut microbiome signature in Crohn's disease and ulcerative colitis; a comprehensive metagenomic analysis of IBDMDB datasets. Front Cell Infect Microbiol 12. https://doi.org/10.3389/fcimb.2022.1015890

Kraemer MUG, Tegally H, Pigott DM, Dasgupta A, et al (2022) Tracking the 2022 monkeypox outbreak with epidemiological data in real-time. Lancet Infect Dis 22(7):941–942. https://doi.org/10.1016/S1473-3099(22)00359-0

Lubrano P, Smollich F, Schramm T, Lorenz E, Alvarado A, Eigenmann SC, Stadelmann A, Thavapalan S, Waffenschmidt N, Glatter T, Hoffmann N, Müller J, Peter S, Drescher K, Link H (2025) Metabolic mutations reduce antibiotic susceptibility of E. coli by pathway-specific bottlenecks. Mol Syst Biol 21:274–293. https://doi.org/10.1038/s44320-024-00084-z

Ma J et al (2022) Engineered probiotics. Microb Cell Fact. BioMed Central Ltd. https://doi.org/10.1186/s12934-022-01799-0

Metcalfe-Roach A et al (2024) Metagenomic analysis reveals large-scale disruptions of the gut microbiome in Parkinson's disease. Movement Disorders. https://doi.org/10.1002/mds.29959

Monir MM et al (2023) Genomic attributes of Vibrio cholerae O1 responsible for 2022 massive cholera outbreak in Bangladesh. Nat Commun 14(1). https://doi.org/10.1038/s41467-023-36687-7

Mu Y et al (2022) Development and applications of CRISPR/Cas9-based genome editing in lactobacillus. Int J Mol Sci. MDPI. https://doi.org/10.3390/ijms232112852

Nooij S et al (2025) Metagenomic global survey and in-depth genomic analyses of Ruminococcus gnavus reveal differences across host lifestyle and health status. Nat Commun 16(1). https://doi.org/10.1038/s41467-025-56449-x

Oliveira R et al (2024) Using peptide nucleic acid fluorescence in situ hybridization (PNA-FISH) to detect Campylobacter spp. in food samples. LWT 198. https://doi.org/10.1016/j.lwt.2024.115922

On Y et al (2025) Genomic analysis of carbapenem-resistant Klebsiella pneumoniae blood isolates from nationwide surveillance in South Korea. Front Microbiol 16. https://doi.org/10.3389/fmicb.2025.1562222

Praveschotinunt P et al (2019) Engineered E. coli Nissle 1917 for the delivery of matrix-tethered therapeutic domains to the gut. Nat Commun 10(1). https://doi.org/10.1038/s41467-019-13336-6

Pulivarti R et al (2023) Cybersecurity of genomic data. https://doi.org/10.6028/NIST.IR.8432

Rahmati R et al (2025) CRISPR-assisted probiotic and in situ engineering of gut microbiota: a prospect to modification of metabolic disorders. Probiotics Antimicrob Proteins [Preprint]. Springer. https://doi.org/10.1007/s12602-025-10561-y

Ruden DM (2025) TIGR-Tas and the expanding universe of RNA-guided genome editing systems: a new era beyond CRISPR-Cas. Genes 16(8):896. https://doi.org/10.3390/genes16080896

Saravanan KA et al (2022) Role of genomics in combating COVID-19 pandemic. Gene. Elsevier B.V. https://doi.org/10.1016/j.gene.2022.146387

Sundermann AJ, Kumar P, Griffith MP, Waggle KD, Srinivasa VR, Raabe N, Mills EG, Coyle H, Ereifej D, Creager HM, Ayres A, Van Tyne D, Pless LL, Snyder GM, Roberts M, Harrison LH (2025) Real-Time Genomic Surveillance for Enhanced Healthcare Outbreak Detection and Control: Clinical and Economic Impact. Clin Infect Dis. https://doi.org/10.1093/cid/ciaf216

de Souza EV et al (2024) Large-scale proteogenomics characterization of microproteins in Mycobacterium tuberculosis. Sci Rep 14(1). https://doi.org/10.1038/s41598-024-82465-w

Sroka-Oleksiak A et al (2025) AI-driven rapid identification of bacterial and fungal pathogens in blood smears of septic patients

Sundermann AJ et al (2025) Pathogen genomics in healthcare: overcoming barriers to proactive surveillance. Antimicrob Agents Chemother. American Society for Microbiology. https://doi.org/10.1128/aac.01479-24

Thomas AM et al (2019) Metagenomic analysis of colorectal cancer datasets identifies cross-cohort microbial diagnostic signatures and a link with choline degradation. Nat Med 25(4):667–678. https://doi.org/10.1038/s41591-019-0405-7

Vashisht V et al (2023) Genomics for emerging pathogen identification and monitoring: prospects and obstacles. BioMedInformatics. Multidisciplinary Digital Publishing Institute (MDPI):1145–1177. https://doi.org/10.3390/biomedinformatics3040069

Volkova A, Ruggles KV (2021) Predictive metagenomic analysis of autoimmune disease identifies robust autoimmunity and disease specific microbial signatures. Front Microbiol 12. https://doi.org/10.3389/fmicb.2021.621310

Waddington C et al (2022) Exploiting genomics to mitigate the public health impact of antimicrobial resistance. Genome Med. BioMed Central Ltd. https://doi.org/10.1186/s13073-022-01020-2

Wang J et al (2025) Integrative analysis of multi-omics data and gut microbiota composition reveals prognostic subtypes and predicts immunotherapy response in colorectal cancer using machine learning. Sci Rep 15(1). https://doi.org/10.1038/s41598-025-08915-1

Wang WL et al (2015) Application of metagenomics in the human gut microbiome. World J Gastroenterol. WJG Press:803–814. https://doi.org/10.3748/wjg.v21.i3.803

Wu H et al (2023) Neuroprotective effects of an engineered Escherichia coli Nissle 1917 on Parkinson's disease in mice by delivering GLP-1 and modulating gut microbiota. Bioeng Transl Med 8(5). https://doi.org/10.1002/btm2.10351

Zhang P et al (2021) Genomic analysis of Shiga toxin-producing Escherichia coli O157:H7 from cattle and pork-production related environments. NPJ Sci Food 5(1). https://doi.org/10.1038/s41538-021-00097-0

Single-Cell Analysis and Cancer Research: Implications for Diagnosis and Treatment

Partha Biswas, Md Abdur Rahim, Dipankor Chatterjee, Mohammad Ali, Jannatul Nayeem Tabia, Mohammad Kawsar Sharif Siam, and Mohammad Umer Sharif Shohan

Abstract Single-cell technologies have revolutionized cancer studies by enabling high-resolution granular exploration of individual cells in cancer cell populations. Here, we explored how single-cell sequencing of genomics, epigenomics, transcriptomics, and multi-omics provides deep insight into cancer biology, interactions in tumor microenvironment, dynamics of immune response, and evolution of cancer. We describe the central platforms and computational approaches used in single-cell

---

Partha Biswas, Md Abdur Rahim and Dipankor Chatterjee contributed equally with all other contributors.

P. Biswas
Division of Medical Oncology, Department of Internal Medicine, The Ohio State University College of Medicine, Columbus, OH, USA

Pelotonia Institute for Immuno-Oncology, The Ohio State University Comprehensive Cancer Center, The Arthur G James Cancer Hospital and Richard J Solove Research Institute, Columbus, OH, USA

M. A. Rahim
Human Microbiome Medical Research Center (HM·MRC), School of Medicine, Soonchunhyang University, Asan-si, Republic of Korea

D. Chatterjee
Department of Molecular Genetics, The Ohio State University, Columbus, OH, USA

M. Ali
Nuffield Department of Clinical Medicine, University of Oxford, Oxford, UK

J. N. Tabia
Late Phase Haematology, Oxford University Hospital NHS Foundation Trust, Oxford, UK

M. K. S. Siam
School of Pharmacy, Brac University, Dhaka, Bangladesh

M. U. S. Shohan (✉)
Center for Human Genetics, University of Oxford, Oxford, UK
e-mail: mohammad.sharifshohan@seh.ox.ac.uk

A. H. M. N. Nabi, H. Uddin Shekhar (eds.), *Biotechnological Advances in Healthomics*, Interdisciplinary Biotechnological Advances,
https://doi.org/10.1007/978-981-95-3612-2_3

research, along with data sparsity, integration, and challenges in clinical translation. Key applications are illustrated in early diagnosis of cancer, identification of distinct biomarkers, monitoring of minimal residual disease, and classification of rare cell types. We further explored the identification of resistance mechanism to drugs, blockade of immune checkpoints, and strategies for adaptive immune therapy through single-cell technologies. Finally, we assess the drawbacks and future directions which is shifting toward precision oncology assisted by advanced computational analytics and initiatives taken globally. Overall, this chapter underlines the potential of single-cell technology to transform cancer diagnosis and therapeutics, which paves the way for precision medicine.

Keywords Single-cell sequencing · Transcriptomics · Multi-omics · Tumor microenvironment · Biomarkers · Precision oncology

Abbreviations

G&T-seq	genome and transcriptome sequencing
scCITE-seq	Cellular Indexing of Transcriptomes and Epitopes by Sequencing
scSNARE-seq	single-nucleus chromatin accessibility and mRNA expression sequencing
scVDJseq	single-cell V(D)J sequencing

1 Introduction

Cancer is considered one of the most difficult diseases due to its cellular and molecular complexity. Bulk sequencing approaches have traditionally provided insights into cancer biology, but they are unable to show the diversity present in cells and their functions (Patel et al. 2014; Sun et al. 2021). The breakthrough of single-cell sequencing technology has transformed our way to characterize thousands of cells individually, which provides us with perspectives on transcriptomics, genomics, epigenomics, and metabolomics within a tumor in high resolution (Jovic et al. 2022). Compared to bulk that provides average expression of all the cells, single-cell sequencing provides granular view of cancer dynamics including insights into the circulating malignant tumor cells, tumor microenvironment, immune cell distribution, mechanisms associated with tumorigenesis, metastasis and resistance to therapy (Sun et al. 2021). Single-cell analysis is significant for characterizing the molecular and cellular structure of the tumor microenvironment, delineating high-resolution architecture of the spatial and functional heterogeneity among the different malignant, stromal and immune cell population (Hu et al. 2022). By analyzing cell-type-specific transcriptional programs, the single-cell technologies annotate the dynamic reprogramming of tumor-infiltrating lymphocytes, tumor-mediated macrophages, and myeloid-derived suppressor cells that can induce the immunosuppressive function and tumor progression (Chuntova et al. 2020). A comprehensive visualization of intra-tumoral heterogeneity at the single-cell level provides a foundation for the rational development of precision medicine (Xulu et al. 2023).

In this chapter, we will explore the implications of single-cell technology for diagnosis and treatment in cancer research (Liang and Fu 2017). We will begin by examining the foundation of single-cell analysis, followed by exploring cancer biology, specifically tumor heterogeneity, microenvironment, and evolution (Liang and Fu 2017). We will then explain how single-cell technology is reshaping the diagnostic sector by identifying novel biomarkers through increasing precision (Van de Sande et al. 2023). In this chapter, we will also illustrate the role of this technology in guiding personalized treatment by identifying the therapeutic target and monitoring response (Tang et al. 2023). In the last part of the chapter, we will evaluate the challenges and future direction of single-cell technology in advancing cancer research (Jovic et al. 2022; Liang and Fu 2017).

Through navigating this chapter, the reader will understand that the single-cell technology is not just an advancement of technology, but a paradigm shift in how we explore cancer research in diagnosis and treatment (Liang and Fu 2017). Single-cell analysis is continually proving to be a process that will bring us closer to conquering cancer through precise, personalized, and effective diagnosis and treatment (Jovic et al. 2022; Liang and Fu 2017).

2 Fundamental Technologies for Research on Single-Cell Cancer

This technology has become prevalent in research since the inaugural single-cell mRNA sequencing experiment in 2009 (Tang et al. 2009), succeeded by the first single-cell DNA sequencing experiment in human cancer cells in 2011(Navin et al. 2011), and subsequently the initial single-cell exome sequencing experiment in 2012 (Hou et al. 2012), this technology has entered most of the research area. The breakthrough of our understanding of human biology, including application in cancer research that came from this technology, exceeds any other sequencing technologies (Jovic et al. 2022; Liang and Fu 2017).

2.1 Platforms for Single-Cell Isolation and Sequencing

2.1.1 Single-Cell Isolation and Preparation

Single-cell technologies are evolving rapidly with continuous advancement in processes for isolation, sequencing, cDNA library preparation, and analytical algorithms. The first step in this process is isolation of individual cells from samples (Jovic et al. 2022). There are several methods that have been developed for this purpose:

Serial Dilution: It is a straightforward and affordable method that requires more time and labor. It does not have an automation process, and simultaneous separation of a large number of cells is not possible with this method (Hwang et al. 2018).

Micro-Manipulation: It is a classical method that is suitable for a few cells, but it is low throughput and has potential for cell shearing (Hwang et al. 2018).

Laser Capture Dissection (Optical Tweezers): This method is used to isolate single cells from complex matrices. It utilizes a combination of imaging-based cell selection with optical trapping, allowing cell differentiation based on types without the need for labeling. However, inaccuracy in probe-guided slicing may lead to impurities or loss of important genes (Hwang et al. 2018).

Microfluidics: This method offers several advantages. Here, consumption of cells is low, analysis of cost is low, and control of the experiment is precise. Microfluidic devices can encapsulate a single cell with a single bead inside an oil droplet, creating a sealed noncontaminated environment (Zhou et al. 2021b). The most significant feature of this method is that its high-throughput process captures thousands of cells in a single pass, leading to the identification of rare cells in different biological conditions (Hwang et al. 2018).

Fluorescence Activated Cell Sorting (FACS): This method allows for sorting of cells based on different characteristics using unrestricted sorting gates. Subsequently, antibodies labelled with fluorescence recognize surface markers, leading to the identification of distinct cell populations (Penter et al. 2018).

3 Single-Cell Sequencing Technologies

There are several single-cell sequencing technologies, each having diverse applications and advantages as follows:

3.1 Single-Cell DNA Sequencing (scDNA-seq)

scDNA-seq carries out massive parallel sequencing of thousands to millions of DNA molecules at the same time. The steps involve isolation of a single cell, amplification of whole genomes of specific cells, followed by high-throughput sequencing to decipher biological functions of genes in single-cell resolution (Hong et al. 2020). This is particularly useful for studying variation in genomes, including mutation, copy number variation, and chromosomal rearrangements in cancer cells (Mallory et al. 2020).

After isolating the limited number of DNA (6–7 pg), whole genome amplification (WGA) is needed (Wang et al. 2022b). There are several methods available for this purpose that include degenerate multiple displacement amplification (MDA), oligonucleotide-primed PCR (DOP-PCR), and multiple annealing and looping-based amplification cycles (MALBAC). Each of these methods has some advantages and disadvantages relating to coverage, uniformity, and error rates (Wang et al. 2022b).

3.2 Single-Cell RNA Sequencing (scRNA-seq)

The steps involved in scRNA-seq technologies include isolation of single cells, extraction of RNA, reverse transcription, preamplification, and detection. The initial steps are vital for accuracy and the final amount of amplified material (Haque et al. 2017). The research in this area is evolving rapidly, and several platforms have been developed (Table 1).

Smart-seq/Smart-seq2: This provides full-length coverage of transcripts that help identify allele-specific expression and alternative splicing, but the caveat is low throughput (Picelli et al. 2014; Wang et al. 2021).

Drop-seq, Seq-Well, and inDrop: These methods are based on beads that employ microfluidic systems for encapsulating single cells with barcoded beads. Although this method is high throughput, the capture is only for 3′ or 5′ ends of transcripts (Zhang et al. 2019).

10× Genomics Chromium: This is the most widely used commercial platform that parallelly uses microfluidics and barcoding to analyze thousands of cells with great sensitivity (Wang et al. 2021).

3.3 Single-Cell ATAC Sequencing (scATAC-seq)

scATAC-seq (Assay for Transposase-Accessible Chromatin using sequencing) is developed to identify and map regions that are open in chromatin in single cells. This sequencing method employs Tn5 transposase to fragment and tag accessible DNA simultaneously with sequencing adapters. The main purpose of this method is that it provides a regulatory process of the whole cells, which reveals the enhancers and transcription factors that are involved in the biological processes (De Rop et al. 2024).

3.4 Single-Cell Multi-Omics Technologies

With the advancement of research, there are technologies developed in single-cell space that can simultaneously profile several features from the same single cell (Wu et al. 2024):

Table 1 A descriptive comparison of different single-cell platforms

Platform	Throughput (Cells/ Run)	Transcript Resolution	Cost per Cell	Primary Use Cases	References
Smart-seq2	Low (~100–1000); requires manual handling or FACS	Full-length transcript coverage; captures isoforms and splicing	High (~$2–$10)	Rare cell types, isoform detection, allele-specific expression, high-resolution analysis	Picelli et al. (2014)
10× Genomics	High (~1000–100,000+); droplet-based microfluidics	3′ or 5′ end counting with UMIs; gene-level resolution	Low (~$0.10–$0.50)	Immune profiling, tumor microenvironment, atlas projects, large-scale population studies	Zheng et al. (2017)
Drop-seq	High (~10,000–50,000); academic droplet platform	3′ end with UMI tagging; moderate sensitivity	Very low (<$0.10)	Cost-efficient profiling, novel cell type discovery, proof-of-concept experiments	Macosko et al. (2015)
Seq-Well	Moderate (~10,000); microwell-based and portable	3′ end with UMIs; lower sensitivity than droplet methods	Low (~$0.10)	Point-of-care, low-resource labs, translational/clinical settings	Gierahn et al. (2017)
inDrop	High (~10,000–100,000); droplet encapsulation with hydrogel beads	3′ end with UMIs; customizable protocol	Moderate (~$0.20–$0.60)	Flexible profiling in developmental biology, immunology, and early-stage technology development	Klein et al. (2015)

scCITE-seq: This process employs oligonucleotide-labelled antibodies to profile proteins in combination with transcriptomics profiling in the same single cell (Stoeckius et al. 2017).

scSNARE-seq: This process employs simultaneous profiling of chromatin accessibility along with transcriptomics profiling in the same single cell (Plongthongkum et al. 2021).

G&T-seq: This process allows for sequencing DNA and RNA within the same single cell, revealing the genetic variation that can affect the transcriptomics (Macaulay et al. 2015).

scVDJseq: This process allows profiling the immune receptor repertoire through sequencing variable (V), diversity (D), and joining (J) regions of B-cell and T-cell, along with the transcriptomics, leading to insights into the clonal diversity, specificity of antigen, and immune response across disease conditions in the same cell (Schnormeier and Budeus 2024).

These multi-omics approaches are getting popular in the research area as they provide comprehensive information on the function and state of cells and the relative relationship across modalities in a single cell, which was unprecedented with bulk sequencing (Wu et al. 2024).

4 Data Analysis and Computational Challenges in Single-Cell Sequencing

4.1 Computational Complexity

Due to the nature of the single-cell sequencing technology that provides granular-level information about each cell, it has also introduced significant computational complexity. Compared to bulk sequencing data, single-cell data are noisy, have a higher number of dimensions, and are sparse. As each cell is separately captured, they vary significantly in sequencing depth and detection of genes. As a result of this, we get a large matrix of expression data with genes in rows and cells in columns with many missing or zero values, creating difficulty in interpreting the biological phenomena (Arya et al. 2025).

4.1.1 Key Analytical Workflows in Single-Cell Data Analysis

Preprocessing and Quality Control

The first step in the single-cell data analysis includes filtering of cells with high mitochondrial content, or low in gene count, followed by normalizing for depth of sequencing and correcting the batch effect for different samples. Harmony and SCTransform are widely used to normalize and correct for batches, improving the quality of the data for downstream processing (Arya et al. 2025; Korsunsky et al. 2019; Hafemeister and Satija 2019). Preprocessing and quality control are crucial in single-cell transcriptomic analysis to eliminate low-quality cells, characterized by high mitochondrial gene expression or low transcript counts, followed by normalization and batch correction to guarantee data consistency. SCTransform utilizes a regularized negative binomial regression model to normalize UMI count data successfully while addressing technical variance, whereas Harmony implements strong batch effect correction by matching cells across conditions without undermining biological variability (Hafemeister and Satija 2019; Korsunsky et al. 2019). Azimuth further extends this framework by enabling reference-based cell type annotation through supervised mapping of query datasets to annotated single-cell atlases, facilitating accurate and scalable cellular classification (Hao et al. 2021).

Dimensionality Reduction and Clustering

As the single-cell matrix has large dimensions, dimensionality reduction helps for efficient computation and visualization. Several methods are used for this purpose, including principal component analysis (PCA), t-distributed stochastic neighbor embedding (t-SNE), and uniform manifold approximation and projection (UMAP), creating low dimensions for the whole data while maintaining the main information available in the data. After this, clustering algorithms like Louvain or Leiden are used to identify similar cell types (Arya et al. 2025; Ghojogh et al. 2023).

Differential Expression and Functional Enrichment

One of the targets for conducting single-cell experiments is to identify the cells that are differentially expressed between different cell types of conditions, either using FindMarker function employing Wilcoxon-rank-sum tests or DESeq2 algorithms (Love et al. 2014). This is followed up with a functional enrichment study based on different databases like Gene Ontology (GO), Kyoto Encyclopedia of Genes and Genomes (KEGG) (Kanehisa et al. 2016), or Reactome to provide information about the biological systems and processes (Arya et al. 2025).

Gene Regulatory Network Inference

SCENIC (Single-Cell Regulatory Network Inference and Clustering) algorithm is employed to explore the transcriptional regulators at the single-cell level by providing information about dynamic processes in cell differentiation or progression of tumor (Arya et al. 2025; Aibar et al. 2017).

Cell Type Annotation and Functional Assignment

Assigning the cell types to different clusters is a vital step to check the different cell types present in the data. This process is either done using different markers and manually setting the cell type based on those specific markers or some reference-based cell annotation algorithms like SingleR and Azimuth (Arya et al. 2025; Cheng et al. 2023).

4.2 Computational Bottlenecks and Emerging Solutions

Although single-cell research is evolving and expanding rapidly, there are many areas that would require deep research. Some of these bottlenecks are explained below:

Data sparsity: Single-cell data have a high count of zeros which can obscure the models build on them leading to unreal biological signals and interpretations (Soneson and Robinson 2018).
Scalability: As these sequencing can be done on multiple samples parallelly, the data come out are of high size requiring memory-efficient algorithms for processing (Soneson and Robinson 2018).
Multi-modal integration: Newer studies are being conducted combining data from transcriptomics, repertoire data, and protein data, which requires improved methods that can properly integrate them (Soneson and Robinson 2018; Stanojevic et al. 2022).

5 Single-Cell Profiling Provides Biological Insights

The advancement of single-cell technology has transformed our understanding of cancer biology (Keller and Pantel 2019). Previously, traditional methods such as genomic and transcriptomic technology have profiled the interaction of diverse cancer and noncancer cells, determining the tumor biology. These methods overlooked individual cellular functions, the position of the cells, and the interaction of those cells with the surrounding environment, also known as the tumor microenvironment, creating a knowledge gap in our ultimate understanding of this disease. However, recent advances in single-cell profiling have paved the way for studying tumors at the individual cell level, providing a new strategy to bridge gaps in our understanding of tumors (Tirosh and Suvà 2018).

5.1 Tumor Heterogeneity and the Evolution of Cancer

Tumors are composed of a diverse array of cell types that can vary significantly in their genetic and phenotypic characteristics. Based on the findings on tumor heterogeneity, clinicians and researchers define cancer development as an evolving process. Despite the significant improvement of clinical aspects, the role of compositional tumor heterogeneity and selection pressures stimulated by the tumor microenvironment (TME) in provoking tumor evolution was largely overlooked. In this background, single-cell technologies are able to track clonal evolution within tumors and reveal how tumor cells adapt over time and evolve in response to therapeutic pressures (Zellmer and Zhang 2014). For example, in recent years, single-cell technologies have provided a better understanding of tumor heterogeneity through integrating epigenomic data with multi-omics data at the single-cell resolution (Dawson and Kouzarides 2012; Adams 2017). Additionally, research had been conducted to find key gene mutations that initiate tumor progression utilizing the scRNA-seq assays (Coons et al. 1941). At the same time, it can provide a comprehension between heterogeneity and temporal dynamics (Peng et al. 2019).

5.2 *Immune Dynamics and the Tumor Microenvironment*

A tumor is a heterogeneous collection of infiltrating and resident host cells, secreted factors, and extracellular matrix. Tumor cells provoke molecular, cellular, and physical changes within their host tissues, thereby supporting tumor growth and progression. Actually, TMF is a complex and continuously evolving entity and the hallmark features include immune cells, stromal cells, blood vessels, extracellular matrix, fibroblasts, endothelial cells, and cancer-associated fibroblasts (CAFs), extracellular matrix components, signaling molecules, and metabolites produced by the tumor and surrounding tissue (Anderson and Simon 2020; Zhu et al. 2012; Yin et al. 2010). Therefore, TMF is not only considered a silent bystander but also an active promoter of cancer progression (Truffi et al. 2020). Immune cells are critical components of the tumor microenvironment. Depending on the context, immune cells can either suppress tumor growth or enhance it. In this context, single-cell techniques have enabled the detailed characterization of immune cell infiltration within tumors. Thus, researchers can identify key immune subsets that play a vital role either in a supportive or detrimental manner in the case of tumor growth. In this context, the study shows the beneficial impact of scRNA-seq assays that enabled researchers to analyze tumor-infiltrating immune cells of many cancers, such as melanoma (Durante et al. 2020; Gerber et al. 2016), lymphoma (Roider et al. 2020), glioma (Goswami et al. 2019), breast (Azizi et al. 2018; Chung et al. 2017), and lung (Guo et al. 2018; Lambrechts et al. 2018) cancers. Moreover, a study conducted by Chen et al. showed the pancreatic ductal adenocarcinoma heterogeneity and identified a new cancer-associated fibroblast (CAF) subset utilizing this technology (Chen et al. 2021). Ultimately, these insights can be beneficial in advancing cancer immunotherapy by utilizing the full potential against cancer.

5.3 *Cancer Stem Cells and Circulating Tumor Cells*

In the case of cancer, most diagnosis comes at an advanced stage where dissemination to other organs is already present (Ando et al. 2021). Based on the evidence, scientists have shifted their focus to target the early stage of cancer which will increase the chance to develop the effective treatments for this disease (Dunn 2012). In the early stages, there is a risk of recurrence that seems to be related to cancer stem cells (CSCs) that can be the cause of cancer progression, metastasis, and drug resistance (Aramini et al. 2022). Scientists across the world agreed that these cells will revolutionize our understanding of the cellular and molecular events related to cancer progression (Lambert et al. 2017). One of the most important aspects of these cells is our capacity to identify specific markers, such as superficial antigens, such as CD133, CD44, and CD90, or intracellular enzymes such as aldehyde dehydrogenase (ALDH) (Zhou et al. 2021a; Herreros-Pomares 2022). In this background, scRNA-seq assays play a vital role in revealing mechanisms of cancer stem cells. For instance, a recent study using scRNA-seq showed an enrichment of cancer stem-cell subpopulation during bladder cancer recurrence with elevated expression of EZH2 and also revealed distinct EMT features associated with bladder cancer subtype (Wang et al. 2022a).

Another kind of cells that also play a vital role in cancer biology is circulating tumor cells (CTCs), which are a distinct population of cancer cells that have detached from the primary tumor and flow into the blood circulation (Tibbe et al. 2007; Hermann et al. 2007). Later, they migrate to healthy tissues, and their survival depends on their interaction with the host (Mani et al. 2008). Alongside, many angiogenic factors and signaling agents contribute to their colonization. Thus, CTCs are valued as a prognostic factor in the development of cancer and progression to metastases. Regarding this, a study reported an identification of melanoma biomarkers in circulating tumor cells using single-cell technologies that have significant potential in disease tracking (Fankhauser et al. 2022). This indicates that they might have cancer stem cell (CSC)-like features: the ability to grow new tumors (Figueira et al. 2009; Tien et al. 2004). In this background, single-cell technologies have provided a way to reveal the unique characteristics of CSCs and CTCs, allowing for the identification of specific markers and pathways that govern their behavior and ultimately paving the way for more adaptive treatment approaches.

6 Prognostic and Diagnostic Uses

6.1 Early Detection and Biomarker Discovery

Cancer can be fatal if detected after metastasis has already started. The idea of being able to detect cancer at an early stage would open a new window for cancer treatment (Gerstberger et al. 2023). However, cancers are conventionally diagnosed according to tissue origin and histological characteristics. The clinical and published data indicate that molecular characteristics affect the tumor microenvironment and hence modify clinical behaviors and therapeutic responses (Junttila and De Sauvage 2013). An enhanced understanding of molecular characteristics by identifying biomarkers, mutations, and gene expression profiles would advance the diagnosis and prognosis of cancer in the early stage (Sethi et al. 2013). The advancement of single-cell sequencing approaches has led to a significant advancement in cancer biology, empowering the investigation of cancer heterogeneity with the highest efficacy. The high-throughput methodologies accelerate the identification of distinct cellular cancer subpopulations and identify the gene expressions profiles that was previously impossible to identify through bulk sequencing, hence increasing early-stage cancer detection and biomarker discovery (Hu et al. 2023). A study employing scRNA-seq on primary lung cancer samples discovered differentially elevated genes, including CXCL1 and CXCL2, which were further verified via quantitative real-time PCR. Furthermore, microRNAs such as miR-532-5p and miR-1266-3p exhibited inverse correlations with chemokine mRNAs, indicating their potential as circulating biomarkers for early detection (Kim et al. 2021). In addition, the single-cell sequencing methods have enabled the capability to detect the rare cell types of malignancies along with the gene expression. In ovarian cancer, the single-cell sequencing approach identified two primary sub-populations based on their gene expression profiles and intratumoral heterogeneity with the effective treatment targets (Balan et al. 2024). Moreover, the single-cell immune cell profiling has enhanced the understanding of the tumor and immune interactions in the cancer microenvironment. By analyzing T cell single-cell data and gene expression profiles, the scientists can easily identify the previously unknown immune cell population within the tumor microenvironment, which will increase the efficacy of predictive and prognostic biomarkers (Zhou et al. 2024).

Although single-cell RNA-seq offers unparalleled insights into gene expression, the inherent tissue separation process associated with the technology inevitably impacts the spatial arrangement of tissues (Longo et al. 2021). A novel technique that addresses this restriction is spatial transcriptomics (ST), which utilizes the inherent architecture of tissue to preserve the spatial localization of gene expression. The early-stage cancer initiation can be detected by the spatial transcriptomics (ST), which integrates the transcriptomics data with the tissue histological context to uncover the region-specific gene expression patterns and intracellular connections (Rao et al. 2021). The recent development of spatial transcriptomics platforms like 10× Genomics, Slide-seq, and GeoMX Digital Spatial Profiler has increased the transcriptomics efficiency in the tumor and peritumoral regions. These technologies have been significant in differentiating malignant lesions from benign precursors and in detecting tumor-adjacent fields with molecular aberrations that seem histologically normal, which will be effective in detecting early-stage cancer initiation (Hu et al. 2022). A study conducted by spatial transcriptomics revealed that a unique set of genes including CCl20 and IL1B are upregulated in breast cancer, however, these genes have not been identified through single-cell RNA-seq alone (Dave et al. 2023). Additionally, in pancreatic neoplasia, the spatial transcriptomic analysis has reported the immune-excluded regions with the high expression of CXCL12, a chemokines known to exhausted T-cell infiltration (Karamitopoulou et al. 2024). These research findings suggest that the biomarkers identified by spatial transcriptomics will improve efficacy of early detection of cancer by capturing context-specific gene expression in the tumor-immune interactions.

Single-cell multi-omics approach has revealed its significance to become the next-generation tool to discover biomarkers by combining multiple layers of molecular characterization like as the epigenome, transcriptome, chromatin accessibility, and protein-protein interaction of any distinct cell population. These comprehensive characteristics reveal a multidimensional aspect of the cellular states and transitions that can differentiate malignant cells and their normal counterparts with high efficiency (Wu et al. 2024). By using the multi-omics approach, a research work identified the early activation of WNT and MYC-driven enhancer networks in colorectal cancer, which can be developed as an effective marker of this cancer progression (Koh et al. 2010). In addition, the combination of scRNA-seq with the CITE-seq

(Cellular Indexing of Transcriptomes and Epitopes by Sequencing) provides a potential efficacy to quantify the mRNA and surface protein markers to detect the early immune dysregulation. A study using the CITE-seq in melanoma biopsies detected a higher level of PD-1 expression on the T cells despite the low expression of PDCD1 transcripts, indicating that the RNA-seq alone failed to capture the post-translational regulation (Marmonti et al. 2022).

Moreover, a recent pan-cancer study integrated the scRNA-seq, scATAC-seq, and spatial proteomics and revealed that the early tumor exhibited the chromatin remodeling at interferon response elements in Tregs, which are effectively associated with the loss of MHC-I and some immunosuppressive ligand expression (Zhang et al. 2024a). Furthermore, other high-throughput computational models such as Seurat v5 and MOFA+ (Multi-omics Factor Analysis) can conduct the integration analysis in a diverse cellular population in the cancer microenvironment (Sammut et al. 2021). Consequently, the above-mentioned tools will be more familiar in future oncology research to discover effective cancer biomarkers that will be used in the clinics to detect cancer in the early stage (Fig. 1). The whole mechanistic process is elaborated in Fig. 1 including different steps such as acquiring samples from the patient's body, creating single ell suspension, high-throughput single-cell sequencing, analysis of processed singlecell sequencing data by multiple single-cell analysis approaches to measure the expression of multiple marker genes that can be used to detect the cancer initiation in the early stage.

6.2 *Precision Classification and Minimal Residual Disease*

The tumor heterogeneity exhibits the major challenges in cancer classification and risk assessment because of the failure of the conventional histopathology approach to identify the molecular complexity. The single-cell technologies like scRNA-seq and multi-omics approaches have overcome these limitations by evaluating the cellular hierarchies and functional diversity within the tumor microenvironment, which is really significant for personalized cancer therapy (Peng et al. 2020). A research work conducted by single-cell analysis has identified the malignant subtypes of pancreatic ductal adenocarcinoma (PDAC), such as basal-like and intermediate populations, that are associated with the prognosis and treatment resistance (Pompella et al. 2020). In recent times, the single-cell technologies are revolutionizing the detection and monitoring of minimal residual disease (MRD), a clinical scenario where a specific subclinical population of malignant cells persists after the posttherapy and effectively contributes to the recurrence (Zhang et al. 2024b). These approaches are really effective in hematologic malignancies for precise tracking of the leukemic clones based on their immunoglobulin and T cell receptor rearrangements (Abbas et al. 2021). In addition, the spatial transcriptomics has improved the efficacy of MRD detection by localizing residual tumor cells within the histologically ambiguous or morphologically normal tissues. In the early-stage lung cancer and colorectal cancer, the spatial transcriptomics analysis has reported a

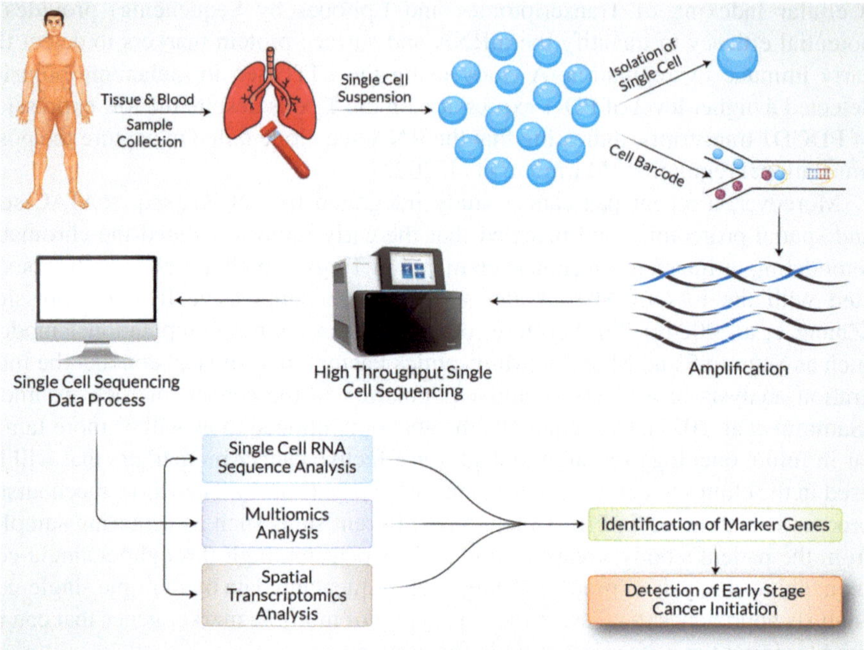

Fig. 1 A diagrammatic representation of the approach of the single-cell technologies to detect cancer in the early stage. Following the collection of the tissue sample and blood from the patient's body, the single-cell suspension is generated to isolate the single cell. The cell barcode is used to amplify the single cells for the high-throughput single-cell sequencing. The single-cell sequencing raw data are processed by the high-configured computational setup. After that, the processed raw single-cell count data can be analyzed by multiple single-cell analysis approaches, such as single-cell RNA sequencing analysis, spatial transcriptomics analysis, and multi-omics analysis, for identifying the marker genes. Finally, any types of cancer can be detected in the early stage by measuring the expression of a group of marker genes responsible for cancer initiation and progression. (Some sections of the figure were adapted from Huang et al. (2021), modified and generated by the authors for better illustration)

molecularly aberrant fields in surgical margin-zones that were previously undetectable through the standard histopathology (Zhang et al. 2024c). These findings prioritize the effectiveness of spatially informed MRD assessment to guide more potential surveillance strategies and surgical decision making. Additionally, the integration of multi-omics approaches enhances the MRD tracking by capturing harmonized changes among the epigenetic, transcriptional, and proteomic layers (Cantafio et al. 2019). For example, the integration of scRNA-seq and scATAC-seq recognizes the regulatory circuitries that sustain quiescent residual clones, which were undetectable in conventional approaches due to low transcriptional activity (Pierce et al. 2021). Furthermore, the single-cell multi-omics approach identified the epigenetically primed cell states in acute lymphoblastic leukemia (ALL) that persisted in remission and anticipated relapse months in advance of clinical indication (Li et al. 2021). In consequence, the integration of single-cell, spatial transcriptomics, and multi-omic technologies is transforming cancer classification and the monitoring of residual disease. These platforms enable an unprecedented level of diagnostic precision, transforming the capacity to stratify patients by risk, personalize therapy, and anticipate disease recurrence with a degree of accuracy.

7 Personalized Oncology and Therapeutic Implications

Single-cell technology has transformed personalized cancer treatment by allowing scientists to study individual tumor cells in detail. This enables the identification of hidden groups of cancer cells (subclones), the discovery of genetic changes that can be targeted with drugs, the monitoring of tumor responses to treatment in real-time, and the design of therapies tailored to each patient's unique cancer.

7.1 Drug Resistance and Targeted Therapies

Drug resistance is still one of the greatest challenges in cancer treatment. Tumors are heterogeneous; they consist of multiple cancer types with many different genetic mutations and gene expressions. This heterogeneity sets the stage for the possibility that even if some small set of cancer cells responds well to a drug, resistant cells in the tumor body will continue to grow and lead to relapse. Standard sequencing focuses on the whole tumor; for most sequencing methods, the signal averaged over many cells will mask small subpopulations of resistant cells. These hidden subpopulations may have relevant mutations or changes in gene expressions that allow them to escape treatment, leading to a very limited possibility of successful long-term treatment (McGranahan and Swanton 2017).

As a result of the application of mutation detection and single-cell population studies methods, we have advanced to studies using single-cell technologies without restrictions related to sequencing depth or throughput. Based on single-cell studies, scientists have shown that even some cancer cells exhibit increased stress response or alter cholesterol metabolism traits that differentiate those cells and allow survival despite aggressive treatment with systemic therapies (Zhang et al. 2022). Other studies have linked resistance to the epithelial-to-mesenchymal transition (EMT), a process where cancer cells change their characteristics to become more mobile and invasive (Dudas et al. 2020). Importantly, single-cell data do not just help explain resistance but also point to ways to overcome it. Aissa et al. (2021) detailed a study in which they used single-cell RNA sequencing to track the evolution of cancer cells in resistance to drug treatment. The authors were able to integrate single-cell RNA sequencing data with a large database on drug response to predict what other drugs could be used to treat the resistant cells (Aissa et al. 2021). In one case, combining an EGFR inhibitor with crizotinib prevented the growth of cells that had previously been resistant, showing how such an approach can guide more effective treatment combinations.

Using this idea, researchers developed a new approach called scTherapy, which applies machine learning and single-cell transcriptome data to suggest specific drug combinations for a patient's tumor (Ianevski et al. 2024). This method identifies co-vulnerabilities targeted at a time, which are distinct pathways or processes that can be attacked to target multiple populations of cancer cells in a tumor. This increases the chances of killing the cancer in the multidimensional space of future therapeutic opportunities, while also potentially protecting healthy cells from their harmful effects by not exposing them to unnecessary treatments.

These approaches represent a major change in the way we think about cancer treatment. Moving from one-size-fits-all strategies to single-cell analysis allows us to see the diversity in each tumor and develop personalized treatment regimens that can deal with the changing nature of resistant tumors. By revealing surprising resistance mechanisms and suggesting drug combinations targeting specific cell states, single-cell technologies will bring us closer to durable, meaningful cancer therapy.

7.2 Immune Checkpoint Therapies and Adaptive Therapy

Immune checkpoint blockade (ICB) therapies, including PD-1/PD-L1 and CTLA-4 antagonists, have revolutionized cancer treatment by reinvigorating the immune system's ability to recognize and kill tumor cells. Despite their increasing application and success in certain malignancies, the major limitation of ICBs to date is that only a subset of patients will benefit from therapy, and many of those who do will ultimately develop resistance. A major driving force behind biomarker research is the significant variability in response to ICBs across patients, and the need for reliable biomarkers to identify which patients will respond and how to monitor response in real time to inform adaptive-treatment decisions. Single-cell technologies, notably single-cell RNA seq (scRNA-seq) and mass cytometry (CyTOF), are leading the charge by providing unparalleled detail regarding the cellular composition and immune microenvironment of the tumor microenvironment (TME).

7.3 Predictive Biomarkers for Immune Checkpoint Therapy

Conventional biomarkers, including PD-L1 expression and tumor mutational burden (TMB), have been used for a long time to gauge the likelihood of response to ICB; while high levels of PD-L1 expression or TMB can provide prognostic advantages, the predictive value of both in different cancer types and patient populations has remained inconsistent. Their limitations underscore the requirement for biomarkers with greater information and dynamism.

Single-cell methodologies are bridging this gap by identifying the immune cell subsets and gene expression patterns that are undetectable by bulk profiling. For instance, scRNA-seq can differentiate between functional and exhausted CD8+ T cells, identify regulatory T cells (Tregs), and detect suppressive myeloid cells, which contribute to ICB response (Krishna et al. 2021). Using scRNA-seq, the researchers uncovered important immune cell populations involved with positive ICB response, specifically effector-like T cells that contained strong cytotoxic capabilities. Furthermore, exhaustion markers (i.e., TIM-3 and LAG-3) identifiable through scRNA-seq may potentially be negative predictors of response or indications of a need for combination therapy (Tietscher et al. 2023). As these technologies continue to evolve, they provide a clearer view of the immune landscape and help develop multidimensional biomarker panels for more accurate patient stratification.

7.4 *Real-Time Monitoring of Therapeutic Efficacy*

In the early phases after initiation of therapy, it remains essential to timely assess the individual response and progression/resistance, to inform clinical decisions. Standard imaging technologies usually lag behind biological responses and may misleadingly indicate pseudo-progression. Liquid biopsies, including circulating tumor DNA (ctDNA), are evolving as a potential strategy to assess tumor burden and provide characterization of response dynamics, in real-time (Boukovala et al. 2024).

In addition to ctDNA analysis, scRNA-seq gives fine-resolution information on the immune landscape during treatment. For example, Parker et al. (2020) applied scRNA-seq to brain mural cells and found that mural cells express CD19, a protein present on B cells (Parker et al. 2020). This finding has significant implications for CAR-T cell therapies targeting CD19, as it suggests potential off-tumor effects that could lead to neurotoxicity.

Moreover, single-cell TCR sequencing can trace T-cell clonal expansion and diversity before and after ICB therapy, providing functional readouts of immune activation. This has been critical in studies showing that newly recruited T cell clones, rather than reactivated resident ones, often drive the antitumor response after PD-1 blockade (Yost et al. 2019).

7.5 *Adaptive Therapy Strategies*

By monitoring immune dynamics in real time, adaptive therapy is a treatment paradigm that is in a constant state of flux based on a patient's current tumor and immune profile. Single-cell techniques are well suited to enable adaptive therapy. By sampling tumors, or peripheral blood, ideally multiple times, and then performing scRNA-seq, clinicians can detect early signs of resistance patterns and immune exhaustion, alongside shifts in the TME that require switching therapies or adding combinations.

For example, in the sarcoma models of Wisdom et al., by using single-cell RNA seq, they compared the immune landscapes of primary (autochthonous) and transplant sarcomas and recognized significant changes in the immune landscape, which predicted a response to immunotherapy. Activation of $CD8^{+}$T cells, as well as pro-inflammatory myeloid cells, suggested that the transplant sarcoma has a highly inflammatory immunological milieu. Ultimately, these transplant sarcomas showed a nice response to radiation and PD-1 checkpoint inhibition. Primary sarcomas, on the other hand, had an immunosuppressive milieu with decreased immune evasion characteristics and neoantigen presentation, which resulted in resistance to the same therapy. In all tumor types, radiotherapy changed the myeloid cell population; however, this change only bolstered antitumor immunity in transplant tumors (Wisdom et al. 2020). These results demonstrate that the immunological environment of a tumor is crucial in influencing therapy results and imply that immune profiling should direct individualized, flexible treatment plans. Precise characterization of these immunological states is made possible by single-cell analysis, which allows real-time monitoring and well-informed choices for the combination or modification of medicines to overcome resistance and enhance patient outcomes.

8 Difficulties, New Developments, and Prospects

While the potential of single-cell technologies such as scRNA-seq, spatial transcriptomics, and multi-omics analysis has effectively increased the understanding of the cancer biology, their broad application in the clinical setup is obstructed by multiple economic, technical, and regulatory challenges. Despite these limitations, the current research and development field is continuing their efforts to enhance the efficiency, cost effectiveness of these platforms and providing significant prospects for their integration into the routine diagnostic and prognostic workflows (Baysoy et al. 2023). Despite of having complexity of single-cell data, the advanced bioinformatics pipelines and valuable computational infrastructure to analyze this massive volume of high-dimensional data generated from multi-omics single-cell profiling (Ma et al. 2020). In addition, the interpretation of data requires potential expertise in genomics, biostatistics, machine learning, and clinical oncology. Standardized algorithms for normalized, batch correction, cell-type annotations, and integrative analysis are most needed to acquire the reproducibility and clinical reliability of these findings (Luecken et al. 2021). Additionally, the high-level price of the high-throughput single-cell technologies reduces their potential for widespread clinical application. The single-cell techniques like scRNA-seq, CITE-seq, and spatial transcriptomics require a high-level cost per sample, which is not only for the reagents but also for the sequencing depth and data storage. However, by optimizing and scaling these techniques, the cost per sample is being reduced compared to the conventional bulk analysis (Farkas 2021). Another important obstacle to the clinical application is the lack of standardized protocols and regulatory frameworks to conduct this analysis because the sample processing protocols, tissue dissociation conditions, and storage protocols are not consistent across the institutions. Furthermore, there is a critical requirement for quality control metrics, reference cell atlas, and clinically validated gene signatures. The regulatory bodies such as the FDA and EMA should implement a proper guideline on how the single-cell data can inform diagnostic or therapeutic decisions, which will boost the efficacy of early-stage cancer diagnostic and therapeutic intervention.

Despite of these limitations and hurdles, the field continues its development, likewise the advancements in the single-cell multi-omics and spatial transcriptomics approaches are increasing their potential to capture more information per well and enhancing spatial resolution down to the subcellular levels. The development in microfluidics, sample barcoding, and library preparation has also increased scalability and cost-effectiveness of these approaches. Most importantly, the international collaborative initiatives such as the Human Cell Atlas and Cancer Grand Challenges are working toward harmonization and data-sharing standards, which accelerate the clinical data integration progress (Horgan et al. 2025). Looking forward, the future of the single-cell technologies in precision oncology is highly promising. By the advancement of analytical frameworks and clinical validation, these technologies are making a transition from bench to bedside, offering a real-time insight into tumor evolution, therapeutic evolution, immune modulation, and therapeutic resistance. Therefore, their integration into personalized medicine pipelines will redefine early detection, risk stratification, and therapeutic intervention in cancer care.

9 Final Thoughts

9.1 A Paradigm Shift in the Treatment of Cancer

The development of single-cell technology is a potential game-changer for cancer research and treatment. Single-cell analysis provides an integrative, high-dimensional view of cellular and molecular heterogeneity in tumors (bulk sequencing provides average signals across millions of cells). The ability to investigate the tumor microenvironment at a single-cell level helped to highlight the complex interactions between stromal, immunological, and malignant cells. Each of the three cell types exists within the complex environment and plays its own discrete, if not interrelated, role in disease evolution, therapeutic response, and resistance. We can now view single-cell methods as not only potential vehicles for research but also enabling a conceptual change in how we think about and deal with cancer. In the future, single-cell sequencing in the precision oncology has a lot of potential. We expect increasingly thorough multi-omics techniques that integrate single-cell-level genomic, transcriptomic, spatial transcriptomics, epigenomic, and proteomic data as technologies develop (Lei et al. 2021). By offering a comprehensive understanding of tumor biology, these integrated studies will make it possible to find new treatment targets and biomarkers for patient classification. In conclusion, single-cell sequencing has truly launched cancer research and treatment into a spree of discovery. Integrating it into precision oncology may revolutionize cancer treatment to allow more effective, precise, and individualized care of each tumor.

9.2 Future Directions in Computational Single-Cell Analysis

The future for the computational single cell will include the use of machine learning and AI models for predicting cell types, feature selections, and automation of the full process. Newer deep generative models are also being employed to offer a probabilistic framework for interpreting the data in a more certain way (Ma and Xu 2022). Apart from this, the single cell technology is also using as a clinical trial real-time prediction tool to measure the treatment response and to identify the reasons for therapy resistance. As the single cell technologies are continuing to grow in use for making clinical decisions, ethical issues, data privacy, data reproducibility and governance regulation will need critical insights so that the best of different singlecell analysis will make significant impact in healthcare system (Kim et al. 2020; Walker et al. 2024).

9.3 Closing Summary

To sum up, single-cell analysis is a paradigm change in the biology and therapy of cancer, not only a technical development. We believe that combined single-cell and computational methods will play a key role in solving long-standing oncology problems as the field develops. Our capacity to decipher tumor complexity, track treatment resistance, and eventually direct the creation of next-generation, precision-guided cancer treatments will be improved by these technologies. In the upcoming 10 years, single-cell biology, multi-omics, artificial intelligence, and clinical medicine will all come together to provide a more accurate and more customized cancer diagnosis and therapy.

References

Abbas HA, Lu W, Marques-Piubelli ML et al (2021) Single cell T cell landscape and T cell receptor repertoire profiling of AML in context of PD-1 blockade therapy. Nat Commun 12. https://doi.org/10.1038/s41467-021-26282-z

Adams PD (2017) Epigenetics of cancer and aging: from mechanisms to therapies. Exp Gerontol 94:123. https://doi.org/10.1016/j.exger.2017.02.057

Aibar S, Wouters J, Aerts J et al (2017) SCENIC: single-cell regulatory network inference and clustering. Nat Methods 14:1083–1086

Aissa AF, Islam ABMMK, Ariss MM et al (2021) Single-cell transcriptional changes associated with drug tolerance and response to combination therapies in cancer. Nat Commun 12. https://doi.org/10.1038/s41467-021-21884-z

Anderson NM, Simon MC (2020) The tumor microenvironment. Curr Biol 30:R921–R925. https://doi.org/10.1016/j.cub.2020.06.081

Ando K, Oki E, Kasagi Y et al (2021) Recent developments in cancer research: expectations for a new remedy. Ann Gastroenterol Surg 5:419–426. https://doi.org/10.1002/ags3.12440

Aramini B, Masciale V, Grisendi G et al (2022) Dissecting tumor growth: the role of cancer stem cells in drug resistance and recurrence. Cancer 14:976. https://doi.org/10.3390/cancers14040976

Arya A, Tripathi P, Dubey N et al (2025) Navigating single-cell RNA-sequencing: protocols, tools, databases, and applications. Genom Informat 23. https://doi.org/10.1186/s44342-025-00044-5

Azizi E, Carr AJ, Plitas G et al (2018) Single-cell map of diverse immune phenotypes in the breast tumor microenvironment. Cell 174:1293–1308. https://doi.org/10.1016/j.cell.2018.05.060

Balan D, Abd Aziz NH, Kampan NC, Plebanski M (2024) Unlocking ovarian cancer heterogeneity: advancing immunotherapy through single-cell transcriptomics. Front Oncol 14. https://doi.org/10.3389/fonc.2024.1388663

Baysoy A, Fan R, Bai Z, Satija R (2023) The technological landscape and applications of single-cell multi-omics. Nat Rev Mol Cell Biol 24:695–713. https://doi.org/10.1038/s41580-023-00615-w

Boukovala M, Westphalen CB, Probst V (2024) Liquid biopsy into the clinics: current evidence and future perspectives. J Liquid Biopsy 4:100146. https://doi.org/10.1016/j.jlb.2024.100146

Cantafio MEG, Arbitrio M, Barbieri V et al (2019) From single level analysis to multi-omics integrative approaches: a powerful strategy towards the precision oncology. High Throughput 7(4):33. https://doi.org/10.3390/ht7040033

Chen K, Wang Q, Guo H et al (2021) Single-cell RNA-Seq reveals dynamic change in tumor microenvironment during pancreatic ductal adenocarcinoma malignant progression. SSRN Electron J. https://doi.org/10.2139/ssrn.3745132

Cheng C, Chen X, Jin H, Chen W (2023) A review of single-cell RNA-Seq annotation, integration, and cell-cell communication. Cells 12:1970. https://doi.org/10.3390/cells12151970

Chung W, Eum HH, Lee H-O et al (2017) Single-cell RNA-seq enables comprehensive tumour and immune cell profiling in primary breast cancer. Nat Commun 8:15081. https://doi.org/10.1038/ncomms15081

Chuntova P, Li MO, Watchmaker PB et al (2020) Unique challenges for glioblastoma immunotherapy-discussions across neuro-oncology and non-neuro-oncology experts in cancer immunology. Meeting report from the 2019 SNO immuno-oncology think tank. Neuro-Oncol 23:356–375. https://doi.org/10.1093/neuonc/noaa277

Coons AH, Creech HJ, Jones RN (1941) Immunological properties of an antibody containing a fluorescent group. Exp Biol Med 47:200–202. https://doi.org/10.3181/00379727-47-13084p

Dave A, Charytonowicz D, Francoeur NJ et al (2023) The breast cancer single-cell atlas: defining cellular heterogeneity within model cell lines and primary tumors to inform disease subtype, stemness, and treatment options. Cell Oncol (Dordrecht, Netherlands) 46:603–628. https://doi.org/10.1007/s13402-022-00765-7

Dawson MA, Kouzarides T (2012) Cancer epigenetics: from mechanism to therapy. Cell 150:12–27. https://doi.org/10.1016/j.cell.2012.06.013

De Rop FV, Hulselmans G, Flerin C et al (2024) Systematic benchmarking of single-cell ATAC-sequencing protocols. Nat Biotechnol 42:916–926. https://doi.org/10.1038/s41587-023-01881-x

Dudás J, Steinbichler TB, Ladányi A et al (2020) Epithelial to mesenchymal transition: a mechanism that fuels cancer radio/chemoresistance. Cells 9:428. https://doi.org/10.3390/cells9020428

Dunn B (2012) Cancer: solving an age-old problem. Nature 483:S2–S6. https://doi.org/10.1038/483s2a

Durante MA, Livingstone AS, Rodriguez DA et al (2020) Abstract 1591: single cell analysis of uveal melanoma reveals new evolutionary complexity. Cancer Res 80:1591. https://doi.org/10.1158/1538-7445.am2020-1591

Fankhauser R, Fuiten A, Depatie N et al (2022) Single-cell identification of melanoma biomarkers in circulating tumor cells. Cancer 14:4921. https://doi.org/10.3390/cancers14194921

Farkas DL (2021) Biomedical applications of translational optical imaging: from molecules to humans. Molecules (Basel, Switzerland) 26:6651. https://doi.org/10.3390/molecules26216651

Figueira RC, Gomes LR, Neto JS et al (2009) Correlation between MMPs and their inhibitors in breast cancer tumor tissue specimens and in cell lines with different metastatic potential. BMC Cancer 9. https://doi.org/10.1186/1471-2407-9-20

Gerber T, Willscher E, Loeffler-Wirth H et al (2016) Mapping heterogeneity in patient-derived melanoma cultures by single-cell RNA-seq. Oncotarget 8:846–862. https://doi.org/10.18632/oncotarget.13666

Gerstberger S, Jiang Q, Ganesh K (2023) Metastasis. Cell 186:1564–1579. https://doi.org/10.1016/j.cell.2023.03.003

Ghojogh B, Karray F, Crowley M, Ghodsi A (2023) Elements of dimensionality reduction and manifold learning. Springer

Gierahn TM, Satija R, Hughes TK et al (2017) Seq-Well: portable, low-cost RNA sequencing of single cells at high throughput. Nat Methods 14:395–398. https://doi.org/10.1038/nmeth.4179

Goswami S, Kong LY, Ott M et al (2019) Immune profiling of human tumors identifies CD73 as a combinatorial target in glioblastoma. Nat Med 26:39–46. https://doi.org/10.1038/s41591-019-0694-x

Guo X, Kang B, Zhang Q et al (2018) Global characterization of T cells in non-small-cell lung cancer by single-cell sequencing. Nat Med 24:978–985. https://doi.org/10.1038/s41591-018-0045-3

Hafemeister C, Satija R (2019) Normalization and variance stabilization of single-cell RNA-seq data using regularized negative binomial regression. Genome Biol 20:296

Haque A, Lönnberg T, Engel J, Teichmann SA (2017) A practical guide to single-cell RNA-sequencing for biomedical research and clinical applications. Genome Med 9. https://doi.org/10.1186/s13073-017-0467-4

Hermann PC, Huber SL, Herrler T et al (2007) Distinct populations of cancer stem cells determine tumor growth and metastatic activity in human pancreatic cancer. Cell Stem Cell 1:313–323. https://doi.org/10.1016/j.stem.2007.06.002

Herreros-Pomares A (2022) Identification, culture and targeting of cancer stem cells. Life 12:184. https://doi.org/10.3390/life12020184

Hong M, Zhang L, Tao S et al (2020) RNA sequencing: new technologies and applications in cancer research. J Hematol Oncol 13. https://doi.org/10.1186/s13045-020-01005-x

Horgan D, Tanner M, Aggarwal C et al (2025) Precision oncology: a global perspective on implementation and policy development. JCO Glob Oncol 11. https://doi.org/10.1200/go-24-00416

Hou Y, Yang L, Yang X et al (2012) Single-cell exome sequencing and monoclonal evolution of a JAK2-negative myeloproliferative neoplasm. Cell 148:873–885. https://doi.org/10.1016/j.cell.2012.02.028

Hu B, Sajid M, Lv R et al (2022) A review of spatial profiling technologies for characterizing the tumor microenvironment in immuno-oncology. Front Immunol 13. https://doi.org/10.3389/fimmu.2022.996721

Hu Y, Shen F, Shen J et al (2023) Single-cell sequencing technology applied to epigenetics for the study of tumor heterogeneity. Clin Epigenetics 15. https://doi.org/10.1186/s13148-023-01574-x

Huang RH, He J, Wang LX, Gao W (2021) Application and prospects of single cell sequencing in tumors. Biomark Res 9. https://doi.org/10.1186/s40364-021-00336-2

Hwang B, Lee JH, Bang D (2018) Single-cell RNA sequencing technologies and bioinformatics pipelines. Exp Mol Med 50:1–14. https://doi.org/10.1038/s12276-018-0071-8

Ianevski A, Nader K, Driva K et al (2024) Single-cell transcriptomes identify patient-tailored therapies for selective co-inhibition of cancer clones. Nat Commun 15. https://doi.org/10.1038/s41467-024-52980-5

Jovic D, Luo Y, Lin L et al (2022) Single-cell RNA sequencing technologies and applications: a brief overview. Clin Transl Med 12. https://doi.org/10.1002/ctm2.694

Junttila MR, De Sauvage FJ (2013) Influence of tumour micro-environment heterogeneity on therapeutic response. Nature 501:346–354. https://doi.org/10.1038/nature12626

Kanehisa M, Furumichi M, Tanabe M, Sato Y, Morishima K (2016) KEGG: new perspectives on genomes, pathways, diseases and drugs. Nucleic Acids Res 45(D1):D353–D361. https://doi.org/10.1093/nar/gkw1092

Karamitopoulou E, Wenning AS, Acharjee A, Aeschbacher P, Marinoni I, Zlobec I, Gloor B, Perren A (2024) Spatial heterogeneity of immune regulators drives dynamic changes of local immune responses, affecting disease outcomes in pancreatic cancer. Clin Cancer Res 30(18):4215–4226. https://doi.org/10.1158/1078-0432.ccr-24-0368

Keller L, Pantel K (2019) Unravelling tumour heterogeneity by single-cell profiling of circulating tumour cells. Nat Rev Cancer 19(10):553–567. https://doi.org/10.1038/s41568-019-0180-2

Kim N, Kim HK, Lee K, Hong Y, Cho JH, Choi JW, Lee J-I, Suh Y-L, Ku BM, Eum HH, Choi S, Choi Y-L, Joung J-G, Park W-Y, Jung HA, Sun J-M, Lee S-H, Ahn JS, Park K, Ahn M-J, Lee H-O (2020) Single-cell RNA sequencing demonstrates the molecular and cellular reprogramming of metastatic lung adenocarcinoma. Nat Commun 11(1). https://doi.org/10.1038/s41467-020-16164-1

Kim J, Xu Z, Marignani PA (2021) Single-cell RNA sequencing for the identification of early-stage lung cancer biomarkers from circulating blood. NPJ Genom Med 6(1). https://doi.org/10.1038/s41525-021-00248-y

Klein AM, Mazutis L, Akartuna I, Tallapragada N, Veres A, Li V, Peshkin L, Weitz DA, Kirschner MW (2015) Droplet barcoding for single-cell transcriptomics applied to embryonic stem cells. Cell 161(5):1187–1201. https://doi.org/10.1016/j.cell.2015.04.044

Koh CM, Bieberich CJ, Dang CV, Nelson WG, Yegnasubramanian S, De Marzo AM (2010) MYC and prostate cancer. Genes Cancer 1(6):617–628. https://doi.org/10.1177/1947601910379132

Korsunsky I, Millard N, Fan J, Slowikowski K, Zhang F, Wei K, Baglaenko Y, Brenner M, Loh P-R, Raychaudhuri S (2019) Fast, sensitive and accurate integration of single-cell data with harmony. Nat Methods 16(12):1289–1296. https://doi.org/10.1038/s41592-019-0619-0

Krishna C, DiNatale RG, Kuo F, Srivastava RM, Vuong L, Chowell D, Gupta S, Vanderbilt C, Purohit TA, Liu M, Kansler E, Nixon BG, Chen Y-B, Makarov V, Blum KA, Attalla K, Weng S, Salmans ML, Golkaram M, Liu L, Zhang S, Vijayaraghavan R, Pawlowski T, Reuter V, Carlo MI, Voss MH, Coleman J, Russo P, Motzer RJ, Li MO, Leslie CS, Chan TA, Hakimi AA (2021) Single-cell sequencing links multiregional immune landscapes and tissue-resident T cells in ccRCC to tumor topology and therapy efficacy. Cancer Cell 39(5):662–677.e6. https://doi.org/10.1016/j.ccell.2021.03.007

Lambert AW, Pattabiraman DR, Weinberg RA (2017) Emerging biological principles of metastasis. Cell 168:670–691

Lambrechts D, Wauters E, Boeckx B, Aibar S, Nittner D, Burton O, Bassez A, Decaluwé H, Pircher A, Van Den Eynde K, Weynand B, Verbeken E, De Leyn P, Liston A, Vansteenkiste J, Carmeliet P, Aerts S, Thienpont B (2018) Phenotype molding of stromal cells in the lung tumor microenvironment. Nat Med 24(8):1277–1289. https://doi.org/10.1038/s41591-018-0096-5

Lei Y, Liu J, Yu X, Wang W, Xu J, Shi S, Zhang B, Tang R (2021) Applications of single-cell sequencing in cancer research: progress and perspectives. J Hematol Oncol 14(1). https://doi.org/10.1186/s13045-021-01105-2

Li J-F, Ma X-J, Ying L-L, Tong Y-H, Xiang X-P (2021) Multi-omics analysis of acute lymphoblastic leukemia identified the methylation and expression differences between BCP-ALL and T-ALL. Front Cell Develop Biol 8. https://doi.org/10.3389/fcell.2020.622393

Liang S-B, Fu L-W (2017) Application of single-cell technology in cancer research. Biotechnol Adv 35(4):443–449. https://doi.org/10.1016/j.biotechadv.2017.04.001

Longo SK, Khavari PA, Ji AL, Guo MG (2021) Integrating single-cell and spatial transcriptomics to elucidate intercellular tissue dynamics. Nat Rev Genet 22(10):627–644. https://doi.org/10.1038/s41576-021-00370-8

Love MI, Huber W, Anders S (2014) Moderated estimation of fold change and dispersion for RNA-seq data with DESeq2. Genome Biol 15:550

Luecken MD, Büttner M, Chaichoompu K, Danese A, Interlandi M, Mueller MF, Strobl DC, Zappia L, Dugas M, Colomé-Tatché M, Theis FJ (2021) Benchmarking atlas-level data integration in single-cell genomics. Nat Methods 19(1):41–50. https://doi.org/10.1038/s41592-021-01336-8

Ma Q, Xu D (2022) Deep learning shapes single-cell data analysis. Nat Rev Mol Cell Biol 23(5):303–304. https://doi.org/10.1038/s41580-022-00466-x

Ma A, Mcdermaid A, Xu J, Chang Y, Ma Q (2020) Integrative methods and practical challenges for single-cell multi-omics. Trends Biotechnol 38(9):1007–1022. https://doi.org/10.1016/j.tibtech.2020.02.013

Macaulay IC, Haerty W, Kumar P, Li YI, Hu TX, Teng MJ, Goolam M, Saurat N, Coupland P, Shirley LM, Smith M, Van Der Aa N, Banerjee R, Ellis PD, Quail MA, Swerdlow HP, Zernicka-Goetz M, Livesey FJ, Ponting CP, Voet T (2015) G&T-seq: parallel sequencing of single-cell genomes and transcriptomes. Nat Methods 12(6):519–522. https://doi.org/10.1038/nmeth.3370

Macosko EZ, Basu A, Satija R, Nemesh J, Shekhar K, Goldman M, Tirosh I, Bialas AR, Kamitaki N, Martersteck EM, Trombetta JJ, Weitz DA, Sanes JR, Shalek AK, Regev A, McCarroll SA (2015) Highly parallel genome-wide expression profiling of individual cells using nanoliter droplets. Cell 161(5):1202–1214. https://doi.org/10.1016/j.cell.2015.05.002

Mallory XF, Nakhleh L, Edrisi M, Navin N (2020) Methods for copy number aberration detection from single-cell DNA-sequencing data. Genome Biol 21(1). https://doi.org/10.1186/s13059-020-02119-8

Mani SA, Guo W, Liao M-J, Eaton EN, Ayyanan A, Zhou AY, Brooks M, Reinhard F, Zhang CC, Shipitsin M, Campbell LL, Polyak K, Brisken C, Yang J, Weinberg RA (2008) The epithelial-mesenchymal transition generates cells with properties of stem cells. Cell 133(4):704–715. https://doi.org/10.1016/j.cell.2008.03.027

Marmonti E, Oliva-Ramirez J, Haymaker C (2022) Dendritic cells: the long and evolving road towards successful targetability in cancer. Cells 11(19):3028. https://doi.org/10.3390/cells11193028

Mcgranahan N, Swanton C (2017) Clonal heterogeneity and tumor evolution: past, present, and the future. Cell 168(4):613–628. https://doi.org/10.1016/j.cell.2017.01.018

Navin N, Kendall J, Troge J, Andrews P, Rodgers L, McIndoo J, Cook K, Stepansky A, Levy D, Esposito D, Muthuswamy L, Krasnitz A, McCombie WR, Hicks J, Wigler M (2011) Tumour evolution inferred by single-cell sequencing. Nature 472(7341):90–94. https://doi.org/10.1038/nature09807

Parker KR, Wilson NE, Maillard I, Reddy R, Montine TJ, Kriegstein AR, Haris M, Mohan S, Perkey E, Migliorini D, Satpathy AT, Scholler J, Posey AD, Bhoj VG, Yost KE, Liu F, June CH, Gabunia K, Chang HY, Bhaduri A, Bagga P (2020) Single-cell analyses identify brain mural cells expressing CD19 as potential off-tumor targets for CAR-T immunotherapies. Cell 183(1):126–142.e17. https://doi.org/10.1016/j.cell.2020.08.022

Patel AP, Trombetta JJ, Gillespie SM, Suvà ML, Cahill DP, Shalek AK, Curry WT, Wakimoto H, Nahed BV, Bernstein BE, Louis DN, Martuza RL, Rozenblatt-Rosen O, Tirosh I, Regev A (2014) Single-cell RNA-seq highlights intratumoral heterogeneity in primary glioblastoma. Science 344(6190):1396–1401. https://doi.org/10.1126/science.1254257

Peng J, Sun B-F, Chen C-Y, Zhou J-Y, Chen Y-S, Chen H, Liu L, Huang D, Jiang J, Cui G-S, Yang Y, Wang W, Guo D, Dai M, Guo J, Zhang T, Liao Q, Liu Y, Zhao Y-L, Han D-L, Zhao Y, Yang Y-G, Wu W (2019) Single-cell RNA-seq highlights intra-tumoral heterogeneity and malignant progression in pancreatic ductal adenocarcinoma. Cell Res 29(9):725–738. https://doi.org/10.1038/s41422-019-0195-y

Peng A, Mao X, Hu Y, Fan S, Zhong J (2020) Single-cell multi-omics and its prospective application in cancer biology. Proteomics 20(13):1900271. https://doi.org/10.1002/pmic.201900271

Penter L, Dietze K, Bullinger L, Westermann J, Rahn H, Hansmann L (2018) FACS single cell index sorting is highly reliable and determines immune phenotypes of clonally expanded T cells. Eur J Immunol 48(7):1248–1250. https://doi.org/10.1002/eji.201847507

Picelli S, Faridani OR, Björklund ÅK, Winberg G, Sagasser S, Sandberg R (2014) Full-length RNA-seq from single cells using smart-seq2. Nat Protoc 9(1):171–181. https://doi.org/10.1038/nprot.2014.006

Pierce SE, Granja JM, Greenleaf WJ (2021) High-throughput single-cell chromatin accessibility CRISPR screens enable unbiased identification of regulatory networks in cancer. Nat Commun 12(1). https://doi.org/10.1038/s41467-021-23213-w

Plongthongkum N, Diep D, Chen S, Lake BB, Zhang K (2021) Scalable dual-omics profiling with single-nucleus chromatin accessibility and mRNA expression sequencing 2 (SNARE-seq2). Nat Protoc 16(11):4992–5029. https://doi.org/10.1038/s41596-021-00507-3

Pompella L, De Vita F, Ventriglia A, Ciardiello F, Orditura M, Pappalardo A, Caterino M, Tirino G, Nacca V (2020) Pancreatic cancer molecular classifications: from bulk genomics to single cell analysis. Int J Mol Sci 21(8):2814. https://doi.org/10.3390/ijms21082814

Rao A, Yanai I, Barkley D, França GS (2021) Exploring tissue architecture using spatial transcriptomics. Nature 596(7871):211–220. https://doi.org/10.1038/s41586-021-03634-9

Roider T, Seufert J, Uvarovskii A, Frauhammer F, Bordas M, Abedpour N, Stolarczyk M, Mallm J-P, Herbst SA, Bruch P-M, Balke-Want H, Hundemer M, Rippe K, Goeppert B, Seiffert M, Brors B, Mechtersheimer G, Zenz T, Peifer M, Chapuy B, Schlesner M, Müller-Tidow C, Fröhling S, Huber W, Anders S, Dietrich S (2020) Dissecting intratumour heterogeneity of nodal B-cell lymphomas at the transcriptional, genetic and drug-response levels. Nat Cell Biol 22(7):896–906. https://doi.org/10.1038/s41556-020-0532-x

Sammut S-J, Crispin-Ortuzar M, Chin S-F, Provenzano E, Bardwell HA, Ma W, Cope W, Dariush A, Dawson S-J, Abraham JE, Dunn J, Hiller L, Thomas J, Cameron DA, Bartlett JMS, Hayward L, Pharoah PD, Markowetz F, Rueda OM, Earl HM, Caldas C (2021) Multi-omic machine learning predictor of breast cancer therapy response. Nature 601(7894):623–629. https://doi.org/10.1038/s41586-021-04278-5

Schnormeier A-K, Budeus B (2024) Single cell VDJ sequencing of normal and malignant B and T cells. Methods Mol Biol:295–346. https://doi.org/10.1007/978-1-0716-4188-0_14

Sethi S, Ali S, Philip PA, Sarkar FH (2013) Clinical advances in molecular biomarkers for cancer diagnosis and therapy. Int J Mol Sci 14(7):14771–14784. https://doi.org/10.3390/ijms140714771

Soneson C, Robinson MD (2018) Bias, robustness and scalability in single-cell differential expression analysis. Nat Methods 15(4):255–261. https://doi.org/10.1038/nmeth.4612

Stanojevic S, Li Y, Ristivojevic A, Garmire LX (2022) Computational methods for single-cell multi-omics integration and alignment. Genomics Proteomics Bioinformatics 20(5):836–849. https://doi.org/10.1016/j.gpb.2022.11.013

Stoeckius M, Hafemeister C, Stephenson W, Houck-Loomis B, Chattopadhyay PK, Swerdlow H, Satija R, Smibert P (2017) Simultaneous epitope and transcriptome measurement in single cells. Nat Methods 14(9):865–868. https://doi.org/10.1038/nmeth.4380

Sun G, Li Z, Rong D, Zhang H, Shi X, Yang W, Zheng W, Sun G, Wu F, Cao H, Tang W, Sun Y (2021) Single-cell RNA sequencing in cancer: applications, advances, and emerging challenges. Mol Ther Oncolytics 21:183–206. https://doi.org/10.1016/j.omto.2021.04.001

Tang F, Barbacioru C, Wang Y, Nordman E, Lee C, Xu N, Wang X, Bodeau J, Tuch BB, Siddiqui A, Lao K, Surani MA (2009) mRNA-Seq whole-transcriptome analysis of a single cell. Nat Methods 6(5):377–382. https://doi.org/10.1038/nmeth.1315

Tang C, Fu S, Jin X, Li W, Xing F, Duan B, Cheng X, Chen X, Wang S, Zhu C, Li G, Chuai G, He Y, Wang P, Liu Q (2023) Personalized tumor combination therapy optimization using the single-cell transcriptome. Genome Med 15(1). https://doi.org/10.1186/s13073-023-01256-6

Tibbe AGJ, Miller MC, Terstappen LWMM (2007) Statistical considerations for enumeration of circulating tumor cells. Cytometry A 71A(3):154–162. https://doi.org/10.1002/cyto.a.20369

Tien Y-W, Jeng Y-M, Hu R-H, Chang K-J, Hsu S-M, Lee P-H (2004) Intravasation-related metastatic factors in colorectal cancer. Tumour Biol 25(1–2):48–55. https://doi.org/10.1159/000077723

Tietscher S, Wagner J, Anzeneder T, Langwieder C, Rees M, Sobottka B, De Souza N, Bodenmiller B (2023) A comprehensive single-cell map of T cell exhaustion-associated immune environments in human breast cancer. Nat Commun 14(1). https://doi.org/10.1038/s41467-022-35238-w

Tirosh I, Suvà ML (2018) Deciphering human tumor biology by single-cell expression profiling. Annu Rev Cancer Biol 3(1):151–166. https://doi.org/10.1146/annurev-cancerbio-030518-055609

Truffi M, Corsi F, Sorrentino L (2020) Fibroblasts in the tumor microenvironment. Adv Exp Med Biol 1234:15–29. https://doi.org/10.1007/978-3-030-37184-5_2

Van De Sande B, Lee JS, Mutasa-Gottgens E, Naughton B, Bacon W, Manning J, Wang Y, Pollard J, Mendez M, Hill J, Kumar N, Cao X, Chen X, Khaladkar M, Wen J, Leach A, Ferran E (2023) Applications of single-cell RNA sequencing in drug discovery and development. Nat Rev Drug Discov 22(6):496–520. https://doi.org/10.1038/s41573-023-00688-4

Walker CR, Li X, Chakravarthy M, Lounsbery-Scaife W, Choi YA, Singh R, Gürsoy G (2024) Private information leakage from single-cell count matrices. Cell 187(23):6537–6549.e10. https://doi.org/10.1016/j.cell.2024.09.012

Wang X, He Y, Zhang Q, Ren X, Zhang Z (2021) Direct comparative analyses of 10X genomics chromium and smart-SEQ2. Genomics Proteomics Bioinformatics 19(2):253–266. https://doi.org/10.1016/j.gpb.2020.02.005

Wang H, Mei Y, Luo C, Huang Q, Wang Z, Lu G-M, Qin L, Sun Z, Huang C-W, Yang Z-W, Chen J, Yin W, Qian C-N, Zeng J, Chen L, Leng Q, Guo Y, Jia G (2022a) Single-cell analyses reveal mechanisms of cancer stem cell maintenance and Epithelial–Mesenchymal transition in recurrent bladder cancer. Clin Cancer Res 27(22):6265–6278. https://doi.org/10.1158/1078-0432.ccr-20-4796

Wang X, Tan Y, He N, Liu Y, Deng Y, Pan W, Chen Z, Zheng X, Ren J, Liu H, Chen H, Li S (2022b) Recent advances and application of whole genome amplification in molecular diagnosis and medicine. MedComm 3(1). https://doi.org/10.1002/mco2.116

Wisdom AJ, Mowery YM, Hong CS, Himes JE, Nabet BY, Qin X, Zhang D, Chen L, Fradin H, Patel R, Bassil AM, Muise ES, King DA, Xu ES, Carpenter DJ, Kent CL, Smythe KS, Williams NT, Luo L, Ma Y, Alizadeh AA, Owzar K, Diehn M, Bradley T, Kirsch DG (2020) Single cell analysis reveals distinct immune landscapes in transplant and primary sarcomas that determine response or resistance to immunotherapy. Nat Commun 11(1). https://doi.org/10.1038/s41467-020-19917-0

Wu X, Yang X, Dai Y, Zhao Z, Zhu J, Guo H, Yang R (2024) Single-cell sequencing to multi-omics: technologies and applications. Biomark Res 12(1). https://doi.org/10.1186/s40364-024-00643-4

Xulu KR, Nweke EE, Augustine TN (2023) Delineating intra-tumoral heterogeneity and tumor evolution in breast cancer using precision-based approaches. Front Genet 14. https://doi.org/10.3389/fgene.2023.1087432

Yin Y, Grabowska AM, Clarke PA, Whelband E, Robinson K, Argent RH, Tobias A, Kumari R, Atherton JC, Watson SA (2010) Helicobacter pylori potentiates epithelial:mesenchymal transition in gastric cancer: links to soluble HB-EGF, gastrin and matrix metalloproteinase-7. Gut 59(8):1037–1045. https://doi.org/10.1136/gut.2009.199794

Yost KE, Satpathy AT, Wells DK, Qi Y, Wang C, Kageyama R, McNamara KL, Granja JM, Sarin KY, Brown RA, Gupta RK, Curtis C, Bucktrout SL, Davis MM, Chang ALS, Chang HY (2019) Clonal replacement of tumor-specific T cells following PD-1 blockade. Nat Med 25(8):1251–1259. https://doi.org/10.1038/s41591-019-0522-3

Zellmer VR, Zhang S (2014) Evolving concepts of tumor heterogeneity. Cell Biosci 4(1):69. https://doi.org/10.1186/2045-3701-4-69

Zhang X, Li T, Liu F, Chen Y, Yao J, Li Z, Huang Y, Wang J (2019) Comparative analysis of droplet-based ultra-high-throughput single-cell RNA-SEQ systems. Mol Cell 73(1):130–142. e5. https://doi.org/10.1016/j.molcel.2018.10.020

Zhang K, Erkan EP, Jamalzadeh S, Dai J, Andersson N, Kaipio K, Lamminen T, Mansuri N, Huhtinen K, Carpén O, Hietanen S, Oikkonen J, Hynninen J, Virtanen A, Häkkinen A, Hautaniemi S, Vähärautio A (2022) Longitudinal single-cell RNA-seq analysis reveals stress-promoted chemoresistance in metastatic ovarian cancer. Sci Adv 8(8). https://doi.org/10.1126/sciadv.abm1831

Zhang D, Zhan D, Zhang R, Sun Y, Duan C, Yang J, Wei J, Li X, Lu Y, Lai X (2024a) Treg-derived TGF-β1 dampens cGAS-STING signaling to downregulate the expression of class I MHC complex in multiple myeloma. Sci Rep 14(1). https://doi.org/10.1038/s41598-024-62298-3

Zhang S, Wang X, Yang Z, Ding M, Zhang M, Young KH, Zhang X (2024b) Minimal residual disease detection in lymphoma: methods, procedures and clinical significance. Front Immunol 15. https://doi.org/10.3389/fimmu.2024.1430070

Zhang Y, Gong S, Liu X (2024c) Spatial transcriptomics: a new frontier in accurate localization of breast cancer diagnosis and treatment. Front Immunol 15. https://doi.org/10.3389/fimmu.2024.1483595

Zheng GXY, Terry JM, Belgrader P, Ryvkin P, Bent ZW, Wilson R, Ziraldo SB, Wheeler TD, McDermott GP, Zhu J, Gregory MT, Shuga J, Montesclaros L, Underwood JG, Masquelier DA, Nishimura SY, Schnall-Levin M, Wyatt PW, Hindson CM, Bharadwaj R, Wong A, Ness KD, Beppu LW, Deeg HJ, McFarland C, Loeb KR, Valente WJ, Ericson NG, Stevens EA, Radich JP, Mikkelsen TS, Hindson BJ, Bielas JH (2017) Massively parallel digital transcriptional profiling of single cells. Nat Commun 8(1). https://doi.org/10.1038/ncomms14049

Zhou H-M, Zhang J-G, Zhang X, Li Q (2021a) Targeting cancer stem cells for reversing therapy resistance: mechanism, signaling, and prospective agents. Signal Transduct Target Ther 6(1). https://doi.org/10.1038/s41392-020-00430-1

Zhou W-M, Yan Y-Y, Pilarsky C, Guo Q-R, Wang H, Xu T-T, Makabel B, He G, Yu X-Y, Zhang J-Y, Ji H (2021b) Microfluidics applications for high-throughput single cell sequencing. J Nanobiotechnol 19(1). https://doi.org/10.1186/s12951-021-01045-6

Zhou Z, Wang J, Wang J, Yang S, Wang R, Zhang G, Li Z, Shi R, Wang Z, Lu Q (2024) Deciphering the tumor immune microenvironment from a multidimensional omics perspective: insight into next-generation CAR-T cell immunotherapy and beyond. Mol Cancer 23(1). https://doi.org/10.1186/s12943-024-02047-2

Zhu Y, Zhu M, Lance P (2012) Stromal COX-2 signaling activated by deoxycholic acid mediates proliferation and invasiveness of colorectal epithelial cancer cells. Biochem Biophys Res Commun 425(3):607–612. https://doi.org/10.1016/j.bbrc.2012.07.137

Navigating Tumor Microenvironments with Spatial Transcriptomics for Effective Therapeutics

Christopher Krolla, Tonmoy Das, Andreas Tsouris, Sajib Chakraborty, and Geoffroy Andrieux

Abstract Spatial transcriptomics is revolutionizing clinical and biomedical research by providing spatially resolved insights into tissue architecture, cellular topology, and morphological organization. This innovative technology enables the simultaneous profiling of gene expression and spatial location within tissues, uncovering multifaceted cellular interactions. In cancer, spatial transcriptomics guides the development of immunotherapy and targeted therapy by identifying spatially distinct expression patterns associated with tumor heterogeneity. This nuanced approach enhances precision medicine strategies by enabling tailored interventions that are informed by the complex spatial architecture and the entangled transcriptomic landscape of individual tissues. As a powerful tool in biomedical research, spatial transcriptomics—named Method of the Year 2020 by *Nature Methods*—is transforming our ability to resolve complex cellular interactions and close-range dynamics within their native spatial context, driving more effective and personalized therapeutic interventions.

Keywords Spatial-Transcriptomics · Tumor-Microenvironment · Immunotherapies · Precision Medicine

Abbreviations

NMF	Non-negative Matrix Factorization (NMF)
DSP	digital spatial profiling
FACS	fluorescence-activated cell sorting
FEA	functional enrichment analysis

C. Krolla · T. Das · A. Tsouris · S. Chakraborty (✉) · G. Andrieux
Institute of Medical Bioinformatics and Systems Medicine, Faculty of Medicine, University of Freiburg, Freiburg, Germany
e-mail: sajib.chakraborty@uniklinik-freiburg.de; geoffroy.andrieux@uniklinik-freiburg.de

A. H. M. N. Nabi, H. Uddin Shekhar (eds.), *Biotechnological Advances in Healthomics*, Interdisciplinary Biotechnological Advances,
https://doi.org/10.1007/978-981-95-3612-2_4

FFPE	formalin-fixed, paraffin-embedded
FISH	fluorescence in situ hybridization
GSEA	gene set enrichment analysis
ISH	in situ hybridization
ISS	in situ sequencing
LCM	laser capture microdissection
MACS	magnetic-activated cell sorting
MERFISH	multiplexed error-robust fluorescence in situ hybridization
NMF	non-negative matrix factorization
PCA	principal component analysis
QC	quality control
RCTD	robust cell type decomposition
scRNA-seq	single-cell RNA sequencing
ST	spatial transcriptomics
SVG	spatially variable genes
TME	tumor microenvironment
UMI	unique molecular identifier

1 Introduction

1.1 Understanding the Tumor Microenvironment: A Spatial Perspective in Oncology

The tumor microenvironment (TME) represents a complex and dynamic ecosystem composed of malignant cells, diverse immune infiltrates, stromal components, and an intricate vascular network. Beyond merely providing a supportive scaffold for tumor growth and metastatic spread, the TME actively shapes the disease course and profoundly influences therapeutic responses (Hanahan and Weinberg 2011). Importantly, the TME is far from a uniform entity. It exhibits marked spatial and cellular heterogeneity, with distinct niches fostering immunosuppression, therapy resistance, and metastatic potential. The spatial organization of these components—the precise positioning, density, and interactions of various cell types within the tissue matrix—plays a pivotal role in dictating the trajectory of tumor evolution and ultimately impacts clinical outcomes (Hinshaw and Shevde 2019). Understanding these spatial dynamics is therefore essential for the development of more effective, targeted therapeutic strategies.

1.2 Limitations of Conventional Transcriptomic Approaches

Traditionally, our understanding of gene expression within tumors has been largely informed by bulk and single-cell RNA sequencing technologies. These approaches have been instrumental in uncovering critical molecular signatures, delineating cellular identities, and characterizing the diversity of cell states within the TME (Grünwald et al. 2021; Andrieux et al. 2023). However, both techniques come with inherent limitations, particularly regarding spatial context. Bulk RNA sequencing, by averaging gene expression across a usually heterogeneous mixture of cells, effectively masks the intricate cellular heterogeneity that is fundamental to tumor biology. This averaging effect can obscure the presence of rare but biologically significant cell populations and the spatially restricted expression of key regulatory genes. To address this limitation, methods such as laser capture microdissection (LCM) and cell sorting techniques like FACS and MACS have been used to isolate more specific and less heterogeneous cell populations, enabling more precise molecular profiling based on cell type. However, these approaches remain limited in throughput and lack the resolution needed to comprehensively capture transcriptional diversity at the single-cell level.

Single-cell RNA sequencing (scRNA-seq) marked a transformative advancement by offering high-resolution profiles at the level of individual cells. Nevertheless, the process requires dissociation of tissues into single-cell suspensions, thereby stripping cells from their native architectural context and exposing them to mechanical and enzymatic stress that can alter their transcriptional profiles. This disruption severs connections between cells and their surrounding microenvironment, revealing critical insights into intercellular communication, signaling gradients, and the physical layout of functional niches within tissues (Rao et al. 2021). As a result, despite its advantages, scRNA-seq leaves a substantial gap in our ability to fully understand the complex spatial dynamics that underpin tumor progression, immune evasion, and therapeutic resistance. Computational inference tools such as CellChat attempt to reconstruct pseudo-spatial interactions or neighborhood relationships from dissociated single-cell data; however, these methods can introduce false positives and lack the precision of direct spatial measurements (Almet et al. 2021). Several resourceful adaptations have been introduced, including pathologist-driven histomorphological annotation of histology images, enabling spatial contextualization integrated with LCM-derived bulk data. While these strategies offer valuable spatial insight, they rely on manual interpretation or indirect integration and thus only partially reconcile molecular identity with spatial tissue organization. Bridging this disconnect is therefore essential for a comprehensive and integrative view of the TME, with direct implications for the development of even more precise and effective treatment strategies.

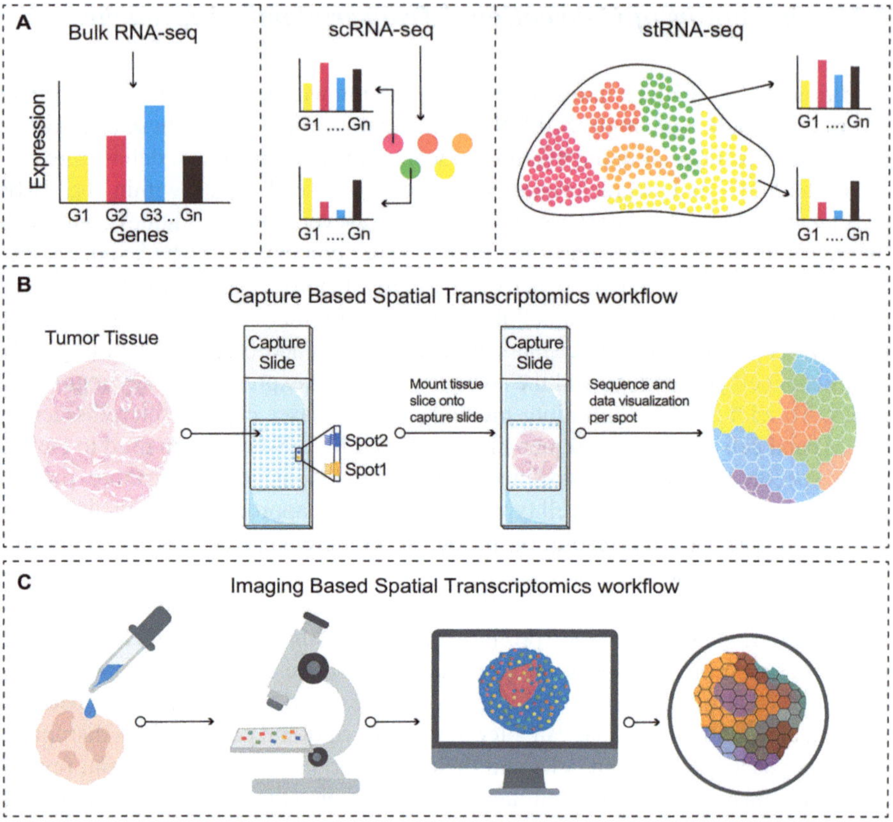

Fig. 1 Overview of spatial transcriptomics method. (**a**) Schematic diagram showing the comparison among bulk RNA-seq, single-cell RNA-seq (scRNA-seq), and spatial transcriptomics (stRNA-seq). (**b**) Capture- and image-based spatial transcriptomics workflow (**c**) Imaging based spatial transcriptomics workflow

1.3 Emergence and Impact of Spatial Transcriptomics

Spatial transcriptomics (ST) has emerged to overcome the lack of spatial information inherent in traditional sequencing methods. By preserving the spatial architecture of tissues while simultaneously capturing comprehensive transcriptomic profiles, ST provides a crucial connection between molecular biology and classical histopathology (Fig. 1) (Moses and Pachter 2022). It enables the investigation of gene expression within the precise topological context of intact tissue sections, allowing researchers to directly link molecular signatures to their corresponding spatial locations (Burgess 2019).

Several cutting-edge technologies have propelled the field forward, including sequencing-based methods that capture spatial gene expression from fresh frozen tissues using barcoded arrays; probe-based approaches that use targeted hybridization panels, enabling transcript detection in FFPE samples; and highly multiplexed imaging-based techniques that visualize RNA molecules directly within intact tissues at subcellular resolution. These methodologies vary in resolution and sensitivity but share a common goal: to generate spatially resolved maps of gene expression that can elucidate how a cell's position within a tissue shapes its phenotype and functional state.

In the context of oncology, ST is proving indispensable for achieving a more nuanced and integrated understanding of tumor biology. It allows researchers to dissect intratumoral heterogeneity with unparalleled clarity, revealing how distinct cell populations are spatially organized and interact within the tumor mass (Fig. 2). Moreover, ST provides critical insights into topographical immune infiltration patterns, identifying not just the presence but also the spatial arrangement of immune cells relative to malignant and stromal elements (Fig. 2) (Lewis et al. 2021). It also illuminates cell–cell communication networks, highlighting how local microenvironments and cellular neighborhoods influence gene expression dynamics (Armingol et al. 2020).

This level of spatially informed resolution is essential for identifying spatially restricted biomarkers—molecular features that may be confined to specific niches or tumor margins and for unraveling mechanisms of immune evasion, often orchestrated through localized signaling (Jin et al. 2024; Sun and Yu 2015). Additionally, spatial transcriptomics holds great promise for the discovery of therapeutic vulnerabilities that remain hidden in bulk or dissociated single-cell datasets, offering new avenues for precision oncology and the design of spatially targeted interventions (Jin et al. 2024).

2 Spatial Transcriptomics Technologies

2.1 General Sample Preparation

Reliable ST data arise from careful attention to the physical specimen before any molecular barcodes are read. This chapter begins with a comparative assessment of FF and FFPE tissue preservation, proceeds through the successive laboratory steps required to generate high-quality libraries, ensuring sample comparability, and concludes with a synopsis of current multiplexing strategies (Frederick et al. 2024).

Rigorous specimen handling from excision to cDNA library construction is the primary determinant of data quality in spatial transcriptomics. Fresh-frozen preservation maximizes transcriptomic breadth, whereas formalin-fixed, paraffin-embedded (FFPE) tissue extends morphological fidelity and clinical reach. Regardless of the modality, best practice begins with prompt, gentle harvesting to minimize ischemic damage, followed by fixation, rapid 4% paraformaldehyde cross-linking for fresh/frozen workflows or standard formalin for FFPE. The sample is then embedded (OCT for cryosectioning, paraffin for FFPE), sectioned at

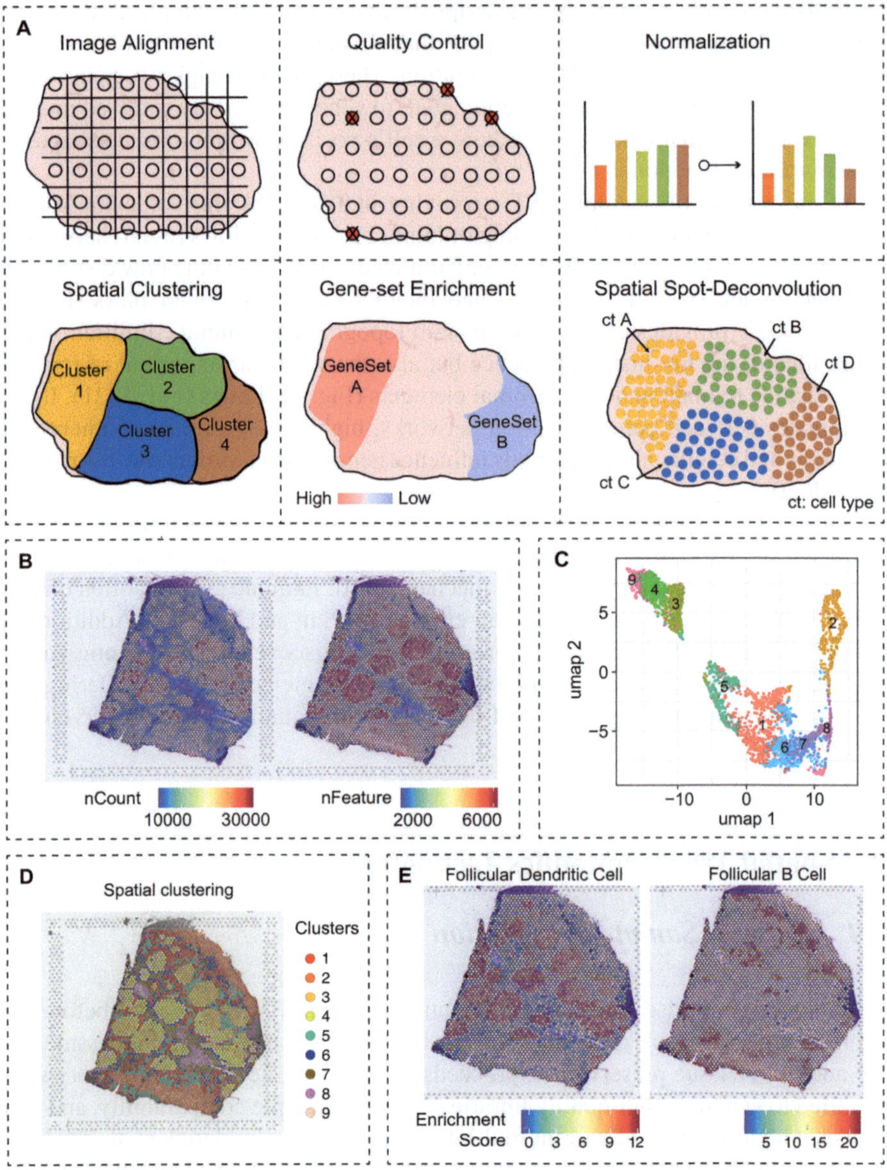

Fig. 2 Quality control and clustering. (**a**) Bioinformatics workflow of spatial transcriptomics analysis. (**b**) The spatial distribution of nCount (UMI count) and nFeature (Gene count) mapped onto the human tonsil tissue H&E image. The analysis shown here is based on the spatial transcriptomics (Visium) data generated by the tonsil cell atlas in the context of the Human Cell Atlas (HCA). Spatial data were processed and visualized through Seurat V5. (**c**) Leiden Clustering of the 10× Visium Spatial Transcriptomics dataset from human tonsil tissue section is shown as UMAP. Color represents different spatial clusters. (**d**) Cluster annotation of the spots is mapped onto spatial space. (**e**) Gene-set enrichment analysis of the human tonsil tissue section using univariate linear model (ULM). Two relevant gene sets were utilized: Follicular Dendritic Cell and Follicular B Cell

5–10 μm, and mounted flat onto capture-chemistry slides within the imaging frame. After securing adhesion by drying or baking, sections are permeabilised or de-cross-linked to expose RNA, with optional autofluorescence quenching or histological staining for quality control. Maintaining RNase-free conditions and appropriate desiccated or ethanol storage preserves transcript integrity until probe hybridization and imaging. Careful calibration of sectioning, permeabilisation, and amplification safeguards spatial resolution and limits the need for aggressive computational correction. Ultimately, a specimen-centered wet-lab workflow is the prerequisite for credible spatial biology.

In essence, FF maximizes transcriptomic depth and minimizes molecular bias, but at the cost of complex logistics and specific cryo-related troubleshooting. FFPE prioritizes robust morphological context and clinical accessibility through established biobanks, with its own set of distinct processing challenges primarily related to water bath behavior and wax interaction. The optimal tissue choice ultimately depends on balancing the need for comprehensive molecular insights with the availability of samples and the specific aims of the spatial transcriptomics investigation.

2.2 *Platform Comparison*

The field of ST is undergoing rapid innovation, with an expanding array of platforms designed to profile gene expression within its native spatial context. These technologies vary significantly in terms of spatial resolution, transcriptome coverage, tissue compatibility, throughput, and the computational demands of downstream analysis (Table 1). Broadly, spatial transcriptomics platforms can be classified into two major categories based on their underlying methodologies: capture-based and imaging-based approaches (Fig. 3). To maintain generality, we use the term "ST units" to refer to both spots (from sequencing-based platforms) and segmented cells (from imaging-based platforms). Where specificity is required, the appropriate term will be used.

2.3 *Capture-Based Platforms*

Capture-based platforms rely on the physical capture of mRNA molecules from tissue sections onto spatially barcoded arrays or slides. These barcodes encode positional information, enabling the reconstruction of spatial gene expression maps following sequencing (Moses and Pachter 2022).

Table 1 Resolution, throughput, and limitations across platforms

Platform	Resolution	Coverage	Throughput	Limitations
Slide-seqV2	~10 μm (near single-cell)	Broad (near whole transcriptome)	Moderate	Lower sensitivity; random bead placement complicates decoding
Stereo-seq	~220 nm (subcellular)	Whole transcriptome	High	Lower capture efficiency; technically demanding setup
Visium	~55 μm (multi-cell)	Whole transcriptome	High	Limited to multi-cell spots; lacks true single-cell resolution
Visium HD	~2 μm (single-cell)	Whole transcriptome	High	Newer platform; protocols still maturing
CosMx SMI	Subcellular (~500 nm)	Targeted (up to 1000 genes)	Moderate	Panel-based; higher complexity in probe and image processing
GeoMX DSP	~10–100 μm (region-based)	Targeted (RNA or protein panels)	High	Region selection may introduce sampling bias; not at single-cell resolution
MERFISH	Subcellular (~1 μm)	Targeted (up to 10,000 genes)	Moderate	Requires complex probe design and sequential imaging
seqFISH	Subcellular (~1 μm)	Targeted (up to 10,000 genes)	Low-moderate	Time-consuming; prone to signal errors over many hybridization cycles
Xenium	Subcellular (~200 nm)	Targeted (5k panel available)	Moderate	Restricted to panel-based detection; requires optimized imaging

2.3.1 Visium

Visium, by 10x Genomics, is one of the most widely adopted spatial transcriptomics platforms, offering a balance between ease-of-use, whole-transcriptome coverage, and moderate spatial resolution. It is a spatial barcoding-based method in which tissue sections are placed on glass slides patterned with oligonucleotide spots (~55 μm diameter) that capture mRNA during permeabilization. The spot-to-spot center distance is 100 μm, which further limits spatial granularity. Visium is compatible with both fresh frozen and FFPE tissues, making it suitable for a wide range of sample types. However, each spot captures transcripts from multiple cells, limiting the resolution to multi-cellular rather than true single-cell spatial analysis.

2.3.2 Visium HD

A next-generation evolution of the Visium platform, Visium HD significantly improves spatial resolution by replacing discrete capture spots with a dense rectangular grid of barcoded probes. This enhancement allows for near single-cell or even

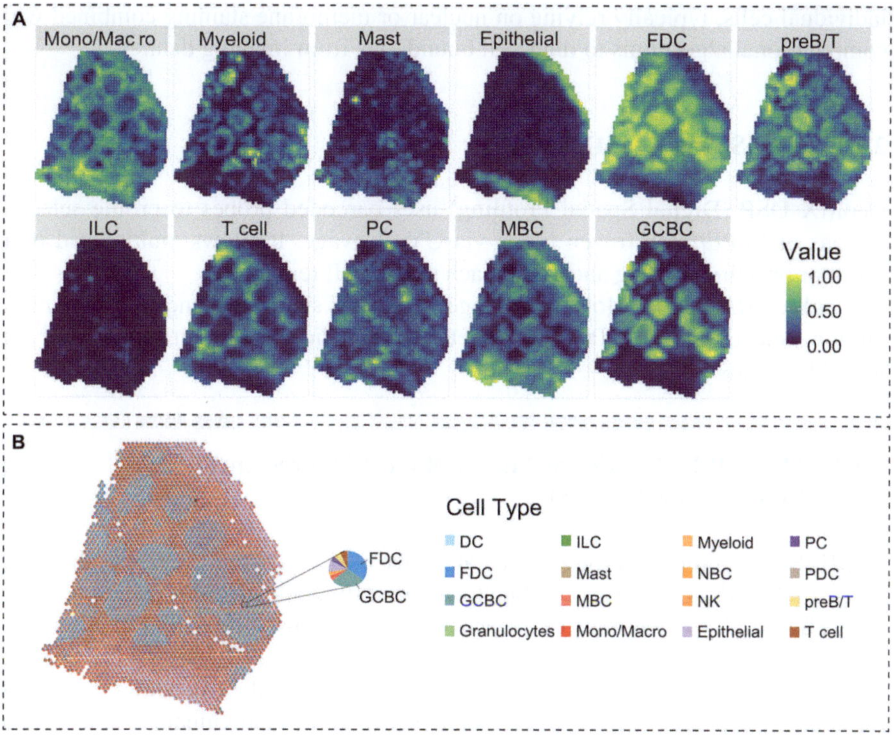

Fig. 3 Deconvolution of spatial spots by the CARD algorithm. (**a**) Spatial distribution of single cells in a human tonsil tissue section. The color represents the score, indicating the probability of occurrence of single cells in a specific spatial location. (**b**) Each spatial spot is deconvoluted, and the proportion of single cells in each spot is illustrated as a pie chart. DC dendritic cells, ILC innate lymphoid cells, PC plasma cells, FDC follicular DCs, NBC naive B cells, PDC plasmacytoid DCs, GCBC germinal center B cells, MBC Memory B cells, NK Natural killer cells, preB/T precursor B and T cells

subcellular resolution while maintaining whole-transcriptome coverage. It offers a powerful upgrade for researchers aiming to study fine tissue architectures such as tumor-immune borders or tissue microanatomy.

2.4 *Imaging-Based Platforms*

Imaging-based platforms utilize in situ hybridization (ISH) or in situ sequencing (ISS) to directly visualize RNA molecules within intact tissue sections. These platforms are capable of single-molecule resolution, providing unmatched spatial precision. However, resolution and throughput vary depending on the platform, and some are more or less limited to targeted gene panels due to probe design constraints. Importantly, accurate cell segmentation is essential for assigning transcripts to individual cells, typically relying on nuclear or membrane staining combined with computational algorithms to infer cell boundaries from imaging data.

2.4.1 NanoString GeoMX/CosMx

GeoMX DSP (Digital Spatial Profiling) uses barcoded probes to profile selected regions of FFPE/fresh frozen tissue via UV cleavage. It allows multiplexed transcript or protein profiling and can't reach single-cell resolution.

CosMx SMI (Spatial Molecular Imager) provides single-cell and subcellular resolution, detecting up to 6000 RNA targets and 64 validated protein analytes per cell in FFPE tissues.

2.4.2 MERFISH (Multiplexed Error-Robust Fluorescence In Situ Hybridization)

MERFISH, developed by the Zhuang lab (Chen et al. 2015) and commercialized via Vizgen's MERSCOPE, uses combinatorial barcoding and error-correcting codes to image hundreds to thousands of transcripts simultaneously. It enables highly multiplexed and error-robust RNA detection through barcoded FISH probes and sequential hybridization rounds. MERFISH achieves subcellular resolution, allowing researchers to precisely localize RNA molecules within individual cells. It is ideal for applications requiring high spatial precision, such as mapping cell–cell communication within tightly packed tissue niches. MERSCOPE is the commercial version supporting FFPE and fresh frozen tissues. It offers subcellular resolution and can detect thousands of genes in situ.

2.4.3 seqFISH (Sequential FISH)

seqFISH enables multiplexed detection of transcripts by repeatedly hybridizing and imaging sets of fluorescent probes (Lubeck et al. 2014). It offers high sensitivity and spatial resolution, making it suitable for studies of tissue organization, neural circuits, and developmental biology.

2.4.4 Xenium In Situ

Also developed by 10× Genomics, Xenium represents a distinct shift toward in situ capture and imaging, integrating elements of both capture- and imaging-based paradigms. It enables subcellular spatial resolution and multiplexed profiling of hundreds to thousands of RNA targets directly in tissue. Xenium is optimized for high-throughput and high-plex experiments. It is compatible with a wide variety of tissue types and sample preparations, including FFPE.

Each platform offers a compromise between resolution and transcriptome breadth. Therefore, the application domain—whole-tissue profiling versus focused cell-type studies—often dictates the platform choice.

3 Bioinformatics Pipelines

The analytical power of ST hinges not only on the experimental platform but also on the computational pipeline used to extract meaningful biological insights from the data generated in the experiment. A robust pipeline for spatial transcriptomics integrates image data, spatial coordinates, and gene expression, enabling multidimensional interpretation of the tissue environment. Below, we detail the key stages of the bioinformatics workflow (Fig. 2).

3.1 Data Preprocessing and Image Alignment

This foundational step prepares raw ST data for downstream analysis and ensures that only high-quality and spatially coherent information is used. Histological images (H&E or immunofluorescence) are aligned with the spatial transcriptomics capture area. This spatial registration ensures that gene expression values can be visually inspected within the tissue section. Tools such as the 10x Space Ranger (https://www.10xgenomics.com/support/software/space-ranger/2.1), STUtility (https://ludvigla.github.io/STUtility_web_site/), and Seurat (https://satijalab.org/seurat/articles/spatial_vignette), Squidpy (Spatial Single Cell Analysis in Python: https://squidpy.readthedocs.io/en/stable/) (Butler et al. 2018) support this alignment step by overlaying expression data on top of the histological image.

3.2 Quality Control

Quality control in spatial transcriptomics involves a series of steps to ensure data quality and interpretability. It begins with global diagnostic assessments to evaluate overall sequencing performance and transcriptome complexity. This is followed by spatial consistency and integration checks to confirm that spatial units align correctly with tissue structure and are properly registered across modalities. Next, ST unit-level filtering, including image-based quality control for imaging platforms, removes unreliable spots or segmented cells based on signal quality, content, and segmentation accuracy. Finally, gene-level filtering eliminates low-quality or potentially uninformative genes, ensuring a clean and biologically meaningful dataset for downstream analysis.

3.2.1 Global Diagnostic Assessments

These initial checks evaluate the overall quality and complexity of the dataset before any filtering is applied. They help detect under-sequencing, technical variability, or sample preparation issues that could compromise downstream analysis. Key assessments include:

- *Saturation curves*: Assess whether additional sequencing would yield more unique RNA molecules.
- *Complexity curves*: Evaluate the relationship between sequencing depth and detected genes or UMIs.
- *Library size distribution*: Examine total UMI counts or reads per ST unit to detect technical outliers.
- *Gene/feature count distribution*: Identify ST units with abnormally low or high numbers of detected genes.
- *Global expression heatmaps or PCA*: Detect batch effects or slide-level variability across samples.

These diagnostics guide parameter selection for later QC steps and help identify problematic datasets early.

3.2.2 Spatial Consistency and Integration Checks

This step ensures that spatial transcriptomics data accurately reflect the physical organization of the tissue and are correctly aligned across data modalities. These checks are crucial for preserving spatial context and avoiding analysis artifacts. Key assessments include:

- *Tissue mask validation*: Confirm that ST units fall within tissue regions, excluding background areas such as glass, folds, or tears.
- *Alignment quality control*: For multi-modal datasets (e.g., transcriptomics + protein abundance), ensure accurate spatial alignment using visual inspection, overlay consistency, or correlation-based methods.
- *Spatial coordinate integrity*: Detect and correct misaligned or corrupted coordinate data that could distort spatial analysis.
- *Spatial outlier detection*: Identify ST units that are spatially isolated or irregularly distributed compared to surrounding units—using visual inspection, spatial density, or distance-based methods—to flag potential misalignments, artifacts, or tissue edge effects.

These checks help preserve biologically meaningful spatial structure and ensure that integrated analyses are spatially coherent.

3.2.3 ST Unit-Level Filtering (Including Image-Based Quality Control)

Not all spots or beads in sequencing-based spatial platforms yield high-quality RNA capture, and not all segmented cells in imaging-based platforms are accurately resolved. Both limitations can compromise the reliability and interpretability of downstream analyses. ST unit-level filtering aims to remove these unreliable units by applying quality control criteria such as:

- *Low RNA content*: ST units with low transcript counts, gene or detected features counts
- *High mitochondrial RNA expression*: It indicates stress, degradation, or cell death
- *Lack of biological signal*: Units located in regions such as fat-rich or necrotic tissue
- *Dominance of uninformative genes*: High proportions of ribosomal, hemoglobin, pseudogenes, or processed transcripts rather than protein-coding genes
- *Segmentation artifacts (for imaging-based platforms)*: Over- or under-segmented cells or misaligned boundaries due to poor staining or low image quality

These filters are critical for maintaining the integrity of the dataset and preventing technical noise from skewing biological interpretation.

3.2.4 Gene-Level Filtering

Gene-level filtering aims to exclude uninformative genes as well as genes that may introduce technical noise to the data. This step refines the dataset, improves the signal-to-noise ratio, and ensures more meaningful downstream analysis.

- *Low detection rate*: Genes expressed in only a small fraction of ST units (e.g., <1–5%
- *Technical artifacts*: Genes affected by platform-specific issues (e.g., cross-hybridization, barcode collisions)
- *Potential biological irrelevance*: Genes with disproportionately high expression from categories such as mitochondrial, ribosomal, hemoglobin, pseudogenes, or processed transcripts
- *Non-coding biotypes*: Genes without protein-coding potential, excluded when the analysis focuses on functional expression

These filters enhance the biological relevance of the gene expression matrix and reduce confounding background signals.

3.3 Normalization

Normalization adjusts for technical variability, such as differences in RNA capture efficiency between spots or tissue sections. The aim is to adjust raw counts for meaningful comparisons while retaining biologically relevant variation, particularly

Table 2 Normalization methods for spatial transcriptomics (ST) and their assumptions and spatial considerations

Method	Key assumptions	Considerations for spatial transcriptomics
Library size normalization and log transformation	Total counts per ST unit primarily reflect technical effects	Simple and interpretable; minimally invasive; preserves spatial structure. Recommended with scaling to median library size except for spot based.
SCTransform	Gene expression follows a negative binomial distribution; sequencing depth drives technical variability	Variance-stabilizing; effective in scRNA-seq; may remove spatially relevant biological variation
Scran	Most genes are not differentially expressed across pooled ST units	Useful for compositional bias correction; assumptions may not hold in spatially heterogeneous tissues
RUVIII-NB	Technical variation can be modeled using negative controls and replicates	Strong adjustment; risks overcorrection; requires well-defined pseudo-replicates and controls
SpaNorm	Library size variation is spatially smooth and can be decomposed	Specifically designed for ST; preserves spatial domains; newer and less widely adopted

where total transcript counts reflect tissue structure. While no gold standard has been established and broader benchmarking is needed, choosing a normalization method remains a practical compromise between technical correction and biological signal preservation, especially for tasks like domain identification.

Several normalization methods are available (Table 2), each based on distinct assumptions. Library size normalization and log transformation is a conservative and widely used method, introducing minimal distortion to underlying expression patterns. The first step consists of normalizing the total number of counts across all genes between ST units. To avoid artificial rescaling, counts should be normalized to the median library size of the ST units rather than a fixed total (e.g., 10,000). This prevents scale shifts that could confound downstream analyses. In the second step, the total counts of each gene in each spot are log transformed.

In spot-based data, library size normalization is often omitted as the number of cells can vary between spots, and therefore cause large fluctuations in the actual total number of counts per spot (Bhuva et al. 2024). In such cases, library size normalization would remove a substantial fraction of the biological information, and analysts tend to directly log-transform the raw counts of each spot.

Accurate normalization is essential for comparing gene expression across ST units, minimizing the loss of spatial context, enabling cross-sample comparisons, and reducing bias in unsupervised clustering. Until broader benchmarks establish best practices, library size normalization to the median library size and log-transformation remains a practical and cautious choice, especially for analyses sensitive to the spatial organization of gene expression (Choudhary and Satija 2022).

3.4 Spatial Clustering and Annotation

The additional spatial dimension in ST can unlock biological insights inaccessible to classical scRNA-seq, enabling not just transcriptional but also topological tissue structure inference. This gives rise to two conceptually distinct clustering strategies: transcriptionally dominated (spatially naïve) clustering versus spatially informed (regionally biased) clustering. While the former aims to identify unbiased molecular phenotypes—e.g., cell types or states—the latter emphasizes spatial coherence, prioritizing anatomical or functional domains. Despite their increasing application, this conceptual distinction remains underrepresented in current literature. Depending on the biological question, both approaches are valid: spatial clustering excels in mapping structured tissue organization, whereas expression-only methods like Leiden, especially after rigorous quality control, remain highly effective for discovering cell-intrinsic phenotypes without any spatial bias. Furthermore, both spatial and non-spatial clustering can be visualized directly in tissue space. This makes it easy to explore and intuitively validate patterns against histological or morphological features.

3.4.1 Unsupervised Spatial Clustering

Spatial clustering segments the tissue into biologically meaningful regions based on similarities in gene expression profiles while respecting the spatial arrangement of the spots or beads. This process captures spatial dependencies, meaning that nearby spots are more likely to share characteristics, thereby reflecting true tissue micro-anatomy and avoiding spurious associations that could arise from purely gene-based clustering.

3.4.2 Exemplary Tools for Spatial Clustering

Several computational frameworks have been developed to perform spatial clustering, each with its own strengths:

3.4.3 Graph-Based Clustering

Seurat integrates spatial information by embedding spot coordinates alongside gene expression values during dimensionality reduction (e.g., PCA, UMAP) and subsequent graph-based clustering (e.g., Louvain, Leiden algorithms) (Hao et al. 2023) (Fig. 2).

- *Features*:
  - Compatible with multiple spatial transcriptomics platforms (e.g., Visium, Slide-seq).
  - Supports spatial visualization of cluster assignments directly overlaid on tissue images.
  - Facilitates downstream annotation by combining known marker genes with spatial domains.
- *Use Case*: Ideal for projects already embedded in a Seurat-based scRNA-seq analysis pipeline.

3.4.4 BayesSpace Clustering

BayesSpace enhances spatial clustering resolution by using Bayesian modeling to infer finer-grained clusters from low-resolution data, such as Visium's ~55 μm spots. It performs resolution enhancement by predicting sub-spot clusters, effectively achieving near single-cell spatial resolution without the need for higher-cost experimental data. In addition, it uses a spatial prior to smooth clusters across neighboring spots while allowing sharp transitions where appropriate (e.g., tumor boundaries) (Zhao et al. 2021).

- *Features*:
  - Supports both normal-resolution and enhanced-resolution modes.
  - Accurately models spatial heterogeneity, especially at invasive tumor margins and immune infiltration fronts.
- *Use Case*: Particularly valuable when fine spatial patterning is critical but experimental resolution is limited, e.g., in cancer-immune interface studies or small substructure mapping.

3.5 Spatial Functional Enrichment

Functional enrichment analysis (FEA), exemplified by methods like gene set enrichment analysis (GSEA), is a widely used method to determine whether a predefined set of genes shows statistically significant differences between biological states (e.g., phenotypes, experimental conditions). Standard FEA of bulk and scRNA-seq data typically lacks spatial information about gene expression changes within tissues or cells, focusing solely on overall functional set enrichment. Here, we will list common methods for spatial functional enrichment analysis.

3.6 Spatial Spot Deconvolution

One of the main limitations of sequencing-based ST platforms—such as 10x Genomics Visium and Slide-seq—is their relatively low spatial resolution: each spot typically aggregates RNA from multiple neighboring cells. This cellular mixture obscures the identity of individual cell types at precise locations. To overcome this, spatial spot deconvolution methods computationally infer the cellular composition of each spot, often by integrating scRNA-seq data as a reference, with some approaches additionally mapping individual cells to spatial spots. These methods refine spatial interpretation, generate cell-type-specific spatial maps, and deepen our understanding of tissue architecture, cell–cell interactions, and disease microenvironments (Gaspard-Boulinc et al. 2025).

ST deconvolution has matured into a diverse methodological landscape spanning statistical, machine learning, and deep learning frameworks. Regression-based approaches such as SpatialDWLS (Dong and Yuan 2021) and SPOTlight (Elosua-Bayes et al. 2021) emphasize computational efficiency and interpretability: SpatialDWLS applies dampened weighted least squares with spatial weighting, while SPOTlight combines seeded NMF with NNLS to achieve robust cell-type decomposition, particularly in noisy data. Bayesian methods like Cell2location (Kleshchevnikov et al. 2022) and Robust Cell Type Decomposition (RCTD) (Cable et al. 2021) offer probabilistic modeling tailored to the overdispersed nature of transcriptomic data: Cell2location uses a negative binomial model with hierarchical priors to integrate reference information while controlling for batch and technology effects; RCTD models spot-level transcriptomes as a Poisson mixture, adjusting for platform variation and providing high-resolution deconvolution.

Dimensionality reduction strategies, exemplified by CARD, integrate gene expression with spatial autocorrelation by coupling NMF with a conditional autoregressive prior, enabling inference of domain-specific composition patterns. Deep learning frameworks like Tangram (Biancalani et al. 2021) align scRNA-seq profiles to spatial locations by optimizing spatial gene correlations in a shared latent space; Tangram remains one of the most widely benchmarked models, and has set a standard for reference-based mapping.

There are more than 60 different methods published. Here, we will only focus on the most prominent in terms of benchmarking results and the number of publications using the method (Table 3).

SPOTlight, SpatialDWLS, Cell2location, RCTD, CARD, and Tangram represent some of the most frequently used and cited tools in the field. Their widespread adoption reflects both methodological maturity and practical performance across varied biological and technical scenarios. Notably, most of these methods perform exceptionally well in independent benchmarks, consistently ranking among the top performers in accuracy and robustness. The main exception is SPOTlight, which, despite its popularity and conceptual simplicity, tends to show lower performance in comprehensive evaluations compared to its peers (Gaspard-Boulinc et al. 2025).

Table 3 Overview of prominent spatial transcriptomics deconvolution methods

Method	Modeling approach	Spatial prior	Programming language	Notable features
SPOTlight	Seeded non-negative matrix factorization (NMF)	No	R	Utilizes marker genes; integrates with Seurat; supports topic-based decomposition.
RCTD	Probabilistic modeling with platform effect correction	No	R	Handles doublets and low-UMI spots; platform-aware normalization.
Cell2location	Bayesian hierarchical modeling	Yes	Python	Incorporates spatial priors; estimates cell abundance and uncertainty.
CARD	Conditional Autoregressive (CAR) model	Yes	R	Models spatial correlation; spatial smoothing through autoregressive terms.
Tangram	Deep learning-based spatial alignment	Optional	Python	Probabilistic cell-to-spot mapping; supports multi-resolution and multimodal data.
SpatialDWLS	Weighted least squares regression	No	R	Fast and lightweight; integrated with Giotto suite for spatial visualization.

3.7 Neighborhood Analysis

Spatial transcriptomics has catalyzed numerous methods to characterize local neighborhood relationships, ranging from gene co-expression patterns and cell-type co-localization to more direct cell–cell interactions. Understanding these neighborhood effects is key to deciphering tissue organization and intercellular communication. This section reviews major computational approaches for neighborhood analysis in spatial omics, arranged chronologically, highlighting each method's assumptions, strategy, and focus (gene co-expression, cell co-localization, or true cell–cell interaction).

Across the landscape of neighborhood analysis, most methods aim to quantify how the spatial context of a cell influences cellular states or transcriptomic programs. A unifying assumption is that proximity implies biological relevance: neighboring cells are more likely to interact, share microenvironmental cues, or coordinate gene expression. Methods diverge in how they operationalize this assumption, which can be broadly categorized by their computational paradigm and biological signal of interest.

Some tools rely on spatial adjacency graphs constructed from tissue coordinates or image segmentation, assuming direct physical contact as a proxy for interaction. Early tools like histoCAT (Schapiro et al. 2017) and Giotto (Dries et al. 2021) use descriptive statistics and permutation-based tests to detect cell-type co-localization, often interpreted as indicative of functional organization. These models are typically rule-based or non-parametric, emphasizing interpretable spatial statistics over inference.

By contrast, methods such as SpaGCN (Hu et al. 2021) and Squidpy (Palla et al. 2022) introduce graph-based modeling and permutation frameworks, respectively. SpaGCN leverages graph convolutional neural networks (GCNs) to model the influence of neighboring gene expression on local profiles, learning spatially smoothed embeddings for domain detection. Squidpy adopts a statistical route, using null model permutation testing to identify non-random cell-type co-occurrence or spatial autocorrelation, and integrates network centrality measures for structural insights.

More recent approaches shift toward predictive or generative modeling to capture causal or mechanistic effects of neighborhood context. MISTy (machine learning framework for extracting relationships from any spatial omics data) (Tanevski et al. 2022) implements multi-view machine learning, where separate models are trained on intra-cellular and inter-cellular (neighbor-derived) features to decompose the contribution of each to gene expression. This allows estimation of neighbor influence on a per-gene, per-cell basis. CellNeighborEX (Kim et al. 2023), on the other hand, uses a comparative differential expression framework, contrasting transcriptional profiles of cells with heterotypic versus homotypic neighbors, to pinpoint context-induced gene programs that emerge from direct neighbor interactions.

At the highest resolution, Intracellular Spatial Transcriptomic Analysis Toolkit (InSTAnT) (Kumar et al. 2024) operates at the subcellular level, applying spatial point pattern statistics to detect transcript colocalization within or between cells, reflecting possible RNA co-regulation or functional interactions. Finally, DeepCOLOR (Kojima et al. 2024) represents an example of a tool leveraging deep learning, using variational autoencoders to map single-cell RNA-seq profiles into spatial locations, thereby reconstructing colocalization networks of inferred cell types from mixed-resolution data.

In sum, neighborhood analysis in spatial omics spans a spectrum from descriptive spatial statistics to sophisticated graph based, machine learning, and deep generative models. These tools aim to decode how local environments shape cellular identity and interaction, with assumptions ranging from spatial proximity as structure, to neighbor identity as functional regulator, and finally to molecular co-localization as a proxy for interaction (Table 4). This table summarizes modeling approaches, spatial assumptions, interaction focus, and implementation details of these methods.

Each of these methods contributes a unique perspective on spatial neighborhoods—from identifying spatial gene modules and enriched cell–cell contacts to quantifying how neighboring cells actively modulate molecular phenotypes. Together, they form a toolkit for interrogating co-expression and interactions in complex tissues, helping to link molecular patterns to cellular crosstalk and tissue architecture. The choice of method depends on the biological question (gene-centric vs. cell-centric, descriptive vs. predictive interaction analysis) and data modality (imaging-based single-cell vs. lower-resolution spots). In practice, these approaches

Table 4 Comparison of neighborhood analysis methods

Method (year)	Modeling approach	Interaction focus	Implementation
histoCAT (2017) (Schapiro et al. 2017)	Interactive toolbox; neighbor frequency & network analysis	Cell phenotype co-localization; adjacency networks.	GUI toolbox (open-source; MATLAB-based).
SpatialDE (2018) (Svensson et al. 2018)	Gaussian Process regression per gene to detect spatial variance	Gene-level spatial expression patterns (SVGs); spatial gene clustering ("expression histology")	Python package; open-source.
Giotto (2021) (Dries et al. 2021)	Pipeline with neighbor graph construction, co-occurrence enrichment	Cell-type co-occurrence; cell–cell communication and neighbor-effects on gene expression	R package (Bioconductor); interactive viewer included.
SpaGCN (2021) (Hu et al. 2021)	Graph Convolutional Network integrates expression & histology	Spatial domains (spot clusters) and domain-specific genes; co-expression in local tissue regions.	Python (PyTorch); GitHub release.
Squidpy (2021) (Palla et al. 2022)	Statistical library (permutation tests, spatial stats like Ripley's K).	Cluster/cell-type co-occurrence; spatial autocorrelation of gene expression.	Python library; part of Scanpy ecosystem.
MISTy (2022) (Tanevski et al. 2022)	Multiview ML (e.g., ensemble regressors) isolating intra- vs. inter-cell effects	Cell–cell interaction influence on gene expression; identifies neighbor-contributed gene regulation	R package (mistyR on Bioconductor)
SpaGene (2022) (Liu et al. 2022)	Moran's I-like SVG detection + NMF on SVG expression matrix	Spatial gene modules (co-expressed gene sets localized in space); gene co-expression in neighborhoods.	Open-source (R/Python)
SpatialPCA (2022) (Shang and Zhou, 2022)	PCA with spatial covariance penalty (spatially aware latent factors)	Spatially correlated gene sets (factors); tissue domain delineation and gradients (trajectory)	R package (code on authors' site)
SpatialCorr (2022) (Bernstein et al. 2022)	Differential correlation analysis for gene sets across regions	Gene–gene co-expression variation by location; detects context-specific gene network activity	Open-source (likely R);
CellNeighborEX (2023) (Kim et al. 2023)	Comparative gene expression between heterotypic vs. homotypic neighbors	Neighbor-induced gene expression changes; cell-type interaction signatures (contact-dependent genes)	Python/R tool; open-source

(continued)

Table 4 (continued)

Method (year)	Modeling approach	Interaction focus	Implementation
DeepCOLOR (2024) (Kojima et al. 2024)	Deep generative model (VAE) fusing scRNA-seq with ST to infer single-cell placement	Intercellular colocalization networks; identifies which cell types/states are adjacent (single-cell resolution)	Python (deep learning; TensorFlow/ PyTorch)
DESpace (2024) (Cai et al. 2024)	Two-step: spatial domain clustering then differential expression testing (edgeR)	Spatially variable genes (overall SVGs) across tissue; consistent SVGs across replicates (multi-sample)	R package (Bioconductor);
InSTAnT (2024) (Kumar et al. 2024)	Single-molecule spatial analysis; point pattern statistics for transcript colocalization	Gene co-localization at subcellular and cell–cell level; finds gene pairs/ modules concentrated in the same locales	Python Package

are often complementary, and integrating their insights can paint a comprehensive picture of spatial organization and neighborhood dynamics in tissues.

Inference in spatial transcriptomics is inherently challenging due to various technical and biological complexities, leading to a high risk of false-positive discoveries if rigorous validation is not implemented. For example, misleading differential expression results often arise from inadequate handling of replicate variability—methods that ignore biological replicates have been shown to identify differentially expressed genes even when no true biological differences exist (Squair et al. 2021). Clustering-based workflows further compound the issue when the same data is used both to define clusters and to perform marker gene discovery, a practice known as "double dipping" that inflates false-positive marker identification (Song et al. 2023). Predictions of cell–cell communication using dissociated single-cell transcriptomics data frequently ignore spatial proximity constraints, leading to biologically implausible interaction networks unless spatial validation is incorporated (Cang et al. 2023). Additionally, imputation and data smoothing techniques, including both classical and deep learning–based methods, have been found to generate artificial gene–gene correlations and false expression signals, misleading downstream biological inference if not carefully cross-validated (Andrews and Hemberg 2018). Collectively, these pitfalls underscore that computational inferences in spatial and single-cell transcriptomics must always be substantiated by independent and biologically grounded validation. This is especially critical for machine learning and deep learning approaches, which, due to their complexity and flexibility, are particularly prone to overfitting and may yield results that are statistically compelling but biologically irrelevant without robust validation (Murphy et al. 2023).

3.8 Spatial Multi-Omics Integration

Spatial transcriptomics and proteomics each offer powerful insights into complex biological processes, but experimental methods to combine these modalities remain limited. To address this, Spatial PrOtein and Transcriptome Sequencing (SPOTS), a high-throughput method for simultaneous spatial profiling of both RNA and proteins, has been developed (Ben-Chetrit et al. 2023). Compared to single-modality approaches, SPOTS significantly enhances signal resolution, improves cell clustering, and increases the sensitivity of differential gene expression analysis across tissue regions. Ravi et al. performed integration of tissue-matched spatially resolved metabolomics (MALDI-FTICR-MSI) and spatial transcriptomics across six patients. The authors applied a workflow including metabolic and transcriptomic data alignment, tissue-wise analysis, and reciprocal principal-component analysis (RPCA) integration (Ravi et al. 2021).

4 Application of Spatial Transcriptomics

One of the most promising applications of ST lies in its ability to resolve spatial and cellular complexity of TME and reveal spatially distinct molecular patterns that correlate with therapeutic outcomes. Traditional biomarkers derived from bulk or single-cell data often overlook localized variations within the TME. In contrast, ST enables the identification of spatial biomarkers, gene expression signatures confined to specific anatomical niches that can serve as predictors of treatment response or resistance.

For example, tertiary lymphoid structures (TLSs) within TME have been proposed to be a determining factor for the ICI response. A recent study leveraged spatial transcriptomics to demonstrate that TLSs can act as in situ sites for B cell maturation, clonal expansion, isotype switching, and plasma cell (PC) differentiation (Meylan et al. 2022). The authors further revealed that TLS-generated PCs migrate into tumor regions, guided by CXCL12^{+} fibroblasts, and contribute to local IgG antibody production against tumor cells (Meylan et al. 2022). IgG binding to tumor cells correlated with increased macrophage proximity and markers of tumor cell apoptosis, suggesting a role for antibody-dependent cellular cytotoxicity (ADCC) (Meylan et al. 2022). Furthermore, higher IgG-tumor labeling was associated with better responses to immune checkpoint inhibitors (ICI) and longer progression-free survival in patients with renal cell carcinoma (Meylan et al. 2022). These findings highlight the critical role of TLS-driven humoral immunity in antitumor responses and suggest new therapeutic strategies aimed at promoting TLS formation or leveraging tumor-specific antibodies to enhance immunotherapy efficacy.

Similarly, in non-small cell lung cancer (NSCLC), the presence of mature tertiary lymphoid structures (mTLSs) within the tumor microenvironment (TME) has been recognized as a predictive biomarker for favorable responses to immune checkpoint inhibitors (ICIs) (Peyraud et al. 2025). However, a subset of mTLS-positive patients remains resistant to therapy. Recent spatial transcriptomics and multiplex imaging analyses identified two fibroblast populations, $FAP^{+}\alpha SMA^{+}$ and $MYH11^{+}\alpha SMA^{+}$ cancer-associated fibroblasts (CAFs), associated with immune exclusion, T cell exhaustion, and therapy resistance in mTLS-positive NSCLC (Peyraud et al. 2025). These CAF subsets promote immunosuppressive environments by restricting $CD8^{+}$ T cell infiltration and enhancing regulatory T cell (Treg) recruitment (Peyraud et al. 2025). Although direct targeting of fibroblasts has shown limited success, emerging strategies focusing on mitigating CAF-mediated immunosuppression and targeting T cell exhaustion, such as combined PD-L1 and TIGIT inhibition, represent promising avenues to improve immunotherapy outcomes in this patient population (Peyraud et al. 2025).

Through integration of ST, scRNA-seq, and multiplex imaging, another study mapped the cellular organization and dynamic interactions between tertiary lymphoid structures (TLS), tumor cell aggregates (TCA), and stromal regions in nasopharyngeal carcinoma (NPC) (Liu et al. 2024). Spatial analysis identified key TLS-associated cell types, including $CXCL13^{+}$ CAFs, $CXCL13^{+}CD8^{+}$ T cells, and plasma cells, revealing their roles in B cell maturation, antibody production, and tumor immune responses (Liu et al. 2024). ST enabled the discovery of distinct TCA subtypes linked to EBV infection, plasma cell localization, and immunotherapy response. Furthermore, ST exposed critical spatial relationships, such as B cell interactions with fibroblasts and stem-like T cells, that drive immune activation or suppression within tumors (Liu et al. 2024). These findings underscore the power of spatial transcriptomics in unraveling the complex immune landscape of tumors and highlight potential strategies to enhance cancer immunotherapy by targeting TLS formation, plasma cell responses, and T cell differentiation pathways.

Apart from the TLS structures, ST-centric studies have the potential to identify and explore other spatial features of TMEs, such as the tumor immune barrier (TIB). Recent investigations driven by spatial transcriptomics have revealed that the tumor immune barrier (TIB) structure, formed through interactions between $SPP1^{+}$ macrophages and cancer-associated fibroblasts (CAFs), is a major contributor to immunotherapy resistance in hepatocellular carcinoma (HCC) (Liu et al. 2023). $SPP1^{+}$ macrophages, associated with poor prognosis, accumulate at tumor boundaries and, alongside CAFs, promote extracellular matrix (ECM) formation, thereby excluding T lymphocyte infiltration into the tumor core. Targeting SPP1 disrupted the TIB structure, enhanced $CD8^{+}$ T cell infiltration, and improved the efficacy of PD-1 blockade in preclinical models (Liu et al. 2023). These findings highlight the spatial architecture of the tumor microenvironment (TME) as a critical determinant of immune response and propose SPP1 inhibition as a promising therapeutic strategy to enhance immune checkpoint blockade efficacy in HCC.

Recently, deep visual proteomics (DVP), a spatial proteomics technology combining AI-driven image analysis with unbiased proteomics, was applied to toxic epidermal necrolysis—a severe and often fatal skin blistering condition. By profiling protein expression in keratinocytes and immune cells at high resolution, DVP revealed upregulation of the JAK–STAT pathway (Making space for spatial biology in the clinic 2025). Treatment with JAK inhibitors promoted skin healing and recovery in seven patients, demonstrating the therapeutic potential of spatial technologies.

5 Challenges in Spatial Biology

ST has revolutionized transcriptomics by enabling the in situ mapping of gene expression/transcript abundance within intact tissue architecture. However, its full potential is currently constrained by three major categories of challenges: technical limitations, data analysis complexities, and difficulties in biological interpretation.

5.1 Technical Challenges

Current spatial transcriptomics technologies face limitations in both resolution and sensitivity. This hinders the accurate capture of gene expression at single-cell resolution and the comprehensive profiling of the transcriptome across entire tissue sections. These constraints often impose a trade-off between achieving high spatial resolution and capturing a broad set of gene targets.

5.1.1 Resolution

Many current spatial transcriptomics technologies offer lower resolution compared to single-cell RNA sequencing, meaning that each ST unit often represents the aggregated transcripts of multiple cells. This limits the ability to resolve individual cell behaviors and interactions. Although newer image-based platforms can achieve subcellular resolution, they typically involve trade-offs, such as limited gene panels and a reliance on accurate cell segmentation, which introduces its own ongoing problem.

5.1.2 Sensitivity and Transcript Coverage

Some spatial transcriptomics methods fail to capture all transcripts present within a cell or tissue, resulting in incomplete gene expression profiles. Variability in RNA capture efficiency and the conversion of RNA to cDNA further contributes to signal loss and coverage gaps, which can obscure lowly expressed genes and affect downstream analyses such as cell type identification or pathway enrichment.

5.1.3 Illumination-Based Biases

Imaging-based spatial transcriptomics methods are susceptible to optical artifacts arising from the physical properties of light and the complex nature of biological tissues. Variability in fluorophore emission intensity, combined with light scattering, refraction, and absorption, leads to uneven illumination and depth-dependent signal attenuation. As light propagates through the tissue, shorter wavelengths scatter more strongly (Rayleigh scattering), while longer wavelengths penetrate deeper but with reduced resolution. Autofluorescence from endogenous biomolecules, such as lipofuscin, collagen, or NADH, adds further background noise, particularly in fixed or aged tissues. Additionally, overlapping emission spectra from different fluorophores can result in spectral crosstalk, complicating signal separation. These effects necessitate careful optical calibration, background subtraction, and spectral unmixing to ensure accurate quantification and localization of transcriptional activity.

5.1.4 FFPE Versus Fresh Frozen Tissue

Most spatial transcriptomics protocols were initially optimized for fresh frozen (FF) tissue, which preserves RNA integrity well but necessitates rapid processing and strict cold-chain logistics, conditions not always feasible in clinical settings. FF tissue can also exhibit spatial RNA degradation at exposed surfaces, ice crystal artifacts, and sample variability due to freezing quality, which may introduce subtle spatial biases.

In contrast, formalin-fixed paraffin-embedded (FFPE) tissue is more widely available in clinical workflows and offers long-term stability and morphological preservation. However, FFPE fixation causes RNA fragmentation and chemical crosslinking, which complicates transcript detection. FFPE protocols are typically probe-based and limited to predefined gene panels. Although recent designs use multiple probes per gene to improve detection sensitivity, true whole-transcriptome coverage remains limited compared to FF-based methods. These technical differences make direct comparisons between FF and FFPE samples challenging and often require tailored normalization strategies.

5.1.5 Standardization

The spatial transcriptomics field currently lacks standardized protocols for sample preparation, data processing, normalization, and interpretation. This variability across platforms, assays, and analytical pipelines makes it difficult to compare results across studies, replicate findings, or integrate datasets. Differences in resolution, gene panel coverage, and output formats further complicate cross-platform benchmarking. The absence of unified standards also hampers the development of broadly applicable analytical tools and hinders clinical translation where reproducibility is essential.

5.1.6 Reagent Costs

Many spatial transcriptomics methods involve expensive reagents and require specialized, often proprietary, equipment. Image-based platforms often utilize costly fluorophores, barcoded probes, and automated imaging systems, while capture-based methods often rely on custom slides or capture arrays. These costs can limit accessibility, especially for smaller labs or clinical settings, and may constrain the number of samples that can be processed, reducing experimental throughput and scalability.

5.2 Data Analysis Challenges

Analyzing spatial transcriptomics data presents unique computational difficulties due to its size, complexity, and spatial structure. These challenges span from raw data processing to spatial modeling, integration, and visualization.

5.2.1 Large and Complex Datasets

Spatial transcriptomics generates high-dimensional datasets that integrate gene expression, spatial coordinates, and frequently imaging data. Imaging-based protocols can profile over a million cells in a single sample, far surpassing the typical scale of single-cell RNA-seq, which requires multiple datasets to reach similar counts. With newer 5000+ gene panels, the resulting count matrices, neighborhood graphs, and pairwise distance matrices become computationally intensive to process, requiring efficient data structures and access to large-memory computing environments. Specialized tools and infrastructure are needed to manage, query, and analyze such data at scale.

5.2.2 Spatial Effects

Spatial effects in spatial transcriptomics refer to structured technical or biological variation that correlates with physical location and distorts transcriptional abundance measurements. These effects can arise from uneven RNA capture efficiency across tissue regions, localized sequencing depth variation, illumination artifacts in imaging-based protocols, or tissue-dependent optical properties. Such biases can generate artificial spatial patterns or obscure real biological signals, complicating downstream analyses such as clustering, spatial domain identification, or cell–cell interaction inference. Traditional batch correction methods from bulk or single-cell RNA-seq often fail to address these spatial dependencies.

5.2.3 Image Analysis and Segmentation

In imaging-based spatial transcriptomics, accurately segmenting individual cells from tissue images remains a major unresolved challenge. Segmentation affects the assignment of transcripts to cells and has a direct impact on downstream analyses. There is a fundamental trade-off: maximizing transcript capture per cell often leads to inflated cell boundaries that include transcripts from neighboring cells, while strict boundary definitions may result in under-segmented cells with fewer transcripts, particularly missing rare or lowly expressed genes. Imperfect or inconsistent segmentation can distort gene expression profiles, compromise cell-type classification, and mislead spatial inference methods. Despite advances in deep learning and computational imaging, no universally robust solution has yet emerged.

5.2.4 Integration with Other Data Modalities

Integrating spatial transcriptomics with other omics data, such as single-cell RNA-seq, proteomics, or epigenomics, offers the potential for a more comprehensive view of tissue organization and function. However, this integration presents significant computational and analytical challenges due to mismatches in resolution, modality-specific noise, and differences in data structure and scale. For example, aligning single-cell RNA-seq with spatial data is non-trivial: spatial datasets often lack full transcriptome coverage, while scRNA-seq lacks spatial context. As a result, inferring spatially resolved single-cell information requires sophisticated computational models and remains an area of active research.

5.2.5 Lack of User-Friendly Tools

Although numerous computational methods for spatial transcriptomics analysis are being actively developed, many lack user-friendly interfaces or robust documentation, making them difficult to adopt for researchers without specialized bioinformatics expertise. A substantial number of tools emerge from academic research efforts with limited long-term support, leading to broken dependencies or difficulties reproducing published results. Furthermore, the growing diversity of available tools is not yet matched by comprehensive, independent benchmarking, making it difficult for users to assess performance and select appropriate methods for their specific analytical goals. This hampers interoperability and slows the development of reliable best practices.

5.2.6 Computational Infrastructure

The analysis of large-scale spatial transcriptomics datasets often demands high-performance computing (HPC) environments with extensive memory and processing capabilities—resources that may not be accessible to all research groups. While cloud-based infrastructure offers scalability in principle, it can be impractical in this context due to the substantial size of raw imaging and sequencing data. Uploading such datasets introduces significant bandwidth and latency bottlenecks, which can hinder cloud-based workflows and make local HPC environments a more feasible option for many users.

5.2.7 Reproducibility

Reproducibility remains a major concern in spatial transcriptomics due to the absence of standardized workflows, frequent updates to analytical tools, and inconsistent documentation. As new methods are rapidly developed and existing ones evolve, reproducing published results becomes increasingly difficult, particularly when code, parameters, or data preprocessing steps are not clearly reported. This limits the reliability of comparative studies and slows the development of robust, generalizable analytical pipelines.

5.3 Biological Interpretation

A key challenge in spatial transcriptomics lies in interpreting the complex biological signals embedded within spatially resolved gene expression data. This includes disentangling mixed cellular signals, understanding spatial organization, and inferring cell–cell interactions based solely on transcriptomics and spatial proximity.

5.3.1 Deconvolution of Multi-Cellular Spots

For lower-resolution technologies, each ST unit may capture transcripts from multiple cells. Deconvolving these composite signals to infer the contributions of individual cell types is a non-trivial task that often relies on reference-based methods, such as integrating with single-cell RNA-seq. However, this process is inherently uncertain due to variability in cell composition, differences in gene coverage, and the absence of the single-cell ground truth.

5.3.2 Understanding Spatial Relationships

Effectively analyzing spatial relationships between distinct cell types or tissue regions remains a methodological difficulty. While spatial proximity can suggest potential interactions or shared functions within cellular neighborhoods, accurately capturing these relationships requires advanced spatial statistics, graph-based modeling, or neighborhood enrichment analyses, many of which are still under active development.

5.3.3 Inferring Cell–Cell Interactions

Although ST provides crucial positional context, inferring direct cell–cell communication from transcriptomic data alone is difficult. Most existing methods rely on co-expression of known ligand-receptor pairs across neighboring cells, but this inference is indirect and sensitive to technical noise, resolution limits, and incomplete ligand-receptor annotations. Multimodal approaches that combine spatial transcriptomics with proteomics or imaging mass cytometry may offer more reliable insights but introduce additional integration problems.

6 Future Trends

Spatial transcriptomics is entering a new phase of innovation where its potential extends far beyond transcript mapping alone. As technologies mature, researchers are beginning to envision a future in which spatial data are integrated seamlessly with other molecular layers, proteins, metabolites, and epigenetic markers, to create richly annotated, multidimensional tissue atlases. Alongside these experimental advances, computational methods, particularly those involving machine learning, are rapidly evolving, enabling the interpretation of spatial complexity at scale. These trends point toward transformative applications, from basic biological discovery to clinical diagnostics. This chapter explores the emerging directions, open challenges, and future possibilities that will shape the next generation of spatial transcriptomics.

6.1 Spatial Multi-Omics Integration

One of the key frontiers in spatial biology is the integration of diverse omics layers (transcriptomic, proteomic, epigenomic, and metabolomic) along with spatial and histological context. Effective integration enables the construction of comprehensive tissue atlases that capture both molecular identity and spatial relationships.

However, harmonizing these data types, often generated at different resolutions and from different platforms, is technically complex and analytically challenging.

However, a previous study exemplifies the power of multi-omics spatial data integration in unraveling the complex, heterogeneous, and adaptive landscape of glioblastoma (GBM) (Ravi et al. 2022). Through the combination of spatial transcriptomics, MALDI imaging (Gessel et al. 2014), DNA copy number variation inference (Patel et al. 2014) and imaging mass cytometry (IMC) (Windhager et al. 2023), the authors elucidate how microenvironmental factors, particularly hypoxia and immune infiltration, drive transcriptional, genomic and epigenetic remodeling within spatially defined tumor regions (Ravi et al. 2022).

By analyzing regions defined as either hypoxia-core (co-enriched for hypoxia signatures via MALDI and spatial transcriptomics) or hypoxia-associated, the study uncovered a spatial link between hypoxic metabolism and genomic instability (Ravi et al. 2022). Hypoxia-core regions displayed consistent and significant chromosomal aberrations, including losses of 14q and 15p and gains in chromosome 7, which were not evident in control regions. These findings suggest that metabolic stress in hypoxic niches fuels subclonal evolution through increased DNA damage and chromosomal instability, highlighting a dynamic interplay between tumor metabolism and genome architecture (Ravi et al. 2022).

To place these molecular changes within their immunological context, the study employed IMC-based single-cell proteomics on spatially defined regions from the same tissue blocks. This enabled the quantification of over 82,000 segmented cells across immune-reactive, hypoxia-reactive, and neuronal differentiation zones (Ravi et al. 2022). Reactive-immune areas exhibited significantly higher infiltration of myeloid and lymphoid cells, with a corresponding enrichment of glioma cells expressing reactive-immune markers (e.g., EGFR+CHI3L1+VIM+), which were absent in neuronal regions. These findings demonstrate that the immune landscape shapes and is shaped by tumor cell transcriptional programs, reinforcing the concept of spatially compartmentalized tumor-immune co-evolution (Ravi et al. 2022).

6.2 3D Reconstruction of the Tissue/Tumor Microenvironment

An exciting advancement in this space is the reconstruction of three-dimensional tissue architecture from consecutive tissue sections. In a landmark study by Mo et al. (2024), the authors examined over 130 tumor sections across six cancer types using Visium spatial transcriptomics, matched single-nucleus RNA-seq, and CODEX (Kuswanto et al. 2022) proteomics. By co-registering 48 consecutive spatial transcriptomics sections from 16 samples, they were able to reconstruct 3D tumor structures, revealing spatial subclones with distinct mutational and metabolic profiles. These reconstructions uncovered not only how tumor cells are spatially organized but also how they interact with immune cells and stromal components in three-dimensional space (Mo et al. 2024).

Importantly, the study identified metabolically active tumor cores surrounded by antigen-presenting and immune-exhausted environments at the periphery, features that would be obscured in a purely 2D analysis. The integration of spatial transcriptomics and multiplexed imaging using deep learning further enabled the detection of immune "hot" and "cold" zones within the same tumor, offering new insights into tumor immune evasion strategies. Such 3D modeling approaches are poised to become integral in future spatial studies, enabling researchers to unravel the dynamics of tissue evolution and cellular interactions with unprecedented clarity (Mo et al. 2024).

Looking ahead, spatial biology is rapidly moving toward multi-omics integration and 48 serial spatial transcriptomics sections to reconstruct the 3D model of the tissue. Combining consecutive spatial sections of tissue for transcriptomics, spatial proteomics (e.g., CODEX, IMC), and metabolomics (MALDI) assays will allow for a more holistic view of tissue function and disease. These integrative efforts require new statistical frameworks capable of modeling cross-modality relationships within a spatial framework, as well as unified data standards and visualization platforms.

6.3 Application of Machine Learning in Spatial Biology

Another transformative trend is the application of artificial intelligence and machine learning to spatial data interpretation. From unsupervised cell type classification to predictive modeling of spatial patterns and tissue dynamics, ML methods are becoming an indispensable part of analytical pipelines. Models trained on large spatial datasets can learn latent representations of tissue architecture and predict functional zones or disease-associated alterations. Importantly, explainability and biological interpretability must remain key goals in the development of spatial AI tools, especially in translational contexts.

6.4 Clinical Translation

For spatial transcriptomics to transition from a research tool to a clinical mainstay, future advances must focus on speed, affordability, standardization, interpretability, and data stewardship. Clinical settings demand rapid turnaround, often within 24 hours, necessitating streamlined protocols, automated workflows, and targeted gene panels that can deliver actionable results within tight timeframes. To enable widespread use, especially beyond major research centers, costs must be significantly reduced through scalable chemistries, efficient readouts, and modular assay designs.

Equally important is the development of standardized, reproducible procedures, from tissue handling to quality control and analysis, supported by validated algorithms capable of linking spatial patterns to clinical outcomes. Making spatial data clinically usable will require abstraction and simplification: future tools must distill complex transcriptomic maps into clear, interpretable formats, such as risk scores, annotated regions of interest, or AI-generated decision-support overlays tailored to the needs of pathologists and oncologists. Finally, as spatial omics becomes embedded in patient records, robust frameworks for secure data storage, ethical sharing, and consent management will be essential. Together, these efforts will lay the groundwork for spatial transcriptomics to play a transformative role in precision diagnostics and personalized medicine.

7 Conclusion

Spatial transcriptomics is quickly evolved from a novel concept to a transformative toolkit for life sciences. In this Chapter, we explored how the field is advancing on multiple fronts: experimental technologies are achieving finer resolution, greater throughput, and multi-omic breadth; computational methods are rising to meet the ever-increasing data analysis demands with integration and AI; and biological applications are yielding new insights across cancer, neuroscience, development, and so on. The commercial landscape is dynamic, fueling innovation while beginning to consolidate expertise for robust product offerings.

Looking to the future, we can expect spatial transcriptomics to become even more powerful and prevalent. Integration with proteomics, metabolomics, and epigenomics will move us closer to a true molecular atlas of tissues, where one can examine any layer of biology in situ. Emerging areas like 3D spatial mapping and spatiotemporal studies will provide a more complete picture of how cells function in context over time. These developments, however, will also require surmounting current challenges such as improving cost-effectiveness, ensuring data reliability, and developing user-friendly analysis tools.

One exciting vision is that in the next decade, spatial omics could be a routine part of large projects like the Human Cell Atlas, yielding an annotated 3D map of the human body at the cellular level, essentially an Open Street Maps of the body, where you could zoom into an organ, then a tissue, then a cell, and see which genes and proteins are active there. Such a resource would be invaluable for science and medicine.

Clinically, while not widespread yet, the trajectory suggests that spatial profiling may find its way into pathology labs for certain high-value applications (for example, guiding immunotherapy by mapping immune cells in a tumor). The first evidence of this translational momentum is visible in specialized centers and clinical research programs investing in spatial omics. Realizing this clinical potential will depend on faster assays, lower costs, standardized protocols, and intuitive tools that translate complex spatial data into actionable insights.

In conclusion, the future of spatial transcriptomics is bright and expansive; it stands at the intersection of genomics, microscopy, image analysis, spatial modeling, big data, and AI—a true multidisciplinary frontier. As cutting-edge as it is now, the field is still young and not fully matured, which means the coming years will likely bring many surprises and breakthroughs. By addressing current limitations and fostering collaboration between tool developers, computational scientists, biologists, and clinicians, spatial transcriptomics will continue to unlock new dimensions of biomedical knowledge. Ultimately illuminating how the intricate tapestry of life is woven in space.

References

Almet AA, Cang Z, Jin S, Nie Q (2021) The landscape of cell–cell communication through single-cell transcriptomics. Curr Opin Syst Biol 26:12–23. https://doi.org/10.1016/j.coisb.2021.03.007

Andrews TS, Hemberg M (2018) False signals induced by single-cell imputation. F1000Res 7:1740. https://doi.org/10.12688/f1000research.16613.1

Andrieux G, Das T, Griffin M, Straehle J, Paine SML, Beck J, Boerries M, Heiland DH, Smith SJ, Rahman R, Chakraborty S (2023) Spatially resolved transcriptomic profiles reveal unique defining molecular features of infiltrative 5ALA-metabolizing cells associated with glioblastoma recurrence. Genome Med 15(1). https://doi.org/10.1186/s13073-023-01207-1

Armingol E, Officer A, Harismendy O, Lewis NE (2020) Deciphering cell–cell interactions and communication from gene expression. Nat Rev Genet 22(2):71–88. https://doi.org/10.1038/s41576-020-00292-x

Ben-Chetrit N, Niu X, Swett AD, Sotelo J, Jiao MS, Stewart CM, Potenski C, Mielinis P, Roelli P, Stoeckius M, Landau DA (2023) Integration of whole transcriptome spatial profiling with protein markers. Nat Biotechnol 41(6):788–793. https://doi.org/10.1038/s41587-022-01536-3

Bernstein MN, Ni Z, Prasad A, Brown J, Mohanty C, Stewart R, Newton MA, Kendziorski C (2022) SpatialCorr identifies gene sets with spatially varying correlation structure. Cell Rep Methods 2(12):100369. https://doi.org/10.1016/j.crmeth.2022.100369

Bhuva DD, Tan CW, Salim A, Marceaux C, Pickering MA, Chen J, Kharbanda M, Jin X, Liu N, Feher K, Putri G, Tilley WD, Hickey TE, Asselin-Labat M-L, Phipson B, Davis MJ (2024) Library size confounds biology in spatial transcriptomics data. Genome Biol 25(1). https://doi.org/10.1186/s13059-024-03241-7

Biancalani T, Scalia G, Buffoni L, Avasthi R, Lu Z, Sanger A, Tokcan N, Vanderburg CR, Segerstolpe Å, Zhang M, Avraham-Davidi I, Vickovic S, Nitzan M, Ma S, Subramanian A, Lipinski M, Buenrostro J, Brown NB, Fanelli D, Zhuang X, Macosko EZ, Regev A (2021) Deep learning and alignment of spatially resolved single-cell transcriptomes with tangram. Nat Methods 18(11):1352–1362. https://doi.org/10.1038/s41592-021-01264-7

Burgess DJ (2019) Spatial transcriptomics coming of age. Nat Rev Genet 20(6):317. https://doi.org/10.1038/s41576-019-0129-z

Butler A, Hoffman P, Smibert P, Papalexi E, Satija R (2018) Integrating single-cell transcriptomic data across different conditions, technologies, and species. Nat Biotechnol 36(5):411–420. https://doi.org/10.1038/nbt.4096

Cable DM, Murray E, Zou LS, Goeva A, Macosko EZ, Chen F, Irizarry RA (2021) Robust decomposition of cell type mixtures in spatial transcriptomics. Nat Biotechnol 40(4):517–526. https://doi.org/10.1038/s41587-021-00830-w

Cai P, Robinson MD, Tiberi S (2024) DESpace: spatially variable gene detection via differential expression testing of spatial clusters. Bioinformatics 40(2). https://doi.org/10.1093/bioinformatics/btae027

Cang Z, Zhao Y, Almet AA, Stabell A, Ramos R, Plikus MV, Atwood SX, Nie Q (2023) Screening cell–cell communication in spatial transcriptomics via collective optimal transport. Nat Methods 20(2):218–228. https://doi.org/10.1038/s41592-022-01728-4

Chen KH, Boettiger AN, Moffitt JR, Wang S, Zhuang X (2015) Spatially resolved, highly multiplexed RNA profiling in single cells. Science 348(6233). https://doi.org/10.1126/science.aaa6090

Choudhary S, Satija R (2022) Comparison and evaluation of statistical error models for scRNA-seq. Genome Biol 23(1). https://doi.org/10.1186/s13059-021-02584-9

Dong R, Yuan G-C (2021) SpatialDWLS: accurate deconvolution of spatial transcriptomic data. Genome Biol 22(1). https://doi.org/10.1186/s13059-021-02362-7

Dries R, Zhu Q, Dong R, Eng C-HL, Li H, Liu K, Fu Y, Zhao T, Sarkar A, Bao F, George RE, Pierson N, Cai L, Yuan G-C (2021) Giotto: a toolbox for integrative analysis and visualization of spatial expression data. Genome Biol 22(1). https://doi.org/10.1186/s13059-021-02286-2

Elosua-Bayes M, Nieto P, Mereu E, Gut I, Heyn H (2021) SPOTlight: seeded NMF regression to deconvolute spatial transcriptomics spots with single-cell transcriptomes. Nucleic Acids Res 49(9):e50. https://doi.org/10.1093/nar/gkab043

Frederick MJ, Perez-Bello D, Yadollahi P, Castro P, Frederick A, Frederick A, Osman RA, Essien F, Yebra I, Hamlin A, Ow TJ, Skinner HD, Sandulache VC (2024) Reliable RNA-seq analysis from FFPE specimens as a means to accelerate cancer-related health disparities research. bioRxiv (Cold Spring Harbor Laboratory). https://doi.org/10.1101/2024.10.10.617597

Gaspard-Boulinc LC, Gortana L, Walter T, Barillot E, Cavalli FMG (2025) Cell-type deconvolution methods for spatial transcriptomics. Nat Rev Genet. https://doi.org/10.1038/s41576-025-00845-y

Gessel MM, Norris JL, Caprioli RM (2014) MALDI imaging mass spectrometry: spatial molecular analysis to enable a new age of discovery. J Proteomics 107:71–82. https://doi.org/10.1016/j.jprot.2014.03.021

Grünwald BT, Devisme A, Andrieux G, Vyas F, Aliar K, McCloskey CW, Macklin A, Jang GH, Denroche R, Romero JM, Bavi P, Bronsert P, Notta F, O'Kane G, Wilson J, Knox J, Tamblyn L, Udaskin M, Radulovich N, Fischer SE, Boerries M, Gallinger S, Kislinger T, Khokha R (2021) Spatially confined sub-tumor microenvironments in pancreatic cancer. Cell 184(22):5577–5592.e18. https://doi.org/10.1016/j.cell.2021.09.022

Hanahan D, Weinberg RA (2011) Hallmarks of cancer: the next generation. Cell 144(5):646–674. https://doi.org/10.1016/j.cell.2011.02.013

Hao Y, Stuart T, Kowalski MH, Choudhary S, Hoffman P, Hartman A, Srivastava A, Molla G, Madad S, Fernandez-Granda C, Satija R (2023) Dictionary learning for integrative, multimodal and scalable single-cell analysis. Nat Biotechnol 42(2):293–304. https://doi.org/10.1038/s41587-023-01767-y

Hinshaw DC, Shevde LA (2019) The tumor microenvironment innately modulates cancer progression. Cancer Res 79(18):4557–4566. https://doi.org/10.1158/0008-5472.can-18-3962

Hu J, Li X, Coleman K, Schroeder A, Ma N, Irwin DJ, Lee EB, Shinohara RT, Li M (2021) SpaGCN: integrating gene expression, spatial location and histology to identify spatial domains and spatially variable genes by graph convolutional network. Nat Methods 18(11):1342–1351. https://doi.org/10.1038/s41592-021-01255-8

Jin Y, Zuo Y, Li G, Liu W, Pan Y, Fan T, Fu X, Yao X, Peng Y (2024) Advances in spatial transcriptomics and its applications in cancer research. Mol Cancer 23(1). https://doi.org/10.1186/s12943-024-02040-9

Kim H, Kumar A, Lövkvist C, Palma AM, Martin P, Kim J, Bhoopathi P, Trevino J, Fisher P, Madan E, Gogna R, Won KJ (2023) CellNeighborEX: deciphering neighbor-dependent gene expression from spatial transcriptomics data. Mol Syst Biol 19(11). https://doi.org/10.15252/msb.202311670

Kleshchevnikov V, Shmatko A, Dann E, Aivazidis A, King HW, Li T, Elmentaite R, Lomakin A, Kedlian V, Gayoso A, Jain MS, Park JS, Ramona L, Tuck E, Arutyunyan A, Vento-Tormo R, Gerstung M, James L, Stegle O, Bayraktar OA (2022) Cell2location maps fine-grained cell types in spatial transcriptomics. Nat Biotechnol 40(5):661–671. https://doi.org/10.1038/s41587-021-01139-4

Kojima Y, Mii S, Hayashi S, Hirose H, Ishikawa M, Akiyama M, Enomoto A, Shimamura T (2024) Single-cell colocalization analysis using a deep generative model. Cell Syst 15(2):180–192.e7. https://doi.org/10.1016/j.cels.2024.01.007

Kumar A, Schrader AW, Aggarwal B, Boroojeny AE, Asadian M, Lee J, Song YJ, Zhao SD, Han H-S, Sinha S (2024) Intracellular spatial transcriptomic analysis toolkit (InSTAnT). Nat Commun 15(1). https://doi.org/10.1038/s41467-024-49457-w

Kuswanto W, Nolan G, Lu G (2022) Highly multiplexed spatial profiling with CODEX: bioinformatic analysis and application in human disease. Semin Immunopathol 45(1):145–157. https://doi.org/10.1007/s00281-022-00974-0

Lewis SM, Asselin-Labat M-L, Nguyen Q, Berthelet J, Tan X, Wimmer VC, Merino D, Rogers KL, Naik SH (2021) Spatial omics and multiplexed imaging to explore cancer biology. Nat Methods 18(9):997–1012. https://doi.org/10.1038/s41592-021-01203-6

Liu Q, Hsu C-Y, Shyr Y (2022) Scalable and model-free detection of spatial patterns and colocalization. Genome Res 32(9):1736–1745. https://doi.org/10.1101/gr.276851.122

Liu Y, Xun Z, Ma K, Liang S, Li X, Zhou S, Sun L, Liu Y, Du Y, Guo X, Cui T, Zhou H, Wang J, Yin D, Song R, Zhang S, Cai W, Meng F, Guo H, Zhang B, Yang D, Bao R, Hu Q, Wang J, Ye Y, Liu L (2023) Identification of a tumour immune barrier in the HCC microenvironment that determines the efficacy of immunotherapy. J Hepatol 78(4):770–782. https://doi.org/10.1016/j.jhep.2023.01.011

Liu Y, Ye S-Y, He S, Chi D-M, Wang X-Z, Wen Y-F, Ma D, Nie R-C, Xiang P, Zhou Y, Ruan Z-H, Peng R-J, Luo C-L, Wei P-P, Lin G-W, Zheng J, Cui Q, Cai M-Y, Yun J-P, Dong J, Mai H-Q, Xia X, Bei J-X (2024) Single-cell and spatial transcriptome analyses reveal tertiary lymphoid structures linked to tumour progression and immunotherapy response in nasopharyngeal carcinoma. Nat Commun 15(1). https://doi.org/10.1038/s41467-024-52153-4

Lubeck E, Coskun AF, Zhiyentayev T, Ahmad M, Cai L (2014) Single-cell in situ RNA profiling by sequential hybridization. Nat Methods 11(4):360–361. https://doi.org/10.1038/nmeth.2892

Making space for spatial biology in the clinic (2025) Nat Biotechnol 43:1–2. https://doi.org/10.1038/s41587-024-02543-2

Meylan M, Petitprez F, Becht E, Bougoüin A, Pupier G, Calvez A, Giglioli I, Verkarre V, Lacroix G, Verneau J, Sun C-M, Laurent-Puig P, Vano Y-A, Elaïdi R, Méjean A, Sanchez-Salas R, Barret E, Cathelineau X, Oudard S, Reynaud C-A, De Reyniès A, Sautès-Fridman C, Fridman WH (2022) Tertiary lymphoid structures generate and propagate anti-tumor antibody-producing plasma cells in renal cell cancer. Immunity 55(3):527–541.e5. https://doi.org/10.1016/j.immuni.2022.02.001

Mo C-K, Liu J, Chen S, Storrs E, Da Costa ALNT, Houston A, Wendl MC, Jayasinghe RG, Iglesia MD, Ma C, Herndon JM, Southard-Smith AN, Liu X, Mudd J, Karpova A, Shinkle A, Goedegebuure SP, Abdelzaher ATMA, Bo P, Fulghum L, Livingston S, Balaban M, Hill A, Ippolito JE, Thorsson V, Held JM, Hagemann IS, Kim EH, Bayguinov PO, Kim AH, Mullen MM, Shoghi KI, Ju T, Reimers MA, Weimholt C, Kang L-I, Puram SV, Veis DJ, Pachynski R, Fuh KC, Chheda MG, Gillanders WE, Fields RC, Raphael BJ, Chen F, Ding L (2024) Tumour evolution and microenvironment interactions in 2D and 3D space. Nature 634(8036):1178–1186. https://doi.org/10.1038/s41586-024-08087-4

Moses L, Pachter L (2022) Museum of spatial transcriptomics. Nat Methods 19(5):534–546. https://doi.org/10.1038/s41592-022-01409-2

Murphy AE, Fancy N, Skene N (2023) Avoiding false discoveries in single-cell RNA-seq by revisiting the first Alzheimer's disease dataset. elife 12. https://doi.org/10.7554/elife.90214

Palla G, Spitzer H, Klein M, Fischer D, Schaar AC, Kuemmerle LB, Rybakov S, Ibarra IL, Holmberg O, Virshup I, Lotfollahi M, Richter S, Theis FJ (2022) Squidpy: a scalable framework for spatial omics analysis. Nat Methods 19(2):171–178. https://doi.org/10.1038/s41592-021-01358-2

Patel AP, Tirosh I, Trombetta JJ, Shalek AK, Gillespie SM, Wakimoto H, Cahill DP, Nahed BV, Curry WT, Martuza RL, Louis DN, Rozenblatt-Rosen O, Suvà ML, Regev A, Bernstein BE (2014) Single-cell RNA-seq highlights intratumoral heterogeneity in primary glioblastoma. Science 344(6190):1396–1401. https://doi.org/10.1126/science.1254257

Peyraud F, Guégan J-P, Rey C, Lara O, Odin O, Del Castillo M, Vanhersecke L, Coindre J-M, Clot E, Brunet M, Grellety T, Tasseel A, Moulec SL, Johnston RJ, Bessede A, Italiano A (2025) Spatially resolved transcriptomics reveal the determinants of primary resistance to immunotherapy in NSCLC with mature tertiary lymphoid structures. Cell Rep Med:101934. https://doi.org/10.1016/j.xcrm.2025.101934

Rao A, Barkley D, França GS, Yanai I (2021) Exploring tissue architecture using spatial transcriptomics. Nature 596(7871):211–220. https://doi.org/10.1038/s41586-021-03634-9

Ravi V, Will P, Kueckelhaus J, Sun N, Joseph K, Salié H, Vollmer L, Kuliesiute U, Von Ehr J, Benotmane JK, Neidert N, Follo M, Scherer F, Goeldner JM, Behringer SP, Franco P, Khiat M, Zhang J, Hofmann UG, Fung C, Ricklefs F, Lamszus K, Boerries M, Ku M, Beck J, Sankowski R, Schwabenland M, Prinz M, Schüller U, Killmer S, Bengsch B, Walch AK, Delev D, Schnell O, Heiland DH (2021) Spatially resolved multi-omics deciphers bidirectional tumor-host interdependence in glioblastoma. SSRN Electron J. https://doi.org/10.2139/ssrn.3865275

Ravi VM, Will P, Kueckelhaus J, Sun N, Joseph K, Salié H, Vollmer L, Kuliesiute U, Von Ehr J, Benotmane JK, Neidert N, Follo M, Scherer F, Goeldner JM, Behringer SP, Franco P, Khiat M, Zhang J, Hofmann UG, Fung C, Ricklefs FL, Lamszus K, Boerries M, Ku M, Beck J, Sankowski R, Schwabenland M, Prinz M, Schüller U, Killmer S, Bengsch B, Walch AK, Delev D, Schnell O, Heiland DH (2022) Spatially resolved multi-omics deciphers bidirectional tumor-host interdependence in glioblastoma. Cancer Cell 40(6):639–655.e13. https://doi.org/10.1016/j.ccell.2022.05.009

Schapiro D, Jackson HW, Raghuraman S, Fischer JR, Zanotelli VRT, Schulz D, Giesen C, Catena R, Varga Z, Bodenmiller B (2017) histoCAT: analysis of cell phenotypes and interactions in multiplex image cytometry data. Nat Methods 14(9):873–876. https://doi.org/10.1038/nmeth.4391

Shang L, Zhou X (2022) Spatially aware dimension reduction for spatial transcriptomics. Nat Commun 13(1). https://doi.org/10.1038/s41467-022-34879-1

Song D, Chen S, Lee C, Li K, Ge X, Li JJ (2023) ClusterDE: a post-clustering differential expression (DE) method robust to false-positive inflation caused by double dipping. bioRxiv (Cold Spring Harbor Laboratory). https://doi.org/10.1101/2023.07.21.550107

Squair JW, Gautier M, Kathe C, Anderson MA, James ND, Hutson TH, Hudelle R, Qaiser T, Matson KJE, Barraud Q, Levine AJ, La Manno G, Skinnider MA, Courtine G (2021) Confronting false discoveries in single-cell differential expression. Nat Commun 12(1). https://doi.org/10.1038/s41467-021-25960-2

Sun X-X, Yu Q (2015) Intra-tumor heterogeneity of cancer cells and its implications for cancer treatment. Acta Pharmacol Sin 36(10):1219–1227. https://doi.org/10.1038/aps.2015.92

Svensson V, Teichmann SA, Stegle O (2018) SpatialDE: identification of spatially variable genes. Nat Methods 15(5):343–346. https://doi.org/10.1038/nmeth.4636

Tanevski J, Flores ROR, Gabor A, Schapiro D, Saez-Rodriguez J (2022) Explainable multiview framework for dissecting spatial relationships from highly multiplexed data. Genome Biol 23(1). https://doi.org/10.1186/s13059-022-02663-5

Windhager J, Zanotelli VRT, Schulz D, Meyer L, Daniel M, Bodenmiller B, Eling N (2023) An end-to-end workflow for multiplexed image processing and analysis. Nat Protoc 18(11):3565–3613. https://doi.org/10.1038/s41596-023-00881-0

Zhao E, Stone MR, Ren X, Guenthoer J, Smythe KS, Pulliam T, Williams SR, Uytingco CR, Taylor SEB, Nghiem P, Bielas JH, Gottardo R (2021) Spatial transcriptomics at subspot resolution with BayesSpace. Nat Biotechnol 39(11):1375–1384. https://doi.org/10.1038/s41587-021-00935-2

Biotechnology-Driven Progress in Liquid Biopsy for Precision Oncology

Md. Raiyan Hosen and Md. Ismail Hosen

Abstract Liquid biopsy is changing how we detect and monitor cancer. By analyzing small traces of tumor DNA and cells in body fluids such as blood and urine, we can now track cancer without needing invasive surgery or tissue samples. This chapter explores how modern biotechnology tools are making liquid biopsy more accurate and useful. Covering a spectrum of cutting-edge methodologies, the chapter explores the isolation and analysis of circulating tumor DNA (ctDNA), circulating tumor cells (CTCs), extracellular vesicles (EVs), and various biomarkers indicative of cancer progression. The key topics for the proposed book chapter include the utilization of next-generation sequencing (NGS) and digital droplet PCR (ddPCR) for the high-throughput profiling of genetic alterations in liquid biopsy samples, enabling comprehensive genomic characterization of tumors. As liquid biopsy continues to evolve as a dynamic field, this book chapter serves as a valuable resource for researchers, clinicians, and biotechnologists seeking a deeper understanding of the biotechnological landscape shaping the liquid biopsy paradigm in cancer diagnostics and monitoring.

Keywords Liquid biopsy · ctDNA · CTCs · ddPCR · NGS

Abbreviations

cfDNA	Cell-free DNA
CSF	Cerebrospinal fluid
CTCs	Circulating tumor cells
ctDNA	Circulating tumor DNA

M. R. Hosen · M. I. Hosen (✉)
Department of Biochemistry and Molecular Biology, University of Dhaka, Dhaka, Bangladesh
e-mail: ismail.hosen@du.ac.bd

A. H. M. N. Nabi, H. Uddin Shekhar (eds.), *Biotechnological Advances in Healthomics*, Interdisciplinary Biotechnological Advances,
https://doi.org/10.1007/978-981-95-3612-2_5

ddPCR	Droplet digital PCR
EVs	Extracellular vesicles
MRD	Minimal residual disease
OSCC	Oral squamous cell carcinoma
POC	Point-of-care
TEPs	Tumor-educated platelets
VOC	Volatile organic compounds

1 Introduction

Liquid biopsy offers a new, less invasive way to detect and monitor cancer. Instead of relying on tissue samples from surgeries, clinicians can now examine tumor-related materials in fluids such as blood, urine, and saliva. This allows for earlier detection, better tracking of treatment response, and more personalized therapies. While the potential of liquid biopsy is enormous, the technological requirement to sensitively detect the tumor-borne materials in body fluids is also a major point of concern. In this chapter, we will first discuss the different facets of liquid biopsy biomarkers, including the circulating tumor DNA (ctDNA), circulating tumor cells (CTCs), and extracellular vesicles (EVs). Subsequently, we will explore recent biotechnological breakthroughs that have made liquid biopsy more powerful, especially in analyzing these biomarkers in body fluids.

2 A Comprehensive Definition of Liquid Biopsy

Liquid biopsy is defined as a non-invasive diagnostic method that involves the analysis of circulating tumor components found in bodily fluids, primarily blood, but also urine or cerebrospinal fluid (Vaidyanathan et al. 2019). This innovative approach allows for the detection of cancer, monitoring of disease progression, and assessment of treatment response by examining biomarkers, such as circulating tumor DNA (ctDNA), circulating tumor cells (CTCs), and extracellular vesicles (EVs) (Vaidyanathan et al. 2019; Alix-Panabieres 2020; Yu et al. 2022). Liquid biopsy has the potential to revolutionize cancer care by providing insights into tumor evolution and therapy response, ultimately guiding personalized treatment strategies for patients. Moreover, advancements in technologies such as NGS and ddPCR have significantly enhanced the sensitivity and accuracy of liquid biopsy, facilitating the detection of rare genetic mutations associated with cancer progression (Ma et al. 2024). The innovations in liquid biopsy are crucial for the diagnosis of cancers originating from organs, where tissue samples are difficult to obtain, such as the pancreas, lung, and colon tissues (Santarpia et al. 2018; Buscail et al. 2019; Vacante et al. 2020).

3 Liquid Biopsy Biomarkers

The success of liquid biopsy depends largely on the detection and analysis of tumor-derived biomarkers, primarily circulating tumor DNA (ctDNA), circulating tumor cells (CTCs), and extracellular vesicles (EVs) (Ma et al. 2024). This section delves into different types of biomarkers detected through liquid biopsy (see Fig. 1 and Table 1).

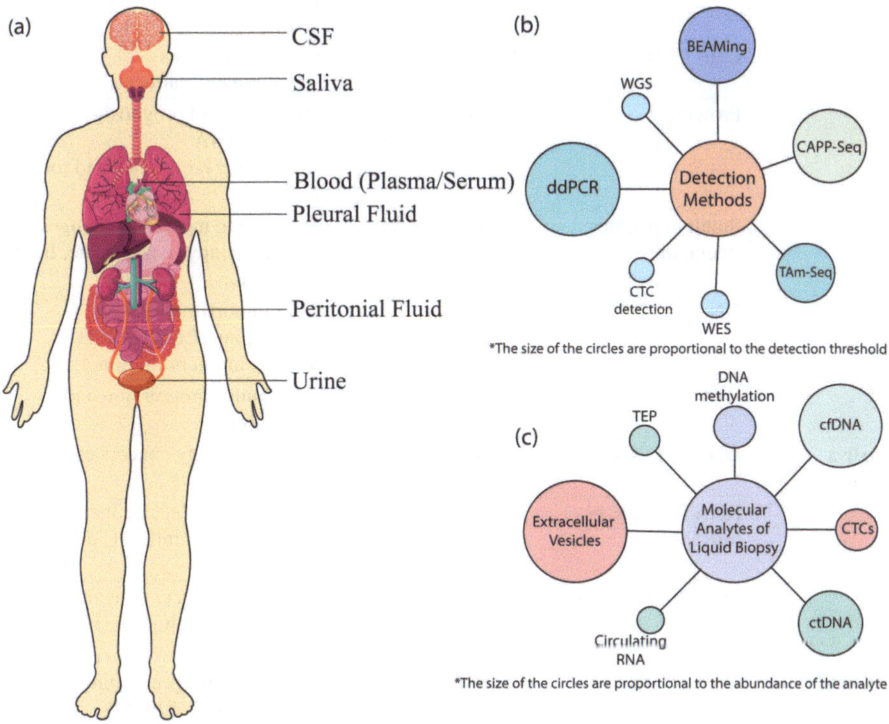

Fig. 1 Outline of major sources of liquid biopsy, detection methods, and primary molecular analytes. (**a**) A diagram of human organs showing sources of each type of liquid biopsy. (**b**) Molecular detection technologies are used for detection. The size of the circle is proportional to the detection threshold. (**c**) Primary molecular analytes that are found in a liquid biopsy. Circle size is proportional to the abundance of each type of analyte. Abbreviations: ddPCR droplet digital PCR, WGS whole genome sequencing, BEAMing beads, emulsion, amplification, and magnetics, CAPP-seq cancer personalized profiling by sequencing, TAm-seq tagged amplicon deep sequencing, WES whole exome sequencing, CTCs circulating tumor cells

Table 1 Analyte types, detection methods, and clinical applications of liquid biopsy across different body fluids in cancer

Liquid biopsy source	Analyte type	Methods used for detection	Clinical application in cancer
Blood	DNA (ctDNA, cfDNA, methylated DNA)	ddPCR, bisulfite sequencing	Tumor genotyping & resistance mutation detection MRD prediction & recurrence detection Tumor burden & early detection
	EVs (exosomal RNA/ protein)	ELISA, ddPCR	Treatment response monitoring Multicancer screening pancreatic, ovarian, breast cancer screening
	Cells (CTCs, TEPs)	CTC-chips	Cancer prognosis Response to treatment
	Proteins	Immunoassays	Diagnosis and prognosis—CEA, CA15-3, PSA Monitoring recurrence and therapy response
	Small molecules (metabolites)	Immunoassays	Monitoring metabolic changes during therapy (e.g., lactate, lipid metabolites)
CSF	DNA (ctDNA)	ddPCR	CNS tumors/metastasis Genotyping (EGFR, IDH1) Monitoring in brain tumors.
	EVs	ELISA	CNS tumor progression and monitoring
Saliva	DNA (ctDNA)	ddPCR	Head/neck/lung cancers early screening
	RNA (cfRNA, miRNAs)	ddPCR	OSCC & head/neck cancer early detection, e.g., miR-21, miR-200a, miR-125a
	Proteins (HER2, p53)	Immunoassays	Breast/oral cancer detection and recurrence monitoring (HER2, CA15-3, p53)
Pleural fluid	DNA (ctDNA)	ddPCR	Diagnosis and genotyping in lung cancer (EGFR/T790M)
	Exosomal miRNAs	ddPCR	Pleural effusion (exosomal miRNA panels) lung cancer
	Cells (CTCs)	CTC-chips	Prognosis and metastasis assessment in lung cancers
	Proteins (VEGF, IL-6, CEA, mesothelin)	Immunoassays	Detecting effusions Therapy response monitoring
Peritoneal fluid	DNA (ctDNA)	ddPCR	Mutation profiling in gastric, ovarian, endometrial cancers
	RNA (EV miRNAs)	ddPCR	Diagnosis and staging in ovarian/ endometrial/colorectal cancers
	Cells (immune cells)	CTC-chips	Immune landscape profiling

(continued)

Table 1 (continued)

Liquid biopsy source	Analyte type	Methods used for detection	Clinical application in cancer
Urine	DNA (ctDNA, cfDNA)	ddPCR	Non-invasive genotyping Early detection & MRD monitoring in breast, colorectal, bladder cancers
	RNA (exosomal miRNAs)	ddPCR	Prostate and bladder cancer detection via miRNA panels POC biosensor applications (miR-21)
	Proteins	Immunoassays	Bladder cancer markers (NMP22, TopoIIA)
	Small molecules (metabolites)	Immunoassays, ms-based	Metabolomic profiling for kidney, bladder cancer detection VOC analysis for non-invasive screening

Data sources: Alix-Panabières and Pantel (2016), Hicks et al. (2017), Oellerich et al. (2017, 2019), Sorber et al. (2016); Volckmar et al. (2018), de Miranda et al. (2021), Martins et al. (2021), Salfer et al. (2022)

3.1 Circulating Tumor DNA (ctDNA)

Circulating tumor DNA (ctDNA) is fragmented DNA released into the bloodstream by cancer cells as they die. It's a type of cell-free DNA (cfDNA) specifically originating from tumor cells. Analyzing ctDNA through liquid biopsies can provide insights into tumor characteristics, aiding in diagnosis, treatment selection, and monitoring for recurrence (Hicks et al. 2017; Oellerich et al. 2019). Recent advancements in the detection of ctDNA have significantly enhanced the sensitivity and specificity of liquid biopsies, facilitating more accurate monitoring of cancer progression and treatment responses (Turner et al. 2020; Amato et al. 2024; Bartolomucci et al. 2025). These innovations are crucial for the early identification of mutations and the adaptation of treatment plans, ultimately leading to improved patient outcomes.

Moreover, the integration of technologies such as NGS and ddPCR has revolutionized the analysis of ctDNA, enabling the detection of rare genetic mutations that are pivotal for personalized cancer therapy (Fig. 1). This progress underscores the potential of liquid biopsy to transform standard clinical practices in oncology. Furthermore, ongoing research is essential to address the challenges related to sensitivity, specificity, and standardization, ensuring that liquid biopsy can be effectively integrated into routine clinical practice.

3.2 Detection Methods for ctDNA

Recent advancements in ctDNA detection have the potential to significantly enhance early cancer diagnosis and treatment strategies. Various methods for detecting ctDNA include polymerase chain reaction (PCR), next-generation sequencing (NGS), and emerging biosensor technologies, each offering unique advantages for clinical applications (Yi et al. 2023; Kumar et al. 2024). Electrochemical biosensors have shown promise due to their sensitivity, specificity, and portability, making them suitable for point-of-care cancer diagnostics (Kumar et al. 2024). The integration of these technologies into routine clinical practice could revolutionize cancer screening and monitoring, ultimately improving patient outcomes. Further research is essential to optimize these detection methods, ensuring they can be effectively implemented in diverse clinical settings and populations.

Analysis of ctDNA can detect cancer with a sensitivity ranging from 40% to 70% depending on the cancer stage. Using NGS or ddPCR, sensitivity can be improved to 85%. Specificity of ctDNA-based cancer detection is reportedly 90%. However, there are several key factors that can severely affect the outcome, such as tumor location and shedding rate of the cancer type (de Abreu et al. 2025).

3.3 Circulating Tumor Cells (CTCs)

Circulating tumor cells (CTCs), which are cancer cells shed from primary tumors into the bloodstream, offer a non-invasive alternative to traditional tissue biopsies, enabling real-time insights into tumor dynamics and heterogeneity. This section aims to synthesize current knowledge on CTCs as liquid biopsy biomarkers, emphasizing their potential in clinical applications. Recent advancements in CTC isolation technologies have significantly enhanced their utility in cancer diagnostics, allowing for improved monitoring of disease progression and treatment response.

These technologies not only facilitate the capture of CTCs but also enable the analysis of their molecular characteristics, which can provide critical information regarding patient prognosis and therapeutic strategies (Woo and Yu 2018; Smit and Pantel 2024). The integration of advanced microfluidic techniques has enhanced CTC isolation, improving both efficiency and purity while allowing for the simultaneous capture of diverse CTC populations. The ongoing evolution of CTC detection methods continues to pave the way for personalized medicine, enhancing the ability to tailor treatments based on individual tumor profiles.

Combining multiple methods CTC have been proved to have sensitivity of 20% and specificity of 80% in lung cancer (Li et al. 2025). CTC detection is arguably harder as it depends on marker selection, cancer type and stage, isolation technology, and sample handling.

3.4 *Extracellular Vesicles (EVs)*

Extracellular vesicles (EVs) have emerged as a promising class of biomarkers in the field of liquid biopsy, offering a non-invasive approach for disease diagnosis, prognosis, and monitoring. These nanoscale membrane-bound particles, which are secreted by various cell types into body fluids, encapsulate a diverse array of biomolecules, including proteins, lipids, and nucleic acids, reflecting the physiological and pathological states of their parent cells. The ability to analyze EVs provides valuable insights into cellular communication and disease mechanisms, making them significant in oncology, neurology, and other medical disciplines. Recent advancements in biotechnological methods for detecting and characterizing EVs have enhanced their utility as biomarkers, with techniques ranging from ultracentrifugation and immunocapture to cutting-edge microfluidics and nanopore sequencing. As the landscape of EV research evolves, it becomes increasingly clear that standardization of isolation and analysis techniques is paramount for clinical applicability. Variability in sample handling, such as the choice of anticoagulants, can significantly impact the yield and molecular content of EVs, potentially leading to inconsistent results across studies (Yu et al. 2022; Taha 2023). Furthermore, the integration of advanced analytical platforms, such as mass spectrometry and high-throughput sequencing, not only enhances the detection capabilities but also allows for the identification of disease-specific EV signatures, which could revolutionize early diagnosis and personalized treatment strategies (Becker et al. 2016; Tamura et al. 2021). Ultimately, addressing these challenges will not only improve the reproducibility of EV biomarker studies but also facilitate the transition of EV-based diagnostics from research to routine clinical practice, thereby unlocking their full potential in disease management. In conclusion, the establishment of standardized protocols and the adoption of innovative technologies will be critical for the successful integration of EVs into clinical workflows, enhancing their role in precision medicine.

EV-based assays can have sensitivity ranging from 70% to 90% and specificity of up to 95% depending on detection method, cancer type, and stage. EVs are very heterogeneous and can vary widely in size, content, and origin, which makes the development of a standard detection method difficult (Rayamajhi et al. 2024).

3.5 *Exosome and Circulating RNA*

Among the myriad components examined in liquid biopsies, exosomes and circulating RNA have attracted considerable scholarly attention owing to their prospective utility as biomarkers for the early detection and prognostication of cancer. Exosomes, which are diminutive extracellular vesicles that transport a diverse array of proteins, lipids, and nucleic acids, assume a pivotal role in facilitating intercellular communication and serve as a reflection of the molecular characteristics of their progenitor cells (Yu et al. 2022; Liu et al. 2024). In a parallel vein, circulating RNA, comprising microRNAs and messenger RNAs, provides valuable insights into the pathophysiological conditions of tumors (Salehi and Sharifi 2018; Lin et al. 2022; Preethi et al. 2022).

Importantly, the integration of nanotechnology into these detection methodologies has surfaced as a transformative paradigm, markedly enhancing the sensitivity and specificity associated with biomarker identification (Li et al. 2020). Moreover, the emergence of microfluidic technologies has facilitated the creation of portable devices capable of executing rapid and multiplexed liquid biopsies, thereby propelling advancements in point-of-care testing and real-time cancer surveillance. Use of nanoparticles possess the capacity to amplify the sensitivity and specificity of biomarker detection signal enhancement and optimized capture efficiency of exosomes and RNA molecules derived from biological fluids. For instance, recent scholarly investigations have illustrated that gold nanoparticles can be effectively employed for the selective isolation and quantification of exosomal miRNAs, thereby facilitating earlier cancer detection with superior accuracy compared to conventional methodologies. Such advancements not only hold promise for augmenting diagnostic capabilities but also lay the groundwork for personalized therapeutic strategies, customized to align with the unique profiles of individual patients.

Exosomes and miRNAs show a high degree of specificity and sensitivity and can detect cancer with an accuracy of up to 75%. But working with RNA has its downsides, such as a lack of standardization and low abundance in early-stage cancer (Wang et al. 2022) (Table 2).

Table 2 Comparison of sensitivity, specificity, and limitations of key analytes used in liquid biopsy for cancer detection

Analyte	Sensitivity[a]	Specificity[a]	Key limitations
Circulating tumor DNA (ctDNA)	40–85% (varies by stage)	>90%	Low abundance in early cancer; requires high-depth sequencing
Circulating tumor cells (CTCs)	50–80%	>90%	Rare in circulation; challenging isolation method
Extracellular vesicles (EVs)	70–90%	85–95%	Heterogeneity; lack of standardization; sensitive to sample handling
Exosomes & circulating RNA	75–95%	80–98%	RNA degradation risk; complex isolation and analysis

[a]Values are approximate ranges based on recent literature and may vary depending on cancer type, stage, and detection methodology

4 Recent Advancements in ctDNA Detection Techniques

4.1 Next-Generation Sequencing (NGS)

Recent studies have demonstrated that NGS can provide comprehensive insights into ctDNA mutations, aiding in the identification of therapeutic targets and monitoring treatment response in cancer patients. Moreover, the application of NGS in ctDNA analysis has been shown to enhance the detection of tumor heterogeneity, thus informing personalized treatment strategies for improved patient management (Bai et al. 2019). However, despite these advancements, challenges such as the sensitivity of detection techniques and variations in ctDNA levels among patients remain significant hurdles in clinical implementation. Addressing these issues through standardized protocols and innovative technologies will be crucial for the successful integration of ctDNA analysis into routine cancer care.

4.2 Digital PCR Technologies

The development of digital PCR technologies aims to enhance the sensitivity and specificity of ctDNA detection, addressing the critical variations in cfDNA levels among patients and improving clinical reliability (Fig. 2). Recent studies highlight the utility of ddPCR for quantifying low-frequency mutations in ctDNA, enhancing the sensitivity and specificity of cancer detection methods. Moreover, the integration of ddPCR technology in clinical settings could significantly improve patient outcomes by enabling earlier detection of mutations and more personalized treatment strategies (Hosen et al. 2020). The ongoing development and refinement of ddPCR technologies are crucial for overcoming existing challenges in ctDNA analysis, such as sensitivity and specificity, ultimately facilitating better cancer management. The continued evolution of ddPCR methodologies will likely lead to more reliable and standardized approaches for ctDNA analysis in clinical practice, addressing current limitations in sensitivity and specificity (Table 3).

5 Discussion

Even though conventional tissue biopsy remains the gold standard for cancer diagnosis, a rapid shift toward liquid biopsy can be observed with evolving technology. With increasing sensitivity of methods such as ddPCR and NGS, early detection of cancer along with mass surveillance can become feasible. Cancer traces such as circulating tumor DNA, circulating tumor cells and extracellular vesicles provide evidence of cancer and enable us to not only detect cancer but also monitor it when the patient is under treatment. One of the biggest fears of cancer is recurrence, which can have serious consequences on the patient's mental well-being. With liquid biopsy, we can measure and monitor minimal residual disease (MRD) in a non-invasive manner.

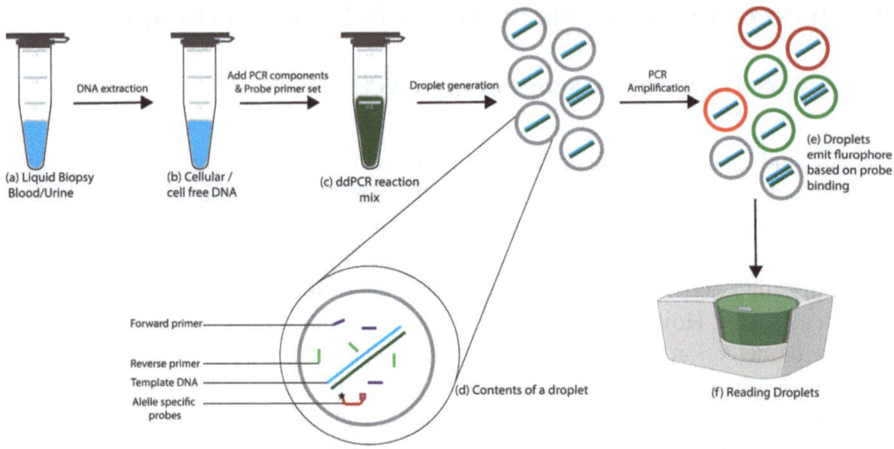

Fig. 2 Principle of droplet digital PCR (ddPCR) for liquid biopsy. (**a**) Liquid biopsy samples such as blood or urine are collected. (**b**) Cellular or cell-free DNA is extracted from the samples. (**c**) The DNA is mixed with PCR components, including forward and reverse primers and allele-specific fluorescent probes, forming the ddPCR reaction mix. (**d**) This mixture is partitioned into thousands of nanoliter-sized droplets, each containing zero, one, or two DNA molecules. (**e**) PCR amplification occurs within each droplet, and target-specific probe binding leads to fluorescence emission. (**f**) The droplets are analyzed using a droplet reader, where fluorescence signals are measured to identify and quantify mutant and wild-type alleles with single-molecule sensitivity

Table 3 A description of state-of-the-art molecular detection methods

Abbreviation	Full form	Pros	Cons
ddPCR	Droplet digital PCR	Extremely sensitive, absolute quantification	Limited multiplexing, Targets must be known, Not suitable for large panels
BEAMing	Beads, emulsion, amplification, magnetics	Ultra-sensitive, quantitative, good for rare allele detection	Technically complex, limited throughput, Requires prior mutation knowledge
CAPP-seq	Cancer personalized profiling by sequencing	High sensitivity and specificity, allows for monitoring disease burden	Requires design of capture panel, only targets known mutations
TAm-seq	Tagged amplicon deep sequencing	High sensitivity, suitable for low-frequency variant detection	Limited to targeted regions, requires known mutations

With great potential, the real-world applications come with a few obstacles. The technologies that are sensitive enough to detect cancer are still in their testing ground with varying degrees of success. However, some applications of liquid biopsy are already changing the lives of cancer patients, such as EGFR mutation detection in lung cancer (Joubert et al. 2022), resistance monitoring, and MRD surveillance in colorectal and breast cancer (Reinert et al. 2019). It is becoming commonplace to find kits that can detect cancer from a liquid biopsy. In the future, integration of multiple types of analytes using genetic as well as epigenetic markers can replace tissue biopsy altogether. Applications of artificial intelligence is also a possibility, which can be used to analyze the data for clinicians for easier interpretation and personalized care, and is a topic that would take another whole chapter to discuss.

6 Key Conclusions on Biotechnology Advancements in Liquid Biopsy for Cancer

Liquid biopsy is revolutionizing cancer detection and monitoring by offering a less invasive alternative to traditional tissue biopsies. This approach analyzes small traces of tumor-related materials in body fluids, enabling earlier detection, better treatment response tracking, and personalized therapies. Despite significant advancements, challenges remain in the sensitivity, specificity, and standardization of liquid biopsy techniques. Continued research and the establishment of standardized protocols are essential for the successful integration of liquid biopsy into routine clinical practice, ultimately improving patient outcomes and enabling personalized therapeutic strategies. In summary, liquid biopsy, supported by cutting-edge biotechnologies and the analysis of various biomarkers, is poised to transform cancer diagnostics and monitoring, moving toward more precise and personalized cancer care.

Acknowledgement of AI Assistance We would like to acknowledge the use of AI assistance in the preparation of this chapter. Specifically, an AI large language model was utilized to assist with structuring the content, synthesizing information, and refining the language for clarity and conciseness. This assistance was instrumental in enhancing the efficiency and quality of the writing process.

References

Alix-Panabières C (2020) The future of liquid biopsy. Nature 579(7800):S9. https://doi.org/10.1038/d41586-020-00844-5

Alix-Panabières C, Pantel K (2016) Clinical applications of circulating tumor cells and circulating tumor DNA as liquid biopsy. Cancer Discov 6(5):479–491. https://doi.org/10.1158/2159-8290.cd-15-1483

Amato O, Giannopoulou N, Ignatiadis M (2024) Circulating tumor DNA validity and potential uses in metastatic breast cancer. NPJ Breast Cancer 10(1). https://doi.org/10.1038/s41523-024-00626-6

Bai Y, Wang Z, Liu Z, Liang G, Gu W, Ge Q (2019) Technical progress in circulating tumor DNA analysis using next generation sequencing. Mol Cell Probes 49:101480. https://doi.org/10.1016/j.mcp.2019.101480

Bartolomucci A, Nobrega M, Ferrier T, Dickinson K, Kaorey N, Nadeau A, Castillo A, Burnier JV (2025) Circulating tumor DNA to monitor treatment response in solid tumors and advance precision oncology. NPJ Precis Oncol 9(1). https://doi.org/10.1038/s41698-025-00876-y

Becker A, Thakur BK, Weiss JM, Kim HS, Peinado H, Lyden D (2016) Extracellular vesicles in cancer: cell-to-cell mediators of metastasis. Cancer Cell 30(6):836–848. https://doi.org/10.1016/j.ccell.2016.10.009

Buscail E, Alix-Panabières C, Quincy P, Cauvin T, Chauvet A, Degrandi O, Caumont C, Verdon S, Lamrissi I, Moranvillier I, Buscail C, Marty M, Laurent C, Vendrely V, Moreau-Gaudry F, Bedel A, Dabernat S, Chiche L (2019) High clinical value of liquid biopsy to detect circulating tumor cells and tumor exosomes in pancreatic ductal adenocarcinoma patients eligible for Up-Front surgery. Cancer 11(11):1656. https://doi.org/10.3390/cancers11111656

De Abreu AR, Wyninckx A, Vandamme T, De Beeck KO, Van Camp G, Peeters M, Laurent-Puig P, Taieb J, Taly V, Benhaim L (2025) CTDNA Detection in Cancer: a comprehensive overview of current detection methods and prospects. Oncologist. https://doi.org/10.1093/oncolo/oyaf204

De Miranda FS, Barauna VG, Santos LD, Costa G, Vassallo PF, Campos LCG (2021) Properties and application of cell-free DNA as a clinical biomarker. Int J Mol Sci 22(17):9110. https://doi.org/10.3390/ijms22179110

Hicks JK, Saller J, Wang E, Boyle T, Gray JE (2017) Cell-free circulating tumor DNA supplementing tissue biopsies for identification of targetable mutations: implications for precision medicine and considerations for reconciling results. Lung Cancer 111:135–138. https://doi.org/10.1016/j.lungcan.2017.06.015

Hosen MI, Forey N, Durand G, Voegele C, Bilici S, Avogbe PH, Delhomme TM, Foll M, Manel A, Vian E, Meziani S, De Tilly B, Polo G, Lole O, Francois P, Boureille A, Pisarev E, Salas AROSE, Monteiro-Reis S, Henrique R, Byrnes G, Jeronimo C, Scelo G, McKay JD, Calvez-Kelm FL, Zvereva M (2020) Development of sensitive droplet digital PCR assays for detecting urinary TERT promoter mutations as non-invasive biomarkers for detection of urothelial cancer. Cancer 12(12):3541. https://doi.org/10.3390/cancers12123541

Joubert V, Chalabreysse L, Gamondes D, Tronc F, Maury J-M, Girard N (2022) Clinical practices for non-neoplastic thymic lesions: results from a multidisciplinary cohort. Lung Cancer 166:76–83. https://doi.org/10.1016/j.lungcan.2022.02.004

Kumar S, Poria R, Kala D, Nagraik R, Dhir Y, Dhir S, Singh B, Kaushik NK, Noorani MS, Kumar D, Gupta S, Kaushal A (2024) Recent advances in ctDNA detection using electrochemical biosensor for cancer. Discover Oncol 15(1). https://doi.org/10.1007/s12672-024-01365-7

Li G, Tang W, Yang F (2020) Cancer liquid biopsy using integrated microfluidic exosome analysis platforms. Biotechnol J 15(5). https://doi.org/10.1002/biot.201900225

Li Y, Nahas M, Stephens D, Froburg K, Hintz E, Champagne D, Lochab A, Brown M, Braun J, Fortuño MA, Ocón M-D-M, Pasquier A, Luque-Vázquez I, Moudgalya H, Kivlehan S, Gjeci I, Korle SL, Campo A, Rodriguez M, Seder CW, Lizotte PH, Bueno R, Borgia JA, Seijo LM, Montuenga LM, Yelensky R (2025) Circulating T-cell receptor repertoire for cancer early detection. NPJ Precis Oncol 9(1). https://doi.org/10.1038/s41698-025-01036-y

Lin B, Jiang J, Jia J, Zhou X (2022) Recent advances in exosomal miRNA biosensing for liquid biopsy. Molecules 27(21):7145. https://doi.org/10.3390/molecules27217145

Liu Q, Zheng J, Xie A, Chen M, Gong R-Y, Sheng Y, Chen H-L, Qi C-B (2024) Exosome, a rising biomarkers in liquid biopsy: advances of label-free and label strategy for diagnosis of cancer. Crit Rev Anal Chem:1–12. https://doi.org/10.1080/10408347.2024.2339961

Ma L, Guo H, Zhao Y, Liu Z, Wang C, Bu J, Sun T, Wei J (2024) Liquid biopsy in cancer current: status, challenges and future prospects. Signal Transduct Target Ther 9(1). https://doi.org/10.1038/s41392-024-02021-w

Martins I, Ribeiro IP, Jorge J, Gonçalves AC, Sarmento-Ribeiro AB, Melo JB, Carreira IM (2021) Liquid Biopsies: applications for cancer diagnosis and monitoring. Genes 12(3):349. https://doi.org/10.3390/genes12030349

Oellerich M, Schütz E, Beck J, Kanzow P, Plowman PN, Weiss GJ, Walson PD (2017) Using circulating cell-free DNA to monitor personalized cancer therapy. Crit Rev Clin Lab Sci 54(3):205–218. https://doi.org/10.1080/10408363.2017.1299683

Oellerich M, Schütz E, Beck J, Walson PD (2019) Circulating Cell-Free DNA—Diagnostic and Prognostic applications in personalized cancer therapy. Ther Drug Monit 41(2):115–120. https://doi.org/10.1097/ftd.0000000000000566

Preethi KA, Selvakumar SC, Ross K, Jayaraman S, Tusubira D, Sekar D (2022) Liquid biopsy: Exosomal microRNAs as novel diagnostic and prognostic biomarkers in cancer. Mol Cancer 21(1). https://doi.org/10.1186/s12943-022-01525-9

Rayamajhi S, Sipes J, Tetlow AL, Saha S, Bansal A, Godwin AK (2024) Extracellular vesicles as liquid biopsy biomarkers across the cancer journey: from early detection to recurrence. Clin Chem 70(1):206–219. https://doi.org/10.1093/clinchem/hvad176

Reinert T, Henriksen TV, Christensen E, Sharma S, Salari R, Sethi H, Knudsen M, Nordentoft I, Wu H-T, Tin AS, Rasmussen MH, Vang S, Shchegrova S, Johansen AFB, Srinivasan R, Assaf Z, Balcioglu M, Olson A, Dashner S, Hafez D, Navarro S, Goel S, Rabinowitz M, Billings P, Sigurjonsson S, Dyrskjøt L, Swenerton R, Aleshin A, Laurberg S, Madsen AH, Kannerup A-S, Stribolt K, Krag SP, Iversen LH, Sunesen KG, Lin C-HJ, Zimmermann BG, Andersen CL (2019) Analysis of plasma cell-free DNA by ultradeep sequencing in patients with stages I to III colorectal cancer. JAMA Oncol 5(8):1124. https://doi.org/10.1001/jamaoncol.2019.0528

Salehi M, Sharifi M (2018) Exosomal miRNAs as novel cancer biomarkers: challenges and opportunities. J Cell Physiol 233(9):6370–6380. https://doi.org/10.1002/jcp.26481

Salfer B, Li F, Wong DTW, Zhang L (2022) Urinary Cell-Free DNA in liquid biopsy and cancer management. Clin Chem 68(12):1493–1501. https://doi.org/10.1093/clinchem/hvac122

Santarpia M, Liguori A, D'Aveni A, Karachaliou N, Gonzalez-Cao M, Daffinà MG, Lazzari C, Altavilla G, Rosell R (2018) Liquid biopsy for lung cancer early detection. J Thorac Dis 10(S7):S882–S897. https://doi.org/10.21037/jtd.2018.03.81

Smit DJ, Pantel K (2024) Circulating tumor cells as liquid biopsy markers in cancer patients. Mol Asp Med 96:101258. https://doi.org/10.1016/j.mam.2024.101258

Sorber L, Zwaenepoel K, Deschoolmeester V, Van Schil PEY, Van Meerbeeck J, Lardon F, Rolfo C, Pauwels P (2016) Circulating cell-free nucleic acids and platelets as a liquid biopsy in the provision of personalized therapy for lung cancer patients. Lung Cancer 107:100–107. https://doi.org/10.1016/j.lungcan.2016.04.026

Taha HB (2023) Plasma versus serum for extracellular vesicle (EV) isolation: a duel for reproducibility and accuracy for CNS-originating EVs biomarker analysis. J Neurosci Res 101(11):1677–1686. https://doi.org/10.1002/jnr.25231

Tamura T, Yoshioka Y, Sakamoto S, Ichikawa T, Ochiya T (2021) Extracellular vesicles as a promising biomarker resource in liquid biopsy for cancer. Extracell Vesicles Circ Nucleic Acids. https://doi.org/10.20517/evcna.2021.06

Turner NC, Kingston B, Kilburn LS, Kernaghan S, Wardley AM, Macpherson IR, Baird RD, Roylance R, Stephens P, Oikonomidou O, Braybrooke JP, Tuthill M, Abraham J, Winter MC, Bye H, Hubank M, Gevensleben H, Cutts R, Snowdon C, Rea D, Cameron D, Shaaban A, Randle K, Martin S, Wilkinson K, Moretti L, Bliss JM, Ring A (2020) Circulating tumour DNA analysis to direct therapy in advanced breast cancer (plasmaMATCH): a multicentre, multicohort, phase 2a, platform trial. Lancet Oncol 21(10):1296–1308. https://doi.org/10.1016/s1470-2045(20)30444-7

Vacante M, Ciuni R, Basile F, Biondi A (2020) The liquid biopsy in the management of colorectal cancer: an overview. Biomedicine 8(9):308. https://doi.org/10.3390/biomedicines8090308

Vaidyanathan R, Soon RH, Zhang P, Jiang K, Lim CT (2019) Cancer diagnosis: from tumor to liquid biopsy and beyond. Lab Chip. https://doi.org/10.1039/c8lc00684a

Volckmar A, Sültmann H, Riediger A, Fioretos T, Schirmacher P, Endris V, Stenzinger A, Dietz S (2018) A field guide for cancer diagnostics using cell-free DNA: from principles to practice and clinical applications. Genes Chromosom Cancer 57(3):123–139. https://doi.org/10.1002/gcc.22517

Wang X, Tian L, Lu J, Ng IO-L (2022) Exosomes and cancer - diagnostic and prognostic biomarkers and therapeutic vehicle. Oncogene 11(1). https://doi.org/10.1038/s41389-022-00431-5

Woo D, Yu M (2018) Circulating tumor cells as "liquid biopsies" to understand cancer metastasis. Transl Res 201:128–135. https://doi.org/10.1016/j.trsl.2018.07.003

Yi K, Wang X, Filippov SK, Zhang H (2023) Emerging ctDNA detection strategies in clinical cancer theranostics. Smart Med 2(4). https://doi.org/10.1002/smmd.20230031

Yu D, Li Y, Wang M, Gu J, Xu W, Cai H, Fang X, Zhang X (2022) Exosomes as a new frontier of cancer liquid biopsy. Mol Cancer 21(1). https://doi.org/10.1186/s12943-022-01509-9

Advances of Genomic Medicine in Neurodevelopmental Disorders

Hosneara Akter, Mohammed Uddin, and A. H. M. Nurun Nabi

Abstract Neurodevelopmental disorders (NDDs) are a group of conditions with developmental onset, characterized by abnormalities in cognition, language, behavior, and motor function. These disorders significantly impair personal, social, academic, and occupational functioning. NDDs are mostly diagnosed by clinical evaluation and using psychological assessment tools. Due to the complex overlapping clinical presentation of NDDs, diagnosis of this disorder is difficult by clinical evaluation and psychological assessment. Early diagnosis of NDD can lead to better outcomes by facilitating educational planning and therapeutics. The integration of genomic medicine is a revolution in the discovery of new genetic causes of NDDs and has become a crucial tool for the diagnosis, understanding, and treatment of NDDs. The advent of advanced genomic technologies, particularly exome sequencing (ES) and genome sequencing (GS), has significantly improved the diagnostic yield for these conditions. Accurate diagnosis enables early intervention by utilizing available pharmacogenomic biomarkers and appropriate developmental therapies. Additionally, targeted gene therapies like engineered viral vectors, antisense oligonucleotides (ASOs), and clustered regulatory interspaced short palindromic repeats (CRISPR)-based genome editing show promise as potential curative approaches.

H. Akter
Genetics and Genomic Medicine Centre (GGMC), NeuroGen Healthcare, Dhaka, Bangladesh

Laboratory of Population Genetics, Department of Biochemistry and Molecular Biology, University of Dhaka, Dhaka, Bangladesh

M. Uddin
Center for Applied and Translational Genomics (CATG), Mohammed Bin Rashid University of Medicine and Health Sciences, Dubai Health, Dubai, UAE

GenomeArc Inc., Mississauga, ON, Canada

College of Medicine, Mohammed Bin Rashid University of Medicine and Health Sciences, Dubai, UAE

A. H. M. Nurun Nabi (✉)
Genetics and Genomic Medicine Centre (GGMC), NeuroGen Healthcare, Dhaka, Bangladesh
e-mail: nabi@du.ac.bd

A. H. M. N. Nabi, H. Uddin Shekhar (eds.), *Biotechnological Advances in Healthomics*, Interdisciplinary Biotechnological Advances,
https://doi.org/10.1007/978-981-95-3612-2_6

While each of these technologies presents distinct limitations, their thoughtful design and strategic implementation may help to overcome these challenges and improve their clinical applicability.

Keywords Neurodevelopmental disorders · Exome Sequencing · Genome Sequencing · Antisense Oligonucleotide · Clustered Regulatory Interspaced Short Palindromic Repeats

Abbreviations

AADC	Aromatic L-amino acid decarboxylase
AAV9	Adeno-associated virus 9
ADHD	Attention-deficit/hyperactive disorder
AS	Angelman syndrome
ASD	Autism spectrum disorder
ASOs	Antisense oligonucleotides
CA	Congenital anomalies
CNVs	Copy number variants
CP	Cerebral palsy
crRNA	CRISPR RNA
CSF	Cerebrospinal fluid
DEEs	Developmental and epileptic encephalopathies
DSB	Double-strand break
ES	Exome sequencing
eTF	Engineered transcription factor
EXACT	Expression Attenuation via Construct Tuning
FDA	U.S. Food and Drug Administration
FXS	Fragile X syndrome
gRNA	Guide RNA
GS	Genome sequencing
ID	Intellectual disability
INDEL	Insertions/deletions
MA	Microarray
MPH	Methylphenidate
NDDs	Neurodevelopmental disorders
PAM	Protospacer adjacent motif
PWS	Prader Willi syndrome
RTT	Rett syndrome
sgRNA	Single guide RNA
SNVs	Single-nucleotide variants
SSRIs	Selective serotonin reuptake inhibitors
tracrRNA	trans-activating crRNA
TSC	Tuberous sclerosis

1 Introduction

Neurodevelopmental disorders (NDDs) are a group of conditions that begin in the developmental period, typically from conception to early adulthood, and are characterized by impairments in cognition, language, behavior, and motor skills, significantly affecting personal, social, academic, and occupational functioning (DSM-5. 5th edn. 2013; Thapar et al. 2016). NDDs include intellectual disability (ID), autism spectrum disorder (ASD), epilepsy, attention deficit/hyperactivity disorder (ADHD), and more specific difficulties of learning, such as dyslexia and language disorders (DSM-5. 5th edn. 2013). The term syndromic is often used to describe people with NDDs who also present with craniofacial dysmorphism or other congenital anomalies (CA) (Jamra et al. 2011) which includes Down's syndrome, Rett syndrome (RTT), Fragile X syndrome (FXS), Prader-Willi syndrome (PWS), Angelman syndrome (AS), Tuberous sclerosis (TSC), etc. The population prevalence of ID is 3% (Shevell et al. 2003), while ASD affects up to 1 in 36 children (Maenner et al. 2023).

The completion of the Human Genome Project in 2003 marked a significant milestone in the field of genomics, and since then, genomic medicine has made tremendous progress in improving our understanding and management of NDDs. This field utilizes advanced sequencing technologies, including next-generation sequencing and long-read sequencing, known as third-generation sequencing, to identify the genetic causes of NDDs. The ultimate goal is to achieve personalized treatments that improve patient outcomes. Large-scale genomic studies using exome sequencing (ES) and genome sequencing (GS) have identified a range of genetic variants associated with the risk of NDDs, including single-nucleotide variants (SNVs), small insertions and deletions (indels), and structural variants, which encompass both rare and common variants (Liu et al. 2015; Woodbury-Smith et al. 2017; Rahman et al. 2019; Uddin et al. 2019; Satterstrom et al. 2020; Begum et al. 2021; Xiang et al. 2021; Mim et al. 2024). Moreover, epigenetic modifications, including DNA methylation and histone modifications, regulate critical brain development processes such as neuronal differentiation and synaptogenesis. Deregulation of epigenetic mechanisms is associated with NDDs (Jakovcevski and Akbarian 2012; Zoghbi and Bear 2012).

The identification of bona fide genetic risk factors, along with the application of functional genomics to link genetic variants to phenotypes, provides a strong foundation for developing personalized therapeutic approaches. In this chapter, we shall start by discussing complex phenotypes of NDDs and highlight the significance of genomic medicine in understanding their genetic etiology and personalized treatment.

2 Overlapping Phenotypes in NDDs

Many neurodevelopmental disorders often co-occur with other conditions (Akter et al. 2023; Mim et al. 2024; Akter et al. 2024; Akter et al. 2025), making diagnosis and treatment more complex. This comorbidity can lead to a range of challenges, including overlapping symptoms, conflicting treatment approaches, and increased burden on individuals and families. These comorbid conditions can be neurological, psychiatric, medical, or behavioral (see Figure 1A). It is common for individuals with autism and ADHD to also have epilepsy, with the occurrence of epilepsy increasing with age and being strongly linked to the co-occurrence of intellectual disability (ID) (Danielsson et al. 2005; Tuchman 2017). Additionally, about 30% of children and adolescents with childhood-onset schizophrenia also have autism spectrum disorder (Canitano and Pallagrosi 2017; Rapoport et al. 2009). It was observed that 45% of children with cerebral palsy (CP) met the criteria for ADHD, ASD, or both. Of those children, 51% also had ID. Two-thirds of the children had ADHD, ASD, and/or ID (Påhlman et al. 2020; Hendson et al. 2022).

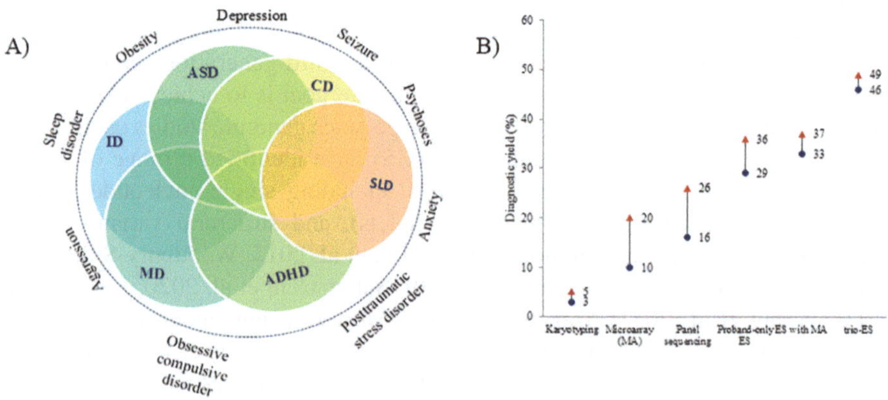

Fig. 1 Comorbidities in NDDs and genetic testing yield. (**a**) NDDs exhibit considerable phenotypic heterogeneity and are frequently characterized by a spectrum of comorbidities spanning multiple clinical domains. These include co-occurring neurological impairments (e.g., epilepsy, motor dysfunction, sleep disorder), psychiatric conditions (e.g., anxiety, depression, aggression, OCD, etc.), and systemic medical complications. (**b**) The diagnostic yield varies across different genetic testing modalities depending on cohort characteristics and analytical approaches. Conventional karyotyping identifies pathogenic variants in approximately 3–5% of cases, while chromosomal microarray (MA) analysis offers a yield of 10–20%. Targeted gene panel sequencing achieves a diagnostic rate ranging from 16% to 26%, whereas exome sequencing (ES) demonstrates a higher yield, typically between 29% and 36%. A slight increase in diagnostic outcome is found from the combined MA and ES test. Moreover, diagnostic rates differ between trio-ES and proband-only ES approaches, with trio-ES generally offering a better yield, 46–49%. Here ASD: Autism spectrum disorder; CD: Communication disorder; SLD: Specific language disorder; ADHD: Attention-deficit/hyperactive disorder; MD: Motor and tic disorder; ID: Intellectual disability; OCD: Obsessive-compulsive disorder

3 Molecular Etiology of NDDs

The etiology of NDDs is mainly attributed to both environmental and genetic factors, with genetic causes being the most prevalent (Rahman et al. 2019; Uddin et al. 2019; Begum et al. 2021; Shabestari et al. 2022; Akter et al. 2023; Mim et al. 2024; Akter et al. 2024). The advancement of genetic technologies such as microarray (MA), next-generation, and third-generation sequencing has resulted in a much improved insight into their genetic etiology (Zhang et al. 2013; Begum et al. 2021; Akter et al. 2023; Akter et al. 2024). Several genes and genomic regions (Cooper et al. 2011) were found to be implicated in the etiology of NDDs, including specific genetic syndromes associated with ID. Structural genomic variation is a known contributor to a proportion of children diagnosed with NDDs (Liu et al. 2015; Woodbury-Smith et al. 2017; Rahman et al. 2019; Begum et al. 2021). Besides structural variation, single-nucleotide variants (SNVs), small (up to 1 kb) insertions or deletions (Indel), and expanded repeats are also major contributors to the etiology of monogenic forms of NDDs (Akter et al. 2024; Akter et al. 2025). Additionally, mobile DNA elements are also rarely found to be associated with NDDs (Erwin et al. 2014).

Over the past decade, large-scale genomic studies have investigated population-specific genomic variations in NDD cohorts (see Table 1). Some of these studies have shown that *de novo* variants in protein-coding genes are enriched in individuals with NDDs and represent a significant cause of these disorders in outbred populations (de Ligt et al. 2012; O'Roak et al. 2012; Iossifov et al. 2014). It is estimated that 42% to 48% of individuals with a NDD may have a causative *de novo* variant in both known and unknown disease genes (McRae et al. 2017). On average, the effect size of inherited variants is significantly lower than that of *de novo* variants. The overall mutation rate of inherited likely gene-disruptive variants or copy number variants (CNVs) in probands does not differ significantly from that of unaffected siblings. However, private or ultra-rare inherited likely gene-disruptive variants in intolerant genes are significantly more common in individuals with ASD probands (Krumm et al. 2015). The risk of recessive disorders with neurodevelopmental phenotypes is significantly higher in the children of consanguineous parents compared to nonconsanguineous parents (Akter et al. 2024). Wilfert et al. found that children with ASD from multiplex families are more likely to carry multiple private inherited variants compared to those from simplex families, which supports a multi-hit model for ASD (Wilfert et al. 2021). Furthermore, a significant portion of NDD cases shows an association between postzygotic (or somatic) mutations and NDDs (Lim et al. 2017).

Variable neurodevelopmental phenotypes have been reported to be caused by different germline and somatic variants in the *LIS1* and *DCX* genes. In females, missense variants in the *DCX* are associated with ID and cryptogenic epilepsy (Guerrini et al. 2003) whereas, in males, somatic variants in the *LIS1* and *DCX* are associated with double cortex (Gleeson et al. 2000). Variants in the *ARX* gene exhibit marked genotype-phenotype variability. Truncating and null variants in *ARX* are associated

Table 1 Global genomic studies on NDD cohorts representing population-specific data

Category of participant	Country of origin	Cohort size	Genomic technology	PMID with reference
ASD	USA	2508	ES	25,363,768, Iossifov et al. (2014)
ASD	USA	208	GS	26,749,308, Turner et al. (2016)
DD	UK and Republic of Ireland	13,610	ES	37,043,637, Wright et al. (2023)
NDD	UK	489	GS	37,541,188, Sanchis-Juan et al. (2023)
NDD	China	1106	ES	40,131,272, Lan et al. (2025)
ASD	China	112	GS	39,126,614, Chang et al. (2024)
NDD	Bangladesh	576	ES and GS	39,342,494, Akter et al. (2024)
NDD	Korea	553	ES	31,996,704, Lee et al. (2020)
GDD and ID	Ukraine	417	ES or PS	38,539,105, Shchubelka et al. (2024)
NDD	United Arab Emirates	405	ES	38,438,125, Khalaf et al. (2024)
ASD	Japan	309	ES	35,811,316, Kimura et al. (2022)
ASD	Italy	122	ES	39,654,053, Granata et al. (2024)
ID	Pakistan	121 families	ES	27,457,812, Riazuddin et al. (2016)
ID	India	55	ES	38,114,583, Pande et al. (2023)

NDDs neurodevelopmental disorders, *ASD* autism spectrum disorder, *GDD* global developmental delay, *ID* intellectual disability, *ES* exome sequencing, *GS* genome sequencing, *PS* panel sequencing

with the most severe malformations, such as lissencephaly with ambiguous genitalia (OMIM 300215), whereas missense variants and polyalanine expansions are typically associated with ID and seizures in the absence of structural brain abnormalities (Kato et al. 2004; Friocourt et al. 2006).

4 Molecular Diagnosis of NDD

In the context of NDDs, a fundamental aspect of genomic medicine is the precise genetic diagnosis, which reduces the diagnostic odyssey and facilitates timely clinical decision-making. This approach supports the implementation of individualized therapeutic strategies, including multidisciplinary medical management and the application of available targeted pharmacogenomic interventions, thereby optimizing participants' outcomes and long-term care.

No single genetic test can identify all types of genomic aberrations in NDDs simultaneously (Srivastava et al. 2019; Akter et al. 2024). The genetic causes of NDDs, including ID, DD, ASD, or congenital malformations, were traditionally identified by G-banded karyotype analysis. In approximately 3–5% of children with such disorders, Karyotyping can detect disease-associated genetic abnormalities (Miller et al. 2010) (see Fig. 1B). These days, karyotyping has largely been replaced with chromosomal microarray, which can identify both smaller and larger CNVs. This increases the diagnostic yield to about 10–20% of patients (Miller et al. 2010; Akter et al. 2023) (see Fig. 1B).

While the advancement of sequencing technologies such as targeted panel sequencing, exome sequencing (ES), and genome sequencing (GS) has substantially improved the identification of genetic causes underlying NDDs (see Fig. 2), leading to increased diagnostic yields. Panel-based sequencing that includes NDD-related genes can identify pathogenic variants in about 16% of individuals with autism and up to 26% in those with NDDs accompanied by epilepsy. ES focuses on single-nucleotide variants (SNVs) and small insertions or deletions in the coding region of the genome, increasing the diagnostic yield to 29–36% (Srivastava et al. 2019; Akter et al. 2024). Combining chromosomal microarray analysis with ES further enhances diagnostic yield (Kim et al. 2023; Akter et al. 2024). Recent advancements in bioinformatics have enabled the detection of CNVs from ES data, improving clinical applications in genetic disorders (Lemire et al. 2024). Simultaneous analysis of both SNVs and CNVs using ES data in NDD cohorts has been shown to significantly improve diagnostic outcomes (Zhai et al. 2021; Akter et al. 2025). Moreover, diagnostic rates differ between trio-ES and proband-only ES approaches (see Fig. 1B), with trio-ES generally offering a better yield (Lan et al. 2025).

Although the above studies reported variable diagnostic yields, expert guidelines recommend microarray as the first-tier test for cohorts with developmental disabilities or congenital anomalies (Miller et al. 2010), while exome sequencing is advised for broader NDD cohorts (Srivastava et al. 2019). Targeted gene sequencing can be a cost-effective option when the clinical presentation suggests a single-gene disorder such as Rett syndrome. Specifically, targeted *MECP2* sequencing is appropriate for individuals with a typical Rett syndrome phenotype (Akter et al. 2025), while exome sequencing is considered the most cost-effective approach for those with Rett-like or atypical Rett presentations (Akter et al. 2025). Targeted panels are typically designed to include genes known to be associated with NDDs. However, due to the large number of genes implicated, their diagnostic yield is often limited, except in certain monogenic NDD cases. Additionally, as new NDD-related genes are continually being discovered, panel sequencing may fail to capture recently identified genes, potentially leading to missed diagnoses.

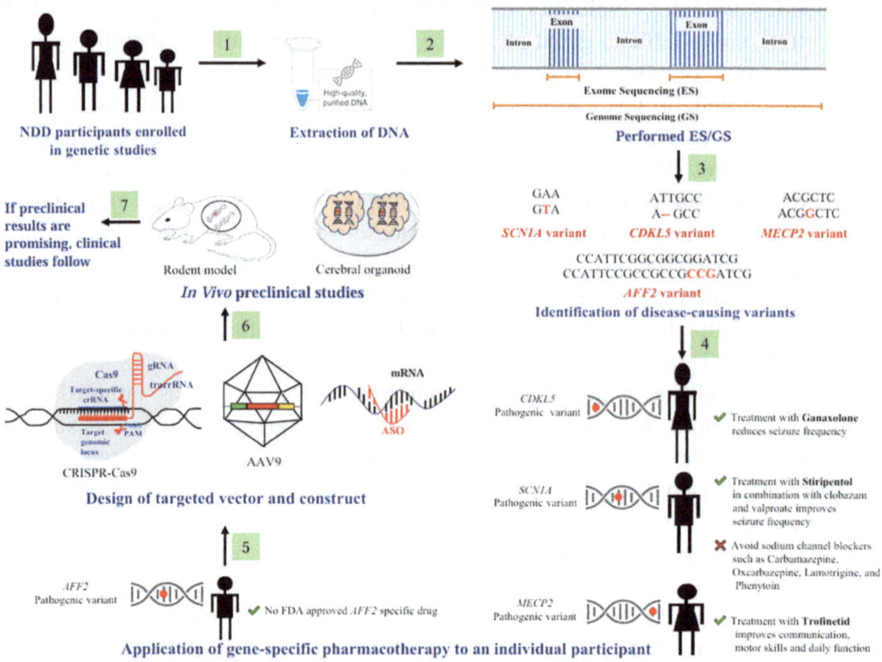

Fig. 2 Translational pipeline of genomic medicine in NDDs. The process begins with the enrollment of NDD participants in genetic studies, followed by the extraction of high-quality genomic DNA (Step 1). Exome sequencing (ES) or genome sequencing (GS) coupled with bioinformatic analysis are performed to identify disease-causing variants (Steps 2 and 3), for example, the identification of pathogenic variants in the *SCN1A*, *CDKL5*, *MECP2*, and *AFF2*. The molecular diagnoses guide therapeutic strategies (Step 4), enabling pharmacogenomic interventions such as Ganaxolone for *CDKL5*-related epilepsy, Stiripentol (with avoidance of sodium channel blockers) for *SCN1A* variants, and Trofinetide for *MECP2*-associated disorders. In cases where no FDA-approved therapy exists (e.g., *AFF2*), gene-specific pharmacological intervention remains a challenge. Subsequently, targeted molecular therapies including ASOs, AAV9 vectors, and CRISPR-Cas9-based genome editing can be developed (Step 5). These constructs can be tested in vivo using rodent models and cerebral organoids (Step 6), and if successful, proceed to clinical trials (Step 7)

However, due to inherent limitations of ES and short-read GS technologies, clinically relevant variants in repetitive genomic regions, such as the CGG repeat expansion in *FMR1* associated with Fragile X syndrome, may go undetected. To address these limitations, long-read sequencing platforms offer a promising alternative, enabling improved read alignment and variant detection within and near repetitive sequences, as well as facilitating *de novo* genome assembly (Mantere et al. 2019; Begum et al. 2021; Akter et al. 2025).

5 Advancement of Precision Medicine for NDDs

Treating individuals with NDDs presents several challenges, including their diverse underlying causes and the necessity for therapeutic agents to cross the blood-brain barrier. Advances in multidisciplinary medical care, including speech and language therapy, occupational therapy, behavioral therapy, and special education (Wolraich et al. 2019; Wood et al. 2019) alongside symptom-targeted drug treatments and pharmacogenomic drugs (see Table 2) have notably improved both lifespan and daily functioning for those affected.

Moreover, emerging gene-based therapies, including antisense oligonucleotides (ASOs), CRISPR/Cas9 genome editing, and gene replacement strategies using Adeno-associated virus 9 (AAV9), offer transformative potential for monogenic NDDs. Clinical trials are underway for several rare syndromic disorders (see Table 3), aiming to correct or modulate the expression of faulty genes. These novel approaches, though still in early stages for most NDDs, represent a major leap toward disease-modifying treatments rather than purely symptomatic management.

Table 2 Pharmacogenomic-driven therapies in NDDs

Target gene	Disorder/ phenotype	Pharmacogenomic drug	Mechanism/effect	References
SCN1A	Dravet syndrome	Stiripentol	Enhances GABAergic activity; used with clobazam and valproate	Bacq et al. (2024)
		Avoid: Carbamazepine, Lamotrigine, Phenytoin	Exacerbate seizures via a sodium channel blocker	
CDKL5	CDKL5 deficiency disorder	Ganaxolone	Neurosteroid GABA-A receptor modulator; reduces seizure burden	Knight et al. (2022)
MECP2	Rett syndrome	Trofinetide	Synthetic IGF-1 analog; improves synaptic and motor function	Neul et al. (2023)
TSC1/ *TSC2*	Tuberous sclerosis complex	Everolimus	mTOR inhibitor; reduces seizure frequency in patients with tuberous sclerosis complex and treatment-resistant seizures	French et al. (2016)
CYP2D6	ADHD	Atomoxetine	Genotype-specific dose adjustments	Brown et al. (2019)
DRD4, *COMT*, *ADRA4*, *SLC6A3*, and *SLC6A2*	ADHD	Methylphenidate (MPH)	Positive response to MPH treatment, including improvement of ADHD symptoms	Myer et al. (2018)
CYP2D6	Irritability/ Aggression in ASD	Risperidone, Aripiprazole	Genotype-specific dose adjustments	Dean (2012)

Table 3 List of gene therapy drugs related to NDDs under clinical trial

Gene name	Associated diseases	Drug name	Mechanism of action of the drug	Root of administration	Drug type	Status of clinical trials	Name of company	Clinical trial number
MECP2	Rett syndrome	NGN-401	A functional copy of the *MECP2* gene is delivered to the CNS using an AAV9 vector to restore normal MeCP2 function in neurons	Intracerebroventricularly	AAV9	Stage 1/2	Neurogene Inc.	NCT05898620
SCN1A	Dravet syndrome	ETX101	An engineered transcription factor (eTF) is delivered to the CNS via an AAV9 vector to upregulate *SCN1A* gene expression	Intracerebroventricularly	AAV9	Stage 1/2	Encoded therapeutics	NCT06112275
SCN1A	Dravet syndrome	STK-001	A mutation-specific ASO is administered into the cerebrospinal fluid, where it binds to the mutated exon and promotes its exclusion from the final mRNA transcript. This exon skipping results in the production of more functional *SCN1A* transcripts and increased expression of the NaV1.1 protein	Intrathecally	ASO	Stage 2	Stoke therapeutics	NCT04740476
MECP2	MECP2 duplication syndrome	ION440	The ASO is specifically designed to target *MECP2* mRNA, modulating gene expression to reduce the overproduction of the MeCP2 protein	Intrathecally	ASO	Stage 1/2	Ionis pharmaceuticals	NCT06430385
UBE3A	Angelman syndrome	ION582	In neurons, the paternal UBE3A allele is silenced by the UBE3A-ATS long non-coding RNA. The engineered ASO inhibits UBE3A-ATS to reactivate the paternal allele and restore UBE3A protein production	Intrathecally	ASO	Stage 3	Ionis pharmaceuticals	NCT05127226
SCN8A	DEEs	N/A	The ASO selectively binds SCN8A mRNA in neurons, triggering RNase H1-mediated degradation and reducing Nav1.6 protein production	Intrathecally	ASO	Pre-clinical	PMID: 31943325	N/A

5.1 Pharmacogenomic-Driven Therapies

Pharmacogenomics holds transformative potential for the management of NDDs. These disorders, characterized by heterogeneous etiologies and complex neurobiological underpinnings, often present with variable responses to pharmacological interventions. Integrating pharmacogenomic data into clinical decision-making enables the tailoring of therapeutic regimens based on individual genetic profiles, thus optimizing efficacy while minimizing adverse effects.

A key example is the *SCN1A* carrying pathogenic variants are associated with Dravet syndrome (OMIM #607208), a severe epileptic encephalopathy. Participants with pathogenic *SCN1A* variants often exhibit paradoxical seizure exacerbation in response to sodium channel blockers such as carbamazepine and phenytoin. Pharmacogenomic insight recommends avoidance of these agents and supports the use of GABAergic modulators like stiripentol in combination with clobazam and valproate, which have been shown to reduce seizure frequency and severity (see Fig. 2 and Table 2) (Bacq et al. 2024). Similarly, *CDKL5* variants, causative of CDKL5 deficiency disorder (OMIM #300672), are now being targeted with ganaxolone (see Fig. 2 and Table 2), a synthetic neurosteroid that positively modulates GABA-A receptors (Knight et al. 2022). In Rett syndrome, caused by pathogenic variants in *MECP2*, trofinetide, a synthetic analog of IGF-1, has demonstrated efficacy in improving motor function, behavior, and communication (Neul et al. 2023) (see Fig. 2 and Table 2). Moreover, Adjunctive everolimus treatment (see Table 2) significantly reduces seizure frequency in participants with tuberous sclerosis complex and treatment-resistant seizures carrying pathogenic variants in the *TSC1*/*TSC2* gene (French et al. 2016).

Beyond monogenic NDDs with epilepsy, pharmacogenomic strategies are increasingly applied in ADHD and ASD. In ASD, risperidone and aripiprazole (Dean 2012), both metabolized by *CYP2D6,* exhibit genotype-dependent pharmacokinetics, with poor metabolizers at higher risk of side effects and ultrarapid metabolizers potentially experiencing reduced efficacy. Similarly, atomoxetine, used in ADHD (Brown et al. 2019), also shows *CYP2D6*-dependent metabolism. FDA guidelines recommend dose adjustments based on *CYP2D6* genotype (see Table 2). Genetic polymorphisms in genes such as *DRD4* (48-bp VNTR), *COMT* (rs4680), *ADRA2A* (rs1800544), *SLC6A3* (40-bp VNTR), and *SLC6A2* (rs28386840 and rs5569) influence individual responses to methylphenidate (see Table 2) and selective serotonin reuptake inhibitors (SSRIs) (Myer et al. 2018).

Despite these improvements, the advancement of gene therapy technologies has been relatively slow (Rubin and Spradling 1982; Hanna et al. 2017). However, recent successes with AAV-based gene replacement therapy (Pattali et al. 2019), ASO therapy (Corey 2017) and CRISPR-Cas9 therapy (Frangoul et al. 2020) have shown promise as targeted treatment strategies. These approaches are discussed below, and the available clinical/preclinical trial information has been summarized in Table 3.

5.2 *Adeno-Associated Virus (AAV)-Based Gene Replacement Therapy*

There has been much progress since the conception of the ground-breaking idea of using viruses as vectors for gene therapy in the twentieth century (Friedmann and Roblin 1972). The small (20–25 nm), nonenveloped single-stranded DNA virus called adeno-associated virus (AAV), initially discovered in the 1960s, has evolved into a widely studied and powerful tool for gene delivery (Wu et al. 2006). There are various types of AAV serotypes used for gene therapy in different tissues. One of these serotypes, AAV9, is the current gold standard for central nervous system (CNS) gene therapy, which can cross the blood-brain barrier with systemic administration and has a tropism for neurons and astrocytes (Foust et al. 2008). To date, several AAV9-based gene therapy products for inherited diseases have received US Food and Drug Administration (FDA) and European Medicines Agency approval for treating inherited disorders such as Aromatic L-amino acid decarboxylase (AADC) deficiency (Kebilidi), Metachromatic leukodystrophy (Lenmeldy), Spinal Muscular Atrophy (Zolgensma), Retinal Dystrophy (Luxturna), Hemophilia A (Roctavian), Hemophilia B (Hemgenix and Beqvez), and Duchenne Muscular Dystrophy (Elevidys).

Although no AAV-based gene therapy has yet received regulatory approval for NDDs, two AAV9-based candidates, ETX101 and NGN-401, are currently undergoing clinical trials (see Table 3). ETX101 is administered directly to the CNS via intracerebroventricular (ICV) infusion to Dravet syndrome (OMIM# 607,208) participants. It employs an AAV9 vector to deliver DNA encoding an engineered transcription factor (eTF) to neurons, where it persists long-term as an episome without integrating into or altering the patient's native genome. The expressed eTF interacts with the endogenous DNA to enhance transcription of the *SCN1A* gene, aiming to increase functional sodium channel (NaV1.1) levels in target neurons. This mechanism is designed to restore proper neuronal signaling and potentially modify the disease course in Dravet syndrome. Preclinical studies have demonstrated that ETX101 improves survival and reduces seizure burden in a Dravet syndrome mouse model and is well-tolerated in nonhuman primates (Tanenhaus et al. 2022). NGN-401 is another investigational AAV9-based gene therapy that utilizes the Expression Attenuation via Construct Tuning (EXACT) platform for precise transgene regulation (https://www.neurogene.com/). It employs an AAV9 capsid to deliver functional copies of the human *MECP2* gene mutated in Rett syndrome (OMIM# 312750), which encodes a critical protein involved in brain development and regulation.

5.3 Antisense Oligonucleotide (ASO) Therapy

ASOs are short, synthetic single-stranded DNA molecules (10–30 bp) that bind to cellular RNAs via complementary base pairing and influence pre-mRNA splicing, mRNA stability, transcription, or RNA-protein interaction. ASO therapy can be personalized to target patient-specific variants (Carroll et al. 2011; Kim et al. 2019). Although several ASOs have received approval from the US Food and Drug Administration (FDA) for conditions such as Duchenne muscular dystrophy (Eteplirsen, Golodirsen, Viltolarsen, Casimersen), spinal muscular atrophy (Nusinersen), familial hypercholesterolemia (Mipomersen), and hereditary transthyretin amyloidosis (Tegsedi), no ASO has yet been approved for the treatment of NDDs. However, a limited number of ASOs are currently in preclinical and clinical trials (see Table 3) for some NDDs.

Currently, three ASOs are actively in human trials in the field of NDDs, named ION582, STK-001, and ION440. ASO for the treatment of Developmental and epileptic encephalopathies (DEEs) caused by *SCN8A* is under preclinical development (see Table 3). ION582 and ION440 ASOs are being developed by Ionis Pharmaceuticals (https://ionis.com/) for the treatment of Angelman syndrome (AS, OMIM #105830) and MECP2 duplication syndrome (MDS, OMIM #300260), respectively. AS is a rare NDD caused by the loss of function of the maternally inherited *UBE3A* gene. In neurons, the paternal *UBE3A* allele is typically silenced by a long non-coding RNA known as UBE3A antisense transcript (UBE3A-ATS). ION582 is designed to inhibit the expression of UBE3A-ATS, thereby activating the paternal *UBE3A* allele and restoring the production of UBE3A protein in neurons. MDS is another rare NDD caused by the overexpression of the *MECP2* allele. ION440 is another investigational antisense oligonucleotide (ASO) therapy developed by Ionis Pharmaceuticals for the treatment of MDS. It employs Ionis' proprietary antisense technology to specifically target and modulate the expression of the *MECP2* gene. By binding to the MECP2 mRNA, ION440 aims to reduce the overproduction of the MeCP2 protein (https://ionis.com/). STK-001 is being developed by Stoke Therapeutics (https://www.stoketherapeutics.com/). STK-001 is administered via intrathecal (IT) injection, delivering the ASO directly into the cerebrospinal fluid (CSF). It binds to the mutated exon, promoting its skipping from the final mRNA transcript. As a result, more productive SCN1A transcripts are generated, leading to increased NaV1.1 protein expression. Another investigational ASO is STK-001, developed by Stoke Therapeutics for the treatment of Dravet syndrome, currently being evaluated in ongoing clinical trials. It is designed to upregulate NaV1.1 protein expression from the wild-type allele of the *SCN1A* gene to restore physiological NaV1.1 levels. There is another ASO that is now under preclinical trial is being developed for the treatment of Developmental and epileptic encephalopathies (DEEs) caused by the pathogenic variant in the *SCN8A* gene (Lenk et al. 2020). In mouse models carrying the *SCN8A* p.Arg1872Trp mutation, which disrupts sodium channel inactivation, treatment with *SCN8A*-targeted ASOs led to a dose-dependent decrease in *SCN8A* mRNA levels. This was associated with the delayed onset of seizures and extended survival, highlighting the therapeutic promise of this strategy (Lenk et al. 2020). All of these ASOs are delivered through intrathecal (IT) injection directly into the cerebrospinal fluid (CSF).

5.4 CRISPR-Cas9 Therapy

The CRISPR-Cas system is a programmable, RNA-guided endonuclease composed of a Cas enzyme and a guide RNA. The Cas enzyme induces a double-strand break (DSB) at a specific DNA site, while the guide RNA (gRNA) directs the Cas enzyme to the target sequence by simple Watson-Crick base pairing (Kim and Kim 2014). Here, the single guide RNA (sgRNA) is a fusion of CRISPR RNA (crRNA) and trans-activating crRNA (tracrRNA). The crRNA identifies the target DNA sequence, known as the protospacer, which can be easily designed, making the system highly customizable. Adjacent to the protospacer is a short DNA sequence called the protospacer adjacent motif (PAM), which is recognized by the Cas enzyme and is essential for its activity. For Cas9, the DSB occurs three nucleotides upstream of the PAM sequence.

Casgevy, the first FDA-approved gene therapy utilizing CRISPR/Cas9 technology, represents a significant milestone in the field of genetic medicine, specifically for the treatment of sickle cell disease and transfusion-dependent β-thalassemia. To date, no CRISPR/Cas9-based therapeutic agents are in clinical trials for the treatment of NDDs.

6 Challenges of Genomic Medicine in NDDs

Despite significant progress in genomic medicine for NDDs, several challenges persist. First, genetic heterogeneity and incomplete penetrance complicate the interpretation of genomic variants, often leading to variants of uncertain significance (VUS) that limit clinical decision-making. Second, the limited availability of large, diverse, and ethnically representative genomic datasets hinders the generalizability of findings across populations, particularly in underrepresented regions. Third, technical challenges such as the detection and interpretation of structural variants, mosaicism, and non-coding regulatory regions remain areas of ongoing development. Moreover, while ES and GS are increasingly accessible, integration into routine clinical practice is uneven due to cost, infrastructure constraints, and a shortage of trained personnel in low-resource settings. Finally, therapeutic translation of genomic discoveries, including pharmacogenomics and gene therapy, is still in early stages for most NDDs, limiting current clinical utility beyond diagnosis and risk stratification.

7 Conclusion

Genomic medicine has revolutionized the landscape of NDD research and clinical care, offering unprecedented opportunities for early diagnosis, individualized risk assessment, and the potential for targeted interventions. Genomic analysis can help identify specific genetic variants associated with NDDs, allowing for more accurate diagnoses and prognostications. This information can also inform treatment decisions, such as choosing the most effective medications or therapies. However, to fully realize its potential, significant challenges must be addressed. These include enhancing data interpretation, expanding equitable access to genomic testing, developing multidisciplinary collaboration, and translating genomic insights into actionable therapeutic strategies. Continued investment in large-scale genomic research, capacity building in clinical genomics, and ethical frameworks will be essential to bridge existing gaps. Ultimately, the integration of genomic medicine, including precision diagnostics, pharmacogenomics, and gene therapy, into a comprehensive, multidisciplinary model of care holds the potential to significantly improve clinical outcomes, optimize therapeutic efficacy, and enhance quality of life for individuals with NDDs and their families.

References

Akter H, Rahman MM, Sarker S et al (2023) Construction of copy number variation landscape and characterization of associated genes in a Bangladeshi cohort of neurodevelopmental disorders. Front Genet 14. https://doi.org/10.3389/fgene.2023.955631

Akter H, Rahaman MA, Eshaque TB et al (2024) Genomic insights from a deeply phenotyped highly consanguineous neurodevelopmental disorders cohort. Genet Med 27:101282. https://doi.org/10.1016/j.gim.2024.101282

Akter H, Rahman MM, Mim RA, Rahaman A, Eshaque TB, Hossain SA, Ganguly A, Omar FB, Taniya MA, Hasan N, Islam A, Jamalalail B, Nassir N, Zehra B, Sarker S, Uddin KMF, Nabi AHMN, Woodbury-Smith M, Uddin M (2025) Genomic insights into Rett syndrome-like features in Bangladeshi participants. Genet Med Open 3:103438. https://doi.org/10.1016/j.gimo.2025.103438

Bacq A, Depaulis A, Castagné V et al (2024) An update on stiripentol mechanisms of action: a narrative review. Adv Ther 41:1351. https://doi.org/10.1007/S12325-024-02813-0

Begum G, Albanna A, Bankapur A, Nassir N, Tambi R, Berdiev BK, Akter H, Karuvantevida N, Kellam B, Alhashmi D, Sung WWL, Thiruvahindrapuram B, Alsheikh-Ali A, Scherer SW, Uddin M (2021) Long-read sequencing improves the detection of structural variations impacting complex non-coding elements of the genome. Int J Mol Sci 22(4):2060. https://doi.org/10.3390/ijms22042060

Brown JT, Bishop JR, Sangkuhl K, Nurmi EL, Mueller DJ, Dinh JC, Gaedigk A, Klein TE, Caudle KE, McCracken JT, De Leon J, Leeder JS (2019) Clinical pharmacogenetics implementation consortium guideline for cytochrome P450 (CYP)2D6 genotype and atomoxetine therapy. Clin Pharmacol Ther 106(1):94–102. https://doi.org/10.1002/cpt.1409

Canitano R, Pallagrosi M (2017) Autism spectrum disorders and schizophrenia spectrum disorders: excitation/inhibition imbalance and developmental trajectories. Front Psych 8. https://doi.org/10.3389/fpsyt.2017.00069

Carroll JB, Warby SC, Southwell AL, Doty CN, Greenlee S, Skotte N, Hung G, Bennett CF, Freier SM, Hayden MR (2011) Potent and selective antisense oligonucleotides targeting single-nucleotide polymorphisms in the Huntington disease gene/allele-specific silencing of mutant huntingtin. Mol Ther 19(12):2178–2185. https://doi.org/10.1038/mt.2011.201

Chang S, Liu JJ, Zhao Y, Pang T, Zheng X, Song Z, Zhang A, Gao X, Luo L, Guo Y, Liu J, Yang L, Lu L (2024) Whole-genome sequencing identifies novel genes for autism in Chinese trios. Sci China Life Sci 67(11):2368–2381. https://doi.org/10.1007/s11427-023-2564-8

Cooper GM, Coe BP, Girirajan S, Rosenfeld JA, Vu TH, Baker C, Williams C, Stalker H, Hamid R, Hannig V, Abdel-Hamid H, Bader P, McCracken E, Niyazov D, Leppig K, Thiese H, Hummel M, Alexander N, Gorski J, Kussmann J, Shashi V, Johnson K, Rehder C, Ballif BC, Shaffer LG, Eichler EE (2011) A copy number variation morbidity map of developmental delay. Nat Genet 43(9):838–846. https://doi.org/10.1038/ng.909

Corey DR (2017) Nusinersen, an antisense oligonucleotide drug for spinal muscular atrophy. Nat Neurosci 20(4):497–499. https://doi.org/10.1038/nn.4508

Danielsson S, Gillberg IC, Billstedt E, Gillberg C, Olsson I (2005) Epilepsy in young adults with autism: a prospective population-based follow-up study of 120 individuals diagnosed in childhood. Epilepsia 46(6):918–923. https://doi.org/10.1111/j.1528-1167.2005.57504.x

De Ligt J, Willemsen MH, Van Bon BWM, Kleefstra T, Yntema HG, Kroes T, Silfhout ATV-V, Koolen DA, De Vries P, Gilissen C, Del Rosario M, Hoischen A, Scheffer H, De Vries BBA, Brunner HG, Veltman JA, Vissers LELM (2012) Diagnostic exome sequencing in persons with severe intellectual disability. N Engl J Med 367(20):1921–1929. https://doi.org/10.1056/nejmoa1206524

Dean L (2012) Risperidone Therapy and CYP2D6 Genotype. In: Medical Genetics Summaries. National Center forBiotechnology Information (US)

Erwin JA, Marchetto MC, Gage FH (2014) Mobile DNA elements in the generation of diversity and complexity in the brain. Nat Rev Neurosci 15(8):497–506. https://doi.org/10.1038/nrn3730

Foust KD, Nurre E, Montgomery CL, Hernandez A, Chan CM, Kaspar BK (2008) Intravascular AAV9 preferentially targets neonatal neurons and adult astrocytes. Nat Biotechnol 27(1):59–65. https://doi.org/10.1038/nbt.1515

Frangoul H, Altshuler D, Cappellini MD, Chen Y-S, Domm J, Eustace BK, Foell J, De La Fuente J, Grupp S, Handgretinger R, Ho TW, Kattamis A, Kernytsky A, Lekstrom-Himes J, Li AM, Locatelli F, Mapara MY, De Montalembert M, Rondelli D, Sharma A, Sheth S, Soni S, Steinberg MH, Wall D, Yen A, Corbacioglu S (2020) CRISPR-CAS9 gene editing for sickle cell disease and B-thalassemia. N Engl J Med 384(3):252–260. https://doi.org/10.1056/nejmoa2031054

French JA, Lawson JA, Yapici Z, Ikeda H, Polster T, Nabbout R, Curatolo P, De Vries PJ, Dlugos DJ, Berkowitz N, Voi M, Peyrard S, Pelov D, Franz DN (2016) Adjunctive everolimus therapy for treatment-resistant focal-onset seizures associated with tuberous sclerosis (EXIST-3): a phase 3, randomised, double-blind, placebo-controlled study. Lancet 388(10056):2153–2163. https://doi.org/10.1016/s0140-6736(16)31419-2

Friedmann T, Roblin R (1972) Gene therapy for human genetic disease? Science 175(4025):949–955. https://doi.org/10.1126/science.175.4025.949

Friocourt G, Poirier K, Rakić S, Parnavelas JG, Chelly J (2006) The role of ARX in cortical development. Eur J Neurosci 23(4):869–876. https://doi.org/10.1111/j.1460-9568.2006.04629.x

Gleeson JG, Minnerath S, Kuzniecky RI, Dobyns WB, Young ID, Ross ME, Walsh CA (2000) Somatic and germline mosaic mutations in the doublecortin gene are associated with variable phenotypes. Am J Hum Genet 67(3):574–581. https://doi.org/10.1086/303043

Granata P, Zito A, Cocciadiferro D, Novelli A, Pessina C, Mazza T, Ferri M, Piccinelli P, Luoni C, Termine C, Fasano M, Casalone R (2024) Unveiling genetic insights: Array-CGH and WES discoveries in a cohort of 122 children with essential autism spectrum disorder. BMC Genomics 25(1). https://doi.org/10.1186/s12864-024-11077-5

Guerrini R, Moro F, Andermann E, Hughes E, D'Agostino D, Carrozzo R, Bernasconi A, Flinter F, Parmeggiani L, Volzone A, Parrini E, Mei D, Jarosz JM, Morris RG, Pratt P, Tortorella G, Dubeau F, Andermann F, Dobyns WB, Das S (2003) Nonsyndromic mental retardation and cryptogenic epilepsy in women with Doublecortin gene mutations. Ann Neurol 54(1):30–37. https://doi.org/10.1002/ana.10588

Hanna E, Rémuzat C, Auquier P, Toumi M (2017) Gene therapies development: slow progress and promising prospect. J Mark Access Health Policy 5(1):1265293. https://doi.org/10.1080/20016689.2017.1265293

Hendson L, Church PT, Banihani R (2022) Follow-up care of the extremely preterm infant after discharge from the neonatal intensive care unit. Paediatr Child Health 27(6):359–364. https://doi.org/10.1093/pch/pxac058

Iossifov I, O'Roak BJ, Sanders SJ, Ronemus M, Krumm N, Levy D, Stessman HA, Witherspoon KT, Vives L, Patterson KE, Smith JD, Paeper B, Nickerson DA, Dea J, Dong S, Gonzalez LE, Mandell JD, Mane SM, Murtha MT, Sullivan CA, Walker MF, Waqar Z, Wei L, Willsey AJ, Yamrom B, Lee YH, Grabowska E, Dalkic E, Wang Z, Marks S, Andrews P, Leotta A, Kendall J, Hakker I, Rosenbaum J, Ma B, Rodgers L, Troge J, Narzisi G, Yoon S, Schatz MC, Ye K, McCombie WR, Shendure J, Eichler EE, State MW, Wigler M (2014) The contribution of de novo coding mutations to autism spectrum disorder. Nature 515(7526):216–221. https://doi.org/10.1038/nature13908

Jakovcevski M, Akbarian S (2012) Epigenetic mechanisms in neurological disease. Nat Med 18(8):1194–1204. https://doi.org/10.1038/nm.2828

Jamra RA, Wohlfart S, Zweier M, Uebe S, Priebe L, Ekici A, Giesebrecht S, Abboud A, Khateeb MAA, Fakher M, Hamdan S, Ismael A, Muhammad S, Nöthen MM, Schumacher J, Reis A (2011) Homozygosity mapping in 64 Syrian consanguineous families with non-specific intellectual disability reveals 11 novel loci and high heterogeneity. Eur J Hum Genet 19(11):1161–1166. https://doi.org/10.1038/ejhg.2011.98

Kato M, Das S, Petras K, Kitamura K, Morohashi K, Abuelo DN, Barr M, Bonneau D, Brady AF, Carpenter NJ, Cipero KL, Frisone F, Fukuda T, Guerrini R, Iida E, Itoh M, Lewanda AF, Nanba Y, Oka A, Proud VK, Saugier-Veber P, Schelley SL, Selicorni A, Shaner R, Silengo M, Stewart F, Sugiyama N, Toyama J, Toutain A, Vargas AL, Yanazawa M, Zackai EH, Dobyns WB (2004) Mutations of ARX are associated with striking pleiotropy and consistent genotype–phenotype correlation. Hum Mutat 23(2):147–159. https://doi.org/10.1002/humu.10310

Khalaf T, Ojaimi MA, Saleh DA, Sulaiman A, Sohal AP, Khan A, El-Hattab AW (2024) The utility of exome sequencing in diagnosing pediatric neurodevelopmental disorders in a highly consanguineous population. Clin Genet 106(1):82–89. https://doi.org/10.1111/cge.14508

Kim H, Kim J-S (2014) A guide to genome engineering with programmable nucleases. Nat Rev Genet 15(5):321–334. https://doi.org/10.1038/nrg3686

Kim J, Hu C, Achkar CME et al (2019) Patient-customized oligonucleotide therapy for a rare genetic disease. N Engl J Med 381:1644–1652. https://doi.org/10.1056/nejmoa1813279

Kim J, Lee J, Jang D-H (2023) Combining chromosomal microarray and clinical exome sequencing for genetic diagnosis of intellectual disability. Sci Rep 13(1). https://doi.org/10.1038/s41598-023-50285-z

Kimura H, Nakatochi M, Aleksic B, Guevara J, Toyama M, Hayashi Y, Kato H, Kushima I, Morikawa M, Ishizuka K, Okada T, Tsurusaki Y, Fujita A, Miyake N, Ogi T, Takata A, Matsumoto N, Buxbaum J, Ozaki N, Sebat J (2022) Exome sequencing analysis of Japanese autism spectrum disorder case-control sample supports an increased burden of synaptic function-related genes. Transl Psychiatry 12(1). https://doi.org/10.1038/s41398-022-02033-6

Knight EMP, Amin S, Bahi-Buisson N, Benke TA, Cross JH, Demarest ST, Olson HE, Specchio N, Fleming TR, Aimetti AA, Gasior M, Devinsky O (2022) Safety and efficacy of ganaxolone in patients with CDKL5 deficiency disorder: results from the double-blind phase of a randomised, placebo-controlled, phase 3 trial. Lancet Neurol 21(5):417–427. https://doi.org/10.1016/s1474-4422(22)00077-1

Krumm N, Turner TN, Baker C, Vives L, Mohajeri K, Witherspoon K, Raja A, Coe BP, Stessman HA, He Z-X, Leal SM, Bernier R, Eichler EE (2015) Excess of rare, inherited truncating mutations in autism. Nat Genet 47(6):582–588. https://doi.org/10.1038/ng.3303

Lan X, Tang X, Weng W, Xu W, Song X, Yang Y, Sun H, Ye H, Zhang H, Yu G, Wu S (2025) Diagnostic utility of Trio–Exome sequencing for children with neurodevelopmental disorders. JAMA Netw Open 8(3):e251807. https://doi.org/10.1001/jamanetworkopen.2025.1807

Lee Y, Park S, Lee JS, Kim SY, Cho J, Yoo Y, Lee S, Yoo T, Lee M, Seo J, Lee J, Kneissl J, Lee J, Jeon H, Jeon EY, Hong SE, Kim E, Kim H, Kim WJ, Kim JS, Ko JM, Cho A, Lim BC, Kim WS, Choi M, Chae J-H (2020) Genomic profiling of 553 uncharacterized neurodevelopment patients reveals a high proportion of recessive pathogenic variant carriers in an outbred population. Sci Rep 10(1). https://doi.org/10.1038/s41598-020-58101-8

Lemire G, Sanchis-Juan A, Russell K et al (2024) Exome copy number variant detection, analysis, and classification in a large cohort of families with undiagnosed rare genetic disease. Am J Hum Genet 111:863–876. https://doi.org/10.1016/j.ajhg.2024.03.008

Lenk GM, Jafar-Nejad P, Hill SF, Huffman LD, Smolen CE, Wagnon JL, Petit H, Yu W, Ziobro J, Bhatia K, Parent J, Giger RJ, Rigo F, Meisler MH (2020) SCN8A antisense oligonucleotide is protective in mouse models of SCN8A encephalopathy and Dravet syndrome. Ann Neurol 87(3):339–346. https://doi.org/10.1002/ana.25676

Lim ET, Uddin M, De Rubeis S, Chan Y, Kamumbu AS, Zhang X, D'Gama AM, Kim SN, Hill RS, Goldberg AP, Poultney C, Minshew NJ, Kushima I, Aleksic B, Ozaki N, Parellada M, Arango C, Penzol MJ, Carracedo A, Kolevzon A, Hultman CM, Weiss LA, Fromer M, Chiocchetti AG, Freitag CM, Church GM, Scherer SW, Buxbaum JD, Walsh CA (2017) Rates, distribution and implications of postzygotic mosaic mutations in autism spectrum disorder. Nat Neurosci 20(9):1217–1224. https://doi.org/10.1038/nn.4598

Liu Y, Zhao D, Dong R, Yang X, Zhang Y, Tammimies K, Uddin M, Scherer SW, Gai Z (2015) De novo exon 1 deletion of AUTS2 gene in a patient with autism spectrum disorder and developmental delay: a case report and a brief literature review. Am J Med Genet A 167(6):1381–1385. https://doi.org/10.1002/ajmg.a.37050

Maenner MJ, Warren Z, Williams AR, Amoakohene E, Bakian AV, Bilder DA, Durkin MS, Fitzgerald RT, Furnier SM, Hughes MM, Ladd-Acosta CM, McArthur D, Pas ET, Salinas A, Vehorn A, Williams S, Esler A, Grzybowski A, Hall-Lande J, Nguyen RHN, Pierce K, Zahorodny W, Hudson A, Hallas L, Mancilla KC, Patrick M, Shenouda J, Sidwell K, DiRienzo M, Gutierrez J, Spivey MH, Lopez M, Pettygrove S, Schwenk YD, Washington A, Shaw KA (2023) Prevalence and characteristics of autism spectrum disorder among children aged 8 years — autism and developmental disabilities monitoring network, 11 sites, United States, 2020. MMWR Surveill Summ 72(2):1–14. https://doi.org/10.15585/mmwr.ss7202a1

Mantere T, Kersten S, Hoischen A (2019) Long-read sequencing emerging in medical genetics. Front Genet 10:432668. https://doi.org/10.3389/FGENE.2019.00426

McRae JF, Clayton S, Fitzgerald TW et al (2017) Prevalence and architecture of de novo mutations in developmental disorders. Underw Nat 542(7642):433–438. https://doi.org/10.1038/nature21062

Miller DT, Adam MP, Aradhya S, Biesecker LG, Brothman AR, Carter NP, Church DM, Crolla JA, Eichler EE, Epstein CJ, Faucett WA, Feuk L, Friedman JM, Hamosh A, Jackson L, Kaminsky EB, Kok K, Krantz ID, Kuhn RM, Lee C, Ostell JM, Rosenberg C, Scherer SW, Spinner NB, Stavropoulos DJ, Tepperberg JH, Thorland EC, Vermeesch JR, Waggoner DJ, Watson MS, Martin CL, Ledbetter DH (2010) Consensus statement: chromosomal microarray is a first-tier clinical diagnostic test for individuals with developmental disabilities or congenital anomalies. Am J Hum Genet 86(5):749–764. https://doi.org/10.1016/j.ajhg.2010.04.006

Mim RA, Soorajkumar A, Kosaji N, Rahman MM, Sarker S, Karuvantevida N, Eshaque TB, Rahaman MA, Islam A, Chowdhury MSJ, Shams N, Uddin KMF, Akter H, Uddin M (2024) Expanding deep phenotypic spectrum associated with atypical pathogenic structural variations overlapping 15q11–q13 imprinting region. Brain Behav 14(4). https://doi.org/10.1002/brb3.3437

Myer NM, Boland JR, Faraone SV (2018) Pharmacogenetics predictors of methylphenidate efficacy in childhood ADHD. Mol Psychiatry 23(9):1929–1936. https://doi.org/10.1038/mp.2017.234

Neul JL, Percy AK, Benke TA, Berry-Kravis EM, Glaze DG, Marsh ED, Lin T, Stankovic S, Bishop KM, Youakim JM (2023) Trofinetide for the treatment of Rett syndrome: a randomized phase 3 study. Nat Med 29(6):1468–1475. https://doi.org/10.1038/s41591-023-02398-1

O'Roak BJ, Vives L, Girirajan S, Karakoc E, Krumm N, Coe BP, Levy R, Ko A, Lee C, Smith JD, Turner EH, Stanaway IB, Vernot B, Malig M, Baker C, Reilly B, Akey JM, Borenstein E, Rieder MJ, Nickerson DA, Bernier R, Shendure J, Eichler EE (2012) Sporadic autism exomes reveal a highly interconnected protein network of de novo mutations. Nature 485(7397):246–250. https://doi.org/10.1038/nature10989

Påhlman M, Gillberg C, Himmelmann K (2020) Autism and attention-deficit/hyperactivity disorder in children with cerebral palsy: high prevalence rates in a population-based study. Dev Med Child Neurol 63(3):320–327. https://doi.org/10.1111/dmcn.14736

Pande S, Majethia P, Nair K, Rao LP, Mascarenhas S, Kaur N, Rosario MCD, Neethukrishna K, Chaurasia A, Hunakunti B, Jadhav N, Xavier S, Kumar J, Bhat V, Bhavani GS, Narayanan DL, Yatheesha BL, Patil SJ, Nampoothiri S, Kamath N, Aroor S, Y RB, Lewis LE, Sharma S, Bajaj S, Sankhyan N, Siddiqui S, Nayak SS, Bielas S, Girisha KM, Shukla A (2023) De novo variants underlying monogenic syndromes with intellectual disability in a neurodevelopmental cohort from India. Eur J Hum Genet 32(10):1291–1298. https://doi.org/10.1038/s41431-023-01513-7

Pattali R, Mou Y, Li X-J (2019) AAV9 vector: a novel modality in gene therapy for spinal muscular atrophy. Gene Ther 26(7–8):287–295. https://doi.org/10.1038/s41434-019-0085-4

Rahman MM, Uddin KF, Jezawi NKA, Karuvantevida N, Akter H, Dity NJ, Rahaman MA, Begum M, Rahaman MA, Baqui MA, Salwa Z, Islam S, Woodbury-Smith M, Basiruzzaman M, Uddin M (2019) Gonadal mosaicism of large terminal de novo duplication and deletion in siblings with variable intellectual disability phenotypes. Mol Genet Genomic Med 7(10). https://doi.org/10.1002/mgg3.954

Rapoport J, Chavez A, Greenstein D, Addington A, Gogtay N (2009) Autism spectrum disorders and childhood-onset schizophrenia: clinical and biological contributions to a relation revisited. J Am Acad Child Adolesc Psychiatry 48(1):10–18. https://doi.org/10.1097/chi.0b013e31818b1c63

Riazuddin S, Hussain M, Razzaq A, Iqbal Z, Shahzad M, Polla DL, Song Y, Van Beusekom E, Khan AA, Tomas-Roca L, Rashid M, Zahoor MY, Wissink-Lindhout WM, Basra M a R, Ansar M, Agha Z, Van Heeswijk K, Rasheed F, Van De Vorst M, Veltman JA, Gilissen C, Akram J, Kleefstra T, Assir MZ, Grozeva D, Carss K, Raymond FL, O'Connor TD, Riazuddin SA, Khan SN, Ahmed ZM, De Brouwer APM, Van Bokhoven H, Riazuddin S (2016) Exome sequencing of Pakistani consanguineous families identifies 30 novel candidate genes for recessive intellectual disability. Mol Psychiatry 22(11):1604–1614. https://doi.org/10.1038/mp.2016.109

Rubin GM, Spradling AC (1982) Genetic transformation of drosophila with transposable element vectors. Science 218(4570):348–353. https://doi.org/10.1126/science.6289436

Sanchis-Juan A, Megy K, Stephens J, Ricaurte CA, Dewhurst E, Low K, French CE, Grozeva D, Stirrups K, Erwood M, McTague A, Penkett CJ, Shamardina O, Tuna S, Daugherty LC, Gleadall N, Duarte ST, Hedrera-Fernández A, Vogt J, Ambegaonkar G, Chitre M, Josifova D, Kurian MA, Parker A, Rankin J, Reid E, Wakeling E, Wassmer E, Woods CG, Raymond FL, Carss KJ (2023) Genome sequencing and comprehensive rare-variant analysis of 465 families with neurodevelopmental disorders. Am J Hum Genet 110(8):1343–1355. https://doi.org/10.1016/j.ajhg.2023.07.007

Satterstrom FK, Kosmicki JA, Wang J et al (2020) Large-scale exome sequencing study implicates both developmental and functional changes in the neurobiology of autism. Cell 180:568–584.e23. https://doi.org/10.1016/j.cell.2019.12.036

Shabestari SAS, Nassir N, Sopariwala S, Karimov I, Tambi R, Zehra B, Kosaji N, Akter H, Berdiev BK, Uddin M (2022) Overlapping pathogenic de novo CNVs in neurodevelopmental disorders and congenital anomalies impacting constraint genes regulating early development. Hum Genet 142(8):1201–1213. https://doi.org/10.1007/s00439-022-02482-5

Shchubelka K, Turova L, Wolfsberger W, Kalanquin K, Williston K, Kurutsa O, Makovetska A, Hasynets Y, Mirutenko V, Vakerych M, Oleksyk TK (2024) Genetic determinants of global developmental delay and intellectual disability in Ukrainian children. J Neurodev Disord 16(1). https://doi.org/10.1186/s11689-024-09528-x

Shevell M, Ashwal S, Donley D, Flint J, Gingold M, Hirtz D, Majnemer A, Noetzel M, Sheth RD (2003) Practice parameter: evaluation of the child with global developmental delay [RETIRED]. Neurology 60(3):367–380. https://doi.org/10.1212/01.wnl.0000031431.81555.16

Srivastava S, Love-Nichols JA, Dies KA, Ledbetter DH, Martin CL, Chung WK, Firth HV, Frazier T, Hansen RL, Prock L, Brunner H, Hoang N, Scherer SW, Sahin M, Miller DT (2019) Meta-analysis and multidisciplinary consensus statement: exome sequencing is a first-tier clinical diagnostic test for individuals with neurodevelopmental disorders. Genet Med 21(11):2413–2421. https://doi.org/10.1038/s41436-019-0554-6

Tanenhaus A, Stowe T, Young A, McLaughlin J, Aeran R, Lin IW, Li J, Hosur R, Chen M, Leedy J, Chou T, Pillay S, Vila MC, Kearney JA, Moorhead M, Belle A, Tagliatela S (2022) Cell-selective adeno-associated virus-mediated SCN1A gene regulation therapy rescues mortality and seizure phenotypes in a Dravet syndrome mouse model and is well tolerated in nonhuman primates. Hum Gene Ther 33(11–12):579–597. https://doi.org/10.1089/hum.2022.037

Thapar A, Cooper M, Rutter M (2016) Neurodevelopmental disorders. Lancet Psychiatry 4(4):339–346. https://doi.org/10.1016/s2215-0366(16)30376-5

The diagnostic and statistical manual of mental disorders (DSM–5), 5th ed (2013) American Psychiatric Association

Tuchman R (2017) What is the relationship between autism spectrum disorders and epilepsy? Semin Pediatr Neurol 24(4):292–300. https://doi.org/10.1016/j.spen.2017.21004

Turner TN, Hormozdiari F, Duyzend MH, McClymont SA, Hook PW, Iossifov I, Raja A, Baker C, Hoekzema K, Stessman HA, Zody MC, Nelson BJ, Huddleston J, Sandstrom R, Smith JD, Hanna D, Swanson JM, Faustman EM, Bamshad MJ, Stamatoyannopoulos J, Nickerson DA, McCallion AS, Darnell R, Eichler EE (2016) Genome sequencing of Autism-Affected families reveals disruption of putative noncoding regulatory DNA. Am J Hum Genet 98(1):58–74. https://doi.org/10.1016/j.ajhg.2015.11.023

Uddin KF, Amin MR, Sultana N, Aleem MA, Sarker SK, Banu LA, Ali MS, Khanam W, Rahman N, Baqui MA, Akter H, Dity NJ, Hossain MAZ (2019) Detection of clinically relevant copy number variation of SEZ6L2 gene in a Bangladeshi autism spectrum disorder cohort. Bangladesh J Med 30(1):24–29. https://doi.org/10.3329/bjmed.v30i1.39919

Wilfert AB, Turner TN, Murali SC et al (2021) Recent ultra-rare inherited variants implicate new autism candidate risk genes. Nat Genet 53(8):1125–1134. https://doi.org/10.1038/s41588-021-00899-8

Wolraich ML, Hagan JF, Allan C et al (2019) Clinical practice guideline for the diagnosis, evaluation, and treatment of attention-deficit/hyperactivity disorder in children and adolescents. Pediatrics 144. https://doi.org/10.1542/peds.2019-2528

Wood JJ, Kendall PC, Wood KS et al (2019) Cognitive behavioral treatments for anxiety in children with autism spectrum disorder. JAMA Psychiatr 77:474. https://doi.org/10.1001/jamapsychiatry.2019.4160

Woodbury-Smith M, Nicolson R, Zarrei M, Yuen RKC, Walker S, Howe J, Uddin M, Hoang N, Buchanan JA, Chrysler C, Thompson A, Szatmari P, Scherer SW (2017) Variable phenotype expression in a family segregating microdeletions of the NRXN1 and MBD5 autism spectrum disorder susceptibility genes. NPJ Genom Med 2(1). https://doi.org/10.1038/s41525-017-0020-9

Wright CF, Campbell P, Eberhardt RY, Aitken S, Perrett D, Brent S, Danecek P, Gardner EJ, Chundru VK, Lindsay SJ, Andrews K, Hampstead J, Kaplanis J, Samocha KE, Middleton A, Foreman J, Hobson RJ, Parker MJ, Martin HC, FitzPatrick DR, Hurles ME, Firth HV (2023) Genomic diagnosis of rare pediatric disease in the United Kingdom and Ireland. N Engl J Med 388(17):1559–1571. https://doi.org/10.1056/nejmoa2209046

Wu Z, Asokan A, Samulski RJ (2006) Adeno-associated virus serotypes: vector toolkit for human gene therapy. Mol Ther 14(3):316–327. https://doi.org/10.1016/j.ymthe.2006.05.009

Xiang J, Ding Y, Yang F, Gao A, Zhang W, Tang H, Mao J, He Q, Zhang Q, Wang T (2021) Genetic analysis of children with unexplained developmental delay and/or intellectual disability by whole-exome sequencing. Front Genet 12. https://doi.org/10.3389/fgene.2021.738561

Zhai Y, Zhang Z, Shi P et al (2021) Incorporation of exome-based CNV analysis makes trio-WES a more powerful tool for clinical diagnosis in neurodevelopmental disorders: a retrospective study. Hum Mutat 42:990–1004. https://doi.org/10.1002/HUMU.24222

Zhang Y, Haraksingh R, Grubert F, Abyzov A, Gerstein M, Weissman S, Urban AE (2013) Child development and structural variation in the human genome. Child Dev 84(1):34–48. https://doi.org/10.1111/cdev.12051

Zoghbi HY, Bear MF (2012) Synaptic dysfunction in neurodevelopmental disorders associated with autism and intellectual disabilities. Cold Spring Harb Perspect Biol 4(3):a009886. https://doi.org/10.1101/cshperspect.a009886

Protein Arginine Methylation in the Nervous System Development, Health and Disease

Misuzu Hashimoto

Abstract Protein arginine methylation is a post-translational modification catalyzed by the protein arginine methyltransferase (PRMT) family, which is highly conserved across species. PRMTs methylate histone and non-histone proteins, thereby regulating protein function, interactions, stability, and localization. These modifications play essential roles in gene expression, signal transduction, cell proliferation, apoptosis, and survival. Aberrant PRMT activity has been linked to various cancers, suggesting arginine methylation is crucial for biological functions in both health and disease, and may serve as a therapeutic target. Recent advances in tissue-specific gene knockout technology in mice have revealed the physiological roles of PRMTs in nervous system development and homeostasis. The mammalian central nervous system (CNS) develops through neural stem cell proliferation and differentiation into neurons and glial cells. This chapter provides an overview of arginine methylation, its biochemical mechanisms, and biological significance. We focus on how PRMTs regulate molecular networks essential for neural development and maintenance. Furthermore, we examine the pathological implications of dysregulated arginine methylation in neuronal disorders, including neurodegeneration, intellectual disabilities, and ALS, offering insights into its role in disease progression. Through this, we aim to clarify the critical functions of PRMTs in both normal and pathological CNS states.

Keywords Protein arginine methylation · Nervous system · Development · Neurodegenerative disease · Signal transduction · Cell proliferation · Apoptosis

---

M. Hashimoto (✉)
Laboratory of Biological Chemistry, Faculty of Applied Biological Sciences, Gifu University, Gifu, Japan
e-mail: hashimoto.misuzu.b9@f.gifu-u.ac.jp

A. H. M. N. Nabi, H. Uddin Shekhar (eds.), *Biotechnological Advances in Healthomics*, Interdisciplinary Biotechnological Advances,
https://doi.org/10.1007/978-981-95-3612-2_7

Abbreviations

AD	Alzheimer's disease
ADMA	Asymmetric dimethyl arginine
ALS	Amyotrophic lateral sclerosis
BioID	Proximity-dependent biotin identification
CNS	The central nervous system
E	Embryonic day
FUS	Fused in sarcoma
GFAP	Glial fibrillary acidic protein
IEG	Immediate early gene
LLPS	Liquid-liquid phase separation
METTL23	Methyltransferase-like 23
MMA	Monomethyl arginine
NSC	Neural stem cell
OL	Oligodendrocyte
OPC	Oligodendrocyte precursor cell
PRMT	Protein arginine methyltransferase
PRMT1-nKO	Nestin-Cre-mediated PRMT1 knockout
SAH	S-adenosylhomocysteine
SAM	S-adenosyl-L-methionine
SCYL1	SCY1-like pseudokinase 1
SDMA	Symmetric dimethyl arginine
SMN	Survival of motor neuron
SOX2	Sex determining region Y box2
XLID	X-linked intellectual disability

1 Introduction

1.1 Overview of Protein Arginine Methylation—Chemical Biology and Its Physiological Role

1.1.1 Protein Modification

Posttranslational modifications (also known as PTMs) are the chemical modification of proteins after its translation. These modifications on proteins play various roles in protein localization, protein degradation, protein–protein interactions and, therefore, change cell differentiation, signal transduction, or gene expression. In other words, PTMs play an important role in the dynamic biological activities of the cell. It is known that 5% of the human proteome consists of more than 200 enzymes that catalyze posttranslational modification on proteins. These contain kinase, phosphatase, transferase (e.g. methyltransferase), ligase, or protease.

Protein arginine methylation is one of the most common posttranslational modifications in eukaryotes (Wu et al. 2021; Blanc & Richard 2017; Fuhrmann et al. 2015). The family of enzyme called PRMTs transfer the methyl group from S-adenosyl-L-methionine (SAM or AdoMet) to the guanidino group of arginines in substrate proteins. The best-known substrates of PRMTs are RNA-binding proteins involved in multiple RNA functions such as RNA processing or RNA transport. Moreover, recent studies added many kinds of other proteins to PRMTs substrate list.

1.1.2 Arginine and Its Modification Type

Arginine residues have a guanidium-containing side chain, which has the highest p*K*a values (p*K*a = 12.5) and thus they tend to have a strong positive charge at physiological pH. These characteristics of arginine residues give them a high potential of binding themselves with molecules that have negative charges (e.g., nucleic acids), resulting in the generation of protein/DNA or protein/RNA complexes (Fuhrmann et al. 2015). Given these characteristics, it is not surprising that posttranslational modifications of arginine result in dramatic changes in physiology and disease pathogenesis. Arginine residues can receive several modifications such as methylation, citrullination (arginine deimination), phosphorylation, ADP-ribosylation, and arginylation. In this chapter, we focus on the role of arginine methylation in health and disease, as it is the most well-studied modification of arginine in mammals.

1.1.3 Protein Arginine Methyltransferase (PRMT) Family and Newly Identified Methyltransferases

Arginine methylation is catalyzed by a family of enzymes called protein arginine methyltransferases (PRMTs), and nine members (PRMT1–9) have been identified in mammals (Fig. 1). The methylation activity is that transferring a methyl group from SAM to the guanidino nitrogen of arginine, resulting in the production of methylarginine and S-adenosylhomocysteine (SAH). The nine PRMTs are categorized into three groups, depending on the modes of methylation. All the PRMTs are capable of monomethylation to produce monomethyl arginine (MMA). Type I PRMTs (PRMT1, PRMT2, PRMT3, PRMT4/CARM1, PRMT6, and PRMT8) catalyze the formation of asymmetric dimethyl arginine (ADMA) by adding the second methyl group to the same guanidino nitrogen. Type II PRMTs (PRMT5 and PRMT9) catalyze the formation of symmetric dimethyl arginine (SDMA) by adding the second methyl group to the opposite side of guanidino nitrogen. Type III PRMT (PRMT7) only forms MMA. The tenth type II arginine methyltransferase, NDUFAF7, has also been characterized as it methylates mitochondrial protein complex I protein NDUFS2. NDUFAF7 has a seven-b-strand fold similarly to PRMTs (Rhein et al. 2013). Methyltransferase-like 23 (METTL23), which has distant homology with PRMTs, is also identified as histone arginine methyltransferase in

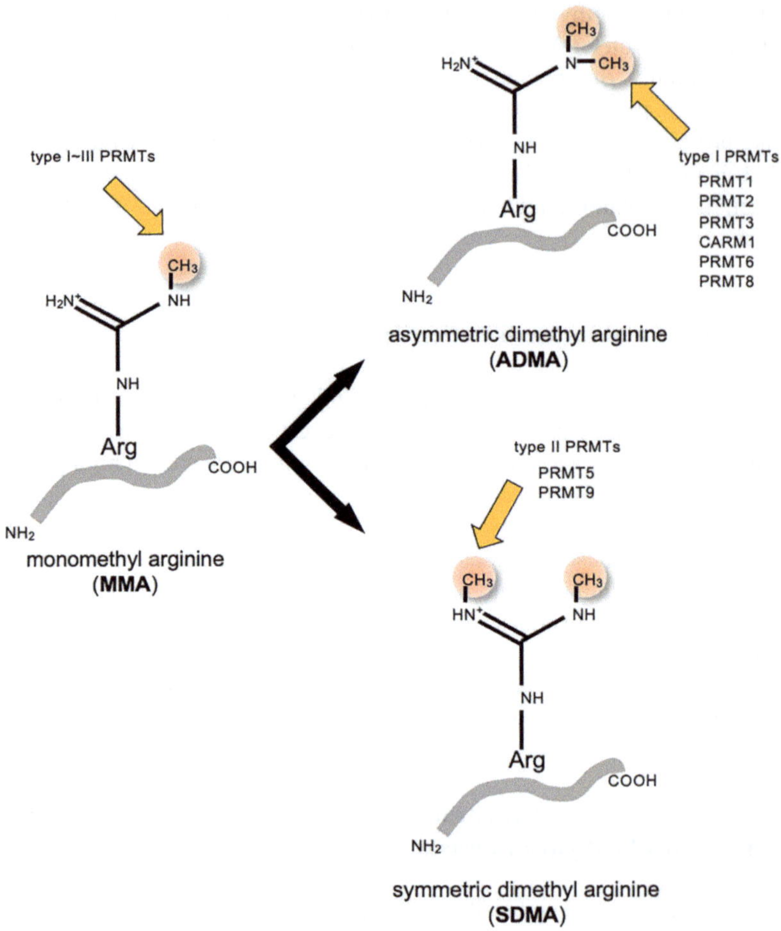

Fig. 1 Protein arginine methylation is catalyzed by PRMT family, which is classified into three types depending on the modes of methylation. PRMT7 only catalyzes monomethylation (type III)

mouse (Hatanaka et al. 2017). From clinical studies, a single nucleotide mutation of *METTL23* was identified in patients with normal-tension glaucoma (NTG) (Pan et al. 2022). NTG is characterized by retinal ganglion cell (RGC) death, resulting in visual field loss. The mutant *METTL23* causes a different splicing pattern of the gene, resulting in a reduction of the methylation level of H3R17me2a. In summary, we now know 11 methyltransferases, including PRMT1–9, NDUFAF7, and METTL23.

It is important to note that the impact of arginine methylation on cellular functions is highly substrate-dependent, and no consistent pattern has been established across different methylation types. As a result, the biological significance of each methylation form—monomethylarginine (MMA), asymmetric dimethylarginine

(ADMA), and symmetric dimethylarginine (SDMA)—is currently being investigated on a substrate-by-substrate basis. This complexity underscores the need for detailed functional studies to elucidate how specific methylation events contribute to cellular processes and disease mechanisms.

The crystal structures of some PRMTs have already been dissolved (Zhang and Cheng 2003). These studies have revealed that most PRMTs form homodimeric complexes that contain two active sites, while PRMT7 forms unusual complexes. In terms of substrate recognition by PRMTs, it is of note that there is no strict conservation of protein residues around arginine methylation sites. One exception is glycine, which is relatively enriched around arginine, such as RG or RGG. Another feature is that substrates of arginine methylation are often found in unstructured (disordered) protein regions such as loops or N- or C-terminal regions (Maron et al. 2021). Disordered regions of the proteins are often related to different cellular processes such as protein-protein/nucleic acid interactions and liquid-liquid phase separation (LLPS) of proteins, which are also a number of neurological disorders. Thus, more recent studies focus on the link between arginine methylation and these characteristics.

1.1.4 Epigenetic Regulator

PRMTs are writers of the methyl mark on histones; thus, the studies on its reader and eraser (demethylase) are ongoing. Tudor domain-containing proteins are known to recognize and bind to methylated lysines and arginines through its aromatic-binding cage structure (Wang and Bedford 2023). One of the representative Tudor domain-containing proteins is SMN (survival of motor neuron), which is the causative gene for spinal muscular atrophy (SMA). The Tudor domain in SMA protein binds to spliceosomal core proteins such as SmD1, SmD3, and SmB/B′, especially when they are symmetrically demethylated at arginine residues. Thus, it is thought that SMN acts as a molecular chaperone for pre-mRNA splicing regulation. Interestingly, SMN Tudor domain prefers SDMA rather than ADMA. In contrast, TDRD3, another Tudor domain-bearing protein, favors ADMA over SDMA or MMA. Demethylation of arginine is poorly understood since it is obvious that arginine methylation is less dynamic compared with other modifications. However, it is also observed that the level of arginine methylation changes depending on the cellular environment, suggesting that cells have demethylating enzymes. There are little studies on the regulatory mechanism for PRMT activity control. In case of PRMT5, MEP50 (a 7-bladed WD40 repeat b-propeller protein) serves as an important cofactor that directly binds to PRMT5 and enhances its activity for histone methylation (Stopa et al. 2015). Phosphorylation of PRMT5 by the oncogenic mutant of JAK2 kinase (JAK2V617F) was shown to impair PRMT5's activity of histone methylation (Liu et al. 2011). Future interactome studies for other PRMTs will identify more PRMTs' interacting partners that potentially regulate their enzymatic activities.

1.1.5 Methylation Substrate

A growing number of studies have tried to make a catalog for PRMTs substrates (Maron et al. 2021; Massignani et al. 2022). These efforts revealed the important functions of PRMTs in various biological processes, including RNA metabolism, RNA processing, gene expression, ribosomal function, and viral processes. Some of the studies use techniques called stable isotope labeling with amino acids in cell culture (SILAC) method, followed by LC-MS/MS to comprehensively search for arginine methylation. Massignani et al. have found that arginine methylation correlates with serine (S)/threonine (T) phosphorylation by discovering that 49% of all arginine methylation sites have phosphorylated residues within a 15-amino acid length, including the arginine (Massignani et al. 2022). Notably, crosstalk between arginine methylation and serine/threonine phosphorylation has been observed in several proteins, suggesting potential functional interdependence rather than mere coincidence. For example, methylation of the Forkhead box family protein FOXO1 promotes its nuclear localization and activation of the apoptotic pathway, while phosphorylation by Akt inhibits this process by retaining FOXO1 in the cytoplasm (Yamagata et al. 2008). Similarly, arginines in the Akt consensus motif (RXRXXS/T) in BCL-2 antagonist of cell death (BAD) has also been shown to be targets of PRMT1 (Sakamaki et al. 2011). This competitive relationship between methylation and phosphorylation at adjacent or overlapping sites highlights a regulatory mechanism in which one modification can directly influence the accessibility or function of another. While such examples support the idea of coordinated regulation, it remains possible that in some cases, the co-occurrence of methylation and phosphorylation may reflect amino acid composition biases in certain protein regions. Therefore, further systematic studies are needed to distinguish functional crosstalk from coincidental proximity.

Another study for PRMTs' target identification performed PRMTs interactome using the proximity-dependent biotin identification (BioID) method (Wei et al. 2021). The BioID method is useful in screening for both direct/indirect interacting proteins as well as neighboring proteins (Roux et al. 2012), and it is likely to be a powerful tool for capturing enzyme-substrate interaction, including PRMTs and their targets.

Type I and II PRMTs share some of the substrates, such as histones. A549, human lung adenocarcinoma cell line, treated with MS023, a type I PRMTs inhibitor, reduced ADMA on the FET family RNA-binding proteins FUS and TAF15. While at the same time, MMA and SDMA levels on FUS and TAF15 increased after MS023 treatment (Maron et al. 2021). Similar findings have been reported by other research groups, showing that inducible knockout of PRMT1 decreases total ADMA levels in mouse embryonic fibroblasts in culture, while MMA and SDMA are significantly upregulated (Dhar et al. 2013). These results suggest that genetic loss of type I PRMTs or inhibition of the activity, their substrates being scavenged by type II and III PRMTs. In other words, type I PRMTs could be protecting their substrates from further modifications by other enzymes.

1.1.6 Protein Arginine Methylation in the Brain

Protein lysine methylation is indeed one of the most prevalent forms of post-translational modification and plays a crucial role in regulating epigenetic outcomes. This modification involves the addition of one, two, or three methyl groups to the lysine residues on proteins, influencing various cellular processes, including gene expression, DNA repair, and protein-protein interactions. An increasing number of studies in the 2000s showed that protein arginine methylation is another essential type of methylation in various types of cells. In late 2010s, a large number of studies demonstrated that arginine methylation by PRMTs play essential roles in developmental and homeostatic control of many of the mammalian tissues. Knockout of PRMT1 and PRMT5 shows embryonic lethality (Pawlak et al. 2000; Tee et al. 2010). CARM1 null mice are born small and shortly die after birth (Yadav et al. 2003). These studies demonstrated that PRMTs are critical in cell differentiation and embryonic development. In terms of the potential role of PRMTs in disease control, PRMT1 and PRMT5 have been shown to be upregulated in multiple types of cancer (Guccione and Richard 2019). The inhibitors of these PRMTs are currently under clinical trials for cancer patients. Thus, it is now established that protein arginine methylation plays vital roles in development, health, and disease.

The most major type I PRMTs, PRMT1, are ubiquitously expressed in mammals. *Prmt1* knockout mice is known to show early embryonic lethality; however, *bgeo* insertion in the *Prmt1* gene locus enabled the expression pattern of *Prmt1* in the whole mouse embryo of the heterozygotes (Pawlak et al. 2000). From the analysis of the whole embryo, it was revealed that *Prmt1* expression starts from embryonic day (E) 7.5. The signals got stronger as the neural tube and the brain tissue develops among E8.5–9.5. These data implicated the importance of *Prmt1* in CNS development.

Remarkably, Immunoaffinity enrichment of arginine methylated proteins and the subsequent mass spectrometry analyses identified over 1000 arginine methylated proteins in human colon cancer cell line (HCT116), adult mouse brain, and whole mouse embryo (Guo et al. 2014). For this purpose, research groups of Cell Signaling Technologies Inc. designed new antibodies against MMA, ADMA, and SDMA. The antibodies of MMA have been produced by serial immunization of rabbits with the first peptide (XXXXXXXR*XXXXXX) and second peptide (XXXXXXXR*GGXXX), both of which contain methylated arginine (R*). This design is based on the knowledge that arginine methylation is frequently observed in RG-rich (especially RGG motif) regions. ADMA antibodies have been produced in a similar method. In HCT116 and whole embryos, both MMA and ADMA-modified major proteins were classified into RNA processing and transcriptional regulation. This finding suggests that many of the MMA and ADMA proteins are localized to the nucleus. However, in the mouse brain, a lot of membrane proteins (e.g., receptor/channel/transporter/ adaptor/scaffold), G proteins, or vesicle proteins were observed. In terms of cellular function, over half of the brain-enriched MMA/ADMA proteins are involved in synaptic transmission. Synapsin I (SYN1) is a member of the synapsin family which is a neuron-specific phosphoprotein family and regulates neurotransmitter release (Ryan et al. 1996). SHANK1 is a member of the Shank (SH3 and multiple ankyrin repeat domains protein) family (SHANK1–3), which is a synaptic scaffold protein. An increasing number of studies highlighted the involvement of mutations of *SHANK* family genes in the etiology of autism spectrum disorder (ASD) (Sato et al. 2012). These notions suggest that arginine methylation is deeply related not only to gene expression regulation but also neurotransmission and its related disorders.

2 Physiological and Pathological Roles of Protein Arginine Methylation in the Nervous System

2.1 Neural Stem Cell

The CNS development in mice starts from early gestational age and continues until after birth. CNS consists mainly of neurons, astrocytes, oligodendrocytes, and microglia. While microglia originate from macrophages that invade the embryonic neural tube during early gestational age, other cells are all originated from neural stem cells (NSCs) (Hippenmeyer 2023). In the murine cortical development, NSCs exist as neuroepithelial stem cells that increase their own population around E8–10. Then, it develops into radial glia progenitor cells that which first generate neurons at around E12–16, then secondarily produce glial cells, including astrocytes and oligodendrocytes, peaking at E16–18. Therefore, it is important to study the underlying molecular control of how and when NSCs proliferate and differentiate. In other words, NSCs are original sources of the CNS tissues, therefore, it is important to clarify the molecular mechanisms that regulate the generation and maintenance of NSCs as well as those that control NSCs' potential for generating neuron and glia.

NSC-specific knockout of PRMT5, the representative member of type II PRMTs, resulted in apoptosis of NSCs, which led to postnatal lethality (Bezzi et al. 2013) (Fig. 2). PRMT5 is known as a regulator of constitutive splicing and the alternative splicing of specific mRNAs (Guccione and Richard 2019). The NSC-specific knockout of PRMT5 also showed aberrant splicing of *Mdm4* mRNA, which leads to activation of p53 signaling that is required for apoptotic response. Notably, additional knockout of *p53* partially rescued the survival rate of the NSC-specific *Prmt5* knockout mice.

Similarly, embryonic NSCs derived from NSC-specific knockout of PRMT1 showed aberrant formation of floating neurospheres in culture as well as massive apoptosis with significant accumulation of p21, an inhibitor of cell cycle progression (Hashimoto et al. 2021) (Fig. 2). In contrast, NSCs in PRMT1-deficient brain tissue are somehow maintained. This in vivo versus in vitro discrepancy could be explained by the upregulation of cytokines that is produced by surrounding cells, which gives NSCs trophic support for their maintenance. It is likely because a pro-inflammatory cytokine, IL-6, induces NSC proliferation in vitro (Wang et al. 2011).

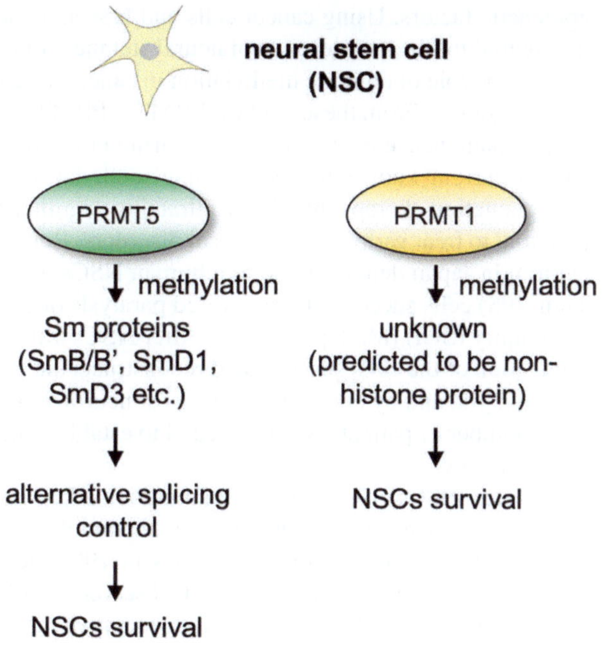

Fig. 2 PRMT5 and 1 are essential for the neural stem cell survival and self-renewal. PRMT5 regulates alternative splicing for NSCs proliferation and survival by methylating Sm proteins

In fact, some of these neurotrophic cytokines, including *IL-6* and *Lif* mRNA, did not change; *Lama1* mRNA was significantly increased in the fetal brain of NSC-specific PRMT1 knockout mice. The results indicate that laminin is one of the potential extracellular cues that salvage dying PRMT1-deficient NSCs.

What is the substrate of PRMT1 in the embryonic NSCs? As discussed before, PRMT1 is also known as one of the epigenetic regulators since it methylates histones and regulates the gene transcriptional network. Asymmetric dimethylation of histone 4 at R3 (H4R3me2as) was examined and found to be distributed in the developing brain cortices (Chittka 2010). However, interestingly, those signals were not observed in the NSCs, suggesting that PRMT1 methylates some other substrates other than histones.

Prmt1–8 mRNA expression has been explored using the mouse fetal cerebral cortex (Honda et al. 2017). This study demonstrated that *Prmt1* expression was the highest among all the *Prmts* tested. In addition, *Prmt1* expression in the neocortex gradually decreased throughout its development from E11.5 to E17.5. Interestingly, this change correlated with Sex determining region Y box2 (SOX2), which is the popular marker of NSCs, suggesting that PRMT1 plays an essential role in NSCs maintenance or their survival.

It is also demonstrated that CARM1 (PRMT4), another type I PRMTs, is specifically upregulated during the differentiation of mouse embryonic stem cells to NSCs (Niu et al. 2020). NSCs share some similarities with cancer cells not only because both of them have high potential for proliferation but also, they have high expression of genes related to protein homeostasis (protein synthesis and degradation) and epigenetic factors. Using cancer cells and NSCs, it was demonstrated that PRMT1 is essential for the maintenance of neural stemness of both types of cells (Chen et al. 2021). The role of arginine methylation in cancer development are well described in other literature. From these studies, PRMT5, PRMT1, and CARM1 are essential for NSCs maintenance or its production during embryonic development. We here discussed current knowledge on arginine methylation and embryonic NSCs. NSC transplantation therapy has been attracting significant public interest due to its potential to treat various neurological disorders. Remarkably, in early 2025, a clinical trial in Japan demonstrated that human NSCs derived from induced pluripotent stem (iPS) cells successfully recovered paralysis of some of the patients with spinal cord injury (SCI) (Mallapaty 2025). After NSCs transplantation, it seems that some population of the cells survives and reconstructs the damage site by producing neurons and glia and by connecting with host neurons. In the future, more trials with a large number of patients will be needed to establish the potential of NSC transplantation therapy.

Besides NSCs' external transplantation, adult mammalian brains even have NSCs that contribute to brain plasticity beyond development (Bond et al. 2015). In concert with the finding of adult NSCs in 1999 (Kempermann and Gage 1999), adult neurogenesis had already been observed in 1965 (Altman and Das 1965). These findings brought new hope in regenerative medicine. Therefore, it is of great importance to understand the molecular network, including protein arginine methylation in survival, proliferation, and differentiation of NSCs.

2.2 Neuron

Neurons comprise neural circuitry that is responsible for high-order CNS functions such as movement, sensation, learning, and memory. In the CNS, neurons make a complex network that is surrounded by glia or blood vessels. In those networks, neurons make synapses that are the physiological space between neurons or gaps between neurons and other cells (e.g., muscle cell), and this space is where neurotransmission occurs. At a synapse, the presynaptic neuron releases a neurotransmitter, and then receptors on the surface of postsynaptic cells receive the molecule to generate a new electrical signal.

Although most cell types in our body have a short lifespan, neurons do not divide after terminal differentiation from their progenitors, and they survive for many decades. This means that neurons have a high ability in proper cell maintenance and mechanisms for counteracting perturbations such as protein misfolding or DNA damage. Therefore, it is of great interest how protein arginine methylation is involved in those cellular processes for neuronal maintenance.

As described in the previous page, the comprehensive analysis of MMA and ADMA marks in the adult mouse brain showed that these modifications are also found in many kinds of membrane-located proteins involved in synaptic functions. This suggests that PRMTs may control neuronal activity not only at the nucleus or cytosol but also at membranous organelles such as synaptic vesicles or Golgi apparatus. Kim et al. have observed mild but spontaneous epileptic seizures in PRMT1 heterozygous knockout mice, which were caused by neuronal hyperactivity in the mutant mice (Kim et al. 2016). Mechanistic analysis demonstrated that methylation of the KCNQ2 protein of the KCNQ channel is essential to maintain healthy neurons.

2.2.1 Control of Neuronal Branching by PRMTs

Neuronal outgrowth (branch formation) consists of axon extension as well as dendrite arborization. The potential involvement of PRMT1 in these processes has been suggested by observation of increased ADMA levels in rat PC12 cells after nerve growth factor (NGF)-treatment (Cimato et al. 2002). While PC12 cells are neuronal model cells, Simandi et al. showed the essential role of PRMT1 in neuronal differentiation using embryonic stem cells (Simandi et al. 2015). More recently, Amano et al. have reported that SCY1-like pseudokinase 1 (SCYL1) as the specific target of PRMT1 for neuron morphological development (Fig. 3). SCYL1 is a catalytically inactive kinase, and it controls Golgi morphology by modulating COPI-mediated retrograde traffic. Amano et al. also observed that knockdown of PRMT1 resulted in enlargement and fragmentation of the Golgi apparatus, demonstrating the vital role of PRMT1 in Golgi morphology (Amano et al. 2020). Defective neuronal maturation and Golgi morphology abnormalities have been associated with mutations in Rab33a, a Rab family small GTPase (Nakazawa et al. 2012). PRMT2

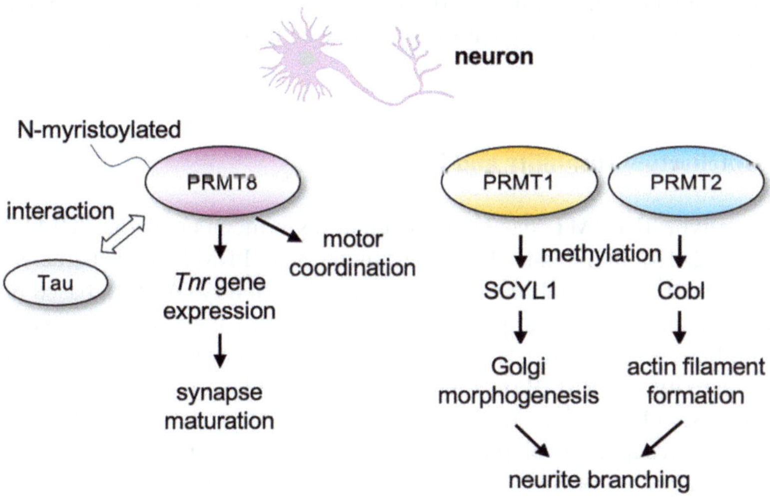

Fig. 3 PRMT8, as a neuronal-specific PRMT member, regulates various neuronal functions. PRMT8 is associated with the plasma membrane as this has a specific myristoylation site at its N-terminal region. PRMT1 and PRMT2 positively regulate neuronal branch number as well as neurite length in different mechanisms

(type I) also shows positive involvement in neurite branching by methylating the actin nucleator protein called Cobl (Hou et al. 2018) (Fig. 3).

2.2.2 PRMT8 in Neuronal Function

Among the PRMT family, PRMT8 expression is restricted to neurons in the overall CNS areas, including the hippocampus, cerebral cortex, and the cerebellum (Kousaka et al. 2009). In addition, the most unique characteristics of PRMT8 are its localization to the plasma membrane through its N-terminal myristoylation motif (Lee et al. 2005). In fact, in the research area of PRMT in neurons, PRMT8 is the top member, which has several evidence on neuronal functions (Fig. 3). Interestingly, PRMT8 is also considered as an epigenetic modifier, and it was found to bind to the Tenascin-R (*Tnr*) gene promoter. This binding is essential for synapse maturation of neurons at visual-somatosensory cortices (Lee et al. 2017).

PRMT8 knockout mice have been generated to analyze the physiological role of PRMT8, especially in the cerebellum as it has been shown to be expressed strongly in Purkinje cells. PRMT8 deletion did not affect the survival of mice, and reproduction was normal. However, the PRMT8-knockout mice exhibited abnormal motor coordination and showed hyperactivity in the open field test. From further biochemical analysis, PRMT8 was shown to act as a phospholipase that hydrolyzes phosphatidylcholine (PC). This suggests that PRMT8 has dual activities as arginine methylation and PC hydrolysis (Kim et al. 2015).

In Alzheimer's disease (AD), tau protein and amyloid-β are known to be important factors for the neurodegeneration; however, the disease causative mechanisms are poorly understood. A recent study identified PRMT8 as one of the interacting proteins of Tau and suggested that PRMT8 may be involved in AD pathology (Ishii et al. 2022) (Fig. 3).

2.2.3 Intellectual Disability and PRMTs

PRMT9, a type II PRMT, has been identified as a member of the PRMT family (Yang et al. 2015). In humans, the G189R mutation in PRMT9 was characterized in association with autosomal recessive intellectual disability (ARID), based on findings from a large-scale whole-genome sequencing study involving 136 consanguineous families. The PRMT9 G189R mutation was shown to completely abolish catalytic activity of PRMT9 on splicing factor SF3B2 R508 methylation. Using a conditional knockout of a mouse model of *Prmt9* demonstrated that PRMT9 is essential for synapse development, learning, and memory (Shen et al. 2024). Mechanistically, loss of SF3B2 R508 methylation impairs mRNA splicing by disrupting the interaction of the spliceosomal complex with RNA. While there seems to be only a few substrates of PRMT9, SF3B2 is the only target of PRMT9, at least in brain development. It is again of note that PRMT5 is the most major type II PRMTs in mammals; however, this study revealed an essential role of PRMT9 in neuronal development and function.

X-linked intellectual disability (XLID) accounts for ~16% of all intellectual disability patients. RNA-binding proteins (RBPs) are closely linked with intellectual disability. For example, Fragile X mental retardation protein (FMRP) is related to fragile X syndrome, which accounts for 3–5% XLID patients. *RBMX*, another RBP coding gene, has been found to have a frameshift mutation in one type of XLID called Shashi-XLID (Shashi et al. 2014). The following analyses identified that the *RBMX* mutation causes aberrant upregulation of p53 pathway, which induces apoptotic cell death. Interestingly, PRMT5 methylates RGG/RG motif of RBMX protein, then it changes the interaction property of RBMX with the SRSF1 protein to change *Mdm4* gene splicing for p53 regulation.

In summary, these studies demonstrate two distinct mechanisms by which PRMTs are involved in intellectual disability, emphasizing the need to differentiate between direct genetic and indirect functional contributions. The PRMT9 G189R mutation represents a direct genetic cause, as it abolishes enzymatic activity and disrupts neuronal development through impaired methylation of SF3B2. This mutation was identified in a large-scale genomic study and validated in animal models, establishing its pathogenic role. In contrast, PRMT5 is not genetically altered in XLID, but its methylation of RBMX affects RNA splicing and p53 pathway regulation, suggesting an indirect mechanism that may contribute to disease progression. These findings highlight the importance of considering both gene mutations and post-translational modifications when investigating the molecular basis of neurodevelopmental disorders.

2.2.4 Involvement of PRMTs in Neuromuscular Function and Neurodegeneration

To investigate the role of PRMT1 in neuromuscular interaction, motoneuron specific *Prmt1* knockout mice were generated by utilizing Hb9-Cre lines. The mn*Prmt1*-KO mice exhibited motor neuron degeneration with mitochondrial dysfunction and subsequent muscle loss (So et al. 2023) (Fig. 4). In addition, specific deletion of *Prmt1* in muscle stem cells by *Pax7*$^{\mathrm{CreERT2}}$ displayed impairment in muscle regeneration. In muscle stem cells, Eya1 (co-activator of Six1) was shown to be methylated by PRMT1, and it regulates *MyoD* gene expression (Blanc et al. 2017) (Fig. 4). These studies show that PRMT1 is essential for myogenesis and functional neuromuscular junction.

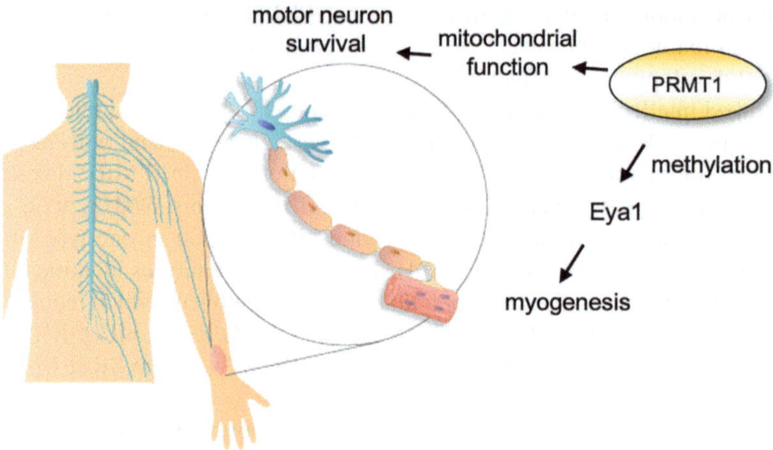

Fig. 4 Physiological roles or arginine methylation events in the peripheral nervous system. PRMT1 has multiple roles in the neuromuscular junction made by motor neurons and the muscle cells. The above image is made from TogoTV with some modifications (https://doi.org/10.7875/togopic.2014.27)

PRMT8, which is expressed in post-mitotic motoneurons of the aging spinal cord, plays a critical role in age-related stress tolerance—such as resistance to oxidative stress and protein misfolding—as well as in neuronal plasticity. This is achieved through the regulation of transcriptional activation of immediate early genes (IEGs) and the DNA damage response (Simandi et al. 2018). Since expression of CREB-dependent IEGs, including *Ier2*, *Fos*, *Myc*, *Egr1,* and *Atf3* is induced upon CNS tissue injury, it is thought that IEGs regulation is linked to neurodegenerative diseases such as Alzheimer's disease, amyotrophic lateral sclerosis (ALS), and frontotemporal dementia (FTD) (Minatohara et al. 2016). While the substrate of PRMT8 in the motoneuron has not been identified, it is likely that PRMT8 positively mediates DNA damage response. There are a large number of studies in search for the substrates of PRMT1 compared with those of PRMT8; however, it is likely that PRMT1 and PRMT8 share at least some of the substrates as they share 80% homology in their amino acid sequence (Scaramuzzino et al. 2013). One of their common substrates is the FET family RNA-binding protein, Fused in sarcoma (FUS). Both PRMT1 and 8 interact with wild-type FUS and ALS-like mutant FUS. Methylation of FUS regulates its nucleus-cytoplasm shuttling (Fig. 5). ALS is a neurodegenerative disease that leads to muscle weakness and early death (Tradewell et al. 2012; Yamaguchi and Kitajo 2012; Dormann et al. 2012). It was found that hypomethylation of FUS enables LLPS, leading to the formation of toxic protein aggregates. Similarly, hypomethylation of other RNA binding proteins, including hnRNPA2 or FMRP1 promotes their LLPS (Ryan et al. 2018; Tsang et al. 2019).

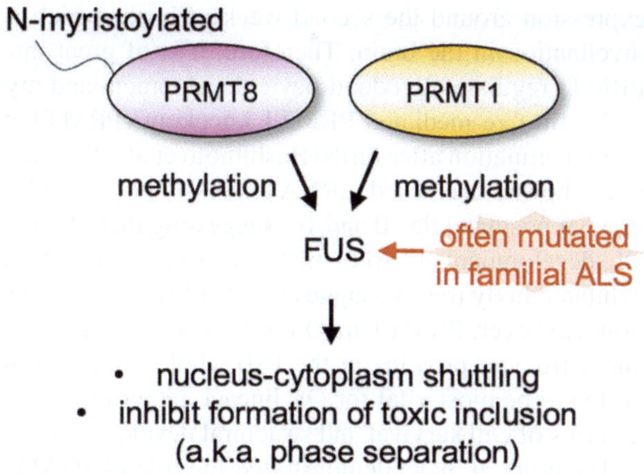

Fig. 5 RNA binding protein FUS, which has three RGG domains, is the methylation target of PRMT1 and 8. ALS-causative FUS mutation leads to its toxic inclusion, while arginine methylation is protective for those events

2.3 Oligodendrocyte (OL), Myelin

2.3.1 Oligodendrocytes and Myelin and Their Development, Disease

The CNS consists of two histologically dependent regions; one is white matter, which is made up of myelinated axons (also called white matter tracts), and the other is grey matter, which is composed of neuronal cell bodies. White matter is named for its white appearance due to the lipid-rich myelin component. Neuronal axons are wrapped by myelin membrane, which is produced by oligodendrocytes (OLs). The myelin structure allows fast and efficient axon conduction as well as providing trophic and metabolic support to axons. Dysregulation in myelination (such as leukodystrophies) or loss of myelin (demyelination), which is seen in multiple sclerosis (MS), or Alzheimer's disease, leads to axonal damage. Clinical studies using human tissues of neuropathology have shown that de novo myelination by differentiation of residual oligodendrocyte precursor cells (OPCs), which is termed remyelination. Studies on developmental myelination and remyelination have revealed that cell-intrinsic systems that control OPCs for its proliferation, migration, and differentiation regulate myelin formation. Moreover, other cell types such as microglia and astrocytes also support OPCs recruitment to the myelination site or differentiation.

2.3.2 PRMT Family and Myelin

PRMT1 and PRMT5 expression has been shown in all the neural cell types, having relatively higher expression in neurons (Zhang et al. 2014). Also, PRMT1-mediated ADMA modification is shown to decrease in the developing mouse brain, suggesting that PRMT1 is important for brain development (Simandi et al. 2018). PRMT5 expression was shown to be increased after birth (Huang et al. 2011). The expression level shift of PRMT1 and 5 seems opposite; however, both show strong expression around the second week of birth, which is the peak of developmental myelination in the brain. Therefore, it is of great interest how type I and type II PRMTs regulate oligodendrocyte development and myelin formation.

Nestin-Cre-mediated PRMT1 knockout (PRMT1-nKO) mice showed dramatic hypomyelination after birth (Hashimoto et al. 2016) (Fig. 6). In addition, the mutant mice showed decreased mRNA expression of *Olig1/2*, *Sox10*, *Nkx2–2* in the whole brain at postnatal day 0 and 10, suggesting that PRMT1 controls genes essential for OL development. PRMT1-nKO mice died around the second week of birth and exhibited body tremor, suggesting that PRMT1 is essential for behavioral coordination. However, PRMT1-nKO mice showed normal expression of neuronal proteins and astrocytic proteins in the P10 whole brain. From this study, PRMT1 is suggested to be most vital for OL lineage rather than neuronal and astrocytic lineages in terms of cell survival and structural development.

Huang et al. have demonstrated the role of PRMT5 in OL development a little earlier (Huang et al. 2011). PRMT5 knockdown in rat primary OPCs and rat C6 glioma cell line caused elevated expression of inhibitors of differentiation/DNA binding, *Id2* and *Id4*, which are known repressors for glial differentiation, resulting in OL maturation. Mechanistically, PRMT5 in normal OPCs was enriched in CpG Island of the promoter regions of *Id2* and *Id4*, while the occupancy level was decreased in the PRMT5 knockdown cells, suggesting that PRMT5 normally inhibits their expression by epigenetic regulation. The epigenome regulatory system by PRMT5 has been shown before the above study using the human β-globin gene

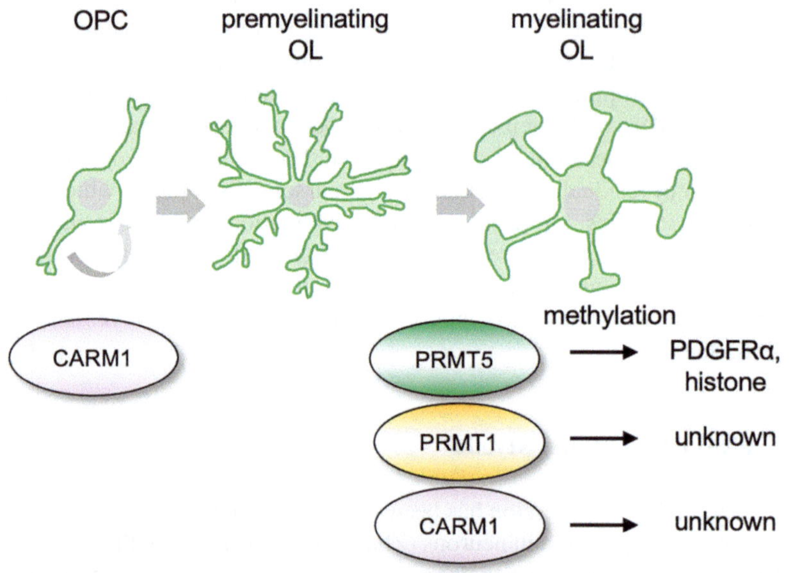

Fig. 6 Oligodendrocyte differentiation system and the related PRMTs for driving the maturation process. Knockout of PRMT1 or 5 is critical in myelin formation in vivo. CARM1 is important for OL differentiation as well as OPC proliferation

locus. Zhao et al. has found that histone H4R3me2s mark mediated by PRMT5 is required for subsequent DNA methylation mark added by the DNA methyltransferase DNMT3A (Zhao et al. 2009) (Fig. 6). This study first demonstrated a new link between histone (protein) and DNA methylation for gene silencing.

Ishino et al. recently identified CARM1 (PRMT4) as an essential methyltransferase for OPC proliferation and differentiation. CARM1-specific inhibitor treatment resulted in decreased OPC proliferation, while apoptotic cell death was not induced. Also, the same inhibitor induced low differentiation potential to OL, which is evaluated by decreased expression of proteolipid protein (PLP), a myelin marker protein, and Caspr, a node of Ranvier protein (Ishino et al. 2022) (Fig. 6).

Remarkably, the role of PRMT5 in OL lineage progression was further specified by the following two research papers. Scaglione et al. made in vivo PRMT5 ablation model by crossing *Prmt5-flox* and *Olig1-Cre* mouse lines (Scaglione et al. 2018) (Fig. 6). These OL-specific PRMT5-knockout mice showed severe hypomyelination at postnatal day 14 and died around the second week of birth. It is of note that PDGFRα-positive OPC proliferation was kept normal in the mutant mice. It is noteworthy that a similar analysis using the Cnp-Cre line, which expresses Cre later than Olig1-Cre, did not affect the level of myelination, suggesting that PRMT5 is required for the OPC differentiation program within a critical time window. From molecular level analyses, the authors found that histone symmetric arginine methylation (H4R3me2s) by PRMT5 prevents acetylation of histone (H4K5ac) which is gene repression mark for genes required for OL differentiation. PRMT5 seemed to have a unique role in orchestrating histone methylation and acetylation in a specific time window during OL development. Another study of the PRMT5 in OL development utilized *Prmt5-flox* and *Olig2-Cre* mouse lines (Calabretta et al. 2018) (Fig. 6). The mutant mice exhibited severe hypomyelination and early death prior to the third postnatal week (almost the same phenotype as the above-mentioned *Prmt5-flox* and *Olig1-Cre* mice). In search for the essential substrate of PRMT5, the authors found that PRMT5 methylates PDGFRα at R554 among the intracellular domain, changing the availability of the protein in the OPC for their differentiation potential. It has been known that PDGFRα trafficking and recycling are controlled by its ubiquitination by Cbl E3 ubiquitin ligase. Inhibition of methylation of PDGFRα by PRMT5 knockdown of PRMT5 inhibitor treatment results in faster PDGFRα degradation mediated by ubiquitination. The crosstalk of arginine methylation by PRMT5 and ubiquitination is an interesting molecular network in myelin development. In addition, beyond OL differentiation, repression of PRMT5 could be an interesting drug target for certain cancers that show elevated expression of PDGFRα, such as glioblastomas. To summarize the role of PRMT5 in OL development, there is no doubt that PRMT5 does not affect OPC proliferation. However, at a specific timing during postnatal stages, PRMT5 methylation of histone or PDGFRα leads to an epigenetic program for driving proper OL progression for normal myelin development.

2.4 Astrocyte

2.4.1 Astrocyte and their Fevelopment

Astrocytes are one of the glial cell types that exceed neuronal number fivefold (Sofroniew and Vinters 2010; Lawrence et al. 2023). Astrocytes play a wide variety of critical roles in supporting neurons and brain homeostasis. They are also part of the blood-brain barrier (BBB), which is a protective cellular layer that wraps blood vessels in the CNS with other cell types, including pericytes and endothelial cells. During brain development, astrocytes support synapse formation. In addition, for proper neurotransmission, cellular processes extended by astrocytes form a complex with pre- and post-synaptic neurons. Astrocytes also support neurotransmission by taking up excess neurotransmitters such as glutamate from the synaptic space for degradation.

Glial fibrillary acidic protein (GFAP), a cytoskeletal intermediate filament protein, has been widely used to label astrocytes in the CNS. Upon various types of damage, astrocytes change their activity to become reactive astrocytes, which are highly proliferative and form a glial scar. This process is called astrogliosis and GFAP expression is elevated in this state. It is of note that GFAP is also expressed by NSCs from embryonic to adult stages. Furthermore, astrocytes are classified into several types, such as fibrous astrocytes and protoplasmic astrocytes, which have different gene expression patterns. Thus, astrocytes are identified as different types of markers, such as glutamine synthetase (GS) or ALDH1L1, depending on the stage of development, brain insult, or brain regions.

2.4.2 Role of PRMTs in Astrocyte Generation and Its Homeostasis

PRMT1 has been shown to be important for astrocyte differentiation from NSCs (Honda et al. 2017). Knockdown of *Prmt1* resulted in reduced level of differentiation to astroglial lineage, while neurons were generated normally. Using luciferase reporter assay for *Gfap* gene promotor, transfection of PRMT1 or its co-transfection with a transcription factor signal transducer and activator of transcription 3 (STAT3) stimulated promoter activity, suggesting that PRMT1 regulates *Gfap* mRNA expression. Biochemical analysis also suggested that STAT3 direct methylation by PRMT1 increased the STAT3 potential of *Gfap* gene expression induction.

Selvi et al. have extracted a natural compound TBBD from pomegranate (*Punica gratum*) (Selvi et al. 2010). The biochemical study revealed that TBBD has activity in repressing CARM1 activity on histone H3R17me2as methylation. The authors tested the biological function of TBBD on neuronal lineage progression using embryonic stem cells and zebrafish embryos (Selvi et al. 2014). The TBBD-treated zebrafish embryos exhibited reduced astroglial lineage progression as shown by GFAP reduction.

These studies highlight the importance of type I PRMTs and astrocyte generation during development. However, there is less evidence on type II PRMTs on the physiological regulation of astrocytes or the involvement of PRMTs in astrogliosis. PRMT1-nKO mice described in the previous section showed massive astrogliosis as well as microglial activation in the developing cerebral cortex (Hashimoto et al. 2021). At the same time, the mutant mice showed increased gene expression of inflammatory cytokines and chemokines. In these mutant mice, PRMT1 is depleted from NSCs and all the cell types, including astrocytes and neurons; therefore, it is still unclear how PRMT1 normally represses the inflammatory state cell autonomously in astrocytes. These results are in stark contrast to the above-mentioned defective astrocytic differentiation capacity of PRMT1 knockdown NSCs (Honda et al. 2017). This discrepancy could be explained by the difference between in vitro and in vivo systems.

References

Altman J, Das GD (1965) Autoradiographic and histological evidence of postnatal hippocampal neurogenesis in rats. J Comp Neurol 124(3):319–335. https://doi.org/10.1002/cne.901240303

Amano G, Matsuzaki S, Mori Y, Miyoshi K, Han S, Shikada S, Takamura H, Yoshimura T, Katayama T (2020) SCYL1 arginine methylation by PRMT1 is essential for neurite outgrowth via Golgi morphogenesis. Mol Biol Cell 31(18):1963–1973. https://doi.org/10.1091/mbc.e20-02-0100

Bezzi M, Teo SX, Muller J, Mok WC, Sahu SK, Vardy LA, Bonday ZQ, Guccione E (2013) Regulation of constitutive and alternative splicing by PRMT5 reveals a role for Mdm4 pre-mRNA in sensing defects in the spliceosomal machinery. Genes Dev 27(17):1903–1916. https://doi.org/10.1101/gad.219899.113

Blanc RS, Richard S (2017) Arginine methylation: the coming of age. Mol Cell 65(1):8–24. https://doi.org/10.1016/j.molcel.2016.11.003

Blanc RS, Vogel G, Li X, Yu Z, Li S, Richard S (2017) Arginine methylation by PRMT1 regulates muscle stem cell fate. Mol Cell Biol 37(3). https://doi.org/10.1128/mcb.00457-16

Bond AM, Ming G-L, Song H (2015) Adult mammalian neural stem cells and neurogenesis: five decades later. Cell Stem Cell 17(4):385–395. https://doi.org/10.1016/j.stem.2015.09.003

Calabretta S, Vogel G, Yu Z, Choquet K, Darbelli L, Nicholson TB, Kleinman CL, Richard S (2018) Loss of PRMT5 promotes PDGFRα degradation during oligodendrocyte differentiation and myelination. Dev Cell 46(4):426–440.e5. https://doi.org/10.1016/j.devcel.2018.06.025

Chen L, Zhang M, Fang L, Yang X, Cao N, Xu L, Shi L, Cao Y (2021) Coordinated regulation of the ribosome and proteasome by PRMT1 in the maintenance of neural stemness in cancer cells and neural stem cells. J Biol Chem 297(5):101275. https://doi.org/10.1016/j.jbc.2021.101275

Chittka A (2010) Dynamic distribution of histone H4 arginine 3 methylation marks in the developing murine cortex. PLoS One 5(11):e13807. https://doi.org/10.1371/journal.pone.0013807

Cimato TR, Tang J, Xu Y, Guarnaccia C, Herschman HR, Pongor S, Aletta JM (2002) Nerve growth factor-mediated increases in protein methylation occur predominantly at type I arginine methylation sites and involve protein arginine methyltransferase 1. J Neurosci Res 67(4):435–442. https://doi.org/10.1002/jnr.10123

Dhar S, Vemulapalli V, Patananan AN, Huang GL, Di Lorenzo A, Richard S, Comb MJ, Guo A, Clarke SG, Bedford MT (2013) Loss of the major type I arginine methyltransferase PRMT1 causes substrate scavenging by other PRMTs. Sci Rep 3(1). https://doi.org/10.1038/srep01311

Dormann D, Madl T, Valori CF, Bentmann E, Tahirovic S, Abou-Ajram C, Kremmer E, Ansorge O, Mackenzie IRA, Neumann M, Haass C (2012) Arginine methylation next to the PY-NLS modulates Transportin binding and nuclear import of FUS. EMBO J 31(22):4258–4275. https://doi.org/10.1038/emboj.2012.261

Fuhrmann J, Clancy KW, Thompson PR (2015) Chemical biology of protein arginine modifications in epigenetic regulation. Chem Rev 115(11):5413–5461. https://doi.org/10.1021/acs.chemrev.5b00003

Guccione E, Richard S (2019) The regulation, functions and clinical relevance of arginine methylation. Nat Rev Mol Cell Biol 20(10):642–657. https://doi.org/10.1038/s41580-019-0155-x

Guo A, Gu H, Zhou J, Mulhern D, Wang Y, Lee KA, Yang V, Aguiar M, Kornhauser J, Jia X, Ren J, Beausoleil SA, Silva JC, Vemulapalli V, Bedford MT, Comb MJ (2014) Immunoaffinity enrichment and mass spectrometry analysis of protein methylation. Mol Cell Proteomics 13(1):372–387. https://doi.org/10.1074/mcp.o113.027870

Hashimoto M, Murata K, Ishida J, Kanou A, Kasuya Y, Fukamizu A (2016) Severe hypomyelination and developmental defects are caused in mice lacking protein arginine methyltransferase 1 (PRMT1) in the central nervous system. J Biol Chem 291(5):2237–2245. https://doi.org/10.1074/jbc.m115.684514

Hashimoto M, Kumabe A, Kim J, Murata K, Sekizar S, Williams A, Lu W, Ishida J, Nakagawa T, Endo M, Minami Y, Fukamizu A (2021) Loss of PRMT1 in the central nervous system (CNS) induces reactive astrocytes and microglia during postnatal brain development. J Neurochem 156(6):834–847. https://doi.org/10.1111/jnc.15149

Hatanaka Y, Tsusaka T, Shimizu N, Morita K, Suzuki T, Machida S, Satoh M, Honda A, Hirose M, Kamimura S, Ogonuki N, Nakamura T, Inoue K, Hosoi Y, Dohmae N, Nakano T, Kurumizaka H, Matsumoto K, Shinkai Y, Ogura A (2017) Histone H3 methylated at arginine 17 is essential for reprogramming the paternal genome in zygotes. Cell Rep 20(12):2756–2765. https://doi.org/10.1016/j.celrep.2017.08.088

Hippenmeyer S (2023) Principles of neural stem cell lineage progression: insights from developing cerebral cortex. Curr Opin Neurobiol 79:102695. https://doi.org/10.1016/j.conb.2023.102695

Honda M, Nakashima K, Katada S (2017) PRMT1 regulates astrocytic differentiation of embryonic neural stem/precursor cells. J Neurochem 142(6):901–907. https://doi.org/10.1111/jnc.14123

Hou W, Nemitz S, Schopper S, Nielsen ML, Kessels MM, Qualmann B (2018) Arginine methylation by PRMT2 controls the functions of the Actin Nucleator COBL. Dev Cell 45(2):262–275.e8. https://doi.org/10.1016/j.devcel.2018.03.007

Huang J, Vogel G, Yu Z, Almazan G, Richard S (2011) Type II arginine methyltransferase PRMT5 regulates gene expression of inhibitors of differentiation/DNA binding Id2 and Id4 during glial cell differentiation. J Biol Chem 286(52):44424–44432. https://doi.org/10.1074/jbc.m111.277046

Ishii A, Matsuba Y, Mihira N, Kamano N, Saito T, Muramatsu S-I, Yokosuka M, Saido TC, Hashimoto S (2022) Tau-binding protein PRMT8 facilitates vacuole degeneration in the brain. J Biochem 172(4):233–243. https://doi.org/10.1093/jb/mvac058

Ishino Y, Shimizu S, Tohyama M, Miyata S (2022) Coactivator-associated arginine methyltransferase 1 controls oligodendrocyte differentiation in the corpus callosum during early brain development. Dev Neurobiol 82(3):245–260. https://doi.org/10.1002/dneu.22871

Kempermann G, Gage FH (1999) New nerve cells for the adult brain. Sci Am 280(5):48–53. https://doi.org/10.1038/scientificamerican0599-48

Kim J-D, Park K-E, Ishida J, Kako K, Hamada J, Kani S, Takeuchi M, Namiki K, Fukui H, Fukuhara S, Hibi M, Kobayashi M, Kanaho Y, Kasuya Y, Mochizuki N, Fukamizu A (2015) PRMT8 as a phospholipase regulates Purkinje cell dendritic arborization and motor coordination. Sci Adv 1(11). https://doi.org/10.1126/sciadv.1500615

Kim H-J, Jeong M-H, Kim K-R, Jung C-Y, Lee S-Y, Kim H, Koh J, Vuong TA, Jung S, Yang H, Park S-K, Choi D, Kim SH, Kang K, Sohn J-W, Park JM, Jeon D, Koo S-H, Ho W-K, Kang J-S, Kim S-T, Cho H (2016) Protein arginine methylation facilitates KCNQ channel-PIP2 interaction leading to seizure suppression. elife 5. https://doi.org/10.7554/elife.17159

Kousaka A, Mori Y, Koyama Y, Taneda T, Miyata S, Tohyama M (2009) The distribution and characterization of endogenous protein arginine N-methyltransferase 8 in mouse CNS. Neuroscience 163(4):1146–1157. https://doi.org/10.1016/j.neuroscience.2009.06.061

Lawrence JM, Schardien K, Wigdahl B, Nonnemacher MR (2023) Roles of neuropathology-associated reactive astrocytes: a systematic review. Acta Neuropathologica Communications 11(1). https://doi.org/10.1186/s40478-023-01526-9

Lee J, Sayegh J, Daniel J, Clarke S, Bedford MT (2005) PRMT8, a new membrane-bound tissue-specific member of the protein arginine methyltransferase family. J Biol Chem 280(38):32890–32896. https://doi.org/10.1074/jbc.m506944200

Lee PKM, Goh WWB, Sng JCG (2017) Network-based characterization of the synaptic proteome reveals that removal of epigenetic regulator Prmt8 restricts proteins associated with synaptic maturation. J Neurochem 140(4):613–628. https://doi.org/10.1111/jnc.13921

Liu F, Zhao X, Perna F, Wang L, Koppikar P, Abdel-Wahab O, Harr MW, Levine RL, Xu H, Tefferi A, Deblasio A, Hatlen M, Menendez S, Nimer SD (2011) JAK2V617F-mediated phosphorylation of PRMT5 downregulates its methyltransferase activity and promotes myeloproliferation. Cancer Cell 19(2):283–294. https://doi.org/10.1016/j.ccr.2010.12.020

Mallapaty S (2025) Paralysed man stands again after receiving 'reprogrammed' stem cells. Nature. https://doi.org/10.1038/d41586-025-00863-0

Maron MI, Lehman SM, Gayatri S, DeAngelo JD, Hegde S, Lorton BM, Sun Y, Bai DL, Sidoli S, Gupta V, Marunde MR, Bone JR, Sun Z-W, Bedford MT, Shabanowitz J, Chen H, Hunt DF, Shechter D (2021) Independent transcriptomic and proteomic regulation by type I and II protein arginine methyltransferases. iScience 24(9):102971. https://doi.org/10.1016/j.isci.2021.102971

Massignani E, Giambruno R, Maniaci M, Nicosia L, Yadav A, Cuomo A, Raimondi F, Bonaldi T (2022) ProMetheusDB: an in-depth analysis of the high-quality human methyl-proteome. Mol Cell Proteomics 21(7):100243. https://doi.org/10.1016/j.mcpro.2022.100243

Minatohara K, Akiyoshi M, Okuno H (2016) Role of immediate-early genes in synaptic plasticity and neuronal ensembles underlying the memory trace. Front Mol Neurosci 8. https://doi.org/10.3389/fnmol.2015.00078

Nakazawa H, Sada T, Toriyama M, Tago K, Sugiura T, Fukuda M, Inagaki N (2012) RAB33A mediates anterograde vesicular transport for membrane exocytosis and axon outgrowth. J Neurosci 32(37):12712–12725. https://doi.org/10.1523/jneurosci.0989-12.2012

Niu H, Xiao J, Ma Z, Chen L (2020) Prmt4-mediated methylation of NF-κB is critical for neural differentiation of embryonic stem cells. Biochem Biophys Res Commun 525(1):27–32. https://doi.org/10.1016/j.bbrc.2020.02.072

Pan Y, Suga A, Kimura I, Kimura C, Minegishi Y, Nakayama M, Yoshitake K, Iejima D, Minematsu N, Yamamoto M, Mabuchi F, Takamoto M, Shiga Y, Araie M, Kashiwagi K, Aihara M, Nakazawa T, Iwata T (2022) METTL23 mutation alters histone H3R17 methylation in normal-tension glaucoma. J Clin Invest 132(21). https://doi.org/10.1172/jci153589

Pawlak MR, Scherer CA, Chen J, Roshon MJ, Ruley HE (2000) Arginine N-methyltransferase 1 is required for early postimplantation mouse development, but cells deficient in the enzyme are viable. Mol Cell Biol 20(13):4859–4869. https://doi.org/10.1128/mcb.20.13.4859-4869.2000

Rhein VF, Carroll J, Ding S, Fearnley IM, Walker JE (2013) NDUFAF7 methylates arginine 85 in the NDUFS2 subunit of human complex I. J Biol Chem 288(46):33016–33026. https://doi.org/10.1074/jbc.m113.518803

Roux KJ, Kim DI, Raida M, Burke B (2012) A promiscuous biotin ligase fusion protein identifies proximal and interacting proteins in mammalian cells. J Cell Biol 196(6):801–810. https://doi.org/10.1083/jcb.201112098

Ryan TA, Li L, Chin LS, Greengard P, Smith SJ (1996) Synaptic vesicle recycling in synapsin I knock-out mice. J Cell Biol 134(5):1219–1227. https://doi.org/10.1083/jcb.134.5.1219

Ryan VH, Dignon GL, Zerze GH, Chabata CV, Silva R, Conicella AE, Amaya J, Burke KA, Mittal J, Fawzi NL (2018) Mechanistic view of HNRNPA2 low-complexity domain structure, interactions, and phase separation altered by mutation and arginine methylation. Mol Cell 69(3):465–479.e7. https://doi.org/10.1016/j.molcel.2017.12.022

Sakamaki J-I, Daitoku H, Ueno K, Hagiwara A, Yamagata K, Fukamizu A (2011) Arginine methylation of BCL-2 antagonist of cell death (BAD) counteracts its phosphorylation and inactivation by Akt. Proc Natl Acad Sci 108(15):6085–6090. https://doi.org/10.1073/pnas.1015328108

Sato D, Lionel AC, Leblond CS, Prasad A, Pinto D, Walker S, O'Connor I, Russell C, Drmic IE, Hamdan FF, Michaud JL, Endris V, Roeth R, Delorme R, Huguet G, Leboyer M, Rastam M, Gillberg C, Lathrop M, Stavropoulos DJ, Anagnostou E, Weksberg R, Fombonne E, Zwaigenbaum L, Fernandez BA, Roberts W, Rappold GA, Marshall CR, Bourgeron T, Szatmari P, Scherer SW (2012) SHANK1 deletions in males with autism spectrum disorder. Am J Hum Genet 90(5):879–887. https://doi.org/10.1016/j.ajhg.2012.03.017

Scaglione A, Patzig J, Liang J, Frawley R, Bok J, Mela A, Yattah C, Zhang J, Teo SX, Zhou T, Chen S, Bernstein E, Canoll P, Guccione E, Casaccia P (2018) PRMT5-mediated regulation of developmental myelination. Nat Commun 9(1). https://doi.org/10.1038/s41467-018-04863-9

Scaramuzzino C, Monaghan J, Milioto C, Lanson NA, Maltare A, Aggarwal T, Casci I, Fackelmayer FO, Pennuto M, Pandey UB (2013) Protein arginine methyltransferase 1 and 8 interact with FUS to modify its sub-cellular distribution and toxicity in vitro and in vivo. PLoS One 8(4):e61576. https://doi.org/10.1371/journal.pone.0061576

Selvi BR, Batta K, Kishore AH, Mantelingu K, Varier RA, Balasubramanyam K, Pradhan SK, Dasgupta D, Sriram S, Agrawal S, Kundu TK (2010) Identification of a novel inhibitor of coactivator-associated arginine methyltransferase 1 (CARM1)-mediated methylation of Histone H3 ARG-17. J Biol Chem 285(10):7143–7152. https://doi.org/10.1074/jbc.m109.063933

Selvi BR, Swaminathan A, Maheshwari U, Nagabhushana A, Mishra RK, Kundu TK (2014) CARM1 regulates astroglial lineage through transcriptional regulation of Nanog and posttranscriptional regulation by miR92a. Mol Biol Cell 26(2):316–326. https://doi.org/10.1091/mbc.e14-01-0019

Shashi V, Xie P, Schoch K, Goldstein DB, Howard TD, Berry MN, Schwartz CE, Cronin K, Sliwa S, Allen A, Need AC (2014) The RBMX gene as a candidate for the Shashi X-linked intellectual disability syndrome. Clin Genet 88(4):386–390. https://doi.org/10.1111/cge.12511

Shen L, Ma X, Wang Y, Wang Z, Zhang Y, Pham HQH, Tao X, Cui Y, Wei J, Lin D, Abeywanada T, Hardikar S, Halabelian L, Smith N, Chen T, Barsyte-Lovejoy D, Qiu S, Xing Y, Yang Y (2024) Loss-of-function mutation in PRMT9 causes abnormal synapse development by dysregulation of RNA alternative splicing. Nat Commun 15(1). https://doi.org/10.1038/s41467-024-47107-9

Simandi Z, Czipa E, Horvath A, Koszeghy A, Bordas C, Póliska S, Juhász I, Imre L, Szabó G, Dezso B, Barta E, Sauer S, Karolyi K, Kovacs I, Hutóczki G, Bognár L, Klekner Á, Szucs P, Bálint BL, Nagy L (2015) PRMT1 and PRMT8 regulate retinoic acid-dependent neuronal differentiation with implications to neuropathology. Stem Cells 33(3):726–741. https://doi.org/10.1002/stem.1894

Simandi Z, Pajer K, Karolyi K, Sieler T, Jiang L-L, Kolostyak Z, Sari Z, Fekecs Z, Pap A, Patsalos A, Contreras GA, Reho B, Papp Z, Guo X, Horvath A, Kiss G, Keresztessy Z, Vámosi G, Hickman J, Xu H, Dormann D, Hortobagyi T, Antal M, Nógrádi A, Nagy L (2018) Arginine methyltransferase PRMT8 provides cellular stress tolerance in aging motoneurons. J Neurosci 38(35):7683–7700. https://doi.org/10.1523/jneurosci.3389-17.2018

So H-K, Kim H, Lee J, You C-L, Yun C-E, Jeong H-J, Jin E-J, Jo Y, Ryu D, Bae G-U, Kang J-S (2023) Protein arginine methyltransferase 1 ablation in motor neurons causes mitochondrial dysfunction leading to age-related motor neuron degeneration with muscle loss. Research 6. https://doi.org/10.34133/research.0158

Sofroniew MV, Vinters HV (2010) Astrocytes: biology and pathology. Acta Neuropathol 119(1):7–35. https://doi.org/10.1007/s00401-009-0619-8

Stopa N, Krebs JE, Shechter D (2015) The PRMT5 arginine methyltransferase: many roles in development, cancer and beyond. Cell Mol Life Sci 72(11):2041–2059. https://doi.org/10.1007/s00018-015-1847-9

Tee W-W, Pardo M, Theunissen TW, Yu L, Choudhary JS, Hajkova P, Surani MA (2010) Prmt5 is essential for early mouse development and acts in the cytoplasm to maintain ES cell pluripotency. Genes Dev 24(24):2772–2777. https://doi.org/10.1101/gad.606110

Tradewell ML, Yu Z, Tibshirani M, Boulanger M-C, Durham HD, Richard S (2012) Arginine methylation by PRMT1 regulates nuclear-cytoplasmic localization and toxicity of FUS/TLS harbouring ALS-linked mutations. Hum Mol Genet 21(1):136–149. https://doi.org/10.1093/hmg/ddr448

Tsang B, Arsenault J, Vernon RM, Lin H, Sonenberg N, Wang L-Y, Bah A, Forman-Kay JD (2019) Phosphoregulated FMRP phase separation models activity-dependent translation through bidirectional control of mRNA granule formation. Proc Natl Acad Sci 116(10):4218–4227. https://doi.org/10.1073/pnas.1814385116

Wang Y, Bedford MT (2023) Effectors and effects of arginine methylation. Biochem Soc Trans 51(2):725–734. https://doi.org/10.1042/bst20221147

Wang F-W, Hao H-B, Zhao S-D, Zhang Y-M, Liu Q, Liu H-J, Liu S-M, Yuan Q-H, Bing L-J, Ling E-A, Hao A-J (2011) Roles of activated astrocyte in neural stem cell proliferation and differentiation. Stem Cell Res 7(1):41–53. https://doi.org/10.1016/j.scr.2011.03.004

Wei H-H, Fan X-J, Hu Y, Tian X-X, Guo M, Mao M-W, Fang Z-Y, Wu P, Gao S-X, Peng C, Yang Y, Wang Z (2021) A systematic survey of PRMT interactomes reveals the key roles of arginine methylation in the global control of RNA splicing and translation. Sci Bulletin 66(13):1342–1357. https://doi.org/10.1016/j.scib.2021.01.004

Wu Q, Schapira M, Arrowsmith CH, Barsyte-Lovejoy D (2021) Protein arginine methylation: from enigmatic functions to therapeutic targeting. Nat Rev Drug Discov 20(7):509–530. https://doi.org/10.1038/s41573-021-00159-8

Yadav N, Lee J, Kim J, Shen J, Hu MCT, Aldaz CM, Bedford MT (2003) Specific protein methylation defects and gene expression perturbations in coactivator-associated arginine methyltransferase 1-deficient mice. Proc Natl Acad Sci U S A 100(11):6464–6468. https://doi.org/10.1073/pnas.1232272100

Yamagata K, Daitoku H, Takahashi Y, Namiki K, Hisatake K, Kako K, Mukai H, Kasuya Y, Fukamizu A (2008) Arginine methylation of FOXO transcription factors inhibits their phosphorylation by AKT. Mol Cell 32(2):221–231. https://doi.org/10.1016/j.molcel.2008.09.013

Yamaguchi A, Kitajo K (2012) The effect of PRMT1-mediated arginine methylation on the subcellular localization, stress granules, and detergent-insoluble aggregates of FUS/TLS. PLoS One 7(11):e49267. https://doi.org/10.1371/journal.pone.0049267

Yang Y, Hadjikyriacou A, Xia Z, Gayatri S, Kim D, Zurita-Lopez C, Kelly R, Guo A, Li W, Clarke SG, Bedford MT (2015) PRMT9 is a Type II methyltransferase that methylates the splicing factor SAP145. Nat Commun 6(1). https://doi.org/10.1038/ncomms7428

Zhang X, Cheng X (2003) Structure of the predominant protein arginine methyltransferase PRMT1 and analysis of its binding to substrate peptides. Structure 11(5):509–520. https://doi.org/10.1016/s0969-2126(03)00071-6

Zhang Y, Chen K, Sloan SA, Bennett ML, Scholze AR, O'Keeffe S, Phatnani HP, Guarnieri P, Caneda C, Ruderisch N, Deng S, Liddelow SA, Zhang C, Daneman R, Maniatis T, Barres BA, Wu JQ (2014) An RNA-sequencing transcriptome and splicing database of glia, neurons, and vascular cells of the cerebral cortex. J Neurosci 34(36):11929–11947. https://doi.org/10.1523/jneurosci.1860-14.2014

Zhao Q, Rank G, Tan YT, Li H, Moritz RL, Simpson RJ, Cerruti L, Curtis DJ, Patel DJ, Allis CD, Cunningham JM, Jane SM (2009) PRMT5-mediated methylation of histone H4R3 recruits DNMT3A, coupling histone and DNA methylation in gene silencing. Nat Struct Mol Biol 16(3):304–311. https://doi.org/10.1038/nsmb.1568

Plant-Derived Compounds as Modulators of the Genetic and Epigenetic Landscape of Polycystic Ovary Syndrome

Neervana Rambaran and Md. Shahidul Islam

Abstract Polycystic ovary syndrome (PCOS) is a complex endocrine disorder in women of reproductive age. It is characterized by hyperandrogenism, menstrual disorders, and polycystic ovarian morphology. Insulin resistance (IR) is commonly associated with PCOS, and patients are at heightened risk for developing metabolic and endocrine comorbidities such as type 2 diabetes mellitus, cardiovascular disease, and endometrial cancer. Globally, PCOS has intensified rapidly, making it a major public health issue with a significant economic burden on the healthcare system. The heterogeneous nature of the syndrome has produced considerable scientific debate, and the current treatment strategies are often unsatisfactory. The application of omics research in PCOS has contributed to a greater understanding and identification of novel biomarkers and targets. To provide a focused, mechanistic, and in-depth discussion, this chapter focuses specifically on the genetic expression and epigenetic mechanisms implicated in PCOS progression and advancement. Special insight has been given into how plant-derived compounds may modulate these molecular pathways to offer innovative therapeutic interventions.

Keywords Hyperandrogenism · Phytochemicals · Insulin resistance · Ovary · Folliculogenesis

Abbreviations

3β-HSD	3β-hydroxysteroid dehydrogenase
17β-HSD	17β hydroxysteroid dehydrogenase
AdipoR1	Adiponectin receptor 1

N. Rambaran · M. S. Islam (✉)
Discipline of Biochemistry, School of Agriculture and Science, University of KwaZulu-Natal, Durban, South Africa
e-mail: islamd@ukzn.ac.za

A. H. M. N. Nabi, H. Uddin Shekhar (eds.), *Biotechnological Advances in Healthomics*, Interdisciplinary Biotechnological Advances,
https://doi.org/10.1007/978-981-95-3612-2_8

AKT	Protein kinase B
AMPK	Activated protein kinase
AMH	Anti-Müllerian hormone
AMHR	Anti-Müllerian hormone receptor 2
AR	Androgen receptor
Arg	Arginine
ATPase	Adenosine triphosphatase
CMIP	c-maf-inducing protein
CRH	Corticotropin-releasing hormone
CYP11A1	Cytochrome P450 11A1
CYP17A1	Cytochrome P450 17A1
CYP19A1	Cytochrome P450 19A1 (aromatase)
CYP21	Cytochrome P450 21A2
CpG	Cytosine-phosphate-guanine dinucleotide
DENND1A	DENN domain-containing protein 1A
DHEA	Dehydroepiandrosterone
DHT	5α-dihydrotestosterone
DMSO	Dimethyl sulfoxide
DNA	Deoxyribonucleic acid
FSH	Follicle-stimulating hormone
FSHR	FSH receptor
GATA1	GATA-binding factor 1
Gly	Glycine
GnRH	Gonadotropin-releasing hormone
GPx	Glutathione peroxidase
GSH	Glutathione (reduced form, containing a sulfhydryl group)
GWAS	Genome-wide association studies
HA	Hyperandrogenism
HFD	High-fat diet
hGCs	Human granulosa cells
HMW	High-molecular-weight adiponectin
HPO	Hypothalamus-pituitary-ovary
IL	Interleukin
IP	Intraperitoneally
IR	Insulin resistance
IRS1	Insulin receptor substrate 1
ITT	Insulin tolerance test
KITL	KIT ligand
LH	Luteinizing hormone
LHCGR	Luteinizing hormone/human chorionic gonadotrophin receptor
LPS	Lipopolysaccharide
MDA	Malondialdehyde
miRNA	microRNA
mRNA	Messenger RNA
ncRNA	Non-coding RNA

PAI-1	Plasminogen activator inhibitor-1
PCK1	Phosphoenolpyruvate carboxykinase
PCOM	Polycystic ovary morphology
PCOS	Polycystic ovary syndrome
PCSK9	Proprotein convertase subtilisin/kexin type 9
PGC-1α	Peroxisome proliferator-activated receptor gamma coactivator-1 alpha
PK	Pharmacokinetics
PPAR-γ	Peroxisome proliferator-activated receptor gamma
Pro12Ala	Proline-to-alanine amino acid substitution
PTGS2	Prostaglandin-endoperoxide synthase 2
RCT	Randomized controlled trials
RNA	Ribonucleic acid
SHBG	Sex hormone-binding globulin
SIRT	Sirtuin or silent information regulator
SNPs	Single-nucleotide polymorphisms
StAR	Steroidogenic acute regulatory protein
SOD	Superoxide dismutase
T2DM	Type 2 diabetes mellitus
TBARS	Thiobarbituric acid reactive substances
TCF7L2	Transcription factor 7-like 2
TLR2	Toll-like receptor 2
TNF-α	Tumor necrosis factor alpha
VEGF	Vascular endothelial growth factor

1 Introduction

Polycystic ovary syndrome (PCOS) is a common gynecological endocrine problem affecting 15–20% of reproductive-aged women (Moosa et al. 2022). The incidence of PCOS has increased significantly, contributing to a substantial reproductive and financial burden on countries and individuals (Liu et al. 2023). Based on the Rotterdam criteria (Rotterdam ESHRE/ASRM-Sponsored PCOS Consensus Workshop Group 2004), PCOS is diagnosed when a patient presents with at least two of the following features: hyperandrogenism, chronic anovulation (oligomenorrhea or amenorrhea), and polycystic ovary morphology (PCOM) (Lentscher and Decherney 2020). Additionally, the majority of patients with PCOS often present with insulin resistance (IR), obesity, increased risk of type 2 diabetes mellitus (T2DM), hypertension, dyslipidemia, cardiovascular disease, and endometrial cancer (Rawat et al. 2022). The disturbance of the hypothalamus-pituitary-ovary (HPO) axis plays a central role in the pathophysiology of PCOS (Haghighat and Mahmoudi 2024). Abnormally high frequency pulsatility from the gonadotropin-releasing hormone (GnRH) pulse generator triggers the hypersecretion of luteinizing hormone (LH), in turn, downregulating the secretion of follicle-stimulating hormone (FSH) (Stener-Victorin 2020). The disproportionally high LH to FSH ratio provokes the

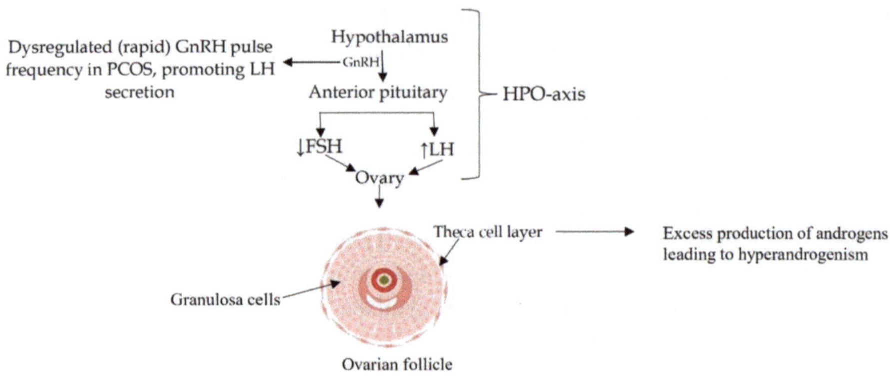

Fig. 1 Neuroendocrine dysregulation of the hypothalamus-pituitary-ovary (HPO) axis in polycystic ovary syndrome (PCOS) results in the gonadotropin-releasing hormone (GnRH) pulse generator producing increased rapid pulses, leading to excess output of luteinizing hormone (LH) that triggers the excess production of androgens

production of excess ovarian androgen, leading to hyperandrogenism that interferes with folliculogenesis and the ovulatory process (Chen et al. 2024) (See Fig. 1). Coincidentally, IR and hyperinsulinemia can impact the neuroendocrine circuit by aggravating the LH production that fuels hyperandrogenism and the arrhythmic reproductive cycle (Fraser et al. 2021).

Conventionally, PCOS has been identified via clinical and biochemical criteria; however, advancements in molecular biology have enhanced the understanding of the pathogenesis of PCOS (Chen et al. 2024). The emergence of health omics encompassing genomics, epigenomics, transcriptomics, proteomics, and metabolomics has enabled the comprehensive profiling of biological pathways associated with the onset and development of the disease (Pei et al. 2021). Genomics and epigenomics have emerged as essential tools for elucidating the heritable and reversible landscape in regulating PCOS (Mimouni et al. 2021; Stener-Victorin 2020). Genetic studies on PCOS have advanced significantly in recent years, with the advent of genome-wide association studies (GWAS) in identifying key susceptibility loci that gives insight into its intricate pathophysiology (Pei et al. 2021). Based on these findings, various single-nucleotide polymorphisms (SNPs) have been linked with heightened PCOS risk, indicating that minor genetic changes can influence the hormonal, metabolic, and reproductive features synonymous with the disorder (Pei et al. 2021).

Concurrently, accumulating evidence highlights the role of epigenetic changes and developmental programming, especially those triggered by hormonal imbalances in the uterine environment, in developing PCOS (Heidarzadehpilehrood et al. 2022). Animal models investigating the induction of PCOS-like traits from prenatal androgen exposure provide evidence that epigenetic modifications, like DNA methylation, can influence susceptibility to PCOS across generations (Dumesic et al. 2020). These epigenetic changes may be established early in life or even in utero and can be amplified by environmental factors such as poor diet, sedentary lifestyle, and exposure to endocrine disruptors (Korolenko and Skinner 2024).

An emerging field of study aims to explore bioactive phytocompounds capable of correcting gene expression abnormalities and modifying epigenetic profiles in PCOS (Muthuswamy et al. 2022; Yuan et al. 2022). Delving into how the compounds can influence genetic expression, together with epigenetic plasticity, will be critical in tailoring personalized therapeutic approaches (Luo et al. 2022).

2 Genetic Architecture of PCOS

Polycystic ovary syndrome has a polygenic nature with a strong genetic component that is aggravated by environmental factors such as obesity (Hiam et al. 2019). The multifactorial manifestation of PCOS may suggest that an array of genes that are associated with hormone regulation, IR, and ovarian function contribute to the pathogenesis of the condition (Bhimwal et al. 2023).

Two popular and widely used approaches employed to study genetic variants include candidate gene association studies and GWAS (Zhao et al. 2016). Both methods offer valuable information for understanding the genetic fundamentals of PCOS, but use different strategies, offering a unique insight into the condition. Genome-wide association studies involve scanning the entire genome for associations between common genetic polymorphisms and the trait or disease in many individuals to elucidate the possible role of genetic variants in the pathophysiology of the disease (Hiam et al. 2019).

Unlike GWAS, candidate gene studies typically focus on investigating specific genes based on their biological or physiological relevance to the condition (Chaudhary et al. 2021). The advantage of merging GWAS and candidate gene studies is rooted in their complementary nature (Hiam et al. 2019). For example, studies identified DENN domain-containing protein 1A (*DENND1A*), a gene involved in gonadotropin regulation, as a potential risk locus for PCOS through GWAS (Hiam et al. 2019). Subsequent candidate gene studies were used to investigate the specific function of DENND1A in hyperandrogenism and ovarian dysfunction in PCOS patients. Kulkarni et al. (2019) found that a truncated splice variant (DENND1A.V2) was expressed at increased levels in PCOS theca cells, contributing to the PCOS phenotype by influencing cytochrome P450 11A1 (CYP11A1) and cytochrome P450 17A1 (CYP17A1) (Kulkarni et al. 2019) (See Fig. 2). The enzymes CYP11A1 and CYP17A1 play vital roles in steroidogenesis, whose dysregulation contributes to the androgen excess observed in PCOS by enhancing the biosynthesis of androgen precursors (Bhimwal et al. 2023).

There have been several reviews that have extensively examined candidate gene studies and GWAS concerning PCOS. Thus, a succinct overview has been provided below for context (Bhandary et al. 2022; Chaudhary et al. 2021; Heidarzadehpilehrood et al. 2022; Hiam et al. 2019; Pei et al. 2021; Zhao et al. 2016).

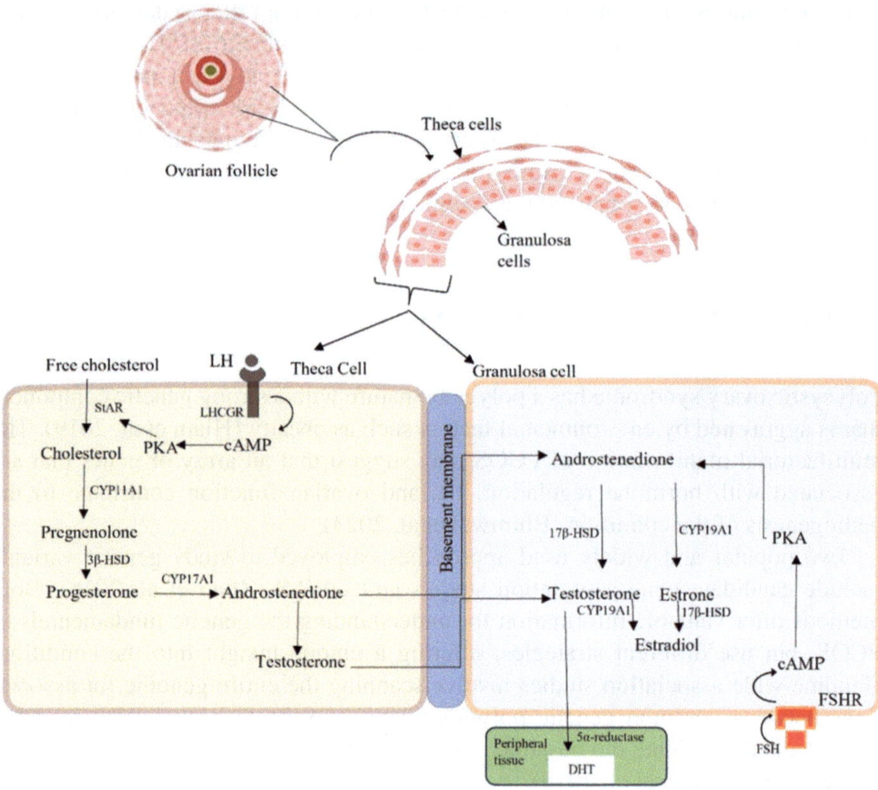

Fig. 2 Ovarian steroid biosynthesis under normal physiological conditions, where theca cell androgens are aromatized to estrogens in the granulosa cells. Abnormality with the cytochrome P450 enzymes, specifically CYP11A1, CYP17A1, and CYP19A1, is associated with increased risk of PCOS manifestation and progression

2.1 Candidate Gene Association Studies to Elucidate PCOS

The most significant genetic variation related to PCOS can be designated into three groups (Zhao et al. 2016):

- Androgens steroidogenesis
- Metabolism
- Inflammatory cytokines

These groups reflect the core pathophysiological mechanisms implicated in PCOS. To probe further, the subsequent Sects. 2.1.1, 2.1.2, and 2.1.3 discuss representative genes within each group, highlighting their roles and supporting evidence from association studies.

2.1.1 Genetic Variations Related to Androgen Steroidogenesis

Steroidogenesis is an enzymatic process by which steroid hormones such as glucocorticoids, mineralocorticoids, and sex hormones (androgens, estrogens, and progesterone) are synthesized from cholesterol, primarily in endocrine tissue, like the adrenal glands and gonads (Heidarzadehpilehrood et al. 2022). Disruptions in the steroidogenic process have been linked to the genetic predisposition of PCOS, predominantly flagged as hyperandrogenism (Wang et al. 2019a) (See Fig. 2).

A notable example is the genetic variations in the expression of proprotein convertase subtilisin/kexin type 9 (PCSK9), which usually plays a critical role in regulating low-density lipoprotein receptor levels (Wang et al. 2019a). Cholesterol is crucial for ovarian hormone production and follicle development (Wang et al. 2019a). Heightened expression of PCSK9 results in abnormal lipid metabolism affecting cholesterol levels, which has been linked to dysregulation of ovarian function and elevated androgen production (Bhandary et al. 2022; Wang et al. 2019a).

Mutations in key genes may influence steroidogenesis in PCOS. These include *CYP11A1*, *CYP17A1*, cytochrome P450 aromatase (*CYP19A1*), cytochrome P450 21A2 (*CYP21*), 3β-hydroxysteroid dehydrogenase (*3β-HSD*), 17β-hydroxysteroid dehydrogenase (*17β-HSD*), sex hormone-binding globulin (*SHBG*), androgen receptor (*AR*), steroidogenic acute regulatory protein (*StAR*), FSH receptor gene (*FSHR*), and luteinizing hormone/human chorionic gonadotrophin receptor (*LHCGR*) gene (Bhandary et al. 2022; Heidarzadehpilehrood et al. 2022).

Given its central role in regulating the bioavailability of circulating androgens, the significance of SHBG warrants attention (Bhandary et al. 2022). Studies have identified that specific polymorphisms in the *SHBG* gene, such as the (TAAAA)n repeat, notably the presence of eight or more TAAAA repeats, have been linked to lower SHBG concentrations in women with PCOS, potentially exacerbating hyperandrogenic symptoms like hirsutism and menstrual irregularities (Ferk et al. 2006). Furthermore, Mendelian randomization analyses have demonstrated that genetically higher SHBG levels are associated with a reduced risk of developing PCOS, suggesting a potential protective role of SHBG against the syndrome (Guo et al. 2024).

2.1.2 Genetic Variations Linked to Metabolism

PCOS is a metabolic disorder intrinsically linked with IR, obesity, T2D, and metabolic syndrome. Of significance is the missense polymorphism in the insulin receptor substrate 1 (*IRS1*) gene, Gly972Arg variant, where glycine (Gly) is replaced by arginine (Arg) at codon 972, which is widely associated with PCOS and clinical features like elevated LH, LH/FSH ratio and ovarian changes (Pei et al. 2021; Zhao et al. 2016). Interestingly, the nuclear receptor, peroxisome proliferator-activated receptor gamma (PPAR-γ), which influences lipid and glucose metabolism, showed that the variants, involving a proline-to-alanine substitution at codon 12, a Pro12Ala polymorphism, may have protective effects on insulin sensitivity (Pei et al. 2021).

Likewise, genes that influence obesity and T2D, such as fat and obesity-associated gene (*FTO*) and transcription factor 7-like 2 (*TCF7L2*), are also regarded as significant candidate genes for the condition (Li et al. 2022). Additionally, the phosphoenolpyruvate carboxykinase 1 (*PCK1*) gene, which encodes for PCK1, an enzyme involved in gluconeogenesis and the regulation of glucose metabolism, may influence insulin sensitivity (Bai et al. 2024). Genetic variants in *PCK1* are linked to the dysregulation of FSH and testosterone levels, making it a vital gene in understanding how metabolic dysregulation intersects with the hormonal irregularities seen in PCOS (Li et al. 2022).

2.1.3 Genetic Variations Linked to Inflammation

PCOS is known as a pro-inflammatory disease, with genes connected with chronic inflammation possibly contributing to its onset (Vasyukova et al. 2023). These genes may play an intermediate role in the interplay between hyperandrogenism, obesity, IR, and anovulation (Zhao et al. 2016). Consequently, genes related to inflammatory cytokines, interleukin 6 (*IL-6*), plasminogen activator inhibitor-1 (*PAI-1*), tumor necrosis factor-alpha (*TNFα*), C-reactive protein, *IL-1β*, *IL-17*, and *IL*-18, have emerged as potential candidate genes for PCOS (Bhimwal et al. 2023; Rudnicka et al. 2021b).

Furthermore, the altered expression of c-maf-inducing protein (CMIP), an immune system regulator, may contribute to the low-grade chronic inflammation characteristically seen in PCOS (Bannigida et al. 2018). Inflammation plays a prominent role in the progression of IR and metabolic abnormalities, suggesting that CMIP may contribute to the metabolic disturbances in PCOS, particularly regarding immune dysfunction and metabolic signalling pathways (Li et al. 2022).

2.2 Genome-Wide Association Studies in PCOS

Genome-wide association studies in populations of Chinese, Korean, and European ancestry have identified several genetic loci linked with PCOS (Hiam et al. 2019). As such, genes identified as PCOS risk include *FSHR*, *LHCGR*, insulin receptor, thyroid adenoma-associated, *DENND1A*, yes-associated protein 1, RAS oncogene family member RAB5B, FSH beta unit, Erb-B2 receptor tyrosine kinase 4, RAD50 double-strand break repair protein, and krr1 small subunit processome component homolog (Bhandary et al. 2022; Hiam et al. 2019). These genes are involved in several biological processes such as gonadotropin action, IR, and folliculogenesis (Bhandary et al. 2022). While several loci are replicated across different populations, further studies with larger sample sizes and in diverse populations are required to confirm these findings and explore novel PCOS susceptibility genes (Hiam et al. 2019).

3 Role of Epigenetics in PCOS

Epigenetics encompasses changes in gene expression without modifying the DNA sequence, but traits can be transgenerationally inherited (Korolenko and Skinner 2024). These changes include DNA methylation, histone modifications, and changes in non-coding RNA (ncRNA) content (Korolenko and Skinner 2024) (See Fig. 3).

Lifestyle and dietary factors play an integral role in the development of PCOS, which may further indicate the strong influence of epigenetic and environmental factors in the pathogenesis of PCOS (Bhingardeve et al. 2025; Szukiewicz et al. 2022). Recent evidence has highlighted the aspect of transgenerational inheritance, suggesting that epigenetic changes linked to PCOS may influence reproductive outcomes across generations (Stener-Victorin and Deng 2021). Maternal health status

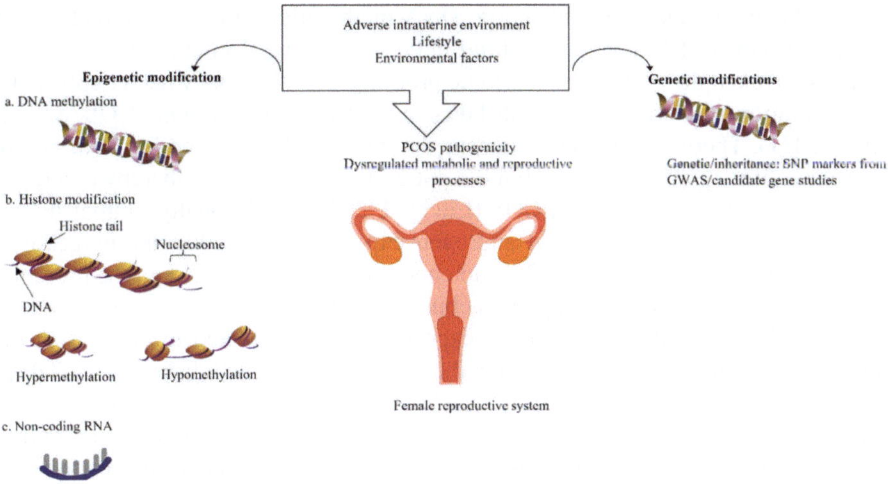

Fig. 3 Interaction of genetic and epigenetic modification in provoking the pathogenesis of polycystic ovary syndrome. Epigenetic changes via DNA methylation, histone modification and non-coding RNA can trigger metabolic and reproductive dysregulation characteristic of polycystic ovary syndrome. (a) DNA methylation involves the selective addition of a methyl group to cytosine residues within the CpG sites, via the enzyme methyltransferase. DNA methylation is associated with gene silencing and aberrations in methylation patterns that catalyze the disease process. (b) Histone modification involves changes to the 'histone code', resulting in structural alterations to the chromatin, thus influencing gene expression. Hypermethylation generally leads to a more condensed chromatin state, restricting access to transcriptional machinery, therefore repressing gene activity. In contrast, hypomethylation is linked with a relaxed chromatin configuration, facilitating the binding of transcription factors, thereby activating gene expression. Hypermethylation or hypomethylation of DNA and histones modifies gene expression in a site-specific manner, leading to dysregulated pathways in PCOS. (c) non-coding RNAs, which are a cluster of functional RNA molecules that do not encode for proteins but regulate epigenetic modifications by targeting mRNAs to degrade or inhibit their translation, as well as by recruiting chromatin-modifying enzymes to interact with specific DNA regions. Aberrant miRNA expression in ovarian granulosa cells affects a wide array of genes related to hormonal regulation, inflammation, cell cycle, and metabolism, all contributing to the multifactorial nature of PCOS

critically impacts the uterine environment and the epigenetic landscape of the developing fetus. Animal studies have shown that exposure to elevated androgen levels during pregnancy can induce PCOS-like traits in offspring, persisting up to the third generation (Stener-Victorin and Deng 2021). This transgenerational transmission indicates the long-lasting effects of an aberrant maternal-fetal environment (Stener-Victorin and Deng 2021).

Thus, expanding on the discussion of epigenetic mechanisms in PCOS, the subsequent section will elaborate on the interaction between epigenetic and hyperandrogenism, considering its critical role in PCOS pathophysiology.

3.1 Hyperandrogenism and PCOS

A significant epigenetic modification, DNA methylation, involves the addition of methyl groups to DNA, often resulting in gene silencing (Liu et al. 2022). Recent studies have employed DNA methylome profiling to investigate whether epigenetic dysregulation of ovarian genes contributes to ovarian dysfunction in PCOS (Salinas et al. 2021). There is growing consensus that abnormally elevated levels of androgen may directly affect the methylation patterns of many genes, thereby contributing to the pathophysiology of PCOS (Liu et al. 2022). Additionally, animal studies have demonstrated that prenatal androgen exposure can modify the epigenome, resulting in long-term reproductive and endocrine complications, i.e., the presentation of PCOS-like symptoms upon reaching adulthood.

Notably, premature androgenized exposure has been linked to changes in the DNA methylation of genes involved in the transforming growth factor beta (TGF-β) signalling pathway (Pei et al. 2021). The TGF-β pathway regulates various cellular functions in the process of folliculogenesis and ovarian development (Bhimwal et al. 2023). Disruptions in this pathway due to methylation may be associated with reproductive abnormalities observed in PCOS (Pei et al. 2021).

To establish a potential link between DNA methylation and gene expression in PCOS, Wang et al. (2014b) detected an association between variations in DNA methylation and mRNA expression in 54 genes, involving 7929 differentially methylated CpG sites. These findings suggest that DNA methylation changes might influence gene expression in PCOS. Importantly, this study showed that differentially methylated genes found in ovarian tissue from PCOS patients in comparison to control subjects were associated with hormone activity, glucose metabolism, insulin signalling, transcriptional regulation, and inflammation (Wang et al. 2014b).

A recent meta-analysis found that the anti-Müllerian hormone (*AMH*) gene was significantly hypomethylated, resulting in altered AMH levels in PCOS women compared to controls (Rawat et al. 2022). Elevated AMH levels in PCOS have been proposed to contribute to anovulation by reducing FSH sensitivity, resulting in the accumulation of small antral follicles and inhibiting dominant follicle growth, thereby disrupting folliculogenesis (Moolhuijsen et al. 2022). Furthermore, the increased AMH levels have been correlated with the severity of hyperandrogenism, indicating a possible role in the androgen excess characterizing this disorder (Dewailly et al. 2016). Slightly contentious, some studies have linked estradiol inhibition to increased AMH expression, suggesting that excess AMH is strongly suspected of playing a central role in the characteristic follicular arrest of PCOS, through the adverse action on aromatase expression and FSH action (Rudnicka et al. 2021a). Additionally, anti-Müllerian hormone receptor 2 (*AMHR*), which is essential for AMH function, was significantly hypomethylated, probably resulting in the increased activity of AMH (Rawat et al. 2022).

Another gene of crucial significance to PCOS hyperandrogenism is the *CYP19A1* gene that is responsible for encoding the steroidogenesis enzyme, aromatase, which catalyzes estrogen biosynthesis from androgens (Panghiyangani et al. 2020). *CYP19A*1 is hypermethylated, resulting in reduced expression and hyperandrogenism (Panghiyangani et al. 2020). This is further complicated by the hypomethylation of the *LHCGR* gene in PCOS, which encodes for the LHCG receptor, leading to LH overactivity (Wang et al. 2014a). Consequently, resulting in the hyperresponse of the LH-stimulated androgen production in the ovarian theca cells (Vázquez-Martínez et al. 2019; Wang et al. 2023).

These findings demonstrate the crucial role of epigenetic modifications in the pathogenesis of PCOS, indicating that hyperandrogenism may induce specific epigenetic changes that contribute to the complex phenotype of PCOS.

4 Role of Plant Compounds in Modulating Genetic and Epigenetic Expression

Plants and their phytochemicals are attracting growing attention as a therapeutic modality in PCOS (Muthuswamy et al. 2022). With fewer side effects and the likelihood of enhanced effectiveness, investigations exploring the role of plants and their phytoconstituents in genetic and epigenetic regulation are being implemented (Włodarczyk et al. 2024). Considering that epigenetic modifications are reversible, a growing spectrum of therapeutic interventions targeting epigenetic alterations are being developed and represents an exciting area of research (Włodarczyk et al. 2024). Phytochemicals such as polyphenols can modulate the expression of genes involved in oxidative stress, metabolism, inflammation, and cancer. By interacting with signalling pathways, plant compounds can potentially prevent or manage diseases linked to genetic and epigenetic alterations, offering a natural approach to support health at the molecular level. The subsequent section highlights key studies that illustrate the impact of plants and their compounds on genetic regulation and possible epigenetic alterations linked with PCOS.

4.1 *The Role of Plants and Their Bioactive Compounds as Genetic and Epigenetic Modulators in PCOS: Emerging Evidence*

4.1.1 Curcumin

Curcumin, the active compound in turmeric, exhibits significant antioxidant properties by scavenging reactive oxygen species and enhancing the activity of antioxidant enzymes like superoxide dismutase (SOD), glutathione, and catalase, thereby reducing oxidative stress (Muthuswamy et al. 2022). The antioxidant property of curcumin was further exemplified in a study by Heshmati et al. (2020) that explored the impact of curcumin on modifying gene expression related to oxidative stress and metabolism, focusing on two key genes: sirtuin-1 or silent information regulator 1 (*SIRT1*) and peroxisome proliferator-activated receptor gamma coactivator-1 alpha (*PGC-1α*). The PGC-1α, a co-activator of PPARγ, serves to enhance the body's antioxidant defense activities by elevating detoxifying enzymes like SOD and glutathione peroxidase (GPx) (Qian et al. 2024). SIRT1, a histone deacetylase, regulates insulin secretion, antioxidant action, apoptosis, and protects DNA (Wu et al. 2022). SIRT1 also deacetylates PGC-1α, thereby enhancing thermogenesis and lipid oxidation, and it helps manage glucose and insulin homeostasis through its interaction with PGC-1α (Pang et al. 2024).

This randomized, double-blind trial included 72 overweight or obese women (BMI >25 kg/m^2) with PCOS and impaired glucose tolerance, diagnosed by Rotterdam criteria. Participants received either 1500 mg/day of curcumin (500 mg three times daily) or a matching placebo (maltodextrin) for 12 weeks (Heshmati et al. 2020). Based on previous studies, the dose was well tolerated with no adverse events and was considered safe and effective (Heshmati et al. 2020).

The study concluded that curcumin supplementation significantly increased the gene expression of *PGC-1α*, which in turn led to the increased enzyme activity of GPx, reducing oxidative stress (Heshmati et al. 2020). Although the increase in *SIRT1* gene expression and SOD enzyme activity was not statistically significant, the observed pattern may indicate indirect evidence of epigenetic modulation that curcumin may exert on these pathways, potentially influencing cellular stress responses (Heshmati et al. 2020). Thus, the findings from this study highlight the potential of curcumin as a therapeutic agent in modulating gene expression and antioxidant defense mechanisms in PCOS (Heshmati et al. 2020).

4.1.2 Quercetin

Quercetin, a flavonoid with antioxidant, antihyperglycemic, and lipid-lowering properties, has been shown to reduce hyperandrogenism and IR, modulate the HPO axis, and improve ovarian morphology in PCOS models (Shah et al. 2023). Investigations by Shah et al. (2023) showed that quercetin treatment reversed several molecular and histological abnormalities in PCOS-induced mice. Quercetin was supplemented to female Parkes strain mice (4–5 weeks old, 18–21 g), following induction of PCOS via oral gavage of letrozole (6 mg/kg body weight) daily for 21 days. After induction, quercetin was administered orally at a dose of 125 mg/kg body weight per day for 45 days (Shah et al. 2023).

Quercetin treatment resulted in the upregulation of the mRNA expression of androgen receptor (AR) and estrogen receptor 1 (Shah et al. 2023). Restoring these receptor expressions promoted granulosa cell proliferation, a key process for follicular growth and maturation (Shah et al. 2023). This remedied the abnormal follicular development seen in PCOS, characterized by cyst formation, small follicles, and thin granulosa cell layers (Shah et al. 2023). Additionally, quercetin increased the expression of CYP19A1 and CYP11A1, which helped restore aromatase and steroidogenesis activity that were impaired in PCOS, thus balancing estrogen and androgen production (Shah et al. 2023). The treatment also normalized ovarian structure, decreased cyst formation, and rectified the LH/FSH ratio, which is essential for proper ovarian function (Shah et al. 2023).

Notably, these findings contrast with a human study that investigated epigenetic regulation of AR expression in granulosa cells obtained from 10 women with PCOS (aged 20–40 years, diagnosed by Rotterdam criteria) and 8 non-PCOS women undergoing IVF due to male partner infertility (Desmawati et al. 2018). The study reported that the AR promoter was significantly hypomethylated, leading to an increase in AR expression (Desmawati et al. 2018). This epigenetic change was proposed to provoke hyperandrogenism by promoting tissue sensitivity to circulating androgens, thereby increasing the abnormal follicular development, a characteristic PCOS trait (Desmawati et al. 2018).

Contrastingly, the quercetin-induced upregulation of AR rectified the hormonal imbalance, including restoring aromatase activity and balancing estrogen-androgen levels (Shah et al. 2023). This distinction demonstrates that the epigenetic AR upregulation may compound PCOS pathology, while phytochemical intervention of AR via quercetin treatment may aid ovarian recovery. These contrasting modes of action emphasize the curative potential of plant compounds as a comprehensive approach to restoring reproductive and metabolic function in PCOS.

4.1.3 Naringenin

The naturally occurring flavonoid, naringenin, predominantly found in citrus fruits, has revealed potential in modulating molecular pathways relevant to PCOS (Wu et al. 2022). To elucidate the curative potential of naringenin, the bioactive phytocompound was supplemented to letrozole-induced PCOS Sprague Dawley rats at a dose of 50 mg/kg/day, either alone or in combination with metformin (300 mg/kg/day) (Rashid et al. 2023).

In vitro, human granulosa cells (hGCs) were treated with naringenin (2.5 μM) alone or with metformin (1 mM) in the presence of forskolin (10 μM) to determine the effects on steroidogenesis (Rashid et al. 2023). Protein expression levels of

CYP17A1, CYP19A1, and 3β-HSD2 were evaluated in both the rat ovary and hGCs to assess the mode of action of naringenin (Rashid et al. 2023). Additionally, given the pivotal role of protein kinase B (AKT) in insulin signalling and glucose uptake, which are impaired in PCOS, the expression of AKT was assessed to determine the insulin-sensitizing properties of naringenin in PCOS (Rashid et al. 2023).

Naringenin treatment led to the restoration of the expression of key steroidogenic enzymes, specifically CYP17A1, CYP19A1, and 3β-HSD2 (Rashid et al. 2023). These enzymes are critical in ensuring hormonal balance and normal follicular process. Furthermore, a reduction in serum androgen levels and cystic follicles was observed, together with improved ovulatory function, after naringenin intervention (Rashid et al. 2023). In addition, naringenin promoted granulosa cell function and ovarian health by activating AKT signalling pathway (Rashid et al. 2023). Notably, the study did not specifically investigate epigenetic changes. However, the molecular pathways influenced by naringenin treatment, such as the activation of AKT signalling, could indirectly influence epigenetic regulatory networks.

Deeper insights into the potential of naringenin to promote metabolic changes in PCOS patients were illustrated by a recent transcriptomic study (Wu et al. 2022). In this study, oocyte samples were collected from two PCOS patients (aged 25–28 years, diagnosed by Rotterdam criteria) for the single-cell transcriptome analyses (Wu et al. 2022). For the animal study, PCOS was induced in female Sprague-Dawley rats (1 mg/kg/day letrozole orally for 28 days). After induction, rats were treated orally with either Diane-35 (0.2 mg/kg/day, anti-androgenic contraceptive) or naringenin (20 mg/kg/day) (Wu et al. 2022).

The study demonstrated that differentially expressed genes in PCOS were linked to ovarian function, adenosine monophosphate-activated protein kinase (AMPK) signalling, insulin secretion, and steroidogenesis (Wu et al. 2022). Interestingly, the AMPK signalling pathway, which regulates various processes like energy balance, glucose uptake, and lipid metabolism, has been found to be dysregulated in PCOS (Mićić et al. 2023). The study found that naringenin intervention activated AMPK, subsequently triggering SIRT1, in turn (Wu et al. 2022), prompting an increased expression of PGC-1α (Wu et al. 2022). Notably, as previously discussed under curcumin, naringenin similarly targets the SIRT1/PGC-1α axis. Although no direct epigenetic markers were measured, these findings indirectly imply that naringenin may influence epigenetic modifications through the AMPK-SIRT1-PGC-1α signalling pathway.

In another study, naringenin significantly impacted the HPO axis in PCOS rats by regulating the gene expression of key hypothalamic regulators (Haghighat and Mahmoudi 2024). This included corticotropin-releasing hormone (CRH) and nesfatin-1. The dysregulated signalling of these two neuropeptides possibly contributes to PCOS pathophysiology by disrupting the HPO axis, thereby altering GnRH/LH secretion that may result in androgen excess (Haghighat and Mahmoudi 2024). In the animal model, estradiol valerate (2 mg/rat) was used to induce PCOS in female Wistar rats (180–200 g). Post-induction, the experimental group received intraperitoneal naringenin (20 or 50 mg/kg) for 14 days. Hypothalamic tissues were collected for gene expression analysis by real-time PCR (Haghighat and Mahmoudi 2024).

Notably, CRH mRNA levels, which are characteristically diminished in PCOS, reverted to baseline levels after naringenin treatment (Haghighat and Mahmoudi 2024). Naringenin was also found to influence ghrelin receptors, a mechanism linked to the restoration of normal CRH function and further aid reproductive health (Haghighat and Mahmoudi 2024). Furthermore, the study demonstrated that naringenin upregulated nesfatin-1 gene expression in the hypothalamus, suggesting the role of naringenin in improving follicular growth and FSH levels (Haghighat and Mahmoudi 2024). Since IR is a prominent feature of PCOS, the effect of naringenin on stabilizing glucose metabolism could contribute to balancing nesfatin-1 levels, thereby enhancing reproductive function (Haghighat and Mahmoudi 2024). While this study supports the role of naringenin in modulating gene expression, epigenetic endpoints such as DNA methylation and histone modification were not evaluated and should therefore be interpreted as indirect evidence.

Overall, these findings show that naringenin has multifaceted benefits in PCOS, improving ovarian function and reproductive health. Furthermore, naringenin may represent a promising epigenetic-modulating compound for remedying PCOS.

4.1.4 Resveratrol

A few studies have focused on the pharmacological effects of resveratrol in regulating genetic expression in PCOS. In one such study by Liang et al. (2023) focused on IR in PCOS patients and found that resveratrol enhanced insulin sensitivity and promoted glycolysis in KGN cells, a human granulosa-like tumor cell line, which mimics ovarian cells (Liang et al. 2023).

In the experimental model, oral administration of resveratrol (20 mg/kg/day) was given via gavage for 30 days to PCOS letrozole-induced Sprague Dawley rats (1 mg/kg/day) and a high-fat diet. In vitro, KGN cells were treated with insulin (4 μg/mL) to replicate hyperinsulinemic conditions and subsequently were exposed to resveratrol at concentrations of 10, 25, 50, and 100 μM for 24 h to assess cellular responses related to steroidogenesis and apoptosis (Liang et al. 2023).

Evidently, key glycolytic enzymes (hexokinase 2, pyruvate kinase M2 isoform, and lactate dehydrogenase A) were downregulated at both mRNA and protein levels when treated with insulin, but resveratrol intervention restored their expression (Liang et al. 2023). Likewise, resveratrol upregulated SIRT2, which was suppressed by high doses of insulin and a selective SIRT2 inhibitor, AGK2 (Liang et al. 2023). It is essential to distinguish that this study examined SIRT2, whereas the other studies involving curcumin, naringenin, and fisetin focused on SIRT1. Unlike SIRT1, which primarily functions in the nucleus, SIRT2 is predominantly cytoplasmic, where it is directly involved in the deacetylation of key glycolytic enzymes, thus influencing the glycolytic pathway (Liang et al. 2023). The study noted a negative correlation between IR and glycolytic markers. However, resveratrol treatment promoted glycolysis via SIRT2-mediated pathways in PCOS-like conditions (Liang et al. 2023). Thus, this study advocates the hypothesis that resveratrol may therapeutically influence PCOS partly via epigenetic mechanisms, mediated through SIRT2 activation. However, this remains to be confirmed by direct epigenetic assays.

A transcriptomic analysis was conducted using publicly available transcriptomic datasets, GSE34526 and GSE138518, containing granulosa cell gene expression profiles from PCOS patients and healthy controls revealed that toll-like receptor 2 (TLR2) was one of the potential hub genes that may be associated with PCOS (Yuan et al. 2022). Toll-like receptors are a family of innate immune receptors that have a well-known role in defense against pathogens. However, they may be implicated in obesity-related IR and metabolic dysfunction, inferred by the binding of dietary fatty acids to TLR2 and the elevated expression and activity of TLR2 in adipose tissue in obesity (Ojeda-Ojeda et al. 2016).

In the in vitro study, human KGN cells were cultured and treated with lipopolysaccharide (LPS, 1 μg/mL) for 36 h to induce an inflammatory response. This was followed by supplementation of resveratrol at concentrations of 10, 20, or 40 μM for 36 h, to evaluate its anti-inflammatory and modulatory effects on TLR2 expression and downstream signaling (Yuan et al. 2022).

Results indicated that LPS stimulation of granulosa cells increased TLR2 expression, including elevating the mRNA and protein levels of pro-inflammatory cytokines (TNFα, IL-1β, IL-6, IL-8), leading to an inflammatory state characteristic of PCOS (Yuan et al. 2022). Resveratrol intervention reduced TLR2 expression in a dose-dependent manner and suppressed the expression of the cytokines, demonstrating the potent anti-inflammatory and gene-regulatory activity of resveratrol (Yuan et al. 2022).

Additionally, LPS-induced inflammation and oxidative stress were attenuated by TLR2 knockdown using siRNA, thereby restoring the expression of antioxidant genes such as catalase, *GPx,* and *SOD* while reducing malondialdehyde activity (Yuan et al. 2022). The overexpression of TLR2 promoted the inflammatory response and oxidative stress in the granulosa cells, both of which were restored by resveratrol treatment (Yuan et al. 2022). These findings may indicate that resveratrol has gene-regulatory properties in PCOS, primarily by down-streaming the effects of TLR2 as well as inflammatory and oxidative pathways, showcasing the potential of plant compounds in influencing gene expression related to PCOS pathophysiology. Further epigenetic studies may be required to verify if resveratrol reduces inflammation in PCOS by silencing TLR2, possibly through epigenetic changes like methylation.

4.1.5 Baicalin

Baicalin (baicalein 7-D-beta-glucuronate), a flavonoid glycoside isolated from *Scutellaria. baicalensis*, was investigated for its underlying mechanisms in ameliorating hyperandrogenism in PCOS (Yu et al. 2019). Baicalin was supplemented to NCI-H295R human adrenocortical carcinoma cells at concentrations ranging from 0.1 to 400 μmol/L, with 24-h exposure used to evaluate the effect of baicalin on luciferase activity, gene expression, and protein analysis (Yu et al. 2019). In vivo, a hyperandrogenism-induced PCOS rat model was established by daily subcutaneous dehydroepiandrosterone injections (0.6 mg/100 g body weight) for 4 weeks (Yu et al. 2019). Baicalin was administered orally at 20 mg/kg/day for 4 weeks, and its effects on hormone profiles and ovarian histology were evaluated (Yu et al. 2019).

The findings indicated that baicalin significantly reduced testosterone production by decreasing the expression of 3β-HSD2, an enzyme essential for androgen biosynthesis (Yu et al. 2019). Gene expression profiling following baicalin treatment demonstrated significant downregulation of genes involved in androgen and cortisol biosynthesis, among these, *3β-HSD2* and wingless-type MMTV integration site family, member 4 were identified as central genes in steroidogenic regulation (Yu et al. 2019).

Furthermore, the study showed that baicalin suppressed 3β-HSD2 promoter activity via the disruption of GATA-binding factor 1 (GATA1) (Yu et al. 2019). The GATA family of proteins is a group of transcription factors that are important regulators of steroidogenesis and have been shown to be key regulators of 3β-HSD2. To confirm these results, GATA1 knockdown and mutagenesis experiments confirmed its critical role in mediating 3β-HSD2 transcription (Yu et al. 2019). The in vivo experiments revealed that baicalin treatment restored estrous cyclicity, normalized the LH: FSH ratio and testosterone levels, and improved ovarian morphology (Yu et al. 2019). Additionally, a reduction in both mRNA and protein levels of 3β-HSD2 in ovarian tissues was observed, aligning with in vitro findings (Yu et al. 2019). These results reflect the anti-androgenic effects that baicalin exerts in PCOS via the targeted suppression of *3β-HSD2* gene expression through GATA1 modulation. This study highlights an innovative transcriptional mechanism by which plant compounds may modulate steroidogenesis in PCOS. While the study did not directly evaluate classic epigenetic markers, the study showed that baicalin was able to inhibit the recruitment of the transcription factor GATA1 to the 3β-HSD2 promoter, which may be influenced by upstream epigenetic modifications by the plant compound. Thus, this may represent indirect evidence of epigenetic modulation by the baicalin.

4.1.6 Gallic Acid

In PCOS patients, the circulating levels of adiponectin have been shown to be significantly reduced (Xu et al. 2022). Adiponectin has insulin-sensitizing and anti-inflammatory properties that exert their influence via adiponectin receptors, such as adiponectin receptor 1 (AdipoR1) (Orszulak et al. 2025). Thus, a decline in adipokines can contribute to IR and hyperandrogenism, two key characteristics of PCOS (Orszulak et al. 2025).

Notably, a recent study focused on the therapeutic properties of gallic acid, a naturally occurring plant compound that is widely used in the food industry due to its ability to prevent the oxidation and rancidity of oils and fats (Xiang et al. 2024). It was inferred by the study that gallic acid increased adiponectin levels and receptor sensitivity, thereby improving insulin sensitivity and reducing ovarian androgen synthesis (Shah et al. 2022). In the experimental protocol of this study, gallic acid was supplemented to letrozole-induced PCOS female Parkes strain mice (aged 4–5 weeks, weighing 16–21 g) by oral gavage at a dose of 75 mg/kg body weight daily for 60 days. Prior to gallic acid administration, PCOS was induced by oral gavage of letrozole at 6 mg/kg body weight daily for 21 days (Shah et al. 2022).

The study demonstrated that treatment of the PCOS mice with gallic acid restored the expression of CYP11A1 and CYP19A1, which are essential for estrogen synthesis (Shah et al. 2022). Additionally, gallic acid increased expression of AdipoR1 and plasma adiponectin levels, indicating improved metabolic and insulin sensitivity (Shah et al. 2022). Histological analyses revealed restored ovarian tissue architecture with reduced cystic follicles (Shah et al. 2022). These findings suggest the potential of gallic acid in modulating gene expression and inflammatory pathways, offering therapeutic promise for managing PCOS.

Of particular interest, adiponectin is encoded by the highly polymorphic *ADIPOQ* gene and a recent study confirmed that the association of rs182052, rs822393, rs7649121 and rs6773957 ADIPOQ variants was significantly correlated with enhanced risk of PCOS among Bahraini women, providing evidence that genetic variation in the *ADIPOQ* gene impacts PCOS development (Al-Awadi et al. 2022). Although the study was in notable discrepancy with other studies based on ethnicity, it does not rule out the potential of these SNPs to affect adiponectin function and their substantial involvement in the PCOS pathogenicity (Al-Awadi et al. 2022). Thus, it may be plausible that bioactive compounds such as gallic acid could counteract the physiological repercussions of ADIPOQ polymorphisms, particularly in populations with a genetic predisposition to low adiponectin levels, therefore representing a novel, personalized strategy in the management of PCOS.

4.1.7 Ellagic Acid

Ellagic acid, a polyphenolic phytochemical that is found in various fruits, nuts, and vegetables, has attracted attention due to its antioxidant, anti-inflammatory, and anti-apoptotic properties (Gupta et al. 2021). With the aim of expanding the knowledge of the potential therapeutic applications of ellagic acid in PCOS, Khoshvaghti and Rahbari (2023) investigated the effects of ellagic acid supplementation on the expression of miRNA-21 in a PCOS-induced adult female Sprague-Dawley rat model. The induction of PCOS in the relevant groups was by subcutaneous injection of estradiol valerate at 2 mg/kg. Ellagic acid was then supplemented via subcutaneous injection at a dose of 200 mg/kg daily for 14 days (Khoshvaghti and Rahbari 2023).

Notably, there was a significant increase in the expression level of miRNA-21 in the PCOS group compared to the control groups, implying the crucial role miRNA-21 dysregulation may have in the pathogenesis of PCOS (Khoshvaghti and

Rahbari 2023). Ellagic acid intervention in the PCOS group resulted in reduced LH levels, improved ovarian follicle development and morphology, as well as the modulation of miRNA-21, indicating possible epigenetic effects (Khoshvaghti and Rahbari 2023). However, it is noted that although the study showed regulation of miRNA-21, a known epigenetic regulator, the study did not further investigate the downstream gene targets or chromatin changes. Hence, the evidence reflects indirect epigenetic modulation via non-coding RNA.

4.1.8 Fisetin

To expand on the role of fisetin as a therapeutic remedy for PCOS, Chahal and Kabra (2024) examined how fisetin modulated oxidative stress and IR through the mechanistic modification of the AMPK and SIRT1 axis. As discussed earlier in the context of naringenin, the dysregulation in AMPK-SIRT1 interaction is suggested as a common event in PCOS, correlated to IR, increased body weight, and adipogenesis (Mihanfar et al. 2021; Tao et al. 2017).

In the experimental design, fisetin intervention was given to 9–12 week old female Sprague-Dawley rats (220–270 g) via oral administration at doses of 20 mg/kg (low dose) and 40 mg/kg (high dose), daily for 21 days following PCOS induction (Chahal and Kabra 2024). PCOS was induced by oral administration of mifepristone (RU-486) at 20 mg/kg/day for 13 consecutive days. The mechanistic group received dorsomorphin (0.2 mg/kg, intraperitoneally) 30 min prior to fisetin (40 mg/kg, given orally) for 21 days to assess AMPK-SIRT1 pathway involvement. Metformin (300 mg/kg, administered orally) was used as a reference treatment.

Earlier research confirmed the ability of fisetin to amplify the AMPK and SIRT1 interaction (Liou et al. 2018). To further verify this conclusion, dorsomorphin was administered, negating the protective effects of fisetin in mifepristone-treated rats, and a consequential deterioration in all biochemical parameters was observed (Chahal and Kabra 2024). This exacerbated the pathological processes related to PCOS, leading to increased levels of testosterone, estradiol, progesterone, thiobarbituric acid reactive substances (TBARS), TNFα, and IL-6, while decreasing serum progesterone, FSH, glutathione, and SOD levels, including altered ovarian histology abnormalities in PCOS rats (Chahal and Kabra 2024). These findings support the involvement of AMPK and SIRT1 in mediating the beneficial effects of fisetin in alleviating PCOS symptoms (Chahal and Kabra 2024). Although epigenetic modulation was not a primary focus of the study, the involvement of AMPK/SIRT1 signalling, both of which are known to influence chromatin structure and gene expression, suggests that fisetin may exert indirect epigenetic effects in the context of PCOS. Taken together, the study highlights the therapeutic potential of fisetin in PCOS, largely due to its ability to modulate metabolic pathways, reduce inflammation, and restore hormonal balance.

5 Therapeutic Efficacy and Toxicological Considerations

The plant-derived compounds discussed in this chapter have demonstrated various therapeutic effects in PCOS models, warranting a closer look at their dosing parameters and potential risks based on previous clinical trials.

In a triple-blind randomized controlled trial conducted on women with PCOS, curcumin was administered at 1000 mg/day (500 mg two times daily) for 12 weeks, significantly reducing fasting blood sugar and improving menstruation characteristics (amenorrhea, oligomenorrhea, and menstrual irregularities), with no safety concerns reported (Ghanbarzadeh-Ghashti et al. 2023). Quercetin has been evaluated in clinical trials at dosages of 500–1000 mg/day for 6–12 weeks, improving IR, testosterone levels, and inflammatory markers in women with PCOS, with no significant adverse effects reported (Khorshidi et al. 2018; Rezvan et al. 2017). Naringenin, administered at 20–50 mg/kg/day in preclinical studies, improved IR, activated the AMPK-SIRT1 pathway, and restored hormonal imbalance, with no toxic effects observed (Haghighat and Mahmoudi 2024; Rashid et al. 2023; Wu et al. 2022). Notably, there are no clinical trials as yet that have evaluated the effects of naringenin in PCOS patients. However, Rebello et al. (2020) conducted a human pharmacokinetic study that showed that oral doses of naringenin up to 900 mg are well tolerated and result in physiologically relevant plasma concentrations. This finding supports the possibility of future translation research. Resveratrol demonstrated reductions in ovarian androgen biosynthesis at doses of 1500 mg/day in human studies, with no major toxicity reported (Banaszewska et al. 2016). Baicalin, at doses of 20–50 mg/kg in animal models, exerted anti-inflammatory, metabolic regulatory effects, and improved hormonal balance, though human data are lacking (Wang et al. 2019b; Yu et al. 2019). Gallic acid and ellagic acid have shown beneficial effects in PCOS animal models at 75–200 mg/kg/day, targeting hyperandrogenism, IR, and ovarian histology, with a favorable safety profile at these doses (Khoshvaghti and Rahbari 2023; Shah et al. 2022). Fisetin, tested in rodent models at 20–40 mg/kg for 21 days, improved IR, oxidative stress, and ovarian morphology through AMPK-SIRT1 modulation, without observed toxicity (Chahal and Kabra 2024).

6 Clinical Relevance: Challenges and Considerations

Despite promising evidence from phytochemical-based in vitro and animal model experiments, the progression to clinical application remains limited. There are several challenges that hinder the translation to effective human treatments.

One of the main concerns is the pharmacokinetic and bioavailability limitations of phytochemicals. Phytocompounds such as curcumin, practically insoluble in water (0.6 μg/mL), are rapidly metabolized and poorly absorbed, leading to quick systemic clearance that contributes to suboptimal curcumin levels in plasma and tissues (Tabanelli et al. 2021). Likewise, for resveratrol, maintaining a therapeutic level in the bloodstream is hindered due to the rapid elimination from the body. This often necessitates high doses in preclinical studies that can cause side effects or be unsafe for human use (Kemper et al. 2022). Furthermore, variability in formulation, extraction methods, and compound purity can affect both bioavailability and reproducibility of results.

Another challenge lies in the use of animal models. While animal models provide important insights into PCOS pathophysiology and treatment response, they do not fully mimic the complexity and heterogeneity of human PCOS (Basak et al. 2024). Furthermore, the acute methods used to induce PCOS (testosterone, letrozole, or estradiol valerate) may not replicate the chronic progression seen in humans (Corrie et al. 2021). Additionally, rodents are polyovulatory, differing significantly from the mono-ovulatory cycle in humans, which may suggest essential differences in dominant follicle selection (Huang-Doran and Franks 2016).

In terms of human clinical evidence, there is a need for statistically rigorous randomized controlled trials (RCTs) with appropriate sample sizes. A recent meta-analysis identified only 15 small RCTs ($n = 916$), often with heterogeneous outcomes that complicate the interpretation and wider application of the results (Almeida Balieiro et al. 2024). To provide a consolidated overview across studies, Table 1 provides a summary of selected phytochemicals investigated in PCOS models and clinical trials, outlining their dosages, model used, observed effects, and associated limitations to facilitate bridging the gap between preclinical and clinical relevance.

Another hurdle in clinical practice is the attraction of phytotherapy, its perceived safety, and the assumption of a lower risk of side effects compared to conventional drugs (Hlengwa et al. 2020). Valid concerns remain regarding herb-drug interactions, the lack of chemical characterization of the active compounds, inconsistent dosing, and a lack of regulatory oversight (Hlengwa et al. 2020). A deeper understanding of optimal dose ranges, treatment durations, and identifying patient profiles is required for the integration of phytochemicals into PCOS management.

7 Conclusion

Plant-derived bioactive compounds are gaining recognition as modulators of gene expression and epigenetic regulation in the context of PCOS. Despite emerging interest in the epigenetic processes implicated in PCOS, relatively few studies have explored the role of plant-based compounds as targeted modulators of the epigenetic landscape associated with PCOS pathogenesis. The vast biochemical reservoir of plant-derived compounds offers a compelling opportunity to uncover novel agents capable of regulating epigenetic effects, gene expression, and altering disease trajectories in PCOS. Direct investigations that explore the role of plant compounds as epigenetic modulators are limited and represent a promising field for future research.

Table 1 Summary of phytochemicals in PCOS: dose, model, and translational limitations

Compound	Dose and duration	Model	Observed effects	Limitations	References
Curcumin	1000 mg/day, 12 weeks	Human (RCT) (PCOS patients per Rotterdam criteria; $n = 72$)	Reduced fasting blood sugar and improved menstrual irregularities	Small sample size	Ghanbarzadeh-Ghashti et al. (2023)
	1500 mg/day, 12 weeks	Human RCT (PCOS patients per Rotterdam criteria; $n = 72$)	↑ *PGC-1α* gene expression, ↑ GPx activity	Need for larger and longer RCT	Heshmati et al. (2020)
Quercetin	1000 mg/day, 12 weeks	Human RCT PCOS patients per Rotterdam criteria; $n = 80$)	↓ resistin gene expression, ↓ testosterone and LH, modest metabolic effects	No significant placebo comparison for glucose/insulin	Khorshidi et al. (2018)
	1000 mg/day, 12 weeks	Human RCT (PCOS patients per Rotterdam criteria; $n = 84$)	↑ Total adiponectin (+5.56%), ↑ HMW adiponectin (+3.9%), ↓ Testosterone, ↓ LH, ↓ HOMA-IR	Focused on biochemical markers; clinical outcomes not evaluated	Rezvan et al. (2017)
	125 mg/kg body weight / day for ≈ 6.4 weeks	18 Parkes strain mice (4–5 weeks old, 18–21 g)	↑ mRNA expression of androgen receptor and estrogen receptor 1 in quercetin-treated PCOS mice	Small sample size, no assessment of quercetin side effects, single dose tested	Shah et al. (2023)

(continued)

Table 1 (continued)

Compound	Dose and duration	Model	Observed effects	Limitations	References
Naringenin	50 mg/kg/day, orally for 20 days, in letrozole-induced PCOS rats 2.5 μM in human granulosa cells (with/without metformin)	25 Female Sprague-Dawley rats (6 weeks old, weighing 100–129 g)	↑ Ovulation, ↓ cystic follicles, ↓ androgens, restored CYP17A1, CYP19A1, and 3β-HSD2 expression, ↑ phosphorylated-AKT signaling	Short treatment duration (20 days); no dose-ranging studies; estrous cycle only normalized late in treatment; insulin sensitivity not confirmed via ITT or HOMA-IR	Rashid et al. (2023)
	20 and 50 mg/kg IP for 14 days in estradiol valerate-induced PCOS rats	20 Female Wistar rats (180–200 g)	↑ CRH and nesfatin-1 gene expression in the hypothalamus; potential restoration of HPO axis signaling	Small sample size (n = 5/group); short treatment duration (14 days)	Haghighat and Mahmoudi (2024)
	20 mg/kg/day orally for 4 weeks in letrozole-induced PCOS	Sprague-Dawley rats (10 weeks old, 270–300 g)	↓ body weight, improved hormone levels, IR, ovarian pathology; upregulated SIRT1, PGC-1α, occludin, claudin-1 in ovary and colon; modulated gut microbiota composition	The underlying mechanism of how naringenin regulates serum hormone levels and metabolic abnormalities remains unclear	Wu et al. (2022)
	Single oral doses (150, 300, 600, 900 mg); blood sampling over 24 h post-dose	Human PK study (healthy adults)	Tolerability up to 900 mg/day; half-life ~3 h; higher bioavailability from orange extract vs isolated form	Single-dose study; no long-term data; small sample size Note: Naringenin data are limited to preclinical studies; no RCTs in PCOS have been conducted	Rebello et al. (2020)

(continued)

Table 1 (continued)

Compound	Dose and duration	Model	Observed effects	Limitations	References
Resveratrol	1500 mg/day orally, ≈ 13 weeks	Human RCT (PCOS patients per Rotterdam criteria; $n = 30$)	↓ Total testosterone (23.1%), DHEA-S (22.2%), fasting insulin (31.8%); ↑ Insulin Sensitivity Index by 66.3%; no change in gonadotropins, lipid profile, inflammatory or endothelial markers	Small sample size; short duration; no long-term follow-up	Banaszewska et al. (2016)
	20 mg/kg/day intragastric administration for 30 days in letrozole and HFD-induced PCOS rats; 25 μM (KGN cells)	24 Sprague-Dawley rats (5 weeks old, 171 ± 11 g); KGN cells	Improved IR and glycolysis via SIRT2; ↓ glucose/insulin; ↑ glycolytic enzymes	No supporting human data	Liang et al. (2023)
	10–40 μM in KGN granulosa cells stimulated with 1 μg/mL LPS	In vitro model of PCOS inflammation	↓ TLR2 expression; ↓ TNF-α, IL-1β, IL-6, IL-8; ↑ CAT, GPx, SOD; ↑ antioxidant capacity; ↓ MDA activity	Cell line-based (KGN)	Yuan et al. (2022)
Baicalin (from *Scutellaria baicalensis*)	20 mg/kg/day orally for 4 weeks in DHEA-induced PCOS rats 25 μM for 48 h in cell line	80 Wistar rats (3.2 weeks old), NCI-H295R human adrenocortical cells	↓ Testosterone (cells and serum); ↓ 3β-HSD2 expression (mRNA & protein); ↑ Estrous cyclicity recovery; ↓ Cystic follicles, ↑ corpora lutea	No human data, potential side effects still need to be elucidated	Yu et al. (2019)

(continued)

Table 1 (continued)

Compound	Dose and duration	Model	Observed effects	Limitations	References
Gallic acid	75 mg/kg/day orally for 60 days, post-PCOS induction with letrozole (6 mg/kg/day for 21 days)	18 Parkes strain mice (4–5 weeks old, 16–21 g)	↓ Body and ovarian weight, testosterone, LH, LH/FSH ratio, insulin, glucose, inflammatory cytokines (TNF-α, VEGF, IL-6); ↑ estrogen, FSH, antioxidant enzymes (SOD, CAT), ATPase activity; normalized estrous cycle; ↓ MDA; ↑ AdipoR1, CYP19A1, PTGS2, KITL mRNA expression	Only one dose (75 mg/kg) of gallic acid tested; no dose-response or toxicity assessment Lacks standard drug control (e.g., metformin) for efficacy benchmarking	Shah et al. (2022)
Ellagic acid	200 mg/kg/day subcutaneously for 14 days, post-PCOS induction (2 mg/kg estradiol valerate, single dose, 60 days prior)	Estradiol valerate-induced PCOS in 40 adult female Sprague-Dawley rats (≈ 10 weeks, 200 ± 20 g)	↓ Body and ovarian weight; ↓ LH, testosterone, miRNA-21 expression; ↑ estradiol, progesterone; restored primordial, antral, and Graafian follicle counts; ↓ cystic and atretic follicles; ↓ theca layer thickness; ↑ oocyte diameter and granulosa layer thickness; improved ovarian morphology and hormone profiles	Short dosing window (14 days) post-PCOS induction; no positive control like metformin	Khoshvaghti and Rahbari (2023)

(continued)

Table 1 (continued)

Compound	Dose and duration	Model	Observed effects	Limitations	References
Fisetin	20 or 40 mg/kg/day orally for 21 days after 13-day mifepristone (20 mg/kg/day) induction; compared to metformin (300 mg/kg) and AMPK inhibitor dorsomorphin	Mifepristone-induced PCOS in female Sprague-Dawley rats (9–12 weeks old, 220–270 g)	↓ Body weight, testosterone, estradiol, LH, HOMA-IR, TNF-α, IL-6, TBARS; ↑ FSH, progesterone, GSH, SOD; restored estrous cycle and ovarian histology. Effects reversed by dorsomorphin co-treatment, confirming AMPK/SIRT1 involvement	Use of 1% DMSO as a vehicle was noted. However, it is unlikely to have an influence on results	Chahal and Kabra (2024)

The plant-derived compounds reviewed in the chapter demonstrated beneficial effects through diverse mechanisms that may align with specific PCOS phenotypes. Fisetin and resveratrol, through their effects on key metabolic regulators like AMPK and SIRT1, may offer therapeutic benefit in the IR phenotype. Particularly, resveratrol may be beneficial in obesity-related IR due to its modulatory effects on the toll-receptors. Curcumin showed anti-inflammatory, antioxidant activities, and regulation of IR, suggesting potential utility in the inflammatory phenotype and IR PCOS group. The other phytocompounds, quercetin, baicalin, gallic acid, and ellagic acid, appear to influence steroidogenesis and hormone balance, which could be helpful in hyperandrogenic PCOS.

Among the discussed compounds, naringenin stands out due to its ability to restore the expression of key steroidogenic enzymes, activate the AMPK-SIRT1 pathway, and positively impact the neuroendocrine axis. Although current evidence is primarily based on preclinical studies, these findings highlight the potential for targeted phytotherapy aligned with PCOS phenotypes. Notably, there were no major toxicological effects observed at the tested doses in animal models, supporting their safety for further exploration.

As the research advances, elucidating phytochemicals that can modulate molecular targets and pathways will be essential for translating this therapeutic potential into personalized therapeutic strategies. Harnessing plant-based epigenetic modulators may ultimately offer a novel avenue for the long-term prevention and management of PCOS.

Acknowledgments This study was supported by an Incentive Research Grant for Rated Researchers from the National Research Foundation (NRF), Pretoria, South Africa.

References

Al-Awadi AM, Babi A, Finan RR, Atageldiyeva K, Shaimardanova M, Mustafa FE, Mahmood NA, Aimagambetova G, Almawi WY (2022) ADIPOQ gene polymorphisms and haplotypes linked to altered susceptibility to PCOS: a case–control study. Reprod Biomed Online 45(5):995–1005. https://doi.org/10.1016/j.rbmo.2022.06.009

Balieiro CCA, Hespanhol LC, Fonseca LM, Wantowski S, Freitas MAA, Dias YJM, Burlá MM, De Oliveira LML (2024) Effects of polyphenol in women with polycystic ovary syndrome: a systematic review and meta-analysis of randomized controlled trials. Eur J Obst Gynecol Reprod Biol. https://doi.org/10.1016/j.ejogrb.2023.12.038

Bai X, Zheng C, Yu Y, Zhang J, Cao S, Hou C, Wang S (2024) RNA-seq and miRNA-seq profiling analyses reveal molecular mechanisms underlying the progression of polycystic ovary syndrome. Gene Rep 35:101908. https://doi.org/10.1016/j.genrep.2024.101908

Banaszewska B, Wrotyńska-Barczyńska J, Spaczynski RZ, Pawelczyk L, Duleba AJ (2016) Effects of resveratrol on polycystic ovary syndrome: a double-blind, randomized, placebo-controlled trial. J Clin Endocrinol Metab 101(11):4322–4328. https://doi.org/10.1210/jc.2016-1858

Bannigida DM, Nayak BS, Vijayaraghavan R (2018) Insulin resistance and oxidative marker in women with PCOS. Arch Physiol Biochem 126(2):183–186. https://doi.org/10.1080/13813455.2018.1499120

Basak S, Dixit AK, Dey RK, Roy S, Singh R, Nair PG, Kumar S, Babu G (2024) Rodent models in polycystic ovarian syndrome: dissecting reproductive and metabolic phenotypes for therapeutic advancements. Steroids 211:109489. https://doi.org/10.1016/j.steroids.2024.109489

Bhandary P, Shetty PK, Manjeera L, Patil P (2022) Hormonal, genetic, epigenetic and environmental aspects of polycystic ovarian syndrome. Gene Rep 29:101698. https://doi.org/10.1016/j.genrep.2022.101698

Bhimwal T, Puneet N, Priyadarshani A (2023) Understanding polycystic ovary syndrome in light of associated key genes. Egypt J Med Hum Genet 24(1). https://doi.org/10.1186/s43042-023-00418-w

Bhingardeve S, Sagvekar P, Mukherjee S (2025) Altered DNA methylation and miRNA expression: epi-marks for elucidating PCOS pathophysiology. Reprod Biomed Online 51(1):104838. https://doi.org/10.1016/j.rbmo.2025.104838

Chahal SK, Kabra A (2024) Fisetin ameliorates polycystic ovary syndrome in rats via a mechanistic modulation of AMP-activated protein kinase and SIRT1 molecular pathway. Naunyn-Schmiedebergs Arch Pharmacol 397(12):10017–10029. https://doi.org/10.1007/s00210-024-03257-7

Chaudhary H, Patel J, Jain NK, Joshi R (2021) The role of polymorphism in various potential genes on polycystic ovary syndrome susceptibility and pathogenesis. J Ovarian Res 14(1). https://doi.org/10.1186/s13048-021-00879-w

Chen Y, Wang G, Chen J, Wang C, Dong X, Chang H-M, Yuan S, Zhao Y, Mu L (2024) Genetic and epigenetic landscape for drug development in polycystic ovary syndrome. Endocr Rev 45(4):437–459. https://doi.org/10.1210/endrev/bnae002

Corrie L, Gulati M, Singh SK, Kapoor B, Khursheed R, Awasthi A, Vishwas S, Chellappan DK, Gupta G, Jha NK, Anand K, Dua K (2021) Recent updates on animal models for understanding the etiopathogenesis of polycystic ovarian syndrome. Life Sci 280:119753. https://doi.org/10.1016/j.lfs.2021.119753

Desmawati N, Febri RR, Hestiantoro A, Asmarinah N (2018) DNA methylation of the androgen receptor gene promoter in the granulosa cells of polycystic ovary syndrome patients. J Phys Conf Ser 1073:032078. https://doi.org/10.1088/1742-6596/1073/3/032078

Dewailly D, Robin G, Peigne M, Decanter C, Pigny P, Catteau-Jonard S (2016) Interactions between androgens, FSH, anti-Müllerian hormone and estradiol during folliculogenesis in the human normal and polycystic ovary. Hum Reprod Update 22(6):709–724. https://doi.org/10.1093/humupd/dmw027

Dumesic DA, Hoyos LR, Chazenbalk GD, Naik R, Padmanabhan V, Abbott DH (2020) Mechanisms of intergenerational transmission of polycystic ovary syndrome. Reproduction 159(1):R1–R13. https://doi.org/10.1530/rep-19-0197

Ferk P, Teran N, Gersak K (2006) The (TAAAA)n microsatellite polymorphism in the SHBG gene influences serum SHBG levels in women with polycystic ovary syndrome. Hum Reprod 22(4):1031–1036. https://doi.org/10.1093/humrep/del457

Fraser GL, Obermayer-Pietsch B, Laven J, Griesinger G, Pintiaux A, Timmerman D, BCJM F, Lademacher C, Combalbert J, Hoveyda HR, Ramael S (2021) Randomized controlled trial of neurokinin 3 receptor antagonist fezolinetant for treatment of polycystic ovary syndrome. J Clin Endocrinol Metabol 106(9):e3519–e3532. https://doi.org/10.1210/clinem/dgab320

Ghanbarzadeh-Ghashti N, Ghanbari-Homaie S, Shaseb E, Abbasalizadeh S, Mirghafourvand M (2023) The effect of Curcumin on metabolic parameters and androgen level in women with polycystic ovary syndrome: a randomized controlled trial. BMC Endocr Disord 23(1). https://doi.org/10.1186/s12902-023-01295-5

Guo X, Chen L, He J, Zhang X, Xu S (2024) Genetically predicted high sex hormone binding globulin was associated with decreased risk of polycystic ovary syndrome. BMC Womens Health 24(1). https://doi.org/10.1186/s12905-024-03144-6

Gupta A, Singh AK, Kumar R, Jamieson S, Pandey AK, Bishayee A (2021) Neuroprotective potential of ellagic acid: a critical review. Adv Nutr 12(4):1211–1238. https://doi.org/10.1093/advances/nmab007

Haghighat K, Mahmoudi F (2024) Evaluation of the effects of naringenin on the hypothalamic mRNA levels of nesfatin-1 and CRH in a rat model of PCOS. Gene Cell Tissue 11(4). https://doi.org/10.5812/gct-154371

Heidarzadehpilehrood R, Pirhoushiaran M, Abdollahzadeh R, Osman MB, Sakinah M, Nordin N, Hamid HA (2022) A review on CYP11A1, CYP17A1, and CYP19A1 polymorphism studies: candidate susceptibility genes for polycystic ovary syndrome (PCOS) and infertility. Genes 13(2):302. https://doi.org/10.3390/genes13020302

Heshmati J, Golab F, Morvaridzadeh M, Potter E, Akbari-Fakhrabadi M, Farsi F, Tanbakooei S, Shidfar F (2020) The effects of curcumin supplementation on oxidative stress, Sirtuin-1 and peroxisome proliferator activated receptor γ coactivator 1α gene expression in polycystic ovarian syndrome (PCOS) patients: a randomized placebo-controlled clinical trial. Diabet Metabol Syndrome 14(2):77–82. https://doi.org/10.1016/j.dsx.2020.01.002

Hiam D, Moreno-Asso A, Teede HJ, Laven JSE, Stepto NK, Moran LJ, Gibson-Helm M (2019) The genetics of polycystic ovary syndrome: an overview of candidate gene systematic reviews and genome-wide association studies. J Clin Med 8(10):1606. https://doi.org/10.3390/jcm8101606

Hlengwa N, Muller CJF, Basson AK, Bowles S, Louw J, Awortwe C (2020) Herbal supplements interactions with oral oestrogen-based contraceptive metabolism and transport. Phytother Res 34(7):1519–1529. https://doi.org/10.1002/ptr.6623

Huang-Doran I, Franks S (2016) Genetic rodent models of obesity-associated ovarian dysfunction and subfertility: insights into polycystic ovary syndrome. Front Endocrinol 7. https://doi.org/10.3389/fendo.2016.00053

Kemper C, Behnam D, Brothers S, Wahlestedt C, Volmar C-H, Bennett D, Hayward M (2022) Safety and pharmacokinetics of a highly bioavailable resveratrol preparation (JOTROL TM). AAPS Open 8(1). https://doi.org/10.1186/s41120-022-00058-1

Khorshidi M, Moini A, Alipoor E, Rezvan N, Gorgani-Firuzjaee S, Yaseri M, Hosseinzadeh-Attar MJ (2018) The effects of quercetin supplementation on metabolic and hormonal parameters as well as plasma concentration and gene expression of resistin in overweight or obese women with polycystic ovary syndrome. Phytother Res 32(11):2282–2289. https://doi.org/10.1002/ptr.6166

Khoshvaghti A, Rahbari R (2023) The effect of ellagic acid on sex hormones and miRNA-21 expression in rats with polycystic ovary syndrome. Naunyn-Schmiedebergs Arch Pharmacol 397(6):4263–4273. https://doi.org/10.1007/s00210-023-02895-7

Korolenko A, Skinner MK (2024) Generational stability of epigenetic transgenerational inheritance facilitates adaptation and evolution. Epigenetics 19(1). https://doi.org/10.1080/15592294.2024.2380929

Kulkarni R, Teves ME, Han AX, McAllister JM, Strauss JF (2019) Colocalization of polycystic ovary syndrome candidate gene products in the CA cells suggests novel signaling pathways. J Endocr Soc 3(12):2204–2223. https://doi.org/10.1210/js.2019-00169

Lentscher JA, Decherney AH (2020) Clinical presentation and diagnosis of polycystic ovarian syndrome. Clin Obstet Gynecol 64(1):3–11. https://doi.org/10.1097/grf.0000000000000563

Li X, Xiao H, Ma Y, Zhou Z, Chen D (2022) Identifying novel genetic loci associated with polycystic ovary syndrome based on its shared genetic architecture with type 2 diabetes. Front Genet 13. https://doi.org/10.3389/fgene.2022.905716

Liang A, Zhang W, Wang Q, Huang L, Zhang J, Ma D, Liu K, Li S, Chen X, Li S, Lei X (2023) Resveratrol regulates insulin resistance to improve the glycolytic pathway by activating SIRT2 in PCOS granulosa cells. Front Nutr 9. https://doi.org/10.3389/fnut.2022.1019562

Liou C-J, Wei C-H, Chen Y-L, Cheng C-Y, Wang C-L, Huang W-C (2018) Fisetin protects against hepatic steatosis through regulation of the SIRT1/AMPK and fatty acid B-Oxidation signaling pathway in High-Fat Diet-Induced obese mice. Cell Physiol Biochem 49(5):1870–1884. https://doi.org/10.1159/000493650

Liu X, Zhang J, Wang S (2023) Global, regional, and national burden of infertility attributable to PCOS, 1990–2019. Hum Reprod 39(1):108–118. https://doi.org/10.1093/humrep/dead241

Liu Y-N, Qin Y, Wu B, Peng H, Li M, Luo H, Liu L-L (2022) DNA methylation in polycystic ovary syndrome: emerging evidence and challenges. Reprod Toxicol 111:11–19. https://doi.org/10.1016/j.reprotox.2022.04.010

Luo ED et al (2022) Advancements in lead therapeutic phytochemicals polycystic ovary syndrome: A review. Front Pharmacol 13:1065243. https://doi.org/10.3389/fphar.2022.1065243

Mićić B, Djordjevic A, Veličković N, Kovačević S, Martić T, Macut D, Milutinović DV (2023) AMPK activation as a protective mechanism to restrain oxidative stress in the insulin-resistant state in skeletal muscle of rat model of PCOS subjected to postnatal overfeeding. Biomedicines 11(6):1586. https://doi.org/10.3390/biomedicines11061586

Mihanfar A, Nouri M, Roshangar L, Khadem-Ansari MH (2021) Therapeutic potential of quercetin in an animal model of PCOS: possible involvement of AMPK/SIRT-1 axis. Eur J Pharmacol 900:174062. https://doi.org/10.1016/j.ejphar.2021.174062

Mimouni NEH, Paiva I, Barbotin A-L, Timzoura FE, Plassard D, Gras SL, Ternier G, Pigny P, Catteau-Jonard S, Simon V, Prevot V, Boutillier A-L, Giacobini P (2021) Polycystic ovary syndrome is transmitted via a transgenerational epigenetic process. Cell Metab 33(3):513–530.e8. https://doi.org/10.1016/j.cmet.2021.01.004

Moolhuijsen LME, Louwers YV, McLuskey A, Broer L, Uitterlinden AG, Verdiesen RMG, Sisk RK, Dunaif A, Laven JSE, Visser JA (2022) Association between an AMH promoter polymorphism and serum AMH levels in PCOS patients. Hum Reprod 37(7):1544–1556. https://doi.org/10.1093/humrep/deac082

Moosa A, Ghani M, O'Neill HC (2022) Genetic associations with polycystic ovary syndrome: the role of the mitochondrial genome; a systematic review and meta-analysis. J Clin Pathol 75(12):815–824. https://doi.org/10.1136/jcp-2021-208028

Muthuswamy S et al (2022) The credible role of curcumin in oxidative stress-mediated mitochondrial dysfunction in mammals. Biomolecules 12(10):1405. https://doi.org/10.3390/biom12101405

Ojeda-Ojeda M, Martínez-García MÁ, Alpañés M, Luque-Ramírez M, Escobar-Morreale HF (2016) Association of TLR2 S450S and ICAM1 K469E polymorphisms with polycystic ovary syndrome (PCOS) and obesity. J Reprod Immunol 113:9–15. https://doi.org/10.1016/j.jri.2015.09.072

Orszulak D, Niziński K, Matonóg A, Zięba-Domalik M, Stojko R, Drosdzol-Cop A (2025) Adipokines as biochemical marker of polycystic ovary syndrome in adolescents—review. Front Endocrinol 16. https://doi.org/10.3389/fendo.2025.1475465

Pang J, Yin L, Jiang W, Wang H, Cheng Q, Jiang Z, Cao Y, Zhu X, Li B, Qian S, Yin X, Wang T, Lu Q, Yang T (2024) Sirt1-mediated deacetylation of PGC-1α alleviated hepatic steatosis in type 2 diabetes mellitus via improving mitochondrial fatty acid oxidation. Cell Signal 124:111478. https://doi.org/10.1016/j.cellsig.2024.111478

Panghiyangani R, Soeharso P, Andrijono N, Suryandari D, Wiweko B, Kurniati M, Pujianto D (2020) CYP19A1 gene expression in patients with polycystic ovarian syndrome. J Hum Reprod Sci 13(2):100. https://doi.org/10.4103/jhrs.jhrs_142_18

Pei C-Z, Jin L, Baek K-H (2021) Pathogenetic analysis of polycystic ovary syndrome from the perspective of omics. Biomed Pharmacother 142:112031. https://doi.org/10.1016/j.biopha.2021.112031

Qian L, Zhu Y, Deng C, Liang Z, Chen J, Chen Y, Wang X, Liu Y, Tian Y, Yang Y (2024) Peroxisome proliferator-activated receptor gamma coactivator-1 (PGC-1) family in physiological and pathophysiological process and diseases. Signal Transduct Target Ther 9(1). https://doi.org/10.1038/s41392-024-01756-w

Rashid R, Tripathi R, Singh A, Sarkar S, Kawale A, Bader GN, Gupta S, Gupta RK, Jha RK (2023) Naringenin improves ovarian health by reducing the serum androgen and eliminating follicular cysts in letrozole-induced polycystic ovary syndrome in the Sprague Dawley rats. Phytother Res 37(9):4018–4041. https://doi.org/10.1002/ptr.7860

Rawat K, Sandhu A, Gautam V, Saha PK, Saha L (2022) Role of genomic DNA methylation in PCOS pathogenesis: a systematic review and meta-analysis involving case-controlled clinical studies. Mol Hum Reprod 28(8). https://doi.org/10.1093/molehr/gaac024

Rebello CJ, Beyl RA, Lertora JJL, Greenway FL, Ravussin E, Ribnicky DM, Poulev A, Kennedy BJ, Castro HF, Campagna SR, Coulter AA, Redman LM (2020) Safety and pharmacokinetics of naringenin: a randomized, controlled, single-ascending-dose clinical trial. Diabetes Obes Metab 22(1):91–98. https://doi.org/10.1111/dom.13868

Rezvan N, Moini A, Janani L, Mohammad K, Saedisomeolia A, Nourbakhsh M, Gorgani-Firuzjaee S, Mazaherioun M, Hosseinzadeh-Attar M (2017) Effects of quercetin on adiponectin-mediated insulin sensitivity in polycystic ovary syndrome: a randomized placebo-controlled double-blind clinical trial. Horm Metab Res 49(02):115–121. https://doi.org/10.1055/s-0042-118705

Rotterdam ESHRE/ASRM-Sponsored PCOS Consensus Workshop Group (2004) Revised 2003 consensus on diagnostic criteria and long-term health risks related to polycystic ovary syndrome. Fertil Steril 81(1):19–25. https://doi.org/10.1016/j.fertnstert.2003.10.004

Rudnicka E, Kunicki M, Calik-Ksepka A, Suchta K, Duszewska A, Smolarczyk K, Smolarczyk R (2021a) Anti-Müllerian hormone in pathogenesis, diagnostic and treatment of PCOS. Int J Mol Sci 22(22):12507. https://doi.org/10.3390/ijms222212507

Rudnicka E, Suchta K, Grymowicz M, Calik-Ksepka A, Smolarczyk K, Duszewska AM, Smolarczyk R, Meczekalski B (2021b) Chronic low grade inflammation in pathogenesis of PCOS. Int J Mol Sci 22(7):3789. https://doi.org/10.3390/ijms22073789

Salinas I, Sinha N, Sen A (2021) Androgen-induced epigenetic modulations in the ovary. J Endocrinol 249(3):R53–R64. https://doi.org/10.1530/joe-20-0578

Shah MZUH, Shrivastva VK, Mir MA, Sheikh WM, Ganie MA, Rather GA, Shafi M, Bashir SM, Ansari MA, Al-Jafary MA, Al-Qahtani MH, Homeida AM, Al-Suhaimi EA (2023) Effect of quercetin on steroidogenesis and folliculogenesis in ovary of mice with experimentally-induced polycystic ovarian syndrome. Front Endocrinol 14. https://doi.org/10.3389/fendo.2023.1153289

Shah MZUH, Soni M, Shrivastava VK, Mir MA, Muzamil S (2022) Gallic acid reverses ovarian disturbances in mice with letrozole-induced PCOS via modulating Adipo R1 expression. Toxicol Rep 9:1938–1949. https://doi.org/10.1016/j.toxrep.2022.10.009

Stener-Victorin E (2020) Epigenetic and transgenerational transmission of polycystic ovary syndrome. Curr Opin Endocr Metab Res 12:72–77. https://doi.org/10.1016/j.coemr.2020.03.005

Stener-Victorin E, Deng Q (2021) Epigenetic inheritance of polycystic ovary syndrome—challenges and opportunities for treatment. Nat Rev Endocrinol 17(9):521–533. https://doi.org/10.1038/s41574-021-00517-x

Szukiewicz D, Trojanowski S, Kociszewska A, Szewczyk G (2022) Modulation of the inflammatory response in polycystic ovary syndrome (PCOS)—searching for epigenetic factors. Int J Mol Sci 23(23):14663. https://doi.org/10.3390/ijms232314663

Tabanelli R, Brogi S, Calderone V (2021) Improving curcumin bioavailability: current strategies and future perspectives. Pharmaceutics 13(10):1715. https://doi.org/10.3390/pharmaceutics13101715

Tao X, Chen L, Cai L, Ge S, Deng X (2017) Regulatory effects of the AMPKα-SIRT1 molecular pathway on insulin resistance in PCOS mice: an in vitro and in vivo study. Biochem Biophys Res Commun 494(3–4):615–620. https://doi.org/10.1016/j.bbrc.2017.09.154

Vasyukova E, Zaikova E, Kalinina O, Gorelova I, Pyanova I, Bogatyreva E, Vasilieva E, Grineva E, Popova P (2023) Inflammatory and anti-inflammatory parameters in PCOS patients depending on body mass index: a case-control study. Biomedicines 11(10):2791. https://doi.org/10.3390/biomedicines11102791

Vázquez-Martínez ER, Gómez-Viais YI, García-Gómez E, Reyes-Mayoral C, Reyes-Muñoz E, Camacho-Arroyo I, Cerbón M (2019) DNA methylation in the pathogenesis of polycystic ovary syndrome. Reproduction 158(1):R27–R40. https://doi.org/10.1530/rep-18-0449

Wang K, Li Y, Chen Y (2023) Androgen excess: a hallmark of polycystic ovary syndrome. Front Endocrinol 14. https://doi.org/10.3389/fendo.2023.1273542

Wang M, Zhao D, Xu L, Guo W, Nie L, Lei Y, Long Y, Liu M, Wang Y, Zhang X, Zhang L, Li H, Zhang J, Yuan D, Yue L (2019a) Role of PCSK9 in lipid metabolic disorders and ovarian dysfunction in polycystic ovary syndrome. Metabolism 94:47–58. https://doi.org/10.1016/j.metabol.2019.02.002

Wang P, Zhao H, Li T, Zhang W, Wu K, Li M, Bian Y, Liu H, Ning Y, Li G, Chen Z-J (2014a) Hypomethylation of the LH/Choriogonadotropin receptor promoter region is a potential mechanism underlying susceptibility to polycystic ovary syndrome. Endocrinology 155(4):1445–1452. https://doi.org/10.1210/en.2013-1764

Wang W, Zheng J, Cui N, Jiang L, Zhou H, Zhang D, Hao G (2019b) Baicalin ameliorates polycystic ovary syndrome through AMP-activated protein kinase. J Ovarian Res 12(1). https://doi.org/10.1186/s13048-019-0585-2

Wang X-X, Wei J-Z, Jiao J, Jiang S-Y, Yu D-H, Li D (2014b) Genome-wide DNA methylation and gene expression patterns provide insight into polycystic ovary syndrome development. Oncotarget 5(16):6603–6610. https://doi.org/10.18632/oncotarget.2224

Włodarczyk M, Ciebiera M, Nowicka G, Łoziński T, Ali M, Al-Hendy A (2024) Epigallocatechin gallate for the treatment of Benign and Malignant Gynecological Diseases—focus on epigenetic mechanisms. Nutrients 16(4):559. https://doi.org/10.3390/nu16040559

Wu Y-X, Yang X-Y, Han B-S, Hu Y-Y, An T, Lv B-H, Lian J, Wang T-Y, Bao X-L, Gao L, Jiang G-J (2022) Naringenin regulates gut microbiota and SIRT1/ PGC-1α signaling pathway in rats with letrozole-induced polycystic ovary syndrome. Biomedicine & Pharmacotherapy 153:113286. https://doi.org/10.1016/j.biopha.2022.113286

Xiang Z, Guan H, Zhao X, Xie Q, Xie Z, Cai F, Dang R, Li M, Wang C (2024) Dietary gallic acid as an antioxidant: a review of its food industry applications, health benefits, bioavailability, nano-delivery systems, and drug interactions. Food Res Int 180:114068. https://doi.org/10.1016/j.foodres.2024.114068

Xu Y, Zhu H, Li W, Chen D, Xu Y, Xu A, Ye D (2022) Targeting adipokines in polycystic ovary syndrome and related metabolic disorders: from experimental insights to clinical studies. Pharmacol Therap 240:108284. https://doi.org/10.1016/j.pharmthera.2022.108284

Yu J, Liu Y, Zhang D, Zhai D, Song L, Cai Z, Yu C (2019) Baicalin inhibits recruitment of GATA1 to the HSD3B2 promoter and reverses hyperandrogenism of PCOS. J Endocrinol 240(3):497–507. https://doi.org/10.1530/joe-18-0678

Yuan B, Luo S, Feng L, Wang J, Mao J, Luo B (2022) Resveratrol regulates the inflammation and oxidative stress of granulosa cells in PCOS via targeting TLR2. J Bioenerg Biomembr 54(4):191–201. https://doi.org/10.1007/s10863-022-09942-7

Zhao H, Lv Y, Li L, Chen Z-J (2016) Genetic studies on polycystic ovary syndrome. Best Pract Res Clin Obstet Gynaecol 37:56–65. https://doi.org/10.1016/j.bpobgyn.2016.04.002

Advances in Biotechnology for Diabetes Research and Therapy

Manjula Nandakumar, Radwan Darwish, Alexandra E. Butler, and Abu Saleh Md Moin

Abstract Diabetes mellitus is a global health challenge marked by progressive pancreatic β-cell dysfunction, insulin resistance, and systemic metabolic imbalance. Despite pharmacological advances, current therapies remain largely supportive, failing to halt disease progression or restore endogenous insulin secretion. Biotechnology has transformed the landscape of diabetes research and therapy by enabling precise disease modelling, gene and cell-based interventions, and real-time metabolic monitoring. Tools such as CRISPR-Cas9, TALENs, and Cre-lox systems allow for tissue- and time-specific gene manipulation to investigate pathophysiological mechanisms and develop targeted therapies. Gene therapies and antisense oligonucleotides targeting insulin signalling, glucose metabolism, and inflammatory pathways show promise in preclinical and clinical studies for both type 1 and type 2 diabetes. Simultaneously, advances in stem cell biology and 3D bioprinting have enabled the differentiation, maturation, and encapsulation of insulin-producing β-cells from induced pluripotent stem cells (iPSCs), opening pathways for personalized regenerative treatments. Complementing these biological innovations, biosensors, smart insulin pumps, and closed-loop artificial pancreas systems exemplify the integration of engineering with therapeutic strategies. This chapter synthesizes recent biotechnological advances and explores how artificial intelligence, machine learning, and data-driven precision medicine may soon redefine diabetes management—transforming it from chronic disease care to curative intervention.

Keywords Type 2 diabetes · β-cell · Induced pluripotent stem cells · Obesity · Diabetes therapy

M. Nandakumar · A. E. Butler · A. S. M. Moin (✉)
Research Department, Royal College of Surgeons in Ireland - Bahrain, Adliya, Bahrain
e-mail: amoin@rcsi.com

R. Darwish
School of Medicine, Royal College of Surgeons in Ireland - Bahrain, Adliya, Bahrain

A. H. M. N. Nabi, H. Uddin Shekhar (eds.), *Biotechnological Advances in Healthomics*, Interdisciplinary Biotechnological Advances,
https://doi.org/10.1007/978-981-95-3612-2_9

Abbreviations

12-LOX	12-Lipoxygenase
AAV	Adeno-Associated Virus
ACC	Acetyl-CoA Carboxylase
ACT	Adenovirus Carrying Transcription Factors
AGE	Advanced Glycation Endproduct
ALK5i/ALK5iII	ALK5 Inhibitor (TGFβ Receptor Inhibitor)
AP	Artificial Pancreas
APOC3	Apolipoprotein C-III
ApoJ	Apolipoprotein J
ASO	Antisense Oligonucleotide
ATDC	Adipose Tissue Dendritic Cell
ATF	Adipose Tissue Macrophage
ATM	Adipose Tissue Macrophage
BMP	Bone Morphogenetic Protein
CAD	Cysteine-Rich Angiogenesis Determinant
CaMK1D	Calcium/Calmodulin Dependent Protein Kinase 1D
CD47	Cluster of Differentiation 47
CGM	Continuous Glucose Monitor
ChGa	Chromogranin A
ChIR99021	GSK3β Inhibitor (small molecule)
c-MYC	Avian Myelocytomatosis Viral Oncogene Homolog
CNT	Carbon Nanotube
CREB	cAMP Response Element-Binding Protein
CRISPR-Cas9	Clustered Regularly Interspaced Short Palindromic Repeats / CRISPR-associated protein 9
CRTC2	CREB-Regulated Transcription Coactivator 2
CSII	Continuous Subcutaneous Insulin Infusion
CXCL12	C-X-C Motif Chemokine Ligand 12
Cyp	Cytochrome P
DAPT	N-[N-(3,5-Difluorophenacetyl)-L-alanyl]-S-phenylglycine t-butyl ester (gamma-secretase inhibitor)
db/db	Leptin Receptor-Deficient Diabetic Mouse Model
DC	Dendritic Cell
DEP1	Density-Enhanced Phosphatase-1 (also known as PTPRJ)
DGAT2	Diacylglycerol Acyltransferase 2
DIO	Diet-Induced Obese
EBV	Epstein-Barr Virus
ECM	Extracellular Matrix
EGF	Epidermal Growth Factor
EGFP	Enhanced Green Fluorescent Protein
EGR1	Early Growth Response 1
ELISA	Enzyme-Linked Immunosorbent Assay

ER	Estrogen Receptor
ESCs	Embryonic Stem Cells
FAS	Fatty Acid Synthase
FcRn	Neonatal Fc Receptor
FET	Field-Effect Transistor
FGF	Fibroblast Growth Factor
FGF2, FGF7, FGF10	Fibroblast Growth Factor 2, 7, 10
FGFR2b	Fibroblast Growth Factor Receptor 2b
FIKL	Foetal INS^+ K/L Cells?
Flt23k	Fms-Like Tyrosine Kinase-1 Intraceptor
FMT	Fecal microbiota transplantation
FoxO1	Forkhead Box Protein O1
G6P	Glucose-6-Phosphate
G6PC	Glucose-6-Phosphatase Catalytic Subunit
G6PT1	Glucose-6-Phosphate Transporter 1
GCCR	Glucocorticoid Receptor
GCG	Glucagon
GCGR	Glucagon Receptor
GCK	Glucokinase
GLIS3	GLIS Family Zinc Finger 3
GLP-1	Glucagon-Like Peptide-1
GLUT1	Glucose Transporter 1
GLUT4	Glucose Transporter Type 4
GM3S	Ganglioside-Monosialic Acid 3 Synthase
GPI	Glycosylphosphatidylinositol
GSIS	Glucose-Stimulated Insulin Secretion
GSiXX	Gamma-Secretase Inhibitor XX
GSK3β	Glycogen Synthase Kinase 3 Beta
HFD	High-Fat Diet
HGF	Hepatocyte Growth Factor
HGFK1	Kringle 1 Domain of Hepatocyte Growth Factor
HIV-1	Human Immunodeficiency Virus Type 1
HLA	Human Leukocyte Antigen
HNF1A	Hepatocyte Nuclear Factor 1 Alpha
HNF4A	Hepatocyte Nuclear Factor 4 Alpha
HRP	Horseradish Peroxidase
IgG	Immunoglobulin G
IKKβ	IκB Kinase β
INS	Insulin
iPSCs	Induced Pluripotent Stem Cells
IR	Insulin Receptor
IRS	Insulin Receptor Substrate
ITC	Isothermal Titration Calorimetry
JNK1	c-Jun N-terminal Kinase 1
Klf6	Krüppel-Like Factor 6

LDN	LDN-193189 (BMP Inhibitor)
LNP	Lipid Nanoparticle
Lox	Locus of Crossing Over (from Cre-lox system)
Lyz2	Lysozyme 2
MAF	MAF BZIP Transcription Factor A
MAFA	V-Maf Avian Musculoaponeurotic Fibrosarcoma Oncogene Homolog A
MARD	Mean Absolute Relative Difference
MDI	Multiple Daily Injections
MIP	Molecularly Imprinted Polymer
miR-302s, miR-367s	MicroRNAs 302 family, 367 family
MIRKO	Muscle-Specific Insulin Receptor Knockout
MRI-PDFF	Magnetic Resonance Imaging – Proton Density Fat Fraction
mRNA	Messenger Ribonucleic Acid
MSCs	Mesenchymal Stem Cells
NAFLD	Non-alcoholic fatty liver disease
NASH	Non-Alcoholic Steatohepatitis
N-Cys	N-acetylcysteine
NEUROD1	Neuronal Differentiation 1
NF-κB	Nuclear Factor Kappa-light-chain-enhancer of Activated B Cells
Ngn3	Neurogenin 3
NGN3	Neurogenin 3
NKX6.1, NKX2.2	NK Homeobox 6.1, 2.2
NOD	Non-Obese Diabetic
NOGGIN	BMP Antagonist Protein
OCT4	Octamer-binding Transcription Factor 4
OIR	Oxygen-Induced Retinopathy
OKSM/OSKM	OCT4, KLF4, SOX2, c-MYC (Yamanaka Factors)
PAX6	Paired Box 6
PCK1	Phosphoenolpyruvate Carboxykinase 1
Pdx1	Pancreatic and Duodenal Homeobox 1
PDX1	Pancreatic and Duodenal Homeobox 1
PEC-01	Pancreatic endoderm cells
PEDF	Pigment Epithelium-Derived Factor
PI3K	Phosphoinositide 3-Kinase
PID	Proportional-Integral-Derivative
PKC	Protein Kinase C
PKCε	Protein Kinase C Epsilon
PPARγ	Peroxisome Proliferator-Activated Receptor Gamma
PTF1A	Pancreas Specific Transcription Factor 1a
PTP1B	Protein Tyrosine Phosphatase 1B
qRT-PCR	Quantitative Reverse Transcription Polymerase Chain Reaction

R428	AXL Receptor Tyrosine Kinase Inhibitor
RA	Retinoic Acid
RAGE	Receptor for Advanced Glycation Endproducts
RBE	Rep Binding Element
RBP4	Retinol Binding Protein 4
Rep78	Replication Protein 78
RFX6	Regulatory Factor X6
RIP-Cre	Rat Insulin Promoter-Cre
RNAi	RNA Interference
ROCK	Rho-Associated Protein Kinase
RT-PCR	Reverse Transcription Polymerase Chain Reaction
saRNAs	Small activating RNAs
SC	Stem Cell (as in SC beta cells)
SCNT	Somatic Cell Nuclear Transfer
scRNA-seq	Single-cell RNA sequencing
SGLT2	Sodium-Glucose Co-Transporter 2
SHH	Sonic Hedgehog
shRNA	Short Hairpin RNA
shRNAs	Short hairpin RNAs
siRNA	Small Interfering RNA
SLC30A8	Solute Carrier Family 30 Member 8
SREBP-1c	Sterol Regulatory Element-Binding Protein 1c
ssASO	Splice-Switching Antisense Oligonucleotide
SST	Somatostatin
STAT3	Signal Transducer and Activator of Transcription 3
STZ	Streptozotocin
SV40	Simian Virus 40
SUR1	Sulfonylurea Receptor 1 (also known as ABCC8: ATP-Binding Cassette Subfamily C Member 8)
T1D	Type 1 Diabetes
T1DM	Type 1 Diabetes Mellitus
T2DM	Type 2 Diabetes Mellitus
T3	Triiodothyronine (Thyroid Hormone)
TALENs	Transcription Activator-Like Effector Nucleases
TAM	Tamoxifen
TCF7L2	Transcription factor 7-like 2
TGFβ	Transforming Growth Factor Beta
TGFβR1	Transforming Growth Factor Beta Receptor 1
TIGIT	T cell Immunoreceptor with Ig and ITIM domains
TLR2	Toll-Like Receptor 2
TNF-α	Tumor Necrosis Factor Alpha
Tregs	Regulatory T Cells
TRIB3	Tribbles Pseudokinase 3
TTR	Transthyretin
UTR	Untranslated Region

VEGF	Vascular Endothelial Growth Factor
WNT	Wingless/Integrated (signaling pathway)
WT	Wild Type
ZDF	Zucker Diabetic Fatty

1 Introduction

Diabetes mellitus is one of the most pervasive chronic diseases of the twenty-first century, affecting over 580 million people globally as of 2025—a number projected to rise to 850 million world-wide by 2050 (Diabetes Atlas, International Diabetes Federation (IDF), 2025). Characterized by sustained hyperglycemia due to defects in insulin secretion, insulin action, or both, diabetes poses significant medical, economic, and societal burdens. Despite major advancements in pharmacological treatments and glucose-monitoring technologies, existing therapies largely focus on symptom management rather than addressing the root causes of disease, particularly the progressive loss or dysfunction of pancreatic β-cells (Hudish et al. 2019). As such, there is a growing impetus to explore disease-modifying strategies that can restore metabolic homeostasis, prevent complications, and ultimately achieve a functional cure.

Recent breakthroughs in biotechnology have transformed our understanding of diabetes pathophysiology and opened new therapeutic frontiers. Tools such as gene therapy, antisense oligonucleotides (ASOs), genome editing platforms (e.g., CRISPR-Cas9), and inducible recombinase systems like Cre-lox have enabled researchers to investigate gene function with unprecedented precision and to design molecular interventions that go beyond conventional pharmacotherapy (Smith et al. 2020; Bevacqua et al. 2021). These approaches have not only facilitated deeper insights into β-cell biology, immune dysregulation, and metabolic signaling but have also laid the foundation for innovative therapies targeting the pancreas, liver, adipose tissue, and kidney.

In particular, gene-based therapies have shown promise in regenerating β-cell mass, in vivo reprogramming of acinar cells to islet endocrine cells, and delivering therapeutic genes such as insulin, PDX1, and GLP-1 in preclinical models of both type 1 and type 2 diabetes (Li et al. 2014; So and Han 2024). The use of viral and non-viral vectors, although presenting distinct challenges and safety considerations, allows for tailored delivery strategies based on target tissues and treatment goals (Lundstrom 2018). Concurrently, ASOs targeting metabolic regulators such as PTP1B, SGLT2, GCGR, CREB, and PEPCK have demonstrated efficacy in modulating insulin sensitivity, glucose production, and lipid metabolism (Chen et al. 2021a, b). Furthermore, inducible and tissue-specific gene editing using Cre-lox models has become an indispensable tool for dissecting gene function in specific cell types. For example, in diabetes research, this system is particularly valuable for targeting pancreatic β-cells. Leveraging the unique expression of the insulin gene exclusively in β-cells, the rat insulin promoter (RIP) is commonly used to drive Cre recombinase expression, allowing for selective gene deletion in β-cells and facilitating the investigation of their role in glucose homeostasis and diabetes pathogenesis (Ray et al. 2000).

Beyond genetic interventions, biotechnology is also reshaping diabetes care through regenerative medicine. The derivation of insulin-producing β-like cells from induced pluripotent stem cells (iPSCs) offers a renewable source of autologous or allogeneic cells for transplantation, disease modelling, and drug screening (Pagliuca et al. 2014). Techniques such as 3D bioprinting, macro- and microencapsulation, and single-cell transcriptomics have improved the maturation, function, and immunoprotection of these stem cell-derived cells, accelerating their translation into clinical therapies (Du et al. 2022; Ramzy et al. 2023). Complementing these cellular strategies, wearable biosensors and closed-loop insulin delivery systems—enabled by advances in materials science, nanotechnology, and algorithm design—are bridging biotechnology with real-time metabolic management. These devices not only improve glycemic control but also provide continuous physiological feedback, paving the way for integrated artificial pancreas systems (Cappon et al. 2019; Sherr et al. 2023).

This chapter will explore the multifaceted applications of biotechnology in diabetes research and treatment. We will discuss gene and cell-based therapies, genome editing tools, antisense oligonucleotides, disease modelling, and regenerative approaches. Additionally, the chapter will address device-based innovations, encapsulation techniques, and the ethical and technical challenges surrounding translational applications. Together, these developments represent a paradigm shift from supportive care toward durable and potentially curative interventions for diabetes.

2 Gene Therapy and Genetic Tools in Diabetes Research

2.1 Gene Delivery Techniques

Gene therapy depends on two components: (1) a therapeutic transgene capable of transmitting expressable genetic modifications, and (2) a robust delivery system. Currently, there are two main gene delivery systems: viral and non-viral vectors (Sayed et al. 2022). A viral vector takes advantage of natural mechanisms of infection to deliver a transgene, first fusing and entering host cells, disassembling its capsid, and releasing its genome into the cytoplasm. The host cell machinery then transcribes and translates the delivered gene into the therapeutic protein (Kay et al. 2001). The most commonly used viral vectors are lentiviruses, adenoviruses, and adeno-associated viruses (AAVs) (Table 1). Lentiviral vectors are derived from Human Immunodeficiency Virus 1 (HIV-1), supporting long-term gene expression due to host genome integration. Lentiviruses can deliver transgenes to both dividing and non-dividing cells, enhancing their suitability in both in vivo and in vitro

Table 1 Comprehensive comparison of gene delivery platforms across vector type, payload capacity, compatibility with non-dividing cells, genomic integration potential, expression durability, and associated advantages and limitations in therapeutic applications

Vector type	Capacity (kb)	Non-dividing cells	Genome integration	Expression duration	Advantages/ disadvantages
Adenovirus	7–8	Yes	No	Transient	Strong expression; highly immunogenic
AAV	~4.5	Yes	No (episomal)	Long-term	Low immunogenicity; small payload
Retrovirus	~8	No	Yes	Stable (genomic)	Efficient for dividing cells; risk of insertional mutagenesis
Lentivirus	~8	Yes	Yes	Stable (genomic)	Infects non-dividing cells; improved safety vectors
Non-viral (lipid/ polymer)	Unlimited	Yes	No	Transient	Non-immunogenic; lower efficiency; flexible payload
Physical (electroporation, ultrasound)	Unlimited	Yes	No	Transient	No vector required; limited tissue applicability

settings. Further, non-integrating adenoviral vectors transiently deliver relatively large DNA cargos, but often elicit strong immune responses, limiting repeated administration (Mingozzi and High 2013). Finally, AAVs are small, non-enveloped viruses originating from the *Parvoviridae* family (Samulski and Muzyczka 2014). They can only replicate in host cells in the presence of a helper virus (usually adenovirus or herpesvirus). Like adenoviruses, AAVs are non-integrating, thereby minimizing the risk of insertional mutagenesis (Park et al. 2008).

Non-viral vector systems generally include physical methods and chemical carriers. While these systems offer improved safety and scalability compared to viral vectors, they typically exhibit lower transfection efficiency (Huang et al. 2022). First, physical methods involve electroporation, a process that involves creating transient membrane pores using electric field to permit nucleic acid entry, microinjection, and hydrodynamic injections (using high-pressure fluids) (Tang et al. 2021; Yang et al. 2021a, b). Chemical methods include using lipid-based nanoparticles (LNPs) to encapsulate nucleic acids to protect them *en route* to their target and enable cellular uptake via endocytosis (Hayes et al. 2006; Colombani et al. 2017; Kulkarni et al. 2018). Further, tunable polymeric carriers can be used to form complexes with nucleic acid and enhance delivery (Zhang et al. 2021a, b). Non-viral methods are non-integrating and hence permit only transient gene expression, unlike their viral counterparts. However, this makes them more amenable to repeated administration.

2.2 Gene Therapy in Diabetes

Gene therapies to revitalize β-cells are effective in both Type 1 diabetes mellitus (T1DM), and Type 2 diabetes mellitus (T2DM). T1DM results from the autoimmune destruction of β-cells, leading to a significant reduction in β-cell mass. Recent trials have shown that, at best, current immunotherapies only delay the progression of T1DM. In this sense, gene therapy aims to reconstitute the residual β-cell mass. One study by So and colleagues showed that the delivery of AAV-mediated paired box 6 (*Pax6*) into the islets of streptozotocin (STZ)-induced diabetic mice improved glycemic control by stimulating β-cell proliferation, and inhibiting apoptosis (So et al. 2021; So et al. 2023). Building on this, α-cells, which arise from the same endocrine progenitor lineage as β-cells, have emerged as promising targets for β-cell regeneration strategies due to their plasticity and capacity for transdifferentiation (Thorel et al. 2010; Habener and Stanojevic 2012). This may be a particularly pertinent strategy, seeing that α-cell hyperplasia is common in T1DM (and T2DM), providing an abundant supply for reprogramming. The α-cells in both β-cell-toxin-induced diabetic mice and autoimmune non-obese diabetic (NOD) mice were induced to differentiate into β-cells using AAVs carrying pancreatic and duodenal homeobox 1 (*Pdx1*) and MAF bZIP transcription factor A (*MafA*) expression cassettes (Xiao et al. 2018; So and Han 2024). The same transdifferentiation phenomenon has been demonstrated in pancreatic acinar cells, achieved upon in vivo delivery of an adenovirus carrying three essential transcription factors: neurogenin 3 (*Ngn3*), *Pdx1*, and *MafA*. The induced β-cells are morphologically indistinguishable from endogenous β-cells and harbour the capacity to secrete insulin and remodel local vasculature (Zhou et al. 2008). This has also been demonstrated in adult human tissue, albeit in vitro, where the ectopic expression of PDX1 in cultures of freshly isolated adult and fetal human liver tissue activated the otherwise silent insulin promoter. Upon transplantation into diabetic, immunodeficient mice, this led to glucose-regulated insulin production and secretion (Ferber et al. 2000; Miyatsuka et al. 2003; Kaneto et al. 2005; Sapir et al. 2005; Wang et al. 2007). The potential of gene therapy extends beyond these three transcription factors, with one study using a fiber-modified infectivity-enhanced adenoviral vector to deliver constitutively active AKT serine/threonine kinase 1 (*Akt1*) into both freshly isolated pancreatic islets and the β-cells of STZ-induced diabetic mice. At the time, β-cell survival and proliferation had only been demonstrated in vitro, as the in vivo experiments were compromised by limited vector delivery systems (Bone et al. 2012). These effects extend to the treatment of T2DM, where the transgenic expression of *Pdx1* in insulin receptor substrate 2 (IRS-2) deficient mice restored β-cell mass through proliferation, effectively preventing the onset of disease. Similarly, AAV-delivered *MafA* and *Pdx1* in db/db mouse cells ameliorated glycemic perturbations by suppressing β-cell apoptosis and enhancing β-cell identity (Matsuoka et al. 2015; So et al. 2021, 2023). The future lies in producing adaptive gene-delivery systems that modulate β-cell identity in response to metabolic cues. This can be potentially achieved by paracrine self-regulation of the islet niche.

2.3 Gene Therapy for Diabetic Complications

The benefits of gene therapy also extend to complications associated with diabetes. In recent years, gene-based therapeutics addressing hepatic complications (NASH, NAFLD) have emerged, and generally target peroxisome proliferator-activated receptor gamma (*PPARγ*). *PPARγ* is a transcription factor that regulates adipocyte differentiation, and its dysregulation is commonly implicated in insulin resistance. The sterol regulatory element-binding protein 1c (SREBP-1c), a transcription factor responsible for lipid synthesis, often referred to as the hallmark of NAFLD, often rises in states of hyperinsulinemia and insulin resistance.

FGF19 inhibits bile acid synthesis in the liver by regulating Cytochrome p (Cyp) 7A1 (Inagaki et al. 2005; Moschetta and Kliewer 2005; Song et al. 2009), while FGF21 primarily acts on adipose tissue, inhibiting lipolysis (Li et al. 2009; Chen et al. 2011), increasing triglyceride uptake (Schlein et al. 2016). However, perhaps most importantly, both FGFs reduce plasma insulin levels, which, as explored previously, reduces de novo lipogenesis by mitigating SREBP-1c activity (Zhang et al. 2011). FGF19 and FGF21 analogues are currently in clinical development for use in NASH, producing a 70% reduction in steatosis (MRI-PDFF) after only 12 weeks (Henriksson and Andersen 2020). As such, they have garnered quite some attention in the academic community and might be worthwhile exploring in gene-based therapeutics.

Further, chronic wound healing in diabetic patients can be encouraged through gene transfer to the damaged site. In such a case, non-viral vectors would be preferable, considering that their viral counterparts can produce an immunological reaction that impedes healing (Uludag et al. 2019). In that regard, small interfering RNAs (siRNAs) targeting matrix metalloproteinase (MMP), ganglioside-mono sialic acid 3 synthase (GM3S), and tumor necrosis factor alpha (TNF-α), all of which are present in nonhealing chronic wounds, are being developed. It naturally follows that knockdown of the genes encoding these proteins would allay the inflammatory process and promote healing (Rabbani et al. 2017; Li et al. 2018; Shi et al. 2024). Since transfection vectors would be inappropriate in this context, nanoneedles are used to deliver the siRNAs to their designated location. However, given the inflammatory milieu into which these RNAs are injected, many fail to reach their target cells. Conversely, plasmids encoding vascular endothelial growth factor (VEGF), an angiogenesis stimulator, could also be used (Ferraro et al. 2009; Ropper et al. 2009; Barc et al. 2021). However, plasmids share similar drawbacks to siRNAs, with limited tissue penetration and possible off-target effects (Arunadevi and Jone Kirubavathy 2023). These can be reduced using sonoporation, which uses ultrasound to permeate cell membranes and allow nucleic acid transfer. Despite its promise, sonoporation has its own limitations, most notably its restricted applicability to small tissue areas, which may be inadequate for achieving effective therapeutic tissue regeneration (Rychak and Klibanov 2014).

In diabetic nephropathy, intramuscular delivery of hepatocyte growth factor (HGF)-encoding plasmids, combined with electroporation, induced a regression in mesangial expansion and albuminuria, though these effects were less pronounced in early diabetic nephropathy (Cruzado et al. 2004). Meanwhile, Matsui and colleagues prevented advanced glycation endproducts (AGEs) from binding to their receptors (RAGEs) using RAGE-directed aptamers (specific single-strand RNAs or DNAs) in STZ-induced diabetic mice (Matsui et al. 2017). Similarly, DNA aptamers binding periostin, an extracellular matrix (ECM) protein, were used to abrogate increases in proinflammatory, pro-fibrotic ECM proteins (Um et al. 2017). Although these studies approach diabetic nephropathy from different angles, they generally aim to reduce fibrosis and proteinuria through mechanisms that remain elusive (Tavakolidakhrabadi et al. 2024). However, a recent study by Gujarati and colleagues (Gujarati et al. 2024) showed that it is not injury to podocytes, which form part of the glomerular filtration barrier, that precipitates nephropathy, but rather proximal tubular dysfunction. To that extent, the group reported that podocyte-specific induction of Krueppel-like factor 6 (*Klf6*) (Mallipattu et al. 2015; Horne et al. 2018) minimized podocyte loss, improved proximal tubular function, and reduced interstitial fibrosis in a male murine model. It is thought that *Klf6* activates calcium/calmodulin-dependent protein kinase 1D (CaMK1D) signaling in proximal tubular cells by triggering the release of ApoJ. This protects the proximal tubule against injury by inhibiting mitochondrial fission (Knott et al. 2008). Therefore, it is thought that constitutively upregulating ApoJ secretion from podocytes, or bypassing ApoJ and upregulating CaMK1D would reduce diabetic nephropathy progression.

Diabetic retinopathy is characterized by neovascularization and increased vascular permeability, both of which can be targeted using gene therapy. Using transgenes like Soluble fms-like tyrosine kinase-1 (sFlt-1), pigment epithelium-derived factor (PEDF), endostatin, angiostatin, and CAD, endothelial proliferation and angiogenesis can be inhibited. For instance, sFlt-1 is a tyrosine kinase protein that acts extracellularly as a VEGF decoy and has been shown to suppress neovascularization in animal models (Bainbridge et al. 2002; Lai et al. 2002; Gehlbach et al. 2003; Ideno et al. 2007; Lamartina et al. 2007; Pechan et al. 2009). Meanwhile, Flt23k is an anti-VEGF intraceptor, effectively degrading VEGF intracellularly (Prea et al. 2015; Zhang et al. 2015). Endogenous inhibitors of PEDF and angiostatin can also be used to reduce VEGF expression and hence neovascularization (O'Reilly et al. 1994; Haurigot et al. 2012). Further, amino-terminal fragment (ATF) and kringle 1 domain of hepatocyte growth factor (HGFK1) inhibit endothelial cell proliferation. Adenovirus-mediated ATF or endostatin delivery reduced retinal neovascularization in oxygen-induced retinopathy (OIR) mice (Le Gat et al. 2003). The same outcomes were observed in OIR mice when HGFK1 was delivered using AAV (Sun and Liu 2019). Beyond angiogenesis, diabetic retinopathy can be targeted in its earlier stages, offering both neuronal and vascular protection from progression. One in vivo study delivering small hairpin RNAs (shRNAs) targeting early growth response 1 (EGR1) using an AAV vector, demonstrated dramatic reductions in apoptosis levels across retinal endothelial cells (Ao et al. 2019).

2.4 Insulin Gene Delivery

The insulin gene can be delivered to individuals with diabetes to generate an endogenous source of insulin. One study treated streptozocin (STZ)-induced diabetic rats with furin-cleavable insulin gene intrapancreatically using recombinant adenoviral vectors, enabling glucose-stimulated insulin secretion (GSIS) from induced hepatocytes (Shifrin et al. 2001). A similar outcome was observed in both STZ-induced and autoimmune insulin-dependent diabetic rats (LEW.1AR1/Ztm-iddm), an inbred strain that spontaneously develops T1DM due to T-cell-mediated β-cell destruction, following treatment with lentiviral vectors encoding insulin cDNA (Elsner et al. 2012). That said, these approaches remain invasive, and so it may be more appropriate to exploit other accessible organs that harbour the capacity to differentiate into a β-cell phenotype (Beumer and Clevers 2016; Meran et al. 2017; Nair et al. 2020; Collier et al. 2022). To that extent, oral gavages of chitosan nanoparticles (non-viral) carrying the insulin gene were administered to diabetic mice and delivered to both K- and L-cells, of which the former was more successful in reducing blood glucose (Rasouli et al. 2013). Though these experiments were conducted in 2013, it was not until 2020 that a subset of foetal enteroendocrine K- and L-cells, denoted foetal INS^+ K/L (FIKL) cells, expressing high insulin levels and other β-cell genes were discovered. These cells bear great resemblance to pancreatic bi-hormonal cells that, following repression of certain genes, commit to different endocrine lineages (Egozi et al. 2021). In the pancreas, these bi-hormonal cells respond to local cues and transcriptional repression by differentiating into either α- or β-cells. The similarity suggests that FIKL cells retain a comparable plasticity, meaning they may be redirected toward a stable β-cell fate with appropriate gene modulation. This lineage flexibility makes FIKL cells a compelling target for gene therapy, as they offer a naturally insulin-expressing cell population that could be engineered into fully functional β-like cells capable of regulated glucose-stimulated insulin secretion. That said, the main challenge associated with insulin gene therapy is achieving highly regulated glucose-stimulated insulin secretion. This leads us to the idea of molecular switches, which serve as glucose-responsive promoters that control insulin production and secretion.

2.5 Molecular Switches of Glucose-Sensitive Insulin

Recent synthetic biology advances have enabled glucose-sensitive genetic circuits via designer cells that mimic β-cell sensing. These cells express glycolysis-mediated calcium channels coupled to an insulin gene under a calcium-responsive promoter. Under hyperglycemic conditions, glucose metabolism induces calcium ion influx, which in turn activates a calcium-responsive promoter, leading to insulin gene expression. These cells coordinate insulin release in a closed-loop manner that is sufficiently sensitive to glucose fluctuations and quantifies these fluctuations accurately. The idea of creating a built-in switch into synthetic circuits to generate biosimilar insulin is particularly pertinent. To that extent, NNC2215, a glucose-responsive insulin conjugate, is especially fascinating. NNC2215 features an insulin molecule conjugated to both a glucose-binding macrocycle and a glucoside, essentially a dual molecular switch (Fig. 1). At low glucose concentrations, this molecular switch remains closed, limiting insulin receptor binding to prevent hypoglycemia. As glucose rises (4–20 mM), the switch opens, enabling receptor binding. The glucose affinity of NNC2215 is matched to physiological needs, as demonstrated using binding studies (ITC and native MS) (Hoeg-Jensen et al. 2024). Molecular switches can be designed using other approaches. One group added a tether between the A and B chains of insulin to inhibit its activation. Fructose, when present, binds to a sensing element, displacing the tether, and enabling controlled activation (Chen et al. 2021a, b). Though these examples of biosimilar insulin do not directly use gene therapy, the molecular logic supporting them may inspire new approaches, delivering genes encoding glucose-responsive insulin elements.

2.6 *Incretin Gene Delivery*

Incretins such as glucagon-like peptide-1 (GLP-1) and GIP enhance insulin secretion and have protective effects on β-cells. Their short half-lives limit conventional therapy, but gene delivery can induce sustained incretin production (Liu et al. 2010a, b; Riedel and Kieffer 2010). Initial studies engineered plasmids by integrating an active GLP-1 peptide into the Fc regions of mouse IgG. Presumably, this would increase the half-life and potency of GLP-1 in vivo. Using this method, plasmids were detectable in plasma up to 4 weeks post-injection in both STZ-induced and db/db mice (Kumar et al. 2007; Soltani et al. 2007). Similar design principles were also applied to exendin-4 peptides, which improved glycemic control in STZ-induced mice (Soltani et al. 2007). In the years that followed, a non-viral gene therapy for GLP-1/Fc was used, employing a REP78-in-trans and REB-in-cis element system to enable site-specific genomic integration. In this study, GLP-1/Fc expression cassettes were fused with the RBE element and injected alongside a Rep78-expressing plasmid intramuscularly into mice genetically modified to contain the human AAVS1 locus on chromosome 19. Rep78, a regulatory protein encoded by AAV, plays a critical role in mediating targeted integration at the AAVS1 site, thereby enhancing the stability and predictability of transgene expression (19759129). On the HFD challenge, GLP-1/Fc improved lipid profiles and reduced weight gain (Liu et al. 2010a, b). Further, Choi et al. constructed a plasmid encoding the active GLP-1(7–37) peptide (with furin cleavage sites) driven by an SV40 promoter with added NF-κB enhancer elements. This plasmid was then incorporated into a double-stranded, adeno-associated viral (dsAAV) expression vector and delivered into db/db mice. The results revealed a reduction in fasting blood glucose levels that extended up to 4 months after a single injection of the dsAAV GLP-1. Concurrently, circulating GLP-1 levels and glucose-dependent insulin secretion increased (Choi

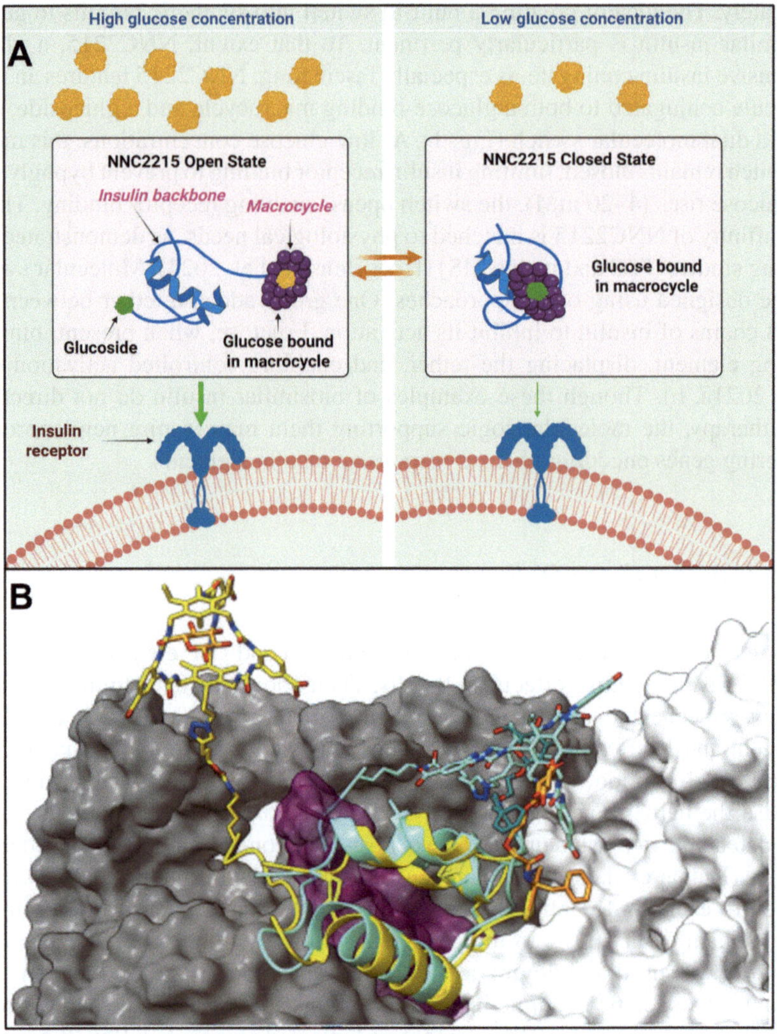

Fig. 1 Functional Principle and 3D Model of NNC2215. (**a**) NNC2215 is an insulin molecule with a glucose-sensitive switch. At high glucose levels, the switch opens, increasing insulin receptor affinity and preventing hyperglycemia. At low glucose levels, it closes, reducing receptor binding and preventing hypoglycemia. (**b**) The 3D models show NNC2215 in open (yellow) and closed (cyan) forms. The insulin backbone appears as ribbons, with switch elements (glucoside and macrocycle) in stick representation. Insulin receptor chains A and C are shown in white and grey. In the open form, free glucose (orange) binds at B29, while the closed form has glucoside bound within the macrocycle, causing a structural clash with the receptor (α-CT, purple). Image **a** was recreated based precisely on an image from the same article, while Image **b** was directly sourced from the article (Hoeg-Jensen et al. 2024). (Written permission was obtained from both the lead author and the publisher)

and Lee 2010). In a more recent study, an oral gene delivery system was designed, using a human IgG-1-Fc antibody modified with cysteine residues. Not only did this system extend the half-life of GLP-1 but also shielded the peptide from degradation in the gastrointestinal tract, facilitating its uptake and transcytosis via the neonatal Fc receptor (FcRn), a receptor responsible for IgG recycling and bidirectional transport across epithelial barriers, and subsequent release in the intestinal region, where it would be absorbed. In db/db mice, this platform improved glucose-dependent insulin secretion and increased GLP-1 peptide localization in the islets (Hasan et al. 2020).

3 Antisense Oligonucleotides (ASOs) in Diabetes Research and Therapy

3.1 ASOs in Gene Silencing

ASOs are short, synthetic nucleic acid sequences designed to bind specific RNA targets through Watson-Crick base pairing (Quemener et al. 2020; Hammond et al. 2021). Once bound, ASOs can modulate gene expression either by promoting transcript degradation or by altering pre-mRNA splicing. For transcript knockdown, the most common mechanism involves RNase H recruitment: DNA-based ASOs form DNA/RNA hybrids with target mRNA, which are then recognized by the endogenous enzyme RNase H, leading to cleavage of the RNA strand (Fig. 2). This results in reduced protein production (Monia et al. 1993). These types of ASOs are known as gapmer antisense oligonucleotides (gapmer ASOs). In addition to gapmers, another class of ASOs called siRNAs also facilitates transcript knockdown but operates through the endogenous RNA interference (RNAi) pathway, which differs mechanistically from RNase H-mediated cleavage (Zhang et al. 2021a, b). Alternatively, chemically modified ASOs can bind to mRNA and sterically hinder ribosome assembly or spliceosome function, thereby modulating splicing. For example, ASOs can target a splice site to exclude a faulty exon from the mature transcript, potentially restoring production of a functional protein (Liang et al. 2017; Sasaki et al. 2019). These are referred to as splice-switching ASOs (ssASOs). Chemical modifications are also incorporated into the structure to increase stability and binding affinity (Lauffer et al. 2024).

3.2 ASOs in Diabetes and Obesity

ASOs targeting PTP1B, such as ISIS 113715, improve insulin sensitivity by (1) modulating protein expression, and (2) enhancing insulin signal transduction (Kenner et al. 1996; Goldstein et al. 2000). ASOs in diabetes and obesity have been

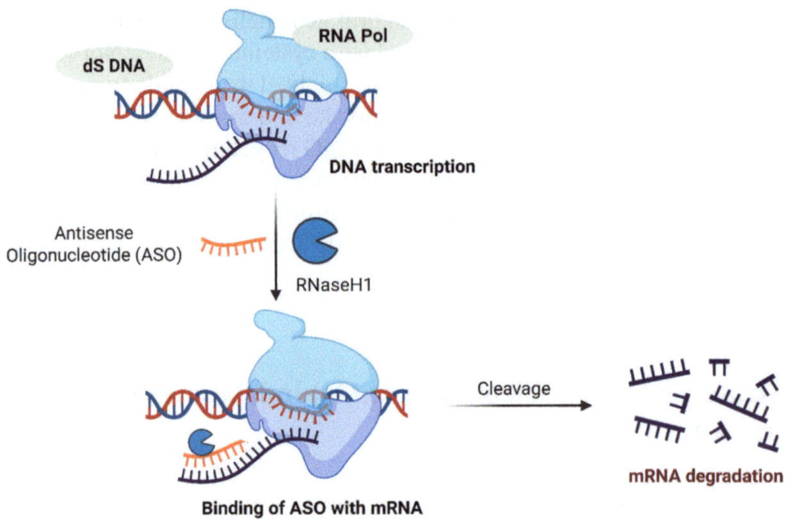

Fig. 2 Mechanism of antisense oligonucleotide (ASO)-mediated mRNA degradation. ASOs are short, synthetic nucleic acid sequences that bind to complementary mRNA transcripts. Following transcription of the target gene by RNA polymerase (RNA Pol), the ASO hybridizes with the mRNA to form a DNA/RNA duplex. This duplex is recognized by the endogenous enzyme RNase H1, which selectively cleaves the RNA strand, resulting in mRNA degradation and subsequent reduction of target gene expression (Sang et al. 2024). ds DNA, double-stranded DNA

summarized in Fig. 3. In both ob/ob and db/db mice, treatment with ISIS 113715 significantly reduced PTP1B expression in the liver and adipose tissue. These changes reduced hyperinsulinemia, while increasing the expression of proteins involved in the insulin signaling pathway, including insulin receptor substrate 2 (IRS2), phosphoinositide 3-kinase (PI3K) p50a, and p55a isoforms (Zinker et al. 2002). ASO treatment also reduces expression of gluconeogenic enzymes (PEPCK, FBP) (Zinker et al. 2002), while increasing adiponectin (Swarbrick et al. 2009), which promotes insulin signaling by activating IRS (Fang and Judd 2018). Further, PTP1B negatively regulates insulin signaling by dephosphorylating insulin receptor (IR) and insulin receptor substrate (IRS). In FAO rat hepatoma cells, PTP1B knockdown enhanced tyrosine phosphorylation of IR, IRS, and Akt, leading to normalized glucose levels (Clampit et al. 2003).

3.2.1 Targeting Glucagon Receptor (GCGR) with ASOs for Metabolic Improvement in T2D

ASOs targeting GCGR, such as ISIS 148359 and ISIS 180475, suppress glucagon signaling to reduce hyperglycemia in T2DM. Normally, glucagon promotes gluconeogenesis and glycogenolysis to promote hepatic glucose (Boden et al. 1990; Pilkis and Granner 1992). In db/db mice treated with ISIS 148359, GCGR knockdown lowered hepatic cAMP, and was reflected in reduced serum glucose, triglycerides, and free fatty acids (Liang et al. 2004). Another study that treated different

ASOs	T2DM patients/ Mouse models; Target tissue	Target Gene	Gene silencing outcome
ISIS 113715	• *ob/ob* and *db/db* mice; • *Liver and fat*	PTP1B	• Reduced hyperinsulinemia • Reduces expression of gluconeogenic enzymes (PEPCK, FBP)
ISIS 148359 ISIS 113715	• *db/db* mice, *db/db* mice ZDF rats ; • *liver*	GCGR	• Lowered hepatic cAMP • Reduction in gluconeogenic enzymes
ISIS 388626	• mice and cynomolgus monkeys ; • *kidneys*	SGLT2	• Increased glucosuria and decreased blood glucose levels
ISIS 180272 ISIS 426115	• *ob/ob*, *db/db*, ZDF, and high-fat diet (HFD); • *Liver and adipose tissue*	GCCR	• suppressed the expression of gluconeogenic enzymes (PEPCK, G6P)
ISIS 338292 ISIS 189594	• Diet-induced rat model of nonalcoholic fatty liver disease (NAFLD); • *Liver*	ACC	• Reduced hepatic triglycerides and diacylglycerol • Improved insulin sensitivity • Lowered glucose production
ISIS 217376 ISIS 369235	• diet-induced obese (DIO) and *ob/ob* mice; • *Liver, adipose tissue*	DGAT2	• Lowered triglyceride, diacylglycerol, and free fatty acid levels • Suppressed lipogenesis-related genes • Promoted expression lipid oxidation genes
ISIS 232697	• HFD rats ; • *Liver*	PKCε	• significant elevation in Akt2 activity and IRS2 tyrosine phosphorylation • Improved glucose handling and reduced hepatic steatosis
ISIS 149008	• *ob/ob* mice ; • *Liver*	G6PT1	• Reduced hepatic glucose production without causing hypoglycemia
ISIS 401723 ISIS 401724	• T2DM patients • *Liver*	RBP4	• Improved muscle glucose uptake, • Lowered hepatic glucose production, • Increased insulin sensitivity

Fig. 3 ASOs in diabetes and obesity associated pre-clinical and clinical trials. List of antisense oligonucleotides (ASOs), their pre-clinical (in obese and diabetic animal models), clinical (in humans with T2DM) studies, target tissue and genes, and the outcome of ASO-based gene modulation. ob/ob mice, Leptin-deficient obese mice; db/db mice, Leptin receptor-deficient diabetic mice; ZDF rats Zucker Diabetic Fatty rats; HFD High Fat Diet; NAFLD Non-alcohol fatty liver disease; PTP1B protein tyrosine phosphatase 1B; GCGR glucagon receptor; SGLT2 Sodium-Glucose Co-Transporter 2; GCCR Glucocorticoid Receptor; ACC Acetyl-CoA Carboxylase; DGAT Diacylglycerol Acyltransferase 2; PKCε Protein Kinase C Epsilon; G6PT1 Glucose-6-Phosphate Transporter 1; RBP4 Retinol Binding Protein 4; PEPCK Phosphoenolpyruvate carboxykinase; G6P glucose 6 phosphate

diabetic rodent models (ob/ob, db/db mouse, Zucker Diabetic Fatty (ZDF rats) with both ISIS 148359 and ISIS 180475, demonstrated a reduction in gluconeogenic enzymes coupled with improvements in GLP-1 and insulin secretion (Sloop et al. 2004). Ultimately, this improved glycemic control in ob/ob and db/db diabetic mouse models without causing hypoglycemia. Additionally, GCGR suppression preserved pancreatic β-cell function (Sloop et al. 2004).

3.2.2 Antisense Inhibition of SGLT2 Induces Glucosuria and Glycemic Reduction in Preclinical Models

ASO ISIS 388626 inhibits renal function, enhancing urinary glucose excretion (Bailey 2011). In one study, mice and cynomolgus monkeys were treated with ISIS 388626, which reduced SGLT2 mRNA expression in kidneys, leading to increased glucosuria and decreased blood glucose levels in mice and cynomolgus monkeys. Dose-dependent effects were observed without impacting renal function, though basophilic granule accumulation occurred (Zanardi et al. 2012). The ASO demonstrated a favorable safety profile and effectiveness, suggesting potential as a non-small molecule alternative to SGLT2 inhibitors like dapagliflozin. This approach has the benefits of reducing hyperglycemia without inducing systemic adverse metabolic effects.

3.2.3 Antisense Therapies Targeting Gluconeogenesis and Lipogenesis: A Metabolic Reprogramming Approach for Insulin Resistance

Glucocorticoids impair insulin action by stimulating gluconeogenesis (Exton 1979, Kraus-Friedmann 1984), reducing glucose uptake (Di Dalmazi et al. 2012; Kuo et al. 2013), and inhibiting glycogen synthesis in skeletal muscle (Stalmans and Laloux 1979; Ruzzin et al. 2005). ASOs such as ISIS 180272 and ISIS 426115 reduce GCCR expression in liver and adipose tissue, reducing fasting glucose, insulin, TNF-α, and triglyceride levels across 4 different rodent models (ob/ob, db/db, ZDF, and high-fat diet (HFD)-fed) (Watts et al. 2005). In ob/ob and db/db mice, ASO-mediated GCCR knockdown also suppressed the expression of gluconeogenic enzymes (PEPCK, G6P), further enhancing insulin sensitivity and mitigating hyperglycemia (Liang et al. 2005). Acetyl-CoA carboxylase (ACC) enzymes regulate lipid metabolism by producing malonyl-CoA, a key intermediate in fatty acid synthesis (Abu-Elheiga et al. 2001; Munday 2002). ASOs targeting ACC1 (ISIS 338292), ACC2 (ISIS 189594), and both simultaneously (ISIS 362037) enhance hepatic fat oxidation and reduce lipogenesis in non-alcoholic fatty liver disease (NAFLD) models (Savage et al. 2006). Dual ACC1/2 knockdown had the most profound effect, reducing hepatic triglycerides and diacylglycerols while improving insulin sensitivity and lowering glucose production (Savage et al. 2006). These results support ASO-based modulation of ACC isoforms as a promising strategy for T2DM by targeting lipid-induced insulin resistance and promoting hepatic metabolic reprogramming, especially in the context of NAFLD.

3.2.4 ASO-Based Modulation of Key Metabolic Enzymes: DGAT2, PEPCK, and PKCε in Insulin Resistance and NAFLD

Diacylglycerol acyltransferase-2 (DGAT2) catalyzes the final step in triglyceride synthesis (Yu et al. 2005). ASOs like ISIS 217376 and ISIS 369235 reduce hepatic and adipose DGAT2 expression, lowering triglyceride, diacylglycerol, and free fatty acid levels (Yu et al. 2005). In diet-induced obese (DIO) and ob/ob mice, ASO treatment also suppressed lipogenesis-related genes (ACC1/2, FAS), promoted expression of lipid oxidation genes (Yu et al. 2005). Further, administration of ISIS 369235 to rats with diet-induced NAFLD also improved hepatic insulin signaling via enhanced PI3K and Akt activity, key components of the insulin signaling pathway where PI3K activation leads to phosphorylation and activation of Akt, promoting glucose uptake and inhibiting gluconeogenesis, coupled with a reduction in hepatic glucose production (Choi et al. 2007). Targeting DGAT2 with ASOs thus offers the dual benefit of alleviating both hepatic steatosis and insulin resistance, much like ACC. Mitochondrial phosphoenolpyruvate carboxykinase (PEPCK-M) catalyzes the conversion of oxaloacetate to phosphoenolpyruvate, a key step in gluconeogenesis (Holyoak et al. 2006). ISIS 421062 targets and suppresses PEPCK-M in rodent models, leading to reductions in blood glucose, insulin, triglycerides, and white adipose tissue mass. This ASO shifts gluconeogenic substrate preference toward glycerol rather than pyruvate, without inducing hypoglycemia or compromising mitochondrial function (Stark et al. 2014). However, in HFD Sprague-Dawley rats, ISIS 113309 reduced hepatic and adipose PEPCK1 expression but impaired hepatic insulin sensitivity and worsened steatosis. Although adiposity decreased, rates of lipolysis, lipogenesis, and gluconeogenesis were unchanged. The hepatic insulin resistance was attributed to reduced GCK expression and impaired glycogen synthesis, associated with increased ATF3, a repressor of GCK. These findings suggest PEPCK1 coordinates metabolic gene regulation rather than directly driving gluconeogenesis in T2DM (Beddow et al. 2019). Seeing that protein kinase C-ε (PKCε) is activated by lipid accumulation, it plays a critical role in fat-induced insulin resistance. ASO ISIS 232697 reduces PKCε in the liver, enhancing insulin receptor kinase activity and IRS2-Akt signaling. In one study, a significant elevation in Akt2 activity and IRS2 tyrosine phosphorylation was observed in HFD rats after treatment, restoring hepatic insulin sensitivity. This translated into improved glucose handling and reduced hepatic steatosis (Samuel et al. 2007).

3.2.5 Dual Control of Gluconeogenesis via CREB, CRTC2 and FoxO1 Suppression: A Transcription-Hormone Interaction Strategy

G6PT1 mediates the final step of gluconeogenesis by transporting glucose-6-phosphate into the endoplasmic reticulum. In ob/ob mice, ASO ISIS 149008 reduced hepatic G6PT1 mRNA by 85% and glucose-6-phosphate transport activity by 75%. This reduced hepatic glucose production without causing hypoglycemia (Sloop et al. 2007). CREB is a transcription factor that activates gluconeogenic genes, including PEPCK, G6Pase, and FBPase (Hanson and Reshef 1997). In HFD-fed rats, treatment with a 2′-O-methoxyethyl (2′-MOE) gapmer ASO targeting CREB significantly reduced expression of PEPCK-C, PEPCK-M, and PGC1α by 43–55%, and suppressed hepatic glucose production by ~45%. Additionally, ASO treatment improved hepatic insulin sensitivity, decreased hepatic diacylglycerol and triglycerides, reduced plasma insulin by 66%, and inhibited PKCε activation (Erion et al. 2009). CRTC2 enhances CREB-mediated transcription of gluconeogenic genes (Koo et al. 2005; Ravnskjaer et al. 2007). In rats, ISIS 384680 decreased the expression of enzymes converting amino acids into gluconeogenic substrates. Alone, CRTC2 knockdown had minimal effect on glucose due to compensatory hyperglucagonemia and impaired glucagon clearance. However, co-administration with a glucagon-neutralizing antibody or somatostatin significantly suppressed glucose production, suggesting that dual inhibition of CRTC2 and glucagon signaling may be more effective in controlling hyperglycemia in T2DM (Erion et al. 2013). FoxO1 regulates hepatic glucose production by promoting gluconeogenesis and glycogenolysis, particularly during fasting or insulin resistance, through transcriptional activation of G6PC, PEPCK, and FBPase (Matsumoto et al. 2007; Rui 2014; I et al. 2015). Insulin normally suppresses FoxO1 via the PI3K/AKT pathway, reducing hepatic glucose output. In T2DM, impaired insulin signaling leads to FoxO1 overactivation and increased glucose production. In diet-induced obese (DIO) mice, ASO ISIS 188764 reduced FoxO1 expression in liver and adipose tissue, lowering plasma glucose by 34% and insulin by 87%, while improving hepatic steatosis and insulin sensitivity (Samuel et al. 2006).

3.2.6 ASOs Targeting Insulin Signaling Inhibitors: A Multi-Faceted Therapeutic Approach in Obesity and T2DM

DEP1 (PTPRJ) dephosphorylates the insulin receptor, dampening insulin signaling (Bandyopadhyay et al. 1997). In HFD-fed mice, ISIS 285564 reduced DEP1 expression in liver and fat, leading to enhanced phosphorylation of IR and Akt. This was reflected in reduction in fasting glucose levels, and significant weight loss (Kruger et al. 2013). RBP4 is an adipokine that promotes insulin resistance via proinflammatory signaling (Yang et al. 2005; Norseen et al. 2012; Moraes-Vieira et al. 2014). In HFD-fed mice, ASO treatment reduced RBP4 expression, normalized glucose and insulin levels, lowered triglycerides, and reduced visceral fat mass. ASO treatment also restored GLUT4, reduced hepatic PEPCK, and prevented hepatic steatosis (Tan et al. 2011). In line with these effects, elevated RBP4 levels are positively correlated with the extent of insulin resistance in T2DM patients with obesity (Cho et al. 2006; Graham et al. 2006; Broch et al. 2007). TTR stabilizes RBP4 in circulation (Zanotti and Berni 2004). When administered subcutaneously twice weekly for 2 weeks, ISIS 401723 and ISIS 401724 reduced RBP4 levels by 80–95%. These changes improved muscle glucose uptake, lowered hepatic glucose production, and increased insulin sensitivity. The effects were sustained for 9 weeks, with reduced insulin and inflammation, supporting TTR knockdown as an indirect but durable strategy to enhance insulin action via RBP4 clearance (Zemany et al. 2015). TRIB3 inhibits Akt (Du et al. 2003; Prudente et al. 2005) and suppresses PPARγ activity, impairing insulin signaling and adipocyte function (Takahashi et al. 2008). In HFD-fed rats, ISIS 391274 enhanced muscle and whole-body glucose uptake by ~50%, increased adiponectin, and upregulated lipid oxidation genes. Further, fat mass expanded significantly, restoring adipose buffering capacity. These effects were diminished with PPARγ antagonism, confirming TRIB3's role in modulating insulin sensitivity via PPARγ (Weismann et al. 2011). IKKβ activates NF-κB signaling and drives inflammatory insulin resistance (Cai et al. 2005; Hayden and Ghosh 2008). In DIO mice, ASO treatment reduced fasting glucose and insulin, increased Akt phosphorylation in adipose tissue, and reduced fat mass by 45%. Hepatic triglyceride accumulation and adipocyte differentiation were also inhibited (Helsley et al. 2016). That said, TLR2 also activates inflammatory pathways via IKKβ and JNK1 (Hirosumi et al. 2002), inducing insulin resistance via IRS1 inhibition. In HFD-fed mice, a single 18mer ASO injection targeting TLR2 reduced IKKβ and JNK1 phosphorylation in skeletal muscle and fat. This was reflected in an increase in IRS1 and Akt phosphorylation, restoring insulin sensitivity (Caricilli et al. 2008).

3.2.7 ASOs in Clinical Trials

Several ASOs are now progressing through early-to-mid phase clinical trials, targeting key regulators of hepatic lipid metabolism, glucose production, and inflammation. One of the most advanced ASOs in the context of metabolic disease targets diacylglycerol acyltransferase 2 (DGAT2), an enzyme catalyzing the final step in triglyceride synthesis. DGAT2 activity is upregulated in nonalcoholic fatty liver disease (NAFLD) and T2DM, and its inhibition has been shown to reduce hepatic steatosis, improve insulin sensitivity, and normalize glucose homeostasis (Choi et al. 2007). ISIS 369235 (also known as IONIS-DGAT2Rx), developed by Ionis Pharmaceuticals, is a second-generation ASO designed to reduce DGAT2 mRNA expression in the liver. In a Phase 1b/2a trial (NCT03334214) in subjects with NAFLD and insulin resistance, treatment with ISIS 369235 produced significant reductions in hepatic fat content as measured by MRI-PDFF, as well as improvements in markers of insulin sensitivity (HOMA-IR) and liver enzyme levels (Loomba et al. 2020). While this agent is not explicitly labeled as a diabetes treatment, its effect on hepatic insulin signaling suggests its utility in broader metabolic syndromes. Further studies are warranted to determine its efficacy in patients with established T2DM.

Another ASO, vupanorsen (formerly AKCEA-ANGPTL3-LRx), inhibits angiopoietin-like protein 3 (ANGPTL3), a liver-derived protein involved in lipid metabolism. ANGPTL3 inhibition reduces plasma levels of triglycerides and low-density lipoprotein cholesterol (LDL-C), and has shown indirect benefits on insulin sensitivity (Kersten 2017). Vupanorsen employs GalNAc (N-acetylgalactosamine) conjugation for liver-specific delivery, enhancing potency while minimizing systemic exposure (Cui et al. 2021). In a randomized, placebo-controlled Phase 2 trial involving patients with T2DM, hypertriglyceridemia, and hepatic steatosis (NCT03360747), vupanorsen significantly reduced fasting triglycerides by up to

53% and demonstrated modest reductions in HbA1c and liver fat content (Gaudet et al. 2020). However, its development for cardiovascular indications was recently halted due to concerns about hepatic enzyme elevations and suboptimal LDL-C lowering. Nevertheless, its insulin-sensitizing potential remains of interest, especially in patients with T2DM and dyslipidemia.

Apolipoprotein C-III (APOC3) regulates triglyceride-rich lipoprotein metabolism by inhibiting lipoprotein lipase (LPL) and hepatic uptake of remnants (Larsson et al. 2013). Elevated APOC3 is associated with insulin resistance, increased cardiovascular risk, and impaired β-cell function (Åvall et al. 2015). Volanesorsen, a first-generation ASO targeting APOC3, received conditional approval for familial chylomicronemia syndrome (FCS) but faced safety concerns due to thrombocytopenia (Cuchel et al. 2012; Witztum et al. 2019). Its next-generation successor, olezarsen (AKCEA-APOC3-LRx), uses a GalNAc-conjugated backbone, offering improved potency and a better safety profile. Though not developed specifically for diabetes, both agents demonstrate marked triglyceride-lowering effects (up to 70% reduction), which can indirectly improve insulin action and reduce ectopic fat deposition in liver and muscle (Bergmark et al. 2024; Karwatowska-Prokopczuk et al. 2024; Stroes et al. 2024).

Pelacarsen (TQJ230), an ASO targeting apolipoprotein(a), aims to reduce plasma Lp(a), a genetically determined lipoprotein associated with atherosclerosis and inflammation (Nordestgaard and Langsted 2024). While not yet trialed specifically in diabetic cohorts, Lp(a) levels are often elevated in individuals with T2DM, particularly those with nephropathy (Li et al. 2023). The Lp(a)HORIZON Phase 3 trial (NCT04023552) has been recently designed to determinte the effect of pelacarsen on major cardiovascular events in individuals with evelated Lp(a) (Cho et al. 2025). Its application to diabetes may evolve as part of an integrated cardiometabolic strategy, especially in patients with mixed vascular and inflammatory risk profiles. Several newer ASO candidates focus on fibroblast growth factor 21 (FGF21) and its downstream pathways, which modulate glucose uptake, lipolysis, and thermogenesis. Although peptide analogs of FGF21 (e.g., pegozafermin, efruxifermin) are being explored more aggressively in clinical trials (Bhatt et al. 2023; Harrison et al. 2023; Loomba et al. 2023; Noureddin et al. 2025), ASO-mediated modulation of FGF21-related pathways remains an attractive alternative for long-term gene suppression with fewer dosing requirements.

4 Other Therapeutics Targets

4.1 RNA-Therapeutics Beyond ASOs

Among the most well-studied RNA therapeutic targets are genes that modulate peripheral insulin sensitivity. One key example is CD36, a fatty acid transporter implicated in lipid accumulation and insulin resistance. In smCd36$^{-/-}$ mice, siRNA-mediated silencing of CD36 reduced fatty acid uptake and restored insulin responsiveness in peripheral tissues. Similarly, inhibition of protein tyrosine phosphatase 1B (*PTP1B*), a known negative regulator of insulin receptor signaling, via siRNA enhanced *AKT* phosphorylation and improved glucose uptake in insulin-responsive tissues (Elchebly et al. 1999). Another strategic target is *FOXO1*, a transcription factor that promotes hepatic gluconeogenesis (Benchoula et al. 2020). Administration of siRNAs silencing *FOXO1* reduced skeletal muscle atrophy, and accordingly, insulin resistance (Du et al. 2023). While improving peripheral insulin action is critical, restoring pancreatic β-cell function is equally important for durable diabetes control. Zhou et al. demonstrated that adenoviral co-delivery of *Pdx1*, *MafA*, and *Ngn3* could convert pancreatic exocrine cells into insulin-producing β-like cells, effectively reversing hyperglycemia in streptozotocin (STZ)-induced diabetic mice (Zhou et al. 2008). Further, small activating RNAs (saRNAs) targeting *MafA* have been used to epigenetically upregulate endogenous *Pdx1*, *NeuroD1*, *Ngn3* and *Nk6x1* in CD34+ progenitor cells. This approach reactivated β-cell-specific transcriptional programs without the need for genome integration (Reebye et al. 2013). In preclinical settings, combination approaches have also demonstrated therapeutic synergy. For example, the administration of short hairpin RNAs (shRNAs) in mice treated with insulin analogs such as glargine has led to improvements in fasting glucose and insulin levels beyond what either therapy achieved alone (Peng et al. 2021). These results indicate that RNA-based tools can be used as adjuncts to existing treatments, as opposed to stand-alone interventions.

4.2 Microbiome-Based Therapies and Nutrigenomics

Microbiome-based therapies for T2DM are an area of active investigation, with interventions primarily focused on modulating the gut microbiota through probiotics, prebiotics, synbiotics, and fecal microbiota transplantation (FMT). Probiotics, live microorganisms that confer health benefits, have been shown to reduce fasting plasma glucose, HbA1c, fasting insulin, and HOMA-IR, as well as central obesity as measured by waist circumference. These effects are most consistently observed with probiotic formulations, while prebiotics and synbiotics show more variable results, with prebiotics also reducing HbA1c and fasting glucose but sometimes increasing fasting insulin levels. The benefits appear to be independent of whether single or multi-strain formulations are used (Vazquez-Marroquin et al. 2023; Keivanlou et al. 2024; Zhou et al. 2025). Prebiotics, such as inulin, fructooligosaccharides, and resistant starch, selectively stimulate the growth of beneficial gut bacteria (e.g., Bifidobacterium, Roseburia, Faecalibacterium) and are associated with improved glycemic indices and insulin sensitivity. These agents may also help restore the balance of butyrate-producing bacteria, which are often depleted in T2DM, and reduce the abundance of potentially pathogenic taxa (Zhou et al. 2025). FMT, which involves the transfer of screened donor stool to a recipient, has shown potential for more rapid and pronounced shifts in gut microbial composition compared to dietary interventions alone. Early clinical studies suggest that FMT, especially when combined with dietary modification, can improve glycemic control, blood pressure, and lipid profiles, likely through enrichment of beneficial taxa (e.g., Bifidobacterium) and reduction of sulfate-reducing bacteria (e.g., Bilophila, Desulfovibrio) that are positively correlated with adverse metabolic indices (Su et al. 2022; Wang et al. 2024a, b). However, the evidence base for FMT in T2DM remains limited, with most studies being small and of short duration. Mechanistically, these microbiome-targeted therapies are thought to act via restoration of gut barrier integrity, modulation of inflammatory pathways (e.g., reduction in CRP, TNF-α, LPS; increase in IL-10), and enhancement of short-chain fatty acid production, which collectively improve insulin sensitivity and reduce systemic inflammation (Chen et al. 2024; Zhou et al. 2025). Despite these promising findings, the observed improvements in glycemic and metabolic parameters are generally modest, and there is significant heterogeneity in study designs, interventions, and outcomes. Long-term safety and efficacy data are lacking, and large-scale, well-controlled trials are needed to clarify optimal formulations, dosing, and patient selection (Qin et al. 2025; Zhou et al. 2025).

Nutrigenomics, the study of how nutrients and dietary patterns influence gene expression and how genetic variation modulates individual responses to diet, helps define the heterogeneity observed in T2DM risk, progression, and therapeutic response (Billings and Florez 2010). Given T2DM's polygenic nature, it is becoming increasingly evident that dietary factors not only act as environmental triggers but also participate in gene modulation via molecular mechanisms that shape metabolic health (Cole and Florez 2020). At the molecular level, various nutrients and bioactive food compounds, notably polyphenols, can regulate gene expression by influencing key metabolic and inflammatory pathways. Polyphenols such as resveratrol, quercetin, genistein, catechins, and curcumin have demonstrated the ability to modulate genes involved in glucose uptake, insulin sensitivity, oxidative stress, and chronic inflammation, hallmarks of T2DM pathophysiology (Goh et al. 2014; Li et al. 2019). These compounds directly modulate transcription factors (e.g., NF-κB, Nrf2), histone acetylation and methylation, and microRNA expression, collectively contributing to epigenetic reprogramming that can either exacerbate or ameliorate diabetic phenotypes (Fenech et al. 2011; Boyanapalli and Kong 2015; Papuc et al. 2022). Furthermore, several genetic variants that modulate dietary response have been identified. Notably, polymorphisms in TCF7L2, PPARγ, FTO, and SLC30A8 have been robustly associated with T2DM susceptibility and variation in clinical outcomes. For instance, the PPARγ Pro12Ala variant influences insulin sensitivity in response to dietary fat, and the metabolic impact of FTO risk alleles can be attenuated through adherence to energy-restricted or Mediterranean-style diets (Ortega-Azorín et al. 2012; Di Renzo et al. 2018). Recent advances in high-throughput technologies have further enabled the identification of novel biomarkers and gene targets for the development of functional foods and nutraceuticals capable of modulating disease-relevant gene expression (Scalbert et al. 2002; Rist et al. 2006; Dimitrov et al. 2016). For example, resveratrol's ability to activate SIRT1 and quercetin's activation of AMPK have been linked to improved mitochondrial function, insulin sensitivity, and reduced hepatic glucose production, offering promising adjuncts to traditional pharmacotherapy (Cantó et al. 2009; Price et al. 2012). These insights support the rationale for genetically tailored nutrition, a cornerstone of precision medicine, to optimize glycemic control, attenuate insulin resistance, and prevent T2DM onset in predisposed individuals (Picó et al. 2019; Marcum 2020). Nutrigenomics also extends into developmental programming, emphasizing how early-life exposures to poor nutrition or adverse intrauterine environments can induce lasting epigenetic alterations that heighten T2DM risk later in life. These developmental origins of health and disease (DOHaD) underscore the importance of maternal and perinatal nutrition as a preventive strategy (Trujillo et al. 2006). Nonetheless, despite promising results, the clinical translation of nutrigenomic findings is hindered by the intricate interplay between genetic variability, dietary complexity, and individual metabolic phenotypes. Moreover, the reproducibility of findings across diverse populations remains a challenge, calling for longitudinal studies with robust designs and ethnically diverse cohorts. The implementation of personalized nutrition approaches also faces logistical, ethical, and economic constraints, especially when scaled for population-wise application (Gohari-Lasaki et al. 2020; Raiten and Bremer 2020).

5 Lox-Cre Recombinase Technology

5.1 Site- and Time-Specific Recombination

The Cre-loxP system is a site-specific recombinase technology used to manipulate genes in mice (Nagy 2000). The system relies two components: a Cre recombinase, initially produced from cyclization recombinase gene of bactriophage 1 (Sternberg et al. 1978; Sternberg and Hamilton 1981; Sternberg et al. 1986; Sauer 1998), and loxP sites, each a 34-base-pair sequence, flanking (floxing) the gene or exon to be manipulated (Sauer and Henderson 1988; O'Neil et al. 1990). Cre binds to two loxP sites and catalyzes recombination between them. The outcome depends on the sites' orientation: if two loxP sites are in the same orientation on a chromosome, Cre excises the intervening DNA as a circular piece, leaving behind one loxP site. If the sites face opposite directions, the DNA segment is inverted. Thus, by designing a floxed allele of a gene, researchers can induce its deletion upon expression of Cre (Nagy 2000; Branda and Dymecki 2004; Meinke et al. 2016). In mouse genetics, one typically creates two strains: (1) a floxed mouse carrying loxP-flanked target genes, and (2) a Cre driver mouse expressing Cre recombinase under a tissue-specific or inducible promoter. Crossing these strains yields progeny in which Cre removes the target gene in the desired cells (Gu et al. 1993; Gu et al. 1994; Rajewsky et al. 1996; McLellan et al. 2017). The strength of Cre-lox lie in its ability to achieve spatial and temporal control of gene knockout. Tissue-specific Cre lines use promoters for genes expressed in particular organs or cell types (Kim et al. 2018). For instance, the Ins1-Cre or Rip-Cre lines drive Cre in pancreatic β-cells, enabling β-cell-specific gene deletion (Thorens et al. 2015). By choosing the appropriate Cre line, a gene's function in target tissues can be studied without affecting its expression elsewhere. For temporal control, inducible Cre systems are used. The most common design fuses Cre to a modified estrogen receptor ligand-binding domain (Cre-ER) (Fig. 4), which is retained in the cytoplasm until tamoxifen is administered. Upon tamoxifen binding, Cre-ER translocates to the nucleus and recombines floxed genes (Brocard et al. 1997; Metzger and Chambon 2001). This allows knockout of the gene at any chosen time (e.g. in adult mice), avoiding developmental effects. Some systems use tetracycline-regulated Cre (Tet-On/Off) to similarly control timing (Gossen and Bujard 1992; Gossen et al. 1995; Kistner et al. 1996). Thus, with these tools, one can generate mice that knock out a gene only in pancreatic β-cells and only after a certain age or upon treatment with an inducer. These models have been crucial in attributing phenotypes to gene function in specific cell types.

5.2 *Applications of Lox-Cre Animal Models in Diabetes*

Lox-Cre models have been instrumental in dissecting the mechanisms of β-cell destruction in T1DM. Conditional knockout of the Ins1 gene in NOD mice using the $Ins1^{Cre}$ driver line provided significant protection from diabetes, confirming Ins1 as a key autoantigen in type 1 diabetes. Importantly, $Ins1^{Cre/WT}$ and $Ins1^{Neo/WT}$ female mice showed marked resistance to disease, with near-complete protection observed in $Ins1^{Cre/WT}$ animals. This effect was associated with reduced insulitis at 12 weeks of age. Notably, Cre knock-in alone also conferred protection, suggesting previously unrecognized immunomodulatory effects of Cre expression in β-cells (Skovso et al. 2022). Also, knockout of the Ins2 gene in NOD mice has been shown to accelerate diabetes onset, likely due to its role in establishing central tolerance within the thymus; its absence may allow more autoreactive T cells to escape into the periphery. In contrast, knockout of Ins1 markedly reduces diabetes and insulitis in NOD mice. Female Ins1 homozygous knockout mice showed no insulitis at 8, 12, or 37 weeks, despite spontaneously producing insulin autoantibodies. These findings suggest that while Ins2 contributes to immune tolerance, Ins1 is critical for driving autoimmune β-cell destruction and disease progression in T1DM (Moriyama et al. 2003).

The role of various immune cell populations in T1DM pathogenesis has also been extensively investigated using Lox-Cre technology Conditional deletion of Cd226, a costimulatory receptor implicated in autoimmune risk, in regulatory T cells (Tregs) using the Foxp3-Cre driver line in NOD mice resulted in reduced insulitis and delayed diabetes onset. This aligns with evidence that CD226 expression impairs Treg stability and suppressive function. CD226-deficient Tregs exhibited increased expression of the inhibitory receptor TIGIT, suggesting that a CD226/TIGIT imbalance contributes to Treg dysfunction in T1DM. Targeting this pathway in Tregs may enhance their protective capacity and delay autoimmune β-cell destruction (Thirawatananond et al. 2023).

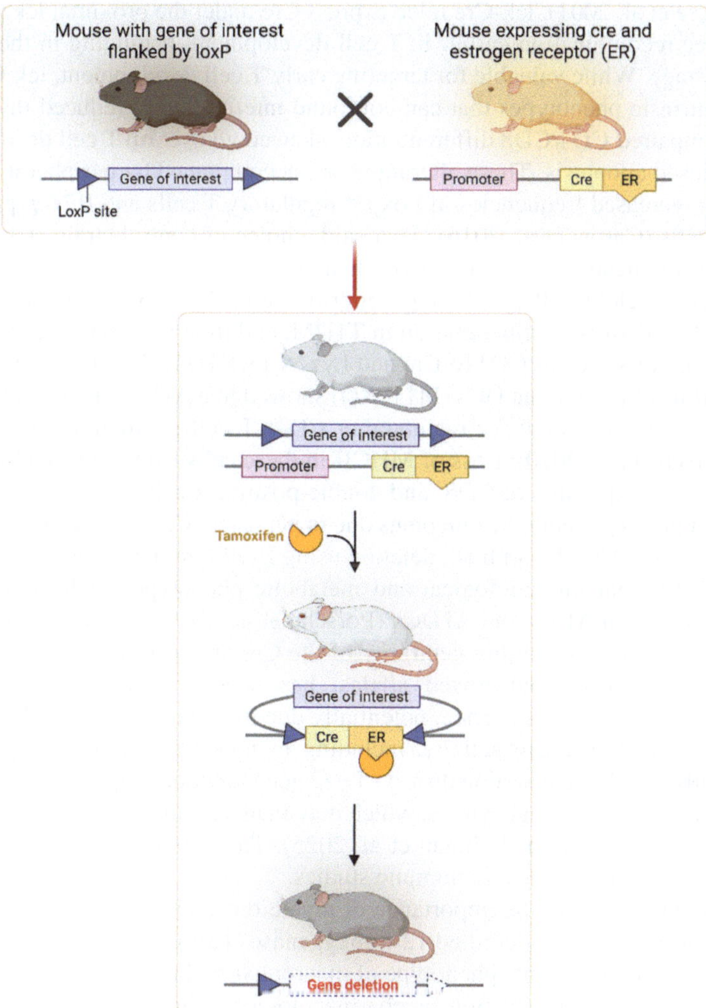

Fig. 4 Tamoxifen-inducible Cre-loxP system for temporal gene deletion in mice. A mouse line carrying a gene of interest flanked by loxP sites is crossed with a transgenic mouse expressing Cre recombinase fused to a modified estrogen receptor (Cre-ER) under the control of a tissue-specific promoter. The resulting offspring harbor both the floxed gene and the Cre-ER transgene. In the absence of tamoxifen, Cre-ER remains inactive in the cytoplasm. Upon tamoxifen administration, Cre-ER translocates to the nucleus, where it excises the floxed gene via recombination at loxP sites, resulting in targeted gene deletion. This system enables spatial and temporal control of gene knockout. (Illustration is created using Biorender.com)

CD4-Cre mice express Cre recombinase under regulatory elements of the *CD4* gene, with recombination occurring in both CD4$^+$ and CD8$^+$ T cells at the double-positive thymocyte stage. This model achieves >99% efficiency in floxed gene deletion and is widely used to study gene function during T cell maturation and in peripheral CD4$^+$ helper T cells, responsible for the autoimmune response. CD8-Cre mice allow parallel investigation of CTLs, which are directly responsible for β-cell killing (Lee et al. 2001). lck-Cre mice express Cre under the proximal lck promoter and induce recombination earlier in T cell development, beginning in the double-negative stage. While valuable for targeting early T cell development, lck-Cre mice exhibit intrinsic phenotypes that can confound interpretation: reduced thymic cellularity, impaired CD4/CD8 differentiation, skewed γδ over αβ T cell development, and increased apoptosis. These alterations are accompanied by peripheral changes, including increased frequencies of $Foxp3^+$ regulatory T cells and IFN-γ–producing effector cells (Carow et al. 2016). That said, choice of control strain (lck-Cre^+ vs. lck-Cre^-) significantly affects reported phenotypes.

Further, myeloid cells, including dendritic cells (DCs) and macrophages, are early mediators of islet inflammation in T1DM, and their roles have been explored using Cre drivers such as CD11c-Cre and Lyz2-Cre. CD11c-Cre–mediated deletion of MHCII in adipose tissue DCs (M11cKO) showed that ATDCs are potent antigen-presenting cells, essential for maintaining $CD4^+$ T cell populations (Caton et al. 2007; Porsche et al. 2020). Loss of MHCII in these cells led to reduced Tconv and Tregs, with an expansion of $CD8^+$ and double-positive T cells in adipose tissue, yet without improving metabolic outcomes due to persistent $CD8^+$-driven inflammation (Porsche et al. 2020). In contrast, deletion using Lyz2-Cre (targeting macrophages) produced different immunological and metabolic phenotypes, indicating distinct functions between ATMs and ATDCs (Porsche et al. 2020). However, recent findings caution against assuming neutrality of the Cre driver itself. Lyz2-Cre expression alone, even without floxed alleles, has been shown to protect against FFA-induced glucose intolerance, potentially due to Cre toxicity, Lyz2 haploinsufficiency, or off-target Cre activity, including in non-myeloid cells like neurons. Importantly, Lyz2-Cre insertion disrupts Lyz2 gene function, impairing macrophage activity and reducing phagocytosis, which may inadvertently improve insulin sensitivity and β-cell function (Rahman et al. 2025). This reinforces the need for Cre-only controls in metabolic and immune studies.

Further emphasizing the importance of myeloid cells in early T1DM pathogenesis, recent studies have identified 12-lipoxygenase (12-LOX) as a key enzyme promoting the prodiabetogenic phenotype of macrophages. In both zebrafish and NOD mouse models, genetic deletion or pharmacological inhibition of 12-LOX in the myeloid lineage significantly reduced islet inflammation, suppressed autoreactive T cell responses, preserved β-cell mass, and protected against T1DM onset. This effect was linked to impaired macrophage migration, driven by loss of the chemokine receptor CXCR3, a defect that could be rescued by CXCR3 overexpression. These findings establish 12-LOX as a crucial regulator of macrophage-mediated islet autoimmunity (Kulkarni et al. 2021).

Lox-Cre models have also been crucial in understanding the mechanisms underlying β-cell dysfunction in T2DM. β-cell specific knockout of PDHA1 (Fig. 5), a key enzyme in glucose metabolism, using an inducible Cre system in db/db mice, impaired glucose-stimulated insulin secretion and glucose tolerance, despite normal body weight and fasting glucose (Wang et al. 2020). Conversely, β-cell-specific deletion of Pten, a negative regulator of insulin signaling, using RIP-Cre, completely protected mice from diabetes in both high-fat diet and db/db models of type 2 diabetes. PTEN expression was upregulated in the islets of both models. In RIP^{cre+}

Pten$^{fl/fl}$ mice, β-cell mass was already increased under basal conditions, with no further increase observed under diabetic conditions. These mice maintained normal β-cell function and intact islet PI3K signaling, unlike wild-type mice fed a high-fat diet or db/db mice, which exhibited impaired β-cell function and reduced PI3K signaling. Notably, the protective effects in β-cells were observed without compromising the response to DNA-damaging stimuli, suggesting that PTEN normally exerts a negative influence on β-cell function and mass, and its inhibition could offer a potential therapeutic strategy (Wang et al. 2010). Additionally, muscle-specific insulin receptor knockout (MIRKO) mice exhibited insulin resistance specifically in skeletal muscle but surprisingly maintained normal glucose tolerance. This finding suggests that while muscle is a major site of insulin-stimulated glucose uptake, other tissues can compensate to maintain overall glucose homeostasis. However, these MIRKO mice did display alterations in fat metabolism, indicating that muscle insulin resistance contributes to lipid dysregulation often observed in T2DM (Bruning et al. 1998). Further, in a study focusing on adipose tissue, a critical insulin-sensitive tissue, adipocyte-specific Kbtbd2 knockout led to an accumulation of p85α, causing insulin resistance (Fig. 5), moderate hyperglycemia, sustained hyperinsulinemia, lipodystrophy, and ectopic lipid accumulation in the liver (Zhang et al. 2016). This was accompanied by adipocyte-extrinsic insulin resistance in the

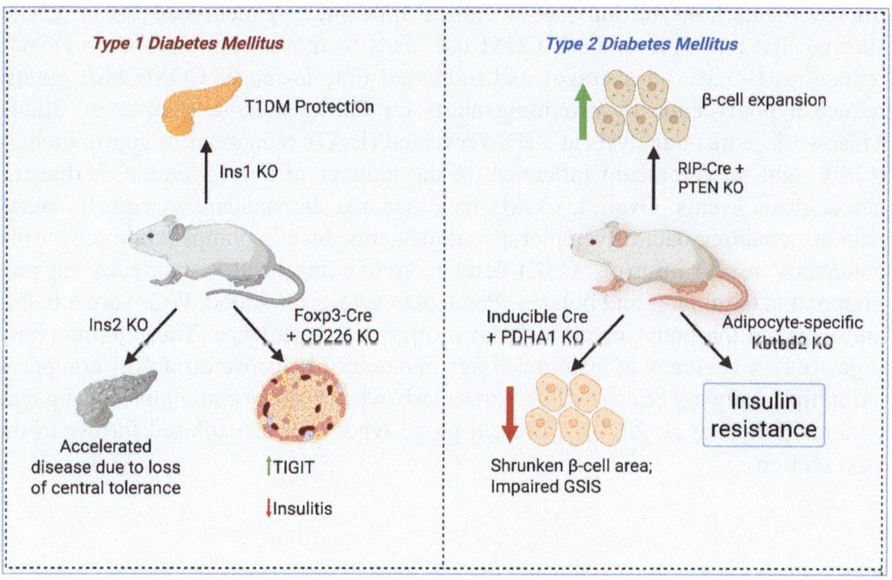

Fig. 5 Lox-Cre mouse models enable tissue-specific gene modifications driving onset of Type 1 diabetes mellitus (T1DM) and Type 2 diabetes mellitus (T2DM). In T1DM models (left), deletion of Ins1 protects against autoimmune diabetes, while Ins2 knockout leads to disease acceleration due to impaired central tolerance. Regulatory T cell-specific deletion of CD226 (via Foxp3-Cre) increases TIGIT expression and reduces insulitis. In T2DM models (right), β-cell–specific PTEN deletion (via RIP-Cre) results in β-cell expansion and diabetes protection, whereas β-cell–specific PDHA1 knockout impairs glucose-stimulated insulin secretion. Adipocyte-specific deletion of Kbtbd2 leads to systemic insulin resistance. (Illustration is created using Biorender.com)

liver and muscle, but no such abnormalities were observed in liver- or muscle-specific Kbtbd2 knockout mice. Additionally, mice with Kbtbd2 knockout in adipocytes, liver, and muscle showed normal growth, suggesting that KBTBD2 may be crucial for proper IGF1 signaling in other tissues (Zhang et al. 2020).

6 Biotechnological Advances in Diabetes Devices

6.1 CGMs and Insulin Pumps

Modern diabetes care is revolutionized by engineering advances in sensing and insulin delivery. Continuous Glucose Monitors (CGMs) (Table 2) now provide real-time glucose readings from the interstitial fluid every few minutes (Lin et al. 2021). State-of-the-art CGM sensors (e.g. Dexcom G6/G7, Abbott FreeStyle Libre 2/3, Medtronic Guardian) use enzyme-coated electrodes or optical fluorescence to measure glucose with high accuracy. Recent CGM generations achieve mean absolute relative differences (MARD) below 10% (versus >20% a decade ago) (Heinemann et al. 2020). Many newer CGMs require no fingerstick calibrations and can be worn 10–14 days per sensor. Abbott's implantable Eversense can last 6 months. These improvements translate into better clinical outcomes. A meta-analysis in T2DM showed that real-time or flash CGM use leads to modest but significant HbA1c reductions (~0.3% on average) and increased time-in-range. CGMs also greatly reduce hypoglycemia by providing alerts for falling glucose (Uhl et al. 2024). Meanwhile, a meta-analysis in T1DM revealed HbA1c reductions of approximately 0.23%, and no significant influence on the number of hypoglycemic or diabetic ketoacidosis events. Overall, CGMs have become the standard of care for many patients, enabling data-driven therapy adjustments. Insulin pumps (continuous subcutaneous insulin infusion, CSII) deliver rapid-acting insulin analogues via programmable basal rates and boluses. Patch pumps (e.g. Omnipod, V-Go) are tubeless and worn on the body for up to 3 days, offering convenience. These pumps have improved the accuracy of insulin delivery and reduced glucose variability compared to multiple daily injections (MDI), particularly when used by compliant and engaged patients (Nimri et al. 2020). Different pump types will be explored further in the next section.

6.2 Closed-Loop Systems

Artificial pancreas systems automate insulin delivery by coupling CGM data with control algorithms (Boughton and Hovorka 2020). In a hybrid closed-loop system, the pump automatically adjusts basal insulin based on CGM readings, but the user still inputs mealtime carbohydrates. Control algorithms are typically either proportional-integral-derivative (PID) or model predictive control (MPC). PID

Table 2 CGMs, insulin pumps, and combined systems. CGMs provide actionable glucose trend data; pumps provide precise insulin dosing control; integration of the two enables closed-loop automation

Device	Examples	Function/Technology	Clinical benefits
CGM (real-time)	Dexcom G6/G7; Libre 2/3; Eversense	Electrochemical or fluorescence sensor; measures interstitial glucose every 1–5 min	↓HbA1c (~0.3–0.5% in T2D); ↑time-in-range; ↑hypoglycemia detection
Insulin Pump (CSII)	Medtronic 770G/780G; Tandem t:slim; Omnipod (patch)	Programmable basal/bolus insulin delivery via subcutaneous cannula or patch pump	More stable basal insulin, greater dosing flexibility, improved glycemic control over MDI
Closed-loop AP (Hybrid)	Medtronic 780G; Tandem Control-IQ; CamAPS FX	Algorithmic insulin dosing driven by CGM feedback; requires meal announcement	Significantly higher time-in-range (often ~70–80% vs ~50–60%), reduced hypoglycemia, lower patient burden
Smart Pens / Algorithms	InPen, apps (e.g. Clarity)	Track insulin doses; suggest doses via bolus calculators	Improved adherence; decision support

CGMs Continuous Glucose Monitors; *MDI* multiple daily injections

algorithms react to current and past glucose deviation, whereas MPC uses a mathematical model of glucose–insulin dynamics to predict future glucose and optimize insulin dosing accordingly (Nwokolo and Hovorka 2023). The MPC approach can anticipate the delayed action of subcutaneous insulin, often yielding tighter control (Kang et al. 2022). Indeed, one clinical trial in humans showed MPC-based controllers achieve higher time-in-range than PID controllers under similar conditions (Pinsker et al. 2016). Commercial systems incorporate these algorithms in proprietary ways. For example, the Tandem Control-IQ uses MPC with user-specific carbohydrate ratios and the Dexcom CGM (Breton et al. 2020), while older Medtronic systems (670G/770G/780G) use PID-based approaches (Bergenstal et al. 2016; Choudhary et al. 2022). All systems aim to maintain glucose within the target range by automatic basal adjustments, with some providing automatic correction boluses. These algorithms also implement safety limits, suspending insulin if hypoglycemia is anticipated. Machine learning enhancements and adaptive features are now being added to personalize control to the individual's insulin sensitivity patterns (Kovatchev et al. 2024).

It might be prudent to also discuss dual-hormone pumps in the context of artifical pancreas, seeing that they help better simulate glucose homeostasis. In T1DM, glucagon secretion is also impaired due to α-cell dysfunction. Some dual-hormone systems incorporate subcutaneous glucagon delivery, acting within 5 minutes of infusion, to counteract hypoglycemia. Preliminary evidence shows that systems containing dual-hormone pumps (insulin + glucagon) are not superior to only insulin (Haidar et al. 2015a, b; Haidar et al. 2016; Abitbol et al. 2018; Haidar 2019). However, a dual hormone artificial pancreas has been shown to reduce nocturnal hypoglycemia in a cohort of T1DM children and adolescents participating in physical activity in a diabetes camp (Haidar et al. 2015a, b). Further, amylin is co-secreted with insulin by β-cells and is involved in postprandial glucose regulation by slowing gastric emptying, suppressing glucagon, and promoting satiety. Therefore, some dual-hormone systems incorporate pramlintide, a synthetic amylin analog. To this extent, two studies on cohorts of T1DM children and adolescents on insulin-only systems showed that the addition of pramlintide both slowed and diminished the peak increase in postprandial glucose levels (Weinzimer et al. 2012; Sherr et al. 2016). Overall, these closed-loop algorithms mimic advanced control in the human body, tightly coupling sensing to delivery. Table 2 contrasts the main features of CGMs, pumps, and combined systems.

7 Biotechnology in Detecting Insulin

Precise measurement of insulin (and related biomarkers) is important for research and for potential diagnostics. Traditional immunoassays (ELISAs) are sensitive but require lab equipment (Even et al. 2007). Newer biosensor technologies (Table 3) promise rapid, point-of-care insulin detection using advanced materials and bio-recognition elements. A major trend is the use of aptamers, short DNA or RNA sequences that bind insulin with high specificity. Aptamers (e.g. the G-quadruplex "IGA3" DNA aptamer) can be selected via SELEX for insulin binding (Yoshida et al. 2009). Unlike antibodies, aptamers are synthetic, stable, and can undergo reversible binding. Many experimental insulin sensors use aptamer probes on electrochemical platforms, seeing that aptamer-based sensors are reported to have faster response and lower detection limits than antibody-based counterparts. For example, an aptamer-conjugated graphene field-effect transistor (FET) has been shown to detect insulin at very low (pM) concentrations, offering fast, label-free readout (Park et al. 2021). Molecularly imprinted polymers (MIPs) are another class of synthetic recognition: polymer matrices with "imprinted" insulin-sized cavities. MIPs act like plastic antibodies, binding insulin reversibly. MIP-coated electrodes or fluorescent sensors have achieved nanomolar insulin sensitivity and good stability (Moein et al. 2014; Chen et al. 2016; Ding et al. 2020).

Recently, nanotechnology has been used to amplify sensor signals. Carbon nanotubes, gold nanoparticles, quantum dots, and nanowires increase surface area and conductivity (Ge et al. 2020; Wardani et al. 2023). For instance, carbon nanotube electrodes have been modified with insulin aptamers to greatly enhance sensitivity (Wardani et al. 2023). Nanoparticle labels (e.g. magnetic beads with HRP enzyme) enable colorimetric detection in low-resource settings. Indeed, a recent colorimetric immunosensor using HRP-labeled antibodies and magnetic nanoparticles achieved a limit of detection of ~10 picomolar insulin, demonstrating that even simple readouts can reach clinically relevant sensitivity (Syed et al. 2024). Looking forward,

Table 3 Selected insulin biosensor technologies

Sensor type	Recognition element	Readout	Sensitivity (LOD)	Notes
Aptamer-based electrochemical	DNA aptamer (IGA3) on electrode	Current (voltammetry)	~pM (in buffer)	Label-free; high specificity; fast response
Field-effect transistor (FET) aptasensor	Insulin aptamer on CNT/ graphene FET	Conductance change	~pM/fM	Ultra-sensitive; real-time capable
Colorimetric immunosensor	Antibody + HRP enzyme on nanoparticle	Color change (optical)	~10 pM (human serum)	Simple, equipment-free; suitable for low-resource settings
Surface plasmon resonance	Antibody or aptamer on chip	Optical resonance shift	~nM–pM (depending on setup)	Label-free; real-time binding kinetics
Molecularly imprinted polymer (MIP) sensor	Synthetic polymer imprinted with insulin	Various (electrochemical/ optical)	~nM	Robust; can be integrated into wearable devices
Enzymatic assay (legacy)	Insulin-specific antibodies	Luminescence/ absorbance	~pM (ELISA)	Gold standard; requires lab; high accuracy

fully wearable microneedle-array sensors are experimental designs that continuously sense two metabolites in interstitial fluid (lactate and glucose, or alcohol and glucose), and relay data onto a mobile device (Tehrani et al. 2022). However, these devices come with some drawbacks; predominantly pertaining to longevity, accuracy, and reliability, and low lag time with a fast, accurate response time. These devices are often tested in controlled environments, which may lead to performance outcomes that do not reflect their effectiveness in everyday use.

8 Biotechnology in Regenerative Medicine and Cell Therapy to Cure Diabetes

Biotechnology has enabled the emergence of regenerative approaches that focus on restoring lost or dysfunctional insulin-producing β-cells in individuals with diabetes. Central to these innovations are induced pluripotent stem cells (iPSCs), which, through gene editing and precise differentiation techniques, are transforming diabetes research and therapy. While stem cells-particularly iPSCs, are invaluable tools in the study of disease pathogenesis and therapeutic development, numerous other biotechnological tools also play critical roles alongside stem cells in advancing research and development in the field of diabetes, which will be discussed in this section.

8.1 Reprogramming Somatic Cells to Pluripotent Stem Cells

This process entails reprogramming specialized cells back into a pluripotent state, allowing them to subsequently differentiate into various cell types for research and therapeutic applications. Achieving this requires not only the induction of pluripotency but also extensive changes at the epigenetic, transcriptional, and chromatin levels, effectively resetting the cells to a state similar to that of embryonic stem cells (Carey et al. 2011). This technological breakthrough offers a significant alternative to the use of embryonic stem cells (ESCs), which is often controversial, as induced pluripotent stem cells (iPSCs) can be generated from adult somatic cells, thus avoiding many of the ethical concerns associated with embryo use (Wang et al. 2011; Palomo et al. 2014).

Early reprogramming methods included somatic cell nuclear transfer (SCNT), a technique in which the nucleus of a differentiated somatic cell is transferred into an enucleated oocyte—an egg cell whose nucleus has been removed. The cytoplasm of the oocyte contains factors that rapidly reprogram the introduced somatic nucleus, erasing its differentiated state and inducing a totipotent or pluripotent state similar to that of a zygote (Campbell et al. 1996). Another approach is cell fusion, where a somatic cell is fused with a pluripotent stem cell, such as an embryonic stem cell. In this process, the resulting hybrid cell contains nuclei from both parent cells, and factors from the pluripotent cell can reprogram the somatic nucleus to a pluripotent state (Matsumura and Tada 2008). Later, the advent of direct reprogramming introduced a method involving the delivery of four transcription factors (OCT4, SOX2, KLF4, and c-MYC)-collectively known as Yamanaka factors or OKSM-using retroviral vectors (Takahashi and Yamanaka 2006).

8.2 Genetic Engineering in Generating iPSCs: Viral and Non-viral Delivery of Genes

Researchers use viral and non-viral methods to deliver reprogramming factors, each with distinct strengths and weaknesses important for optimizing iPSC generation (Fig. 6). Viral vectors are categorized as integrating (e.g. lentiviral, retroviral), which stably express OKSM by inserting genes into the genome but risk insertional mutagenesis and transgene reactivation, or non-integrating, which avoid these risks (Takahashi et al. 2007; Yu et al. 2007). In contrast, non-integrating viral vectors like adenovirus, AAV, and Sendai virus remain episomal and are lost over time, avoiding genomic disruption but with lower efficiency than integrating vectors (Macarthur et al. 2012). Non-viral delivery techniques, such as episomal plasmids (Feng et al. 2020), transposons (Woltjen et al. 2009), and mRNA (Warren et al. 2010) further eliminate the risk of genomic integration while maintaining reasonable reprogramming efficiency. Episomal plasmids use EBV-derived vectors for transient OSKM

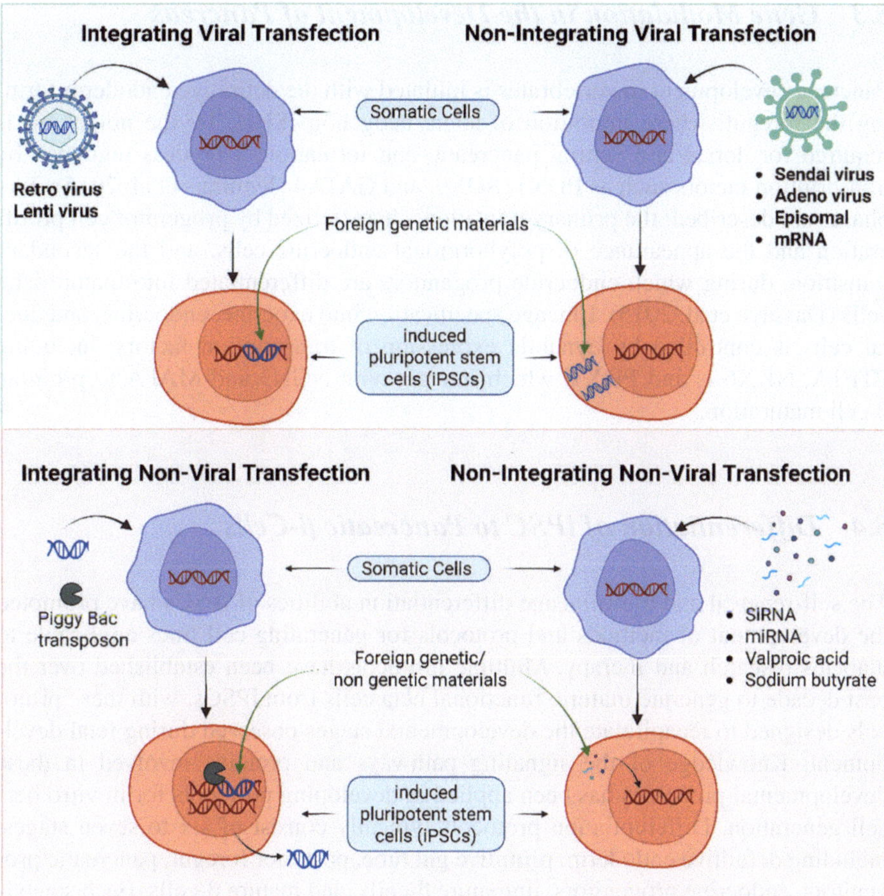

Fig. 6 The four principal methods for delivering reprogramming factors are integrating viral systems-which were initially used to generate stem cells but pose risks of genomic integration and teratoma formation-and three progressively safer alternatives: non-integrating viral vectors, non-viral integrating vectors, and non-integrating non-viral vectors. These improved delivery approaches significantly increase the safety and efficacy of iPSCs, facilitating their broader application in both scientific research and clinical settings

expression without genome integration, offering a low-cost, scalable clinical option. Similarly, mRNA delivery of OKSM enables rapid, efficient reprogramming without integration risks. Recently, small molecules and microRNA overexpression (e.g., miR-302 s, miR-367 s) have also been used to induce pluripotency by modulating cell signaling, providing cost-effective, non-genetic alternatives (Kuo et al. 2012), and signaling through retinoic acid (RA) receptor have also been explored as alternative reprogramming strategies (Yang et al. 2015).

8.3 Gene Modulation in the Development of Pancreas

Pancreas development in vertebrates is initiated with the definitive endoderm forming the foregut, where inhibition of sonic hedgehog (SHH) by the notochord is required for dorsal and ventral pancreatic bud formation, a process regulated by transcription factors such as PDX1, SOX9, and GATA4 (Jennings et al. 2015). Two phases are described: the primary transition, characterized by progenitor cell proliferation and the appearance of polyhormonal endocrine cells, and the secondary transition, during which endocrine progenitors are differentiated into mature islet cells (Dassaye et al. 2015). Lineage specification into exocrine, endocrine, and ductal cells is controlled by dynamic expression of transcription factors, including PTF1A, NKX6.1, and PDX1, which interact with NGN3 and MAFA to promote β-cell maturation.

8.4 Differentiation of iPSC to Pancreatic β-Cells

The self-renewal and multilineage differentiation abilities of iPSCs have prompted the development of methods and protocols for generating cell lines applicable to diabetes research and therapy. Multiple protocols have been established over the past decade to generate mature, functional beta cells from iPSCs, with these protocols designed to recapitulate the developmental stages observed during fetal development. Knowledge of the signaling pathways and proteins involved in these developmental processes has been applied in developing protocols for in vitro beta cell generation. Differentiation protocols typically consist of six to seven stages, including definitive endoderm, primitive gut tube, posterior foregut, pancreatic progenitors, endocrine progenitors, immature β-cells, and mature β-cells. Each stage is transitioned using stage-specific media supplemented with growth factors and small molecules that activate or inhibit the signaling pathways required for β-cell development, and stage-specific markers are analyzed and validated before proceeding to the next step.

8.4.1 Targeting Cellular Signaling Pathways to Generate Insulin Producing β-Cells In Vitro

CHIR99021, a GSK3β inhibitor; Activin A, which stimulates the TGFβ pathway; and Wnt3a, a Wnt signaling activator, have been identified as critical for definitive endoderm formation (Sali et al. 2025). During primitive gut tube formation, the TGFβ pathway is inhibited and SHH is downregulated by adding FGF2, FGF7, FGF10 (activating FGFR2b), and a TGFβR1 inhibitor (Nostro et al. 2015; Kim et al. 2016). In the pancreatic progenitor stage, retinoic acid is used to stimulate PDX1, while BMP inhibition with small molecules such as NOGGIN, LDN, PKC inhibitors, and FGF7 is applied (Rezania et al. 2014; Hogrebe et al. 2021); a combination of NOGGIN, EGF, and nicotinamide has also yielded high numbers of PDX1 and NKX6.1 positive cells (Nostro et al. 2015). Depolymerization of the cytoskeleton with latrunculin A has been reported to improve the yield of SC β-cells (Hogrebe et al. 2021). To induce endocrine progenitors, inhibition of NOTCH, BMP, SHH, and TGFβ is achieved using gamma-secretase inhibitors (DAPT, XXI, SANT1), thyroid hormone (T3), and ALK5i (a TGFβ receptor inhibitor) (Maxwell et al. 2020; Hogrebe et al. 2021). Combinations of BMP and ALK inhibitors, or XXI, ALK5iII, T3, heparin, and betacellulin, have resulted in higher percentages of NKX6.1$^+$ cells co-expressing NGN3, NKX2.2, and NEUROD1 (Pagliuca et al. 2014; Russ et al. 2015; Maxwell et al. 2020; Hogrebe et al. 2021). In the subsequent stage, cocktails of small molecules (GSiXX, ALK5iII, T3, LDN) have been used to enhance expression of mature β-cell markers (NKX6.1, MAFA, NEUROD1, PAX6, SLC30A8, GCK, GLUT1, c-peptide, and INS) and reduce exocrine marker PTF1A expression (Nostro et al. 2015; Hogrebe et al. 2021). In the final stage, a chemical cocktail of R428, ALK5iII, N-Cys, and T3 has been shown to induce a more than ten-fold increase in the maturity marker MAFA (Balboa et al. 2022), while the addition of ROCK inhibitor, WNT signaling molecules, and exendin-4 has improved expression of beta cell maturity markers (Ghazizadeh et al. 2017; Yoshihara et al. 2020).

Early iPSC β-cell differentiation protocols were reported to be less efficient than those using ESCs, with protocols developed for ESCs yielding low numbers of functional beta cells when applied to iPSCs. The resulting cell populations often included both functional (NKX6.1$^+$; cPep$^+$) and polyhormonal (GCG$^+$Cpep$^+$, SST$^+$Cpep$^+$) cells, the latter representing immature, non-functional endocrine cells. Enterochromaffin (ChGa$^+$) cells have recently been identified as off-target populations generated during SC-β cell differentiation, though their functional properties are not clear (Veres et al. 2019).

8.4.2 Harnessing Biotechnology to Mimic In Vivo Architecture: 3D Culture Systems for Functional β-Cell Development

The choice of culture technique significantly influences the differentiation of SC-β cells. Planar 2D cultures are widely used due to their simplicity, cost-effectiveness, reproducibility, and ease of adapting protocols across cell lines. However, 3D cultures more accurately replicate the in vivo environment by supporting enhanced cell-cell interactions, polarity, and tissue-specific architecture. Comparative studies indicate that during definitive endoderm differentiation, lineage marker expression (such as SOX17, FOXA2, and HNF4a) is similar in both 2D and 3D systems, likely because high efficiency is already achieved in 2D at this stage, making 3D unnecessary early on (Yabe et al. 2019). In contrast, 3D suspension cultures maintained throughout the differentiation process yield significantly higher proportions of PDX1+ pancreatic progenitor cells-often exceeding 90%, which is roughly three times greater than in 2D cultures-and also show elevated SOX9 expression (Pagliuca et al. 2014; Veres et al. 2019; Dettmer et al. 2021). Furthermore, the expression of endocrine progenitor and β-cell markers (PDX1, NKX6.1, NGN3, insulin) is markedly higher in 3D systems compared to 2D (Mihara et al. 2017; Wang et al. 2019a, b), with advanced techniques like air-liquid interface further enhancing NGN3 and insulin levels (Rezania et al. 2014). Overall, the spatial organization and microenvironment of 3D cultures better recapitulate the mechanical and biochemical cues present in tissues, resulting in more mature, functional SC-β cells with improved insulin secretion and responsiveness to stimuli (Yin et al. 2016).

8.4.3 Biotechnological Strategies for Safeguarding β-Cells: Advances in Macro-, Micro-, and Nanoencapsulation Systems

After transplantation, SC-β cells or islets-like any allogenic graft-are vulnerable to immune rejection, particularly in type 1 diabetes (T1D), an autoimmune disorder. While immunosuppressive drugs are currently used to address this, their significant side effects can outweigh the benefits of achieving glycemic control post-transplantation. As a result, alternative strategies have been explored, with encapsulation emerging as a promising solution. Encapsulation involves enclosing cells or islets within a semipermeable membrane that allows the exchange of nutrients, oxygen, insulin, and metabolic waste, but blocks immune cells from entering, thereby protecting the transplanted cells from immune attack. Research has demonstrated the application of encapsulation to prevent immune-mediated destruction of allogenic islets, as well as beta cells derived from embryonic stem cells (ESCs), mesenchymal stem cells (MSCs), and induced pluripotent stem cells (iPSCs). Three main encapsulation approaches are commonly utilized: macroencapsulation, microencapsulation, and nanoencapsulation.

8.4.4 Macroencapsulation: Bioengineering Solutions to Safeguard β-Cell Function

Macroencapsulation involves enclosing hundreds of pancreatic islets or SC-β cells whether allogenic, xenogenic, or autologous—within a device larger than 1 mm, such as planar membranes, hydrogel-based chambers, or hollow fibers, which may be implanted intravascularly or extravascularly (Vaithilingam et al. 2017). These devices can accommodate large cell numbers, allow for retrieval in case of graft failure, and enable localized monitoring. In one study, viable cells were successfully retrieved from macrodevices two weeks post-implantation (Scharp et al. 1994). Intravascular designs connect directly to host vasculature, whereas extravascular ones do not, which may impair oxygen diffusion and delay insulin responses due to greater diffusion distances and dense cell packing. This can lead to hypoxia and necrosis, particularly in extravascular devices. To address these limitations, newer designs include integrated oxygen supplies and vascularization features, leading to improved glycemic control and long-term graft survival—up to 10 months in T1D recipients (Ludwig et al. 2013). Notably, one recipient maintained glycemic control for two years post-transplant with macroencapsulated xenogeneic islets, without immunosuppression (Prochorov et al. 2008). Planar macrodevices also support endocrine differentiation, promoting β-cell over α-cell development compared to alginate microcapsules (Motte et al. 2014). A dual-chamber device has further enhanced nutrient delivery and immune protection in preclinical models (Yang et al. 2021a, b).

8.4.5 Microencapsulation: A Bioengineered Barrier for β-Cell Survival

Microencapsulation involves enclosing individual islets or β-cell clusters within semipermeable capsules ranging from 100 to 1000 μm, typically containing one to two islets each (Aijaz et al. 2017). This approach enhances oxygenation and nutrient diffusion due to the high surface-area-to-volume ratio. Alginate-based hydrogels are commonly used; however, environmental changes may affect pore size and permselectivity, compromising long-term stability (Gattas-Asfura et al. 2014). Modified alginate microcapsules loaded with CXCL12 have improved glucose-stimulated insulin secretion (GSIS) and β-cell viability in mouse models (Alagpulinsa et al. 2019). Similarly, encapsulating SC-β cells with zwitterionic alginate-coated nanofibrous membranes has enhanced mass transfer, sustained cell viability, and maintained normoglycemia post-transplantation in diabetic mice (Xiao et al. 2022). Oxygen-generating alginate capsules that react with cellular CO_2 via lithium oxide further improved in vivo viability and glycemic control (Wang et al. 2021). Clinical trials using microencapsulated islets—whether allogenic, xenogenic, or autologous—have shown stable blood glucose, reduced insulin dependence, and minimal immune response in patients not receiving immunosuppression (Soon-Shiong et al. 1994; Calafiore et al. 2006; Basta et al. 2011). A major drawback, however, is the challenge of retrieving dispersed microcapsules after implantation, limiting their clinical reversibility (Dimitrioglou et al. 2019).

8.4.6 Precision Bioengineering: Nanoencapsulation and 3D Printing for β-Cell Survival

Nanoencapsulation coats individual islets or β-cell clusters with an ultrathin biopolymer layer, forming a conformal barrier directly on the cell surface (Stock et al. 2022). This minimizes void space, enhancing nutrient, oxygen, and insulin diffusion, and reduces overall transplant volume—facilitating delivery into sites like the portal vein (Wilson et al. 2008; Wiggins et al. 2020). However, the ultrathin coating lacks long-term stability and may degrade under physiological stress, exposing cells to immune attack. 3D bioprinting is an emerging strategy that mimics pancreatic tissue architecture by embedding β cells or progenitors within bioinks (Yue et al. 2016). Hydrogels such as 3% alginate with 9% methylcellulose have been used to bioprint rat islets, preserving glucose sensitivity in vitro (Duin et al. 2019).

Constructs made of polycaprolactone, and alginate supported SC-β cell clusters, which remained glucose-responsive post-implantation in mice (Marchioli et al. 2016). A polylactic acid-based 3D-printed device coated with platelet-rich plasma enhanced vascularization and islet survival after transplantation in diabetic mice (Farina et al. 2017).

9 Application of Bioengineered iPSC-Derived β-Cells

9.1 Leveraging Stem Cell Engineering to Unravel Disease Mechanism: Modeling Disease in a Dish Using iPSC Derived β-Cells

Mouse models are widely used in diabetes research, but structural differences between mouse and human islets—such as higher β-cell content and central clustering in mice versus dispersed β-cells and a higher α- and δ-cell proportion in humans—limit translatability. Human islets from deceased donors offer valuable insight but are limited by availability, short-term viability, and donor variability. Biotechnological progress in differentiating human pluripotent stem cells (hPSCs), including ESCs and iPSCs, has enabled the generation of glucose-responsive, insulin-producing β-like cells by replicating pancreatic development. iPSCs have been derived from patients with type 2 diabetes (Kudva et al. 2012), cystic fibrosis-related diabetes (Simsek et al. 2016), Wolfram syndrome (Shang et al. 2014), neonatal diabetes (Ma et al. 2018), several forms of monogenic diabetes, maturity-onset diabetes of the young (MODY) (Braverman-Gross et al. 2018) and type 1 diabetes (T1D) (Maehr et al. 2009), providing models for disease-specific research. These cells help dissect disease mechanisms, genetic influences, and β-cell responses to inflammation and apoptosis. Studies show iPSC-derived β-cells from T1D patients function comparably to those from healthy individuals (Leite et al. 2020), and when transplanted into diabetic mice, restore glucose regulation (Dettmer et al. 2022). Coupled with immunoprotective strategies, these cells are now entering clinical trials. Stem cell-derived β-cells thus serve as a powerful biotechnology platform for modeling diabetes pathophysiology and testing personalized therapeutic strategies.

9.2 Biotech-Driven Drug Discovery: SC-β Cells as Next-Gen Screening Tools

Drug screening using SC-β cells enables high-throughput evaluation of drug efficacy and toxicity, particularly when derived from patient-specific iPSCs (Millman et al. 2016; Amin et al. 2018). This personalized approach helps optimize treatment regimens based on individual profiles. Improved bioengineering of SC-β cells has enhanced their functional maturity, increasing their value for drug testing. The FDA's recent policy shift allowing reduced animal testing highlights the promise of SC-β cells as ethical, human-relevant models for early drug assessment (Wang et al. 2024a, b). Furthermore, integrating gene editing with SC-β platforms reveals how genetic variations affect drug response, supporting the development of precision therapies in diabetes.

9.3 Bioengineered β Cells for Transplantation: Toward a Functional Cure for Diabetes

To advance stem cell-derived beta cells (SC-β cells) to the next phase of development, rigorous validation is required. Animal models have been widely employed to assess SC-β cells in studies investigating developmental impairments caused by specific mutations (Saarimaki-Vire et al. 2017; Shi et al. 2017; Balboa et al. 2018; Ma et al. 2018), as well as for evaluating diabetes prevention and reversal (Shang et al. 2014; Zeng et al. 2016). Various forms-including pancreatic progenitors (Aghazadeh et al. 2021), SC-β cells (Saber et al. 2018; Maxwell et al. 2020), SC-β islets, and encapsulated SC-β cell clusters (Vegas et al. 2016) have been transplanted into both murine and primate models. Research has also focused on identifying optimal transplantation sites, with locations such as the subcutaneous space,(Aghazadeh et al. 2021) kidney capsule (Maxwell et al. 2022), epididymal fat (Villa et al. 2017), and intramuscular (Maxwell et al. 2022) being tested in immunocompromised murine models of diabetes induced by streptozotocin or alloxan. Early results from phase 1/2 clinical trials have shown that individuals transplanted with PEC-01 pancreatic endoderm cells encapsulated in the VC-02 device (developed by ViaCyte) were able to produce insulin in response to meals 26 weeks after transplantation, indicating that PEC-01 cells differentiated into mature, insulin-producing cells within the device while eliciting minimal immune reactions (Ramzy et al. 2021; Shapiro et al. 2021).

10 Rewriting the Blueprint: Biotech Tools for iPSC Generation and Gene Editing in Diabetes Research (Illustrated in Fig. 7)

10.1 CRISPR Technologies in SC-β Cell Engineering

CRISPR-Cas9 is a groundbreaking gene-editing technology that enables precise modification of DNA by cutting and altering specific genetic sequences using the Cas9 protein and a guide RNA, allowing for targeted removal, addition, or correction of genes. In diabetes research and therapy, CRISPR is widely used to create

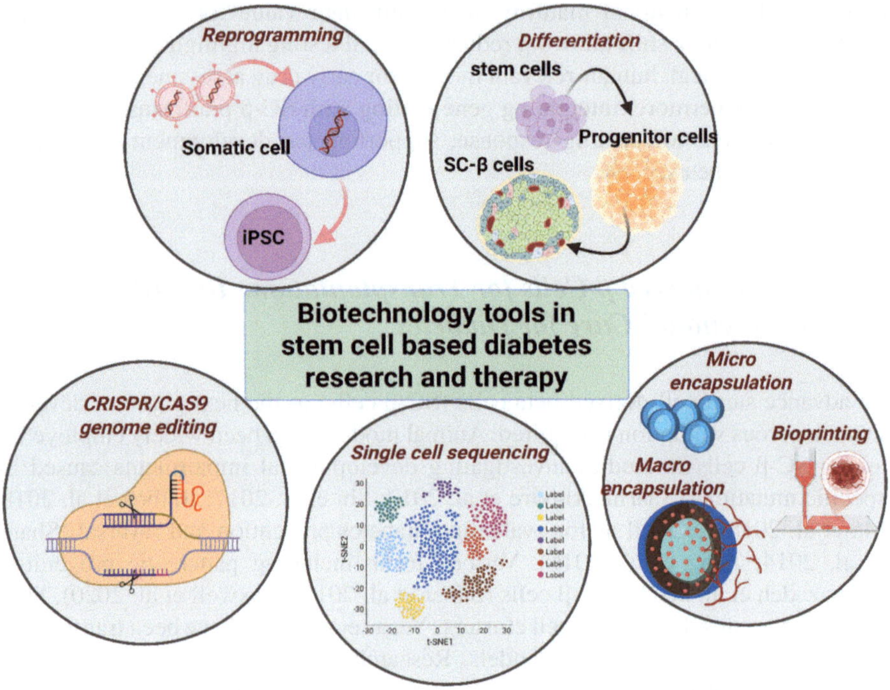

Fig. 7 Various tools and strategies that, in conjunction with stem cells, support diabetes research and therapy

disease models by editing genes in human pluripotent stem cells (hPSCs) at the undifferentiated stage, as efficient knock-in via homology-directed repair (HDR) is only feasible in dividing cells and not in post-mitotic stem cell-derived (SC) beta cells (Smirnikhina et al. 2022) This approach has enabled the generation of various iPSC lines carrying mutations in genes such as HNF1A (Cardenas-Diaz et al. 2019), HNF4A and PDX1 (Wang et al. 2019a, b), GCK and INS (Balboa et al. 2018), GLIS3 (Amin et al. 2018), SUR1 (Lithovius et al. 2021), RFX6 (Aldous et al. 2024) and STAT3 (Saarimaki-Vire et al. 2017) which are then differentiated into SC beta cells to study disease mechanisms and assess the effects of gene correction both in vitro and after transplantation in animal models. CRISPR-mediated correction of mutations in patient-derived iPSCs has been shown to restore impaired function in the resulting beta-like cells, providing a platform to dissect pathogenesis and evaluate therapeutic outcomes. The technology also allows researchers to study the effects of gene knockouts, such as SLC30A8 (Znt8) and ZNF148, revealing enhanced insulin secretion and reduced endoplasmic reticulum (ER) stress responses in edited cells (Ma et al. 2022; de Klerk et al. 2023) Additionally, CRISPR is used to facilitate the purification and tracking of SC beta cells by knocking in fluorescent markers (e.g., EGFP at the PDX1 or INS locus), and to engineer cells with reduced immunogenicity for transplantation by dampening immune responses (Lee et al. 2019). Beyond modeling monogenic diabetes, CRISPR-edited SC beta cells have been used to study and correct genetic defects in disorders such as Wolfram syndrome (Maxwell et al. 2020), Friedreich's ataxia (Miellet et al. 2024) and cystic fibrosis (Simsek et al. 2016) highlighting the broad therapeutic and investigative potential of this technology in diabetes and related diseases.

CRISPR-based hypoimmune engineering strategies are being employed to overcome immune rejection in SC-β cell transplantation. A primary approach involves knocking out B2M and CIITA genes to eliminate HLA class I and II expression, thereby reducing T-cell recognition, while simultaneously overexpressing CD47 to inhibit natural killer (NK) cell-mediated destruction. This combination has enabled hypoimmune SC-β cells to restore normoglycemia in immunocompetent humanized mice and macaque models (Hu et al. 2024). To address antibody-mediated rejection, CD64 has been overexpressed to sequester IgG Fc domains, protecting the cells from antibody-dependent cytotoxicity (Gravina et al. 2023); when combined with HLA ablation and CD47 overexpression, this strategy rendered SC-β cells fully immune-evasive both in vitro and in vivo. Complementary methods include engineering SC-β cells to secrete immunomodulatory cytokines such as IL-10 and TGF-β, which recruit regulatory T cells locally, promoting long-term graft survival in diabetic mice without systemic immunosuppression (Gerace et al. 2023). Genome-wide CRISPR screens have identified targets like CXCL10, whose knockout reduces immune cell recruitment (Sintov et al. 2022), and RNLS, whose deletion confers resistance to endoplasmic reticulum stress and cytokine-induced apoptosis, thereby enhancing graft resilience (Cai et al. 2020). These multiplexed strategies-combining hypoimmune gene edits, antibody evasion, localized immune modulation, and stress-resistance pathways-demonstrate significant potential for developing clinically viable, immune-tolerant SC-β cell therapies for diabetes.

10.2 TALENs as Complementary Tool in β-Cell Engineering

Transcription Activator-Like Effector Nucleases (TALENs) are gene-editing tools that have been utilized in SC-β cell research to enable precise genomic modifications, particularly for applications such as immune evasion and functional tracking. One key application involves targeting the AAVS1 genomic safe harbor locus, a well-characterized region on human chromosome 19 that supports stable transgene expression without disrupting cellular function, allowing for the introduction of therapeutic or reporter genes (Hockemeyer et al. 2011). For instance, TALENs have been used to insert the luciferase gene at the AAVS1 locus, enabling researchers to track SC-β cell survival after transplantation through bioluminescence imaging (Parent et al. 2021). Additionally, TALENs have been combined with CRISPR to enhance immune evasion strategies in SC-β cells, demonstrating the versatility and precision of this technology in advancing cell-based therapies for diabetes (Castro-Gutierrez et al. 2021).

10.3 Single-Cell Sequencing: A Biotechnological Approach to Deciphering Cellular Signatures

Single-cell RNA sequencing (scRNA-seq) technologies are revolutionizing our understanding of pancreatic islet biology by providing high-resolution insights into beta cell development, diversity, and function in both healthy and diseased states. These approaches have demonstrated that beta cells are not a uniform group but display considerable transcriptomic and functional heterogeneity, with multiple subpopulations and developmental pathways identified-including specific progenitor intermediates such as Nkx6.1+ cells that bridge the transition from progenitors to mature beta cells. scRNA-seq has also been instrumental in characterizing the asynchronous maturation of stem cell-derived beta cells, revealing that maturation is a gradual process marked by distinct transcriptomic shifts, which can be enhanced through targeted enrichment (Balboa et al. 2022). Furthermore, these methods have illuminated how different islet cell subtypes respond to metabolic stress, a key insight for refining beta cell replacement therapies. Comparative analyses of in vitro stem cell-derived islets and their in vivo grafts after transplantation show that transplantation promotes the expression of maturation markers and reduces stress signatures, indicating improved function in a physiological environment (Fu et al. 2023). Additionally, longitudinal scRNA-seq studies have revealed that environmental factors, such as caloric restriction, can modulate beta cell metabolic activity, potentially supporting their stability and lifespan-findings that may guide future diabetes treatments (Dos Santos et al. 2023). Overall, single-cell sequencing is providing unprecedented detail on islet cell complexity, developmental dynamics, and adaptive responses, paving the way for more precise and effective therapeutics.

11 Safeguarding Innovation: Ethical and Technical Oversight in β-Cell Bioengineering

Stem cell-based approaches for diabetes research and therapy offer considerable potential, yet they are accompanied by notable limitations, challenges, and ethical dilemmas. Immunological barriers, such as the threat of immune rejection and the necessity for ongoing immunosuppression, as well as the risk of tumor development from residual undifferentiated cells, present significant safety issues. Technical challenges include reliably producing mature, insulin-secreting beta cells, the high costs and complexity of cell manufacturing and transplantation and maintaining cell viability and function over time. Ethical concerns arise from the use of embryonic or fetal tissues, emphasizing the importance of responsible cell sourcing, informed consent, privacy, and fair access to emerging treatments. The advancement of precision 3D islet constructs and genetically engineered cells also requires scalable, reproducible manufacturing to guarantee consistent performance and safety. Rigorous preclinical testing is essential to minimize the tumorigenic potential of pluripotent stem cells, and transitioning to human trials demands well-characterized, safe, and functional cellular products. Long-term animal studies are crucial for demonstrating the absence of malignant transformation post-transplantation, helping to address ethical and safety apprehensions. In gene editing, particularly with CRISPR technology, off-target effects pose a risk of unintended genetic alterations, potentially impacting tumor suppressor genes or activating oncogenes, although strategies such as high-fidelity Cas9 variants and multiple guide RNAs can help mitigate these risks. Additionally, while genetically modifying beta cells to evade immune detection may improve graft survival, it raises concerns about the cells' ability to respond to infections or tumorigenic changes. Therefore, ongoing research, stringent oversight, and innovative solutions are vital to ensure that stem cell-based diabetes therapies are not only effective but also safe and ethically sound for long-term clinical use.

12 Conclusion and Future Perspectives

Biotechnology is revolutionizing diabetes research and treatment, enabling a shift from symptom management to targeted, curative approaches. From gene editing and antisense oligonucleotides to iPSC-derived β-cells and closed-loop insulin delivery systems, these innovations are bridging gaps in understanding, therapy, and patient-specific care. Despite these advances, challenges remain. Translating experimental therapies into clinical practice requires overcoming obstacles related to immune rejection, off-target genetic effects, manufacturing scalability, and long-term functional stability of transplanted cells. Moreover, ethical and regulatory considerations surrounding gene editing and stem cell manipulation must be carefully navigated. As such, sustained interdisciplinary collaboration between molecular biologists, bioengineers, clinicians, data scientists, and ethicists will be essential to fully realize the therapeutic potential of biotechnology in diabetes.

The future of diabetes therapy lies at the intersection of biotechnology, artificial intelligence (AI), and precision medicine. As the volume and complexity of biological data grow—from single-cell transcriptomes and proteomic profiles to real-time sensor outputs—AI will become indispensable in deciphering patterns, predicting disease trajectories, and tailoring individualized treatment plans. Machine learning algorithms can already optimize insulin dosing in closed-loop systems, but future iterations may predict glycemic trends based on multifactorial inputs, including diet, activity, stress, and hormonal cycles.

In gene therapy and regenerative medicine, AI can enhance vector design, predict off-target effects, and guide the differentiation of stem cells into fully functional β-cells. Precision medicine platforms integrating genomic, epigenomic, and metabolomic data will enable stratification of patients based on disease subtype, therapeutic responsiveness, and risk of complications—facilitating the deployment of customized gene or cell-based interventions. Digital twin models of pancreatic islets or systemic metabolism may allow virtual testing of therapies before patient application, reducing trial-and-error in clinical practice.

Additionally, the integration of biosensors and wearable technologies with AI-enabled health monitoring systems can provide continuous feedback loops that not only automate insulin delivery but also inform broader therapeutic decisions. These systems may eventually evolve into intelligent, self-regulating platforms capable of early disease detection and dynamic intervention, shifting the focus from reactive management to proactive, preventative care.

Ultimately, the fusion of biotechnology with AI and data science represents a paradigm shift in diabetes care—from generalized treatment to highly personalized, adaptive solutions that target disease at its source. With continued innovation, regulatory alignment, and ethical stewardship, the vision of a functional cure for diabetes is no longer speculative but increasingly attainable.

References

Abitbol A, Rabasa-Lhoret R, Messier V, Legault L, Smaoui M, Cohen N, Haidar A (2018) Overnight glucose control with dual- and single-hormone artificial pancreas in type 1 diabetes with hypoglycemia unawareness: a randomized controlled trial. Diabetes Technol Ther 20(3):189–196. https://doi.org/10.1089/dia.2017.0353

Abu-Elheiga L, Matzuk MM, Abo-Hashema KAH, Wakil SJ (2001) Continuous fatty acid oxidation and reduced fat storage in mice lacking Acetyl-COA carboxylase 2. Science 291(5513):2613–2616. https://doi.org/10.1126/science.1056843

Aghazadeh Y, Poon F, Sarangi F, Wong FTM, Khan ST, Sun X, Hatkar R, Cox BJ, Nunes SS, Nostro MC (2021) Microvessels support engraftment and functionality of human islets and hESC-derived pancreatic progenitors in diabetes models. Cell Stem Cell 28(11):1936–1949.e8. https://doi.org/10.1016/j.stem.2021.08.001

Aijaz A, Perera D, Olabisi RM (2017) Polymeric materials for cell microencapsulation. Methods Mol Biol:79–93. https://doi.org/10.1007/978-1-4939-6364-5_6

Alagpulinsa DA, Cao JJL, Driscoll RK, Sîrbulescu RF, Penson MFE, Sremac M, Engquist EN, Brauns TA, Markmann JF, Melton DA, Poznansky MC (2019) Alginate-microencapsulation of human stem cell–derived β cells with CXCL12 prolongs their survival and function in immunocompetent mice without systemic immunosuppression. Am J Transplant 19(7):1930–1940. https://doi.org/10.1111/ajt.15308

Aldous N, Elsayed AK, Memon B, Ijaz S, Hayat S, Abdelalim EM (2024) Deletion of RFX6 impairs iPSC-derived islet organoid development and survival, with no impact on PDX1+/NKX6.1+ progenitors. Diabetologia. https://doi.org/10.1007/s00125-024-06232-2

Amin S, Cook B, Zhou T, Ghazizadeh Z, Lis R, Zhang T, Khalaj M, Crespo M, Perera M, Xiang JZ, Zhu Z, Tomishima M, Liu C, Naji A, Evans T, Huangfu D, Chen S (2018) Discovery of a drug candidate for GLIS3-associated diabetes. Nat Commun 9(1). https://doi.org/10.1038/s41467-018-04918-x

Ao H, Liu B, Li H, Lu L (2019) Egr1 mediates retinal vascular dysfunction in diabetes mellitus via promoting p53 transcription. J Cell Mol Med 23(5):3345–3356. https://doi.org/10.1111/jcmm.14225

Arunadevi N, Jone Kirubavathy S (2023) 12 - Three-dimensional approaches based on nanotechnology towards wound management. In: Nanotechnological aspects for next-generation wound management. Elsevier, pp 245–280

Bailey CJ (2011) Renal glucose reabsorption inhibitors to treat diabetes. Trends Pharmacol Sci 32(2):63–71. https://doi.org/10.1016/j.tips.2010.11.011

Bainbridge J, Mistry A, De Alwis M, Paleolog E, Baker A, Thrasher A, Ali R (2002) Inhibition of retinal neovascularisation by gene transfer of soluble VEGF receptor sFlt-1. Gene Ther 9(5):320–326. https://doi.org/10.1038/sj.gt.3301680

Balboa D, Barsby T, Lithovius V, Saarimäki-Vire J, Omar-Hmeadi M, Dyachok O, Montaser H, Lund P-E, Yang M, Ibrahim H, Näätänen A, Chandra V, Vihinen H, Jokitalo E, Kvist J, Ustinov J, Nieminen AI, Kuuluvainen E, Hietakangas V, Katajisto P, Lau J, Carlsson P-O, Barg S, Tengholm A, Otonkoski T (2022) Functional, metabolic and transcriptional maturation of human pancreatic islets derived from stem cells. Nat Biotechnol 40(7):1042–1055. https://doi.org/10.1038/s41587-022-01219-z

Balboa D, Saarimäki-Vire J, Borshagovski D, Survila M, Lindholm P, Galli E, Eurola S, Ustinov J, Grym H, Huopio H, Partanen J, Wartiovaara K, Otonkoski T (2018) Insulin mutations impair beta-cell development in a patient-derived iPSC model of neonatal diabetes. eLife 7. https://doi.org/10.7554/elife.38519

Bandyopadhyay D, Kusari A, Kenner KA, Liu F, Chernoff J, Gustafson TA, Kusari J (1997) Protein-tyrosine phosphatase 1B complexes with the insulin receptor in vivo and is tyrosine-phosphorylated in the presence of insulin. J Biol Chem 272(3):1639–1645. https://doi.org/10.1074/jbc.272.3.1639

Barć P, Antkiewicz M, Śliwa B, Frączkowska K, Guziński M, Dawiskiba T, Małodobra-Mazur M, Witkiewicz W, Kupczyńska D, Strzelec B, Janczak D, Skóra JP (2021) Double VEGF/HGF gene therapy in critical limb ischemia complicated by diabetes mellitus. J Cardiovasc Transl Res 14(3):409–415. https://doi.org/10.1007/s12265-020-10066-9

Basta G, Montanucci P, Kinzer KP, Qi M, Luca G, Barbaro B, Noya G, Oberholzer J, Boselli C, Calafiore R (2011) Long-term metabolic and immunological follow-up of nonimmunosuppressed patients with type 1 diabetes treated with microencapsulated islet allografts. Diabetes Care 34(11):2406–2409. https://doi.org/10.2337/dc11-0731

Beddow SA, Gattu AK, Vatner DF, Paolella L, Alqarzaee A, Tashkandi N, Popov VB, Church CD, Rodeheffer MS, Cline GW, Geisler JG, Bhanot S, Samuel VT (2019) PEPCK1 AntiSense oligonucleotide prevents adiposity and impairs hepatic glycogen synthesis in High-fat male fed rats. Endocrinology 160(1):205–219. https://doi.org/10.1210/en.2018-00630

Benchoula K, Arya A, Parhar IS, Hwa WE (2020) FoxO1 signaling as a therapeutic target for type 2 diabetes and obesity. Eur J Pharmacol 891:173758. https://doi.org/10.1016/j.ejphar.2020.173758

Bergenstal RM, Garg S, Weinzimer SA, Buckingham BA, Bode BW, Tamborlane WV, Kaufman FR (2016) Safety of a hybrid Closed-Loop insulin delivery system in patients with Type 1 diabetes. JAMA 316(13):1407. https://doi.org/10.1001/jama.2016.11708

Bergmark BA, Marston NA, Prohaska TA, Alexander VJ, Zimerman A, Moura FA, Murphy SA, Goodrich EL, Zhang S, Gaudet D, Karwatowska-Prokopczuk E, Tsimikas S, Giugliano RP, Sabatine MS (2024) Olezarsen for hypertriglyceridemia in patients at high cardiovascular risk. N Engl J Med 390(19):1770–1780. https://doi.org/10.1056/nejmoa2402309

Beumer J, Clevers H (2016) Regulation and plasticity of intestinal stem cells during homeostasis and regeneration.Development. Oct 15;143(20):3639–3649. PMID: 27802133. https://doi.org/10.1242/dev.133132

Bevacqua RJ, Dai X, Lam JY, Gu X, Friedlander MSH, Tellez K, Miguel-Escalada I, Bonàs-Guarch S, Atla G, Zhao W, Kim SH, Dominguez AA, Qi LS, Ferrer J, MacDonald PE, Kim SK (2021) CRISPR-based genome editing in primary human pancreatic islet cells. Nat Commun 12(1). https://doi.org/10.1038/s41467-021-22651-w

Bhatt DL, Bays HE, Miller M, Cain JE, Wasilewska K, Andrawis NS, Parli T, Feng S, Sterling L, Tseng L, Hartsfield CL, Agollah GD, Mansbach H, Kastelein JJP (2023) The FGF21 analog pegozafermin in severe hypertriglyceridemia: a randomized phase 2 trial. Nat Med 29(7):1782–1792. https://doi.org/10.1038/s41591-023-02427-z

Billings LK, Florez JC (2010) The genetics of type 2 diabetes: what have we learned from GWAS? Ann N Y Acad Sci 1212(1):59–77. https://doi.org/10.1111/j.1749-6632.2010.05838.x

Boden G, Tappy L, Jadali F, Hoeldtke RD, Rezvani I, Owen OE (1990) Role of glucagon in disposal of an amino acid load. AJP Endocrinol Metab 259(2):E225–E232. https://doi.org/10.1152/ajpendo.1990.259.2.e225

Bone RN, Icyuz M, Zhang Y, Zhang Y, Cui W, Wang H, Peng J-B, Matthews QL, Siegal GP, Wu H (2012) Gene transfer of active Akt1 by an infectivity-enhanced adenovirus impacts β-cell survival and proliferation differentially in vitro and in vivo. Islets 4(6):366–378. https://doi.org/10.4161/isl.22721

Boughton CK, Hovorka R (2020) The artificial pancreas. Curr Opin Organ Transplant 25(4):336–342. https://doi.org/10.1097/mot.0000000000000786

Boyanapalli SSS, Kong A-NT (2015) "Curcumin, the King of Spices": epigenetic regulatory mechanisms in the prevention of cancer, neurological, and inflammatory diseases. Curr Pharmacol Rep 1(2):129–139. https://doi.org/10.1007/s40495-015-0018-x

Branda CS, Dymecki SM (2004) Talking about a revolution. Dev Cell 6(1):7–28. https://doi.org/10.1016/s1534-5807(03)00399-x

Braverman-Gross C, Nudel N, Ronen D, Beer NL, McCarthy MI, Benvenisty N (2018) Derivation and molecular characterization of pancreatic differentiated MODY1-iPSCs. Stem Cell Res 31:16–26. https://doi.org/10.1016/j.scr.2018.06.013

Breton M, Kanapka L, Beck R, Ekhlaspour L, Forlenza G, Cengiz E, Schoelwer M, Ruedy K, Jost E, Carria L, Emory E, Hsu L, Oliveri M, Kollman C, Dokken B, Weinzimer S, Deboer M, Buckingham B, Cherñavvsky D, Wadwa RP, Group ITR, Renard E (2020) A randomized trial of closed-loop control in children with Type 1 diabetes. N Engl J Med 383(9):836–845. https://doi.org/10.1056/nejmoa2004736

Brocard J, Warot X, Wendling O, Messaddeq N, Vonesch J-L, Chambon P, Metzger D (1997) Spatio-temporally controlled site-specific somatic mutagenesis in the mouse. Proc Natl Acad Sci 94(26):14559–14563. https://doi.org/10.1073/pnas.94.26.14559

Broch M, Vendrell J, Ricart W, Richart C, Fernández-Real J-M (2007) Circulating Retinol-Binding protein-4, insulin sensitivity, insulin secretion, and insulin disposition index in obese and non-obese subjects. Diabetes Care 30(7):1802–1806. https://doi.org/10.2337/dc06-2034

Brüning JC, Michael MD, Winnay JN, Hayashi T, Hörsch D, Accili D, Goodyear LJ, Kahn CR (1998) A muscle-specific insulin receptor knockout exhibits features of the metabolic syndrome of NIDDM without altering glucose tolerance. Mol Cell 2(5):559–569. https://doi.org/10.1016/s1097-2765(00)80155-0

Cai D, Yuan M, Frantz DF, Melendez PA, Hansen L, Lee J, Shoelson SE (2005) Local and systemic insulin resistance resulting from hepatic activation of IKK-β and NF-κB. Nat Med 11(2):183–190. https://doi.org/10.1038/nm1166

Cai EP, Ishikawa Y, Zhang W, Leite NC, Li J, Hou S, Kiaf B, Hollister-Lock J, Yilmaz NK, Schiffer CA, Melton DA, Kissler S, Yi P (2020) Genome-scale in vivo CRISPR screen identifies RNLS as a target for beta cell protection in type 1 diabetes. Nat Metab 2(9):934–945. https://doi.org/10.1038/s42255-020-0254-1

Calafiore R, Basta G, Luca G, Lemmi A, Montanucci MP, Calabrese G, Racanicchi L, Mancuso F, Brunetti P (2006) Microencapsulated pancreatic islet allografts into nonimmunosuppressed patients with Type 1 diabetes. Diabetes Care 29(1):137–138. https://doi.org/10.2337/diacare.29.01.06.dc05-1270

Campbell KHS, McWhir J, Ritchie WA, Wilmut I (1996) Sheep cloned by nuclear transfer from a cultured cell line. Nature 380(6569):64–66. https://doi.org/10.1038/380064a0

Cantó C, Gerhart-Hines Z, Feige JN, Lagouge M, Noriega L, Milne JC, Elliott PJ, Puigserver P, Auwerx J (2009) AMPK regulates energy expenditure by modulating NAD+ metabolism and SIRT1 activity. Nature 458(7241):1056–1060. https://doi.org/10.1038/nature07813

Cappon G, Vettoretti M, Sparacino G, Facchinetti A (2019) Continuous glucose monitoring sensors for diabetes management: a review of technologies and applications. Diabetes Metab J 43(4):383. https://doi.org/10.4093/dmj.2019.0121

Cardenas-Diaz FL, Osorio-Quintero C, Diaz-Miranda MA, Kishore S, Leavens K, Jobaliya C, Stanescu D, Ortiz-Gonzalez X, Yoon C, Chen CS, Haliyur R, Brissova M, Powers AC, French DL, Gadue P (2019) Modeling monogenic diabetes using human ESCs reveals developmental and metabolic deficiencies caused by mutations in HNF1A. Cell Stem Cell 25(2):273–289.e5. https://doi.org/10.1016/j.stem.2019.07.007

Carey BW, Markoulaki S, Hanna JH, Faddah DA, Buganim Y, Kim J, Ganz K, Steine EJ, Cassady JP, Creyghton MP, Welstead GG, Gao Q, Jaenisch R (2011) Reprogramming factor stoichiometry influences the epigenetic state and biological properties of induced pluripotent stem cells. Cell Stem Cell 9(6):588–598. https://doi.org/10.1016/j.stem.2011.11.003

Caricilli AM, Nascimento PH, Pauli JR, Tsukumo DML, Velloso LA, Carvalheira JB, Saad MJA (2008) Inhibition of toll-like receptor 2 expression improves insulin sensitivity and signaling in muscle and white adipose tissue of mice fed a high-fat diet. J Endocrinol 199(3):399–406. https://doi.org/10.1677/joe-08-0354

Carow B, Gao Y, Coquet J, Reilly M, Rottenberg ME (2016) LCK-driven CRE expression alters T cell development in the thymus and the frequencies and functions of peripheral T cell subsets. J Immunol 197(6):2261–2268. https://doi.org/10.4049/jimmunol.1600827

Castro-Gutierrez R, Alkanani A, Mathews CE, Michels A, Russ HA (2021) Protecting stem cell derived pancreatic Beta-like cells from diabetogenic T cell recognition. Front Endocrinol 12. https://doi.org/10.3389/fendo.2021.707881

Caton ML, Smith-Raska MR, Reizis B (2007) Notch-RBP-J signaling controls the homeostasis of CD8- dendritic cells inthe spleen. J Exp Med. Jul 9;204(7):1653–64. Epub Jun 25. PMID: 17591855; PMCID: PMC2118632, https://doi.org/10.1084/jem.20062648

Chen K, Wang H, Yang X, Tang C, Hu G, Gao Z (2024) Targeting gut microbiota as a therapeutic target in T2DM: a review of multi-target interactions of probiotics, prebiotics, postbiotics, and synbiotics with the intestinal barrier. Pharmacol Res 210:107483. https://doi.org/10.1016/j.phrs.2024.107483

Chen L, Wang X, Lu W, Wu X, Li J (2016) Molecular imprinting: perspectives and applications. Chem Soc Rev 45(8):2137–2211. https://doi.org/10.1039/c6cs00061d

Chen S, Sbuh N, Veedu RN (2021a) Antisense oligonucleotides as potential therapeutics for type 2 diabetes. Nucleic Acid Ther 31(1):39–57. https://doi.org/10.1089/nat.2020.0891

Chen W, Hoo RL-C, Konishi M, Itoh N, Lee P-C, Ye H-Y, Lam KS-L, Xu A (2011) Growth hormone induces hepatic production of fibroblast growth factor 21 through a mechanism dependent on lipolysis in adipocytes. J Biol Chem 286(40):34559–34566. https://doi.org/10.1074/jbc.m111.285965

Chen Y-S, Gleaton J, Yang Y, Dhayalan B, Phillips NB, Liu Y, Broadwater L, Jarosinski MA, Chatterjee D, Lawrence MC, Hattier T, Michael MD, Weiss MA (2021b) Insertion of a synthetic switch into insulin provides metabolite-dependent regulation of hormone–receptor activation. Proc Natl Acad Sci 118(30). https://doi.org/10.1073/pnas.2103518118

Cho L, Nicholls SJ, Nordestgaard BG, Landmesser U, Tsimikas S, Blaha MJ, Leitersdorf E, Lincoff AM, Lesogor A, Manning B, Kozlovski P, Cao H, Wang J, Nissen SE (2025) Design and rationale of Lp(a)HORIZON trial: assessing the effect of Lipoprotein(a) lowering with Pelacarsen on major cardiovascular events in patients with CVD and elevated Lp(a). Am Heart J. https://doi.org/10.1016/j.ahj.2025.03.019

Cho YM, Youn B-S, Lee H, Lee N, Min S-S, Kwak SH, Lee HK, Park KS (2006) Plasma Retinol-Binding protein-4 concentrations are elevated in human subjects with impaired glucose tolerance and Type 2 diabetes. Diabetes Care 29(11):2457–2461. https://doi.org/10.2337/dc06-0360

Choi CS, Savage DB, Kulkarni A, Yu XX, Liu Z-X, Morino K, Kim S, Distefano A, Samuel VT, Neschen S, Zhang D, Wang A, Zhang X-M, Kahn M, Cline GW, Pandey SK, Geisler JG, Bhanot S, Monia BP, Shulman GI (2007) Suppression of Diacylglycerol Acyltransferase-2 (DGAT2), but not DGAT1, with antisense oligonucleotides reverses diet-induced hepatic steatosis and insulin resistance. J Biol Chem 282(31):22678–22688. https://doi.org/10.1074/jbc.m704213200

Choi SH, Lee HC (2010) Long-term, antidiabetogenic effects of GLP-1 gene therapy using a double-stranded, adeno-associated viral vector. Gene Ther 18(2):155–163. https://doi.org/10.1038/gt.2010.119

Choudhary P, Kolassa R, Keuthage W, Kroeger J, Thivolet C, Evans M, Ré R, De Portu S, Vorrink L, Shin J, Habteab A, Castañeda J, Da Silva J, Cohen O (2022) Advanced hybrid closed loop therapy versus conventional treatment in adults with type 1 diabetes (ADAPT): a randomised controlled study. Lancet Diabetes Endocrinol 10(10):720–731. https://doi.org/10.1016/s2213-8587(22)00212-1

Clampit JE, Meuth JL, Smith HT, Reilly RM, Jirousek MR, Trevillyan JM, Rondinone CM (2003) Reduction of protein-tyrosine phosphatase-1B increases insulin signaling in FAO hepatoma cells. Biochem Biophys Res Commun 300(2):261–267. https://doi.org/10.1016/s0006-291x(02)02839-5

Cole JB, Florez JC (2020) Genetics of diabetes mellitus and diabetes complications. Nat Rev Nephrol. Jul;16(7):377–390. Epub 2020 May 12. PMID: 32398868; PMCID: PMC9639302. https://doi.org/10.1038/s41581-020-0278-5

Collier CA, Mendiondo C, Raghavan S (2022) Tissue engineering of the gastrointestinal tract: the historic path to translation. J Biol Eng 16(1). https://doi.org/10.1186/s13036-022-00289-6

Colombani T, Peuziat P, Dallet L, Haudebourg T, Mével M, Berchel M, Lambert O, Habrant D, Pitard B (2017) Self-assembling complexes between binary mixtures of lipids with different linkers and nucleic acids promote universal mRNA, DNA and siRNA delivery. J Control Release 249:131–142. https://doi.org/10.1016/j.jconrel.2017.01.041

Cruzado JM, Lloberas N, Torras J, Riera M, Fillat C, Herrero-Fresneda I, Aran JM, Alperovich G, Vidal A, Grinyo JM (2004) Regression of advanced diabetic nephropathy by hepatocyte growth factor gene therapy in rats. Diabetes 53(4):1119–1127. https://doi.org/10.2337/diabetes.53.4.1119

Cuchel M, Meagher EA, Du Toit TH, Blom DJ, Marais AD, Hegele RA, Averna MR, Sirtori CR, Shah PK, Gaudet D, Stefanutti C, Vigna GB, Du Plessis AM, Propert KJ, Sasiela WJ, Bloedon LT, Rader DJ (2012) Efficacy and safety of a microsomal triglyceride transfer protein inhibitor in patients with homozygous familial hypercholesterolaemia: a single-arm, open-label, phase 3 study. Lancet 381(9860):40–46. https://doi.org/10.1016/s0140-6736(12)61731-0

Cui H, Zhu X, Li S, Wang P, Fang J (2021) Liver-targeted delivery of oligonucleotides with N-Acetylgalactosamine conjugation. ACS Omega 6(25):16259–16265. https://doi.org/10.1021/acsomega.1c01755

Dassaye R, Naidoo S, Cerf ME (2015) Transcription factor regulation of pancreatic organogenesis, differentiation and maturation. Islets 8(1):13–34. https://doi.org/10.1080/19382014.2015.1075687

De Klerk E, Xiao Y, Emfinger CH, Keller MP, Berrios DI, Loconte V, Ekman AA, White KL, Cardone RL, Kibbey RG, Attie AD, Hebrok M (2023) Loss of ZNF148 enhances insulin secretion in human pancreatic β cells. JCI. Insight 8(11). https://doi.org/10.1172/jci.insight.157572

Dettmer R, Niwolik I, Cirksena K, Yoshimoto T, Tang Y, Mehmeti I, Gurgul-Convey E, Naujok O (2022) Proinflammatory cytokines induce rapid, NO-independent apoptosis, expression of chemotactic mediators and interleukin-32 secretion in human pluripotent stem cell-derived beta cells. Diabetologia 65(5):829–843. https://doi.org/10.1007/s00125-022-05654-0

Dettmer R, Niwolik I, Mehmeti I, Jörns A, Naujok O (2021) New hPSC SOX9 and INS reporter cell lines facilitate the observation and optimization of differentiation into insulin-producing cells. Stem Cell Rev Rep 17(6):2193–2209. https://doi.org/10.1007/s12015-021-10232-9

Di Dalmazi G, Pagotto U, Pasquali R, Vicennati V (2012) Glucocorticoids and type 2 diabetes: from physiology to pathology. J Nutr Metabol 2012:1–9. https://doi.org/10.1155/2012/525093

Di Renzo L, Cioccoloni G, Falco S, Abenavoli L, Moia A, Salimei PS, De Lorenzo A (2018) Influence of FTO rs9939609 and Mediterranean diet on body composition and weight loss: a randomized clinical trial. J Transl Med 16(1). https://doi.org/10.1186/s12967-018-1680-7

Dimitrioglou N, Kanelli M, Papageorgiou E, Karatzas T, Hatziavramidis D (2019) Paving the way for successful islet encapsulation. Drug Discov Today 24(3):737–748. https://doi.org/10.1016/j.drudis.2019.01.020

Dimitrov D, Thiele I, Ferguson LR (2016) Editorial: the human gutome: nutrigenomics of host-microbiome interactions. Front Genet 7. https://doi.org/10.3389/fgene.2016.00158

Ding S, Lyu Z, Niu X, Zhou Y, Liu D, Falahati M, Du D, Lin Y (2020) Integrating ionic liquids with molecular imprinting technology for biorecognition and biosensing: a review. Biosens Bioelectron 149:111830. https://doi.org/10.1016/j.bios.2019.111830

Santos CD, Shrestha S, Cottam M, Perkins G, Lev-Ram V, Roy B, Acree C, Kim K-Y, Deerinck T, Cutler M, Dean D, Cartailler JP, MacDonald PE, Hetzer M, Ellisman M, Drigo RAE (2023) Caloric restriction promotes beta cell longevity and delays aging and senescence by enhancing cell identity and homeostasis mechanisms. bioRxiv (Cold Spring Harbor Laboratory). https://doi.org/10.1101/2023.08.23.554369

Du H, Ma Y, Wang X, Zhang Y, Zhu L, Shi S, Pan S, Liu Z (2023) Advanced glycation end products induce skeletal muscle atrophy and insulin resistance via activating ROS-mediated ER stress PERK/FOXO1 signaling. AJP Endocrinol Metab 324(3):E279–E287. https://doi.org/10.1152/ajpendo.00218.2022

Du K, Herzig S, Kulkarni RN, Montminy M (2003) TRB3: a tribbles homolog that inhibits Akt/PKB activation by insulin in liver. Science 300(5625):1574–1577. https://doi.org/10.1126/science.1079817

Du S, Li Y, Geng Z, Zhang Q, Buhler LH, Gonelle-Gispert C, Wang Y (2022) Engineering islets from stem cells: the optimal solution for the treatment of diabetes? Front Immunol 13. https://doi.org/10.3389/fimmu.2022.869514

Duin S, Schütz K, Ahlfeld T, Lehmann S, Lode A, Ludwig B, Gelinsky M (2019) 3D bioprinting of functional islets of langerhans in an Alginate/Methylcellulose hydrogel blend. Adv Healthc Mater 8(7). https://doi.org/10.1002/adhm.201801631

Egozi A, Llivichuzhca-Loja D, McCourt BT, Halpern KB, Farack L, An X, Wang F, Chen K, Konnikova L, Itzkovitz S (2021) Insulin is expressed by enteroendocrine cells during human fetal development. Nat Med 27(12):2104–2107. https://doi.org/10.1038/s41591-021-01586-1

Elchebly M, Payette P, Michaliszyn E, Cromlish W, Collins S, Loy AL, Normandin D, Cheng A, Himms-Hagen J, Chan CC, Ramachandran C, Gresser MJ, Tremblay ML, Kennedy BP (1999) Increased insulin sensitivity and obesity resistance in mice lacking the protein tyrosine phosphatase-1B gene. Science 283(5407):1544–1548. https://doi.org/10.1126/science.283.5407.1544

Elsner M, Terbish T, Jorns A, Naujok O, Wedekind D, Hedrich HJ, Lenzen S (2012) Reversal of diabetes through gene therapy of diabetic rats by hepatic insulin expression via lentiviral transduction. Mol Ther 20(5):918–926. https://doi.org/10.1038/mt.2012.10

Erion DM, Ignatova ID, Yonemitsu S, Nagai Y, Chatterjee P, Weismann D, Hsiao JJ, Zhang D, Iwasaki T, Stark R, Flannery C, Kahn M, Carmean CM, Yu XX, Murray SF, Bhanot S, Monia BP, Cline GW, Samuel VT, Shulman GI (2009) Prevention of hepatic steatosis and hepatic insulin resistance by knockdown of cAMP response element-binding protein. Cell Metab 10(6):499–506. https://doi.org/10.1016/j.cmet.2009.10.003

Erion DM, Kotas ME, McGlashon J, Yonemitsu S, Hsiao JJ, Nagai Y, Iwasaki T, Murray SF, Bhanot S, Cline GW, Samuel VT, Shulman GI, Gillum MP (2013) cAMP-responsive element-binding protein (CREB)-regulated transcription coactivator 2 (CRTC2) promotes glucagon clearance and hepatic amino acid catabolism to regulate glucose homeostasis. J Biol Chem 288(22):16167–16176. https://doi.org/10.1074/jbc.M112.442277

Even MS, Sandusky CB, Barnard ND, Mistry J, Sinha MK (2007) Development of a novel ELISA for human insulin using monoclonal antibodies produced in serum-free cell culture medium. Clin Biochem 40(1–2):98–103. https://doi.org/10.1016/j.clinbiochem.2006.08.013

Exton JH (1979) Regulation of gluconeogenesis by glucocorticoids. Monogr Endocrinol 12:535–546. https://doi.org/10.1007/978-3-642-67251-9_28

Fang H, Judd RL (2018) Adiponectin regulation and function. Compr Physiol 8(3):1031–1063. https://doi.org/10.1002/cphy.c170046

Farina M, Ballerini A, Fraga DW, Nicolov E, Hogan M, Demarchi D, Scaglione F, Sabek OM, Horner P, Thekkedath U, Gaber OA, Grattoni A (2017) 3D printed vascularized device for subcutaneous transplantation of human islets. Biotechnol J 12(9):1600281. https://doi.org/10.1002/biot.201600281

Fenech M, El-Sohemy A, Cahill L, Ferguson LR, French TA, Tai ES, Milner J, Koh WP, Xie L, Zucker M, Buckley M, Cosgrove L, Lockett T, Fung KY, Head R (2011) Nutrigenetics and nutrigenomics: viewpoints on the current status and applications in nutrition research and practice. J Nutrigenet Nutrigenomics 4(2):69–89. https://doi.org/10.1159/000327772

Feng L, Chao J, Tian E, Li L, Ye P, Zhang M, Chen X, Cui Q, Sun G, Zhou T, Felix G, Qin Y, Li W, Meza ED, Klein J, Ghoda L, Hu W, Luo Y, Dang W, Hsu D, Gold J, Goldman SA, Matalon R, Shi Y (2020) Cell-based therapy for Canavan disease using human iPSC-derived NPCs and OPCs. Adv Sci 7(23):2002155. https://doi.org/10.1002/advs.202002155

Ferber S, Halkin A, Cohen H, Ber I, Einav Y, Goldberg I, Barshack I, Seijffers R, Kopolovic J, Kaiser N, Karasik A (2000) Pancreatic and duodenal homeobox gene 1 induces expression of insulin genes in liver and ameliorates streptozotocin-induced hyperglycemia. Nat Med 6(5):568–572. https://doi.org/10.1038/75050

Ferraro B, Cruz YL, Coppola D, Heller R (2009) Intradermal delivery of plasmid VEGF(165) by electroporation promotes wound healing. Mol Ther 17(4):651–657. https://doi.org/10.1038/mt.2009.1

Fu Q, Jiang H, Qian Y, Lv H, Dai H, Zhou Y, Chen Y, He Y, Gao R, Zheng S, Liang Y, Li S, Xu X, Xu K, Yang T (2023) Single-cell RNA sequencing combined with single-cell proteomics identifies the metabolic adaptation of islet cell subpopulations to high-fat diet in mice. Diabetologia 66(4):724–740. https://doi.org/10.1007/s00125-022-05881-2

Gattas-Asfura K, Valdes M, Celik E, Stabler C (2014) Covalent layer-by-layer assembly of hyperbranched polymers on alginate microcapsules to impart stability and permselectivity. J Mater Chem B 2(46):8208–8219. https://doi.org/10.1039/C4TB01388H

Gaudet D, Karwatowska-Prokopczuk E, Baum SJ, Hurh E, Kingsbury J, Bartlett VJ, Figueroa AL, Piscitelli P, Singleton W, Witztum JL, Geary RS, Tsimikas S, O'Dea LSL, Vupanorsen Study Investigators (2020) Vupanorsen, an N-acetyl galactosamine-conjugated antisense drug to ANGPTL3 mRNA, lowers triglycerides and atherogenic lipoproteins in patients with diabetes, hepatic steatosis, and hypertriglyceridaemia. Eur Heart J 41(40):3936–3945. https://doi.org/10.1093/eurheartj/ehaa872

Ge K, Peng Y, Lu Z, Hu Y, Li G (2020) Aptamer-gold nanoparticle doped covalent organic framework followed by matrix-assisted laser desorption/ionization time-of-flight mass spectrometry for selective enrichment and detection of human insulin. J Chromatogr A 1615:460741. https://doi.org/10.1016/j.chroma.2019.460741

Gehlbach P, Demetriades AM, Yamamoto S, Deering T, Xiao WH, Duh EJ, Yang HS, Lai H, Kovesdi I, Carrion M, Wei L, Campochiaro PA (2003) Periocular gene transfer of sFlt-1 suppresses ocular neovascularization and vascular endothelial growth factor-induced breakdown of the blood-retinal barrier. Hum Gene Ther 14(2):129–141. https://doi.org/10.1089/10430340360532260

Gerace D, Zhou Q, Kenty JH, Veres A, Sintov E, Wang X, Boulanger KR, Li H, Melton DA (2023) Engineering human stem cell-derived islets to evade immune rejection and promote localized immune tolerance. Cell Reports Med 4(1):100879. https://doi.org/10.1016/j.xcrm.2022.100879

Ghazizadeh Z, Kao DI, Amin S, Cook B, Rao S, Zhou T, Zhang T, Xiang Z, Kenyon R, Kaymakcalan O, Liu C, Evans T, Chen S (2017) ROCKII inhibition promotes the maturation of human pancreatic beta-like cells. Nat Commun 8(1):298. https://doi.org/10.1038/s41467-017-00368-1

Goh KP, Lee HY, Lau DP, Supaat W, Chan YH, Koh AF (2014) Effects of resveratrol in patients with type 2 diabetes mellitus on skeletal muscle SIRT1 expression and energy expenditure. Int J Sport Nutr Exerc Metab 24(1):2–13. https://doi.org/10.1123/ijsnem.2013-0082

Gohari-Lasaki S, Sharafshah A, Abbaspour S, Keshavarz P (2020) Single locus and haplotype association of ENPP1 gene variants with the development of retinopathy among type 2 diabetic patients. Int Ophthalmol 40(3):639–647. https://doi.org/10.1007/s10792-019-01222-5

Goldstein BJ, Bittner-Kowalczyk A, White MF, Harbeck M (2000) Tyrosine dephosphorylation and deactivation of insulin receptor substrate-1 by protein-tyrosine phosphatase 1B. Possible facilitation by the formation of a ternary complex with the Grb2 adaptor protein. J Biol Chem 275(6):4283–4289. https://doi.org/10.1074/jbc.275.6.4283

Gossen M, Bujard H (1992) Tight control of gene expression in mammalian cells by tetracycline-responsive promoters. Proc Natl Acad Sci USA 89(12):5547–5551. https://doi.org/10.1073/pnas.89.12.5547

Gossen M, Freundlieb S, Bender G, Muller G, Hillen W, Bujard H (1995) Transcriptional activation by tetracyclines in mammalian cells. Science 268(5218):1766–1769. https://doi.org/10.1126/science.7538621

Graham TE, Yang Q, Bluher M, Hammarstedt A, Ciaraldi TP, Henry RR, Wason CJ, Oberbach A, Jansson PA, Smith U, Kahn BB (2006) Retinol-binding protein 4 and insulin resistance in lean, obese, and diabetic subjects. N Engl J Med 354(24):2552–2563. https://doi.org/10.1056/NEJMoa054288

Gravina A, Tediashvili G, Rajalingam R, Quandt Z, Deisenroth C, Schrepfer S, Deuse T (2023) Protection of cell therapeutics from antibody-mediated killing by CD64 overexpression. Nat Biotechnol 41(5):717–727. https://doi.org/10.1038/s41587-023-01814-8

Gu H, Marth JD, Orban PC, Mossmann H, Rajewsky K (1994) Deletion of a DNA polymerase beta gene segment in T cells using cell type-specific gene targeting. Science 265(5168):103–106. https://doi.org/10.1126/science.8013172

Gu H, Zou YR, Rajewsky K (1993) Independent control of immunoglobulin switch recombination at individual switch regions evidenced through Cre-loxP-mediated gene targeting. Cell 73(6):1155–1164. https://doi.org/10.1016/0092-8674(93)90644-C

Gujarati NA, Frimpong BO, Zaidi M, Bronstein R, Revelo MP, Haley JD, Kravets I, Guo Y, Mallipattu SK (2024) Podocyte-specific KLF6 primes proximal tubule CaMK1D signaling to attenuate diabetic kidney disease. Nat Commun 15(1):8038. https://doi.org/10.1038/s41467-024-25724-9

Habener JF, Stanojevic V (2012) α-cell role in β-cell generation and regeneration. Islets 4(3):188–198. https://doi.org/10.4161/isl.20691

Haidar A (2019) Insulin-and-glucagon artificial pancreas versus insulin-alone artificial pancreas: a short review. Diabetes Spectrum 32(3):215–221. https://doi.org/10.2337/ds19-0026

Haidar A, Legault L, Matteau-Pelletier L, Messier V, Dallaire M, Ladouceur M, Rabasa-Lhoret R (2015a) Outpatient overnight glucose control with dual-hormone artificial pancreas, single-hormone artificial pancreas, or conventional insulin pump therapy in children and adolescents with type 1 diabetes: an open-label, randomised controlled trial. Lancet Diabetes Endocrinol 3(8):595–604. https://doi.org/10.1016/S2213-8587(15)00214-9

Haidar A, Legault L, Messier V, Mitre TM, Leroux C, Rabasa-Lhoret R (2015b) Comparison of dual-hormone artificial pancreas, single-hormone artificial pancreas, and conventional insulin pump therapy for glycaemic control in patients with type 1 diabetes: an open-label randomised controlled crossover trial. Lancet Diabetes Endocrinol 3(1):17–26. https://doi.org/10.1016/S2213-8587(14)70238-9

Haidar A, Rabasa-Lhoret R, Legault L, Lovblom LE, Rakheja R, Messier V, D'Aoust E, Falappa CM, Justice T, Orszag A, Tschirhart H, Dallaire M, Ladouceur M, Perkins BA (2016) Single- and dual-hormone artificial pancreas for overnight glucose control in type 1 diabetes. J Clin Endocrinol Metab 101(1):214–223. https://doi.org/10.1210/jc.2015-3452

Hammond SM, Aartsma-Rus A, Alves S, Borgos SE, Buijsen RAM, Collin RWJ, Covello G, Denti MA, Desviat LR, Echevarria L, Foged C, Gaina G, Garanto A, Goyenvalle AT, Guzowska M, Holodnuka I, Jones DR, Krause S, Lehto T, Montolio M, Van Roon-Mom W, Arechavala-Gomeza V (2021) Delivery of oligonucleotide-based therapeutics: challenges and opportunities. EMBO Mol Med 13(4):e13243. https://doi.org/10.15252/emmm.202013243

Hanson RW, Reshef L (1997) Regulation of phosphoenolpyruvate carboxykinase (GTP) gene expression. Annu Rev Biochem 66:581–611. https://doi.org/10.1146/annurev.biochem.66.1.581

Harrison SA, Frias JP, Neff G, Abrams GA, Lucas KJ, Sanchez W, Gogia S, Sheikh MY, Behling C, Bedossa P, Shao L, Chan D, Fong E, de Temple B, Shringarpure R, Tillman EJ, Rolph T, Cheng A, Yale K, HARMONY Study Group (2023) Safety and efficacy of once-weekly efruxifermin versus placebo in non-alcoholic steatohepatitis (HARMONY): a multicentre, randomised, double-blind, placebo-controlled, phase 2b trial. Lancet Gastroenterol Hepatol 8(12):1080–1093. https://doi.org/10.1016/S2468-1253(23)00305-8

Hasan MN, Hwang YH, An JM, Shahriar SMS, Cho S, Lee YK (2020) Oral GLP1 gene delivery by an antibody-guided nanomaterial to treat type 2 diabetes mellitus. ACS Appl Mater Interfaces 12(35):38925–38935. https://doi.org/10.1021/acsami.0c11201

Haurigot V, Villacampa P, Ribera A, Bosch A, Ramos D, Ruberte J, Bosch F (2012) Long-term retinal PEDF overexpression prevents neovascularization in a murine adult model of retinopathy. PLoS One 7(7):e41511. https://doi.org/10.1371/journal.pone.0041511
Hayden MS, Ghosh S (2008) Shared principles in NF-kappaB signaling. Cell 132(3):344–362. https://doi.org/10.1016/j.cell.2008.01.020
Hayes ME, Drummond DC, Kirpotin DB, Zheng WW, Noble CO, Park JW, Marks JD, Benz CC, Hong K (2006) Genospheres: self-assembling nucleic acid-lipid nanoparticles suitable for targeted gene delivery. Gene Ther 13(7):646–651. https://doi.org/10.1038/sj.gt.3302733
Heinemann L, Schoemaker M, Schmelzeisen-Redecker G, Hinzmann R, Kassab A, Freckmann G, Reiterer F, Del Re L (2020) Benefits and limitations of MARD as a performance parameter for continuous glucose monitoring in the interstitial space. J Diabetes Sci Technol 14(1):135–150. https://doi.org/10.1177/1932296819899000
Helsley RN, Sui Y, Park SH, Liu Z, Lee RG, Zhu B, Kern PA, Zhou C (2016) Targeting IkappaB kinase beta in adipocyte lineage cells for treatment of obesity and metabolic dysfunctions. Stem Cells 34(7):1883–1895. https://doi.org/10.1002/stem.2361
Henriksson E, Andersen B (2020) FGF19 and FGF21 for the treatment of NASH—two sides of the same coin? Differential and overlapping effects of FGF19 and FGF21 from mice to human. Front Endocrinol 11:601349. https://doi.org/10.3389/fendo.2020.601349
Hirosumi J, Tuncman G, Chang L, Gorgun CZ, Uysal KT, Maeda K, Karin M, Hotamisligil GS (2002) A central role for JNK in obesity and insulin resistance. Nature 420(6913):333–336. https://doi.org/10.1038/nature01137
Hockemeyer D, Wang H, Kiani S, Lai CS, Gao Q, Cassady JP, Cost GJ, Zhang L, Santiago Y, Miller JC, Zeitler B, Cherone JM, Meng X, Hinkley SJ, Rebar EJ, Gregory PD, Urnov FD, Jaenisch R (2011) Genetic engineering of human pluripotent cells using TALE nucleases. Nat Biotechnol 29(8):731–734. https://doi.org/10.1038/nbt.1927
Hoeg-Jensen T, Kruse T, Brand CL, Sturis J, Fledelius C, Nielsen PK, Nishimura E, Madsen AR, Lykke L, Halskov KS, Koscova S, Kotek V, Davis AP, Tromans RA, Tomsett M, Penuelas-Haro G, Leonard DJ, Orchard MG, Chapman A, Invernizzi G, Johansson E, Granata D, Hansen BF, Pedersen TA, Kildegaard JK, Pedersen KM, Refsgaard HHF, Alifrangis L, Fels JJ, Neutzsky-Wulff AV, Sauerberg P, Slaaby R (2024) Glucose-sensitive insulin with attenuation of hypoglycaemia. Nature 634(8035):944–951. https://doi.org/10.1038/s41586-024-06489-1
Hogrebe NJ, Maxwell KG, Augsornworawat P, Millman JR (2021) Generation of insulin-producing pancreatic beta cells from multiple human stem cell lines. Nat Protoc 16(9):4109–4143. https://doi.org/10.1038/s41596-021-00547-0
Holyoak T, Sullivan SM, Nowak T (2006) Structural insights into the mechanism of PEPCK catalysis. Biochemistry 45(27):8254–8263. https://doi.org/10.1021/bi060041b
Horne SJ, Vasquez JM, Guo Y, Ly V, Piret SE, Leonardo AR, Ling J, Revelo MP, Bogenhagen D, Yang VW, He JC, Mallipattu SK (2018) Podocyte-specific loss of Kruppel-like factor 6 increases mitochondrial injury in diabetic kidney disease. Diabetes 67(11):2420–2433. https://doi.org/10.2337/db18-0298
Hu X, White K, Olroyd AG, DeJesus R, Dominguez AA, Dowdle WE, Friera AM, Young C, Wells F, Chu EY, Ito CE, Krishnapura H, Jain S, Ankala R, McGill TJ, Lin A, Egenberger KE, Gagnon A, Rukstalis JM, Hogrebe NJ, Gattis C, Basco R, Millman JR, Kievit P, Davis MM, Lanier LL, Connolly AJ, Deuse T, Schrepfer S (2024) Hypoimmune induced pluripotent stem cells survive long term in fully immunocompetent, allogeneic rhesus macaques. Nat Biotechnol 42(3):413–423. https://doi.org/10.1038/s41587-024-01906-2
Huang S, Hao XY, Li YJ, Wu JY, Xiang DX, Luo S (2022) Nonviral delivery systems for antisense oligonucleotide therapeutics. Biomater Res 26(1):49. https://doi.org/10.1186/s40824-022-00305-2
Hudish LI, Reusch JE, Sussel L (2019) β Cell dysfunction during progression of metabolic syndrome to type 2 diabetes. J Clin Invest 129(10):4001–4008. https://doi.org/10.1172/JCI129222
I OS, Zhang W, Wasserman DH, Liew CW, Liu J, Paik J, DePinho RA, Stolz DB, Kahn CR, Schwartz MW, Unterman TG (2015) FoxO1 integrates direct and indirect effects of insulin

on hepatic glucose production and glucose utilization. Nat Commun 6:7079. https://doi.org/10.1038/ncomms8079
Ideno J, Mizukami H, Kakehashi A, Saito Y, Okada T, Urabe M, Kume A, Kuroki M, Kawakami M, Ishibashi S, Ozawa K (2007) Prevention of diabetic retinopathy by intraocular soluble flt-1 gene transfer in a spontaneously diabetic rat model. Int J Mol Med 19(1):75–79. https://doi.org/10.3892/ijmm.19.1.75
Inagaki T, Choi M, Moschetta A, Peng L, Cummins CL, McDonald JG, Luo G, Jones SA, Goodwin B, Richardson JA, Gerard RD, Repa JJ, Mangelsdorf DJ, Kliewer SA (2005) Fibroblast growth factor 15 functions as an enterohepatic signal to regulate bile acid homeostasis. Cell Metab 2(4):217–225. https://doi.org/10.1016/j.cmet.2005.09.002
Jennings RE, Berry AA, Strutt JP, Gerrard DT, Hanley NA (2015) Human pancreas development. Development 142(18):3126–3137. https://doi.org/10.1242/dev.120063
Kaneto H, Nakatani Y, Miyatsuka T, Matsuoka TA, Matsuhisa M, Hori M, Yamasaki Y (2005) PDX-1/VP16 fusion protein, together with NeuroD or Ngn3, markedly induces insulin gene transcription and ameliorates glucose tolerance. Diabetes 54(4):1009–1022. https://doi.org/10.2337/diabetes.54.4.1009
Kang SL, Hwang YN, Kwon JY, Kim SM (2022) Effectiveness and safety of a model predictive control (MPC) algorithm for an artificial pancreas system in outpatients with type 1 diabetes (T1D): systematic review and meta-analysis. Diabetol Metab Syndr 14(1):187. https://doi.org/10.1186/s13098-022-00945-7
Karwatowska-Prokopczuk E, Lesogor A, Yan JH, Hoenlinger A, Margolskee A, Li L, Tsimikas S (2024) Efficacy and safety of olezarsen in lowering apolipoprotein C-III and triglycerides in healthy Japanese Americans. Lipids Health Dis 23(1):329. https://doi.org/10.1186/s12944-024-02045-6
Kay MA, Glorioso JC, Naldini L (2001) Viral vectors for gene therapy: the art of turning infectious agents into vehicles of therapeutics. Nat Med 7(1):33–40. https://doi.org/10.1038/83331
Keivanlou MH, Amini-Salehi E, Sattari N, Hashemi M, Saberian P, Prabhu SV, Javid M, Mirdamadi A, Heidarzad F, Bakhshi A, Letafatkar N, Zare R, Hassanipour S, Nayak SS (2024) Gut microbiota interventions in type 2 diabetes mellitus: An umbrella review of glycemic indices. Diabetes Metab Syndr 18(8):103110. https://doi.org/10.1016/j.dsx.2024.103110
Kenner KA, Anyanwu E, Olefsky JM, Kusari J (1996) Protein-tyrosine phosphatase 1B is a negative regulator of insulin- and insulin-like growth factor-I-stimulated signaling. J Biol Chem 271(33):19810–19816. https://doi.org/10.1074/jbc.271.33.19810
Kersten S (2017) Angiopoietin-like 3 in lipoprotein metabolism. Nat Rev Endocrinol 13(12):731–739. https://doi.org/10.1038/nrendo.2017.91
Kim H, Kim M, Im SK, Fang S (2018) Mouse Cre-LoxP system: general principles to determine tissue-specific roles of target genes. Lab Anim Res 34(4):147. https://doi.org/10.1186/s42826-018-0016-7
Kim Y, Kim H, Ko UH, Oh Y, Lim A, Sohn JW, Shin JH, Kim H, Han YM (2016) Islet-like organoids derived from human pluripotent stem cells efficiently function in the glucose responsiveness in vitro and in vivo. Sci Reports 6:35145. https://doi.org/10.1038/srep35145
Kistner A, Gossen M, Zimmermann F, Jerecic J, Ullmer C, Lübbert H, Bujard H (1996) Doxycycline-mediated quantitative and tissue-specific control of gene expression in transgenic mice. Proc Natl Acad Sci USA 93(20):10933–10938. https://doi.org/10.1073/pnas.93.20.10933
Knott AB, Perkins G, Schwarzenbacher R, Bossy-Wetzel E (2008) Mitochondrial fragmentation in neurodegeneration. Nat Rev Neurosci 9(7):505–518. https://doi.org/10.1038/nrn2417
Koo SH, Flechner L, Qi L, Zhang X, Screaton RA, Jeffries S, Hedrick S, Xu W, Boussouar F, Brindle P, Takemori H, Montminy M (2005) The CREB coactivator TORC2 is a key regulator of fasting glucose metabolism. Nature 437(7062):1109–1111. https://doi.org/10.1038/nature03967
Kovatchev B, Castillo A, Pryor E, Kollar LL, Barnett CL, DeBoer MD, Brown SA (2024) Neural-net artificial pancreas: a randomized crossover trial of a first-in-class automated insulin delivery algorithm. Diabetes Technol Ther 26(6):375–382. https://doi.org/10.1089/dia.2024.0131

Kraus-Friedmann N (1984) Hormonal regulation of hepatic gluconeogenesis. Physiol Rev 64(1):170–259. https://doi.org/10.1152/physrev.1984.64.1.170

Krüger J, Trappiel M, Dagnell M, Stawowy P, Meyborg H, Böhm C, Bhanot S, Ostman A, Kintscher U, Kappert K (2013) Targeting density-enhanced phosphatase-1 (DEP-1) with antisense oligonucleotides improves the metabolic phenotype in high-fat diet-fed mice. Cell Commun Signal 11:49. https://doi.org/10.1186/1478-811X-11-49

Kudva YC, Ohmine S, Greder LV, Dutton JR, Armstrong A, De Lamo JG, Khan YK, Thatava T, Hasegawa M, Fusaki N, Slack JM, Ikeda Y (2012) Transgene-free disease-specific induced pluripotent stem cells from patients with type 1 and type 2 diabetes. Stem Cells Transl Med 1(6):451–461. https://doi.org/10.5966/sctm.2011-0044

Kulkarni A, Pineros AR, Walsh MA, Casimiro I, Ibrahim S, Hernandez-Perez M, Orr KS, Glenn L, Nadler JL, Morris MA, Tersey SA, Mirmira RG, Anderson RM (2021) 12-lipoxygenase governs the innate immune pathogenesis of islet inflammation and autoimmune diabetes. JCI Insight 6(14):e147812. https://doi.org/10.1172/jci.insight.147812

Kulkarni JA, Darjuan MM, Mercer JE, Chen S, van der Meel R, Thewalt JL, Tam YYC, Cullis PR (2018) On the formation and morphology of lipid nanoparticles containing ionizable cationic lipids and siRNA. ACS Nano 12(5):4787–4795. https://doi.org/10.1021/acsnano.8b01697

Kumar M, Huang Y, Glinka Y, Prud'homme GJ, Wang Q (2007) Gene therapy of diabetes using a novel GLP-1/IgG1-Fc fusion construct normalizes glucose levels in db/db mice. Gene Ther 14(2):162–172. https://doi.org/10.1038/sj.gt.3302902

Kuo CH, Deng JH, Deng Q, Ying SY (2012) A novel role of miR-302/367 in reprogramming. Biochem Biophys Res Commun 417(1): 11–16. https://doi.org/10.1016/j.bbrc.2011.11.122

Kuo T, Harris CA, Wang JC (2013) Metabolic functions of glucocorticoid receptor in skeletal muscle. Mol Cell Endocrinol 380(1–2): 79–88. https://doi.org/10.1016/j.mce.2013.04.001

Lai YK, Shen WY, Brankov M, Lai CM, Constable IJ, Rakoczy PE (2002) Potential long-term inhibition of ocular neovascularisation by recombinant adeno-associated virus-mediated secretion gene therapy. Gene Ther 9(12):804–813. https://doi.org/10.1038/sj.gt.3301743

Lamartina S, Cimino M, Roscilli G, Dammassa E, Lazzaro D, Rota R, Ciliberto G, Toniatti C (2007) Helper-dependent adenovirus for the gene therapy of proliferative retinopathies: stable gene transfer, regulated gene expression and therapeutic efficacy. J Gene Med 9(10):862–874. https://doi.org/10.1002/jgm.1073

Larsson M, Vorrsjö E, Talmud P, Lookene A, Olivecrona G (2013) Apolipoproteins C-I and C-III inhibit lipoprotein lipase activity by displacement of the enzyme from lipid droplets. J Biol Chem 288(47):33997–34008. https://doi.org/10.1074/jbc.M113.497178

Lauffer MC, van Roon-Mom W, Aartsma-Rus A, Collaborative N (2024) Possibilities and limitations of antisense oligonucleotide therapies for the treatment of monogenic disorders. Commun Med 4(1):6. https://doi.org/10.1038/s41591-024-02294-1

Le Gat L, Gogat K, Bouquet C, Saint-Geniez M, Darland D, Van Den Berghe L, Marchant D, Provost A, Perricaudet M, Menasche M, Abitbol M (2003) In vivo adenovirus-mediated delivery of a uPA/uPAR antagonist reduces retinal neovascularization in a mouse model of retinopathy. Gene Therapy 10(25): 2098–2103. https://doi.org/10.1038/sj.gt.3302069

Lee PP, Fitzpatrick DR, Beard C, Jessup HK, Lehar S, Makar KW, Makar M, Perez-Melgosa MT, Sweetser MS, Schlissel MS, Nguyen SR, Cherry JH, Tsai SM, Tucker WM, Weaver A, Kelso R, Jaenisch R, Wilson CB (2001) A critical role for Dnmt1 and DNA methylation in T cell development, function, and survival. Immunity 15(5):763–774. https://doi.org/10.1016/S1074-7613(01)00233-4

Lee Y, Choi HY, Kwon A, Park H, Park MH, Kim JW, Kim MJ, Kim YO, Kim S, Kwak S, Koo SK (2019) Generation of a PDX1-EGFP reporter human induced pluripotent stem cell line, KSCBi005-A-3, using the CRISPR/Cas9 system. Stem Cell Res 41:101632. https://doi.org/10.1016/j.scr.2019.101632

Leite NC, Sintov E, Meissner TB, Brehm MA, Greiner DL, Harlan DM, Melton DA (2020) Modeling type 1 diabetes in vitro using human pluripotent stem cells. Cell Rep 32(2):107894. https://doi.org/10.1016/j.celrep.2020.107894

Li J, Wu C, Wang Y, He E, Elkayam L, Joshua-Tor L, Hammond PT (2018) Structurally modulated codelivery of siRNA and Argonaute 2 for enhanced RNA interference. Proc Natl Acad Sci U S A 115(12):E2696–E2705. https://doi.org/10.1073/pnas.1719565115

Li M, Wang Y, Yao Q, Liang Q, Zhang X, Wang Q, Li W, Qiang J, Yang B, Shi B, He M (2023) Association between lipoprotein(a) and diabetic nephropathy in patients with type 2 diabetes. Front Endocrinol (Lausanne) 14:1337469. https://doi.org/10.3389/fendo.2023.1337469

Li R, Zhang Y, Rasool S, Geetha T, Babu JR (2019) Effects and underlying mechanisms of bioactive compounds on type 2 diabetes mellitus and Alzheimer's disease. Oxidative Med Cell Longev 2019:8165707. https://doi.org/10.1155/2019/8165707

Li W, Nakanishi M, Zumsteg A, Shear M, Wright C, Melton DA, Zhou Q (2014) In vivo reprogramming of pancreatic acinar cells to three islet endocrine subtypes. eLife 3: e01846. https://doi.org/10.7554/eLife.01846

Li X, Ge H, Weiszmann J, Hecht R, Li Y, Veniant MM, Xu J, Wu X, Lindberg R, Li Y (2009) Inhibition of lipolysis may contribute to the acute regulation of plasma FFA and glucose by FGF21 in ob/ob mice. FEBS Lett 583(19):3230–3234. https://doi.org/10.1016/j.febslet.2009.09.012

Liang XH, Sun H, Shen W, Wang S, Yao J, Migawa MT, Bui HH, Damle SS, Riney S, Graham MJ, Crooke RM, Crooke ST (2017) Antisense oligonucleotides targeting translation inhibitory elements in 5′ UTRs can selectively increase protein levels. Nucleic Acids Res 45(16):9528–9546. https://doi.org/10.1093/nar/gkx618

Liang Y, Osborne MC, Monia BP, Bhanot S, Gaarde WA, Reed C, She P, Jetton TL, Demarest KT (2004) Reduction in glucagon receptor expression by an antisense oligonucleotide ameliorates diabetic syndrome in db/db mice. Diabetes 53(2):410–417. https://doi.org/10.2337/diabetes.53.2.410

Liang Y, Osborne MC, Monia BP, Bhanot S, Watts LM, She P, DeCarlo SO, Chen X, Demarest K (2005) Antisense oligonucleotides targeted against glucocorticoid receptor reduce hepatic glucose production and ameliorate hyperglycemia in diabetic mice. Metabolism 54(7):848–855. https://doi.org/10.1016/j.metabol.2005.01.027

Lin R, Brown F, James S, Jones J, Ekinci E (2021) Continuous glucose monitoring: a review of the evidence in type 1 and 2 diabetes mellitus. Diabet Med 38(5):e14528. https://doi.org/10.1111/dme.14528

Lithovius V, Saarimaki-Vire J, Balboa D, Ibrahim H, Montaser H, Barsby T, Otonkoski T (2021) SUR1-mutant iPS cell-derived islets recapitulate the pathophysiology of congenital hyperinsulinism. Diabetologia 64(3):630–640. https://doi.org/10.1007/s00125-020-05345-7

Liu MJ, Han J, Lee YS, Park MS, Shin S, Jun HS (2010a) Amelioration of hyperglycemia by intestinal overexpression of glucagon-like peptide-1 in mice. J Mol Med 88(4):351–358. https://doi.org/10.1007/s00109-009-0562-5

Liu R, Li Y, Hu R, Jin T, Deng S, Liang W, Zhang N, Chen J, Prud'homme GJ, Jia WW, Ma D, Wang Q (2010b) A site-specific genomic integration strategy for sustained expression of glucagon-like peptide-1 in mouse muscle for controlling energy homeostasis. Biochem Biophys Res Commun 403(2):172–177. https://doi.org/10.1016/j.bbrc.2010.11.019

Loomba R, Morgan E, Watts L, Xia S, Hannan LA, Geary RS, Baker BF, Bhanot S (2020) Novel antisense inhibition of diacylglycerol O-acyltransferase 2 for treatment of non-alcoholic fatty liver disease: a multicentre, double-blind, randomised, placebo-controlled phase 2 trial. Lancet Gastroenterol Hepatol 5(9):829–838. https://doi.org/10.1016/S2468-1253(20)30166-8

Loomba R, Sanyal AJ, Kowdley KV, Bhatt DL, Alkhouri N, Frias JP, Bedossa P, Harrison SA, Lazas D, Barish R, Gottwald MD, Feng S, Agollah GD, Hartsfield CL, Mansbach H, Margalit M, Abdelmalek MF (2023) Randomized, controlled trial of the FGF21 analogue Pegozafermin in NASH. N Engl J Med 389(11):998–1008. https://doi.org/10.1056/NEJMoa2212585

Ludwig B, Reichel A, Steffen A, Zimerman B, Schally AV, Block NL, Colton CK, Ludwig S, Kersting S, Bonifacio E, Solimena M, Gendler Z, Rotem A, Barkai U, Bornstein SR (2013) Transplantation of human islets without immunosuppression. Proc Natl Acad Sci 110(47):19054–19058. https://doi.org/10.1073/pnas.1315368110

Lundstrom K (2018) Viral vectors in gene therapy. Diseases 6(2):42. https://doi.org/10.3390/diseases6020042

Ma Q, Xiao Y, Xu W, Wang M, Li S, Yang Z, Xu M, Zhang T, Zhang ZN, Hu R, Su Q, Yuan F, Xiao T, Wang X, He Q, Zhao J, Chen ZJ, Sheng Z, Chai M, Wang H, Shi W, Deng Q, Cheng X, Li W (2022) ZnT8 loss-of-function accelerates functional maturation of hESC-derived beta cells and resists metabolic stress in diabetes. Nat Commun 13(1):4142. https://doi.org/10.1038/s41467-022-31900-x

Ma S, Viola R, Sui L, Cherubini V, Barbetti F, Egli D (2018) β cell replacement after gene editing of a neonatal diabetes-causing mutation at the insulin locus. Stem Cell Reports 11(6):1407–1415. https://doi.org/10.1016/j.stemcr.2018.10.020

Macarthur CC, Fontes A, Ravinder N, Kuninger D, Kaur J, Bailey M, Taliana A, Vemuri MC, Lieu PT (2012) Generation of human-induced pluripotent stem cells by a nonintegrating RNA Sendai virus vector in feeder-free or xeno-free conditions. Stem Cells Int 2012:564612. https://doi.org/10.1155/2012/564612

Maehr R, Chen S, Snitow M, Ludwig T, Yagasaki L, Goland R, Leibel RL, Melton DA (2009) Generation of pluripotent stem cells from patients with type 1 diabetes. Proc Natl Acad Sci USA 106(37):15768–15773. https://doi.org/10.1073/pnas.0908235106

Mallipattu SK, Horne SJ, D'Agati V, Narla G, Liu R, Frohman MA, Dickman K, Chen EY, Ma'ayan A, Bialkowska AB, Ghaleb AM, Nandan MO, Jain MK, Daehn I, Chuang PY, Yang VW, He JC (2015) Kruppel-like factor 6 regulates mitochondrial function in the kidney. J Clin Investig 125(3):1347–1361. https://doi.org/10.1172/JCI78547

Marchioli G, Di Luca A, de Koning E, Engelse M, Van Blitterswijk CA, Karperien M, Van Apeldoorn AA, Moroni L (2016) Hybrid polycaprolactone/alginate scaffolds functionalized with VEGF to promote de novo vessel formation for the transplantation of islets of Langerhans. Adv Healthc Mater 5(13):1606–1616. https://doi.org/10.1002/adhm.201600113

Marcum JA (2020) Nutrigenetics/nutrigenomics, personalized nutrition, and precision healthcare. Curr Nutr Reports 9(4):338–345. https://doi.org/10.1007/s13668-020-00337-1

Matsui T, Higashimoto Y, Nishino Y, Nakamura N, Fukami K, Yamagishi SI (2017) RAGE-Aptamer blocks the development and progression of experimental diabetic nephropathy. Diabetes 66(6):1683–1695. https://doi.org/10.2337/db16-1241

Matsumoto M, Pocai A, Rossetti L, DePinho RA, Accili D (2007) Impaired regulation of hepatic glucose production in mice lacking the forkhead transcription factor Foxo1 in liver. Cell Metab 6(3):208–216. https://doi.org/10.1016/j.cmet.2007.08.003

Matsumura H, Tada T (2008) Cell fusion-mediated nuclear reprogramming of somatic cells. Reprod Biomed Online 16(1):51–56. https://doi.org/10.1016/S1472-6483(10)60444-3

Matsuoka TA, Kaneto H, Kawashima S, Miyatsuka T, Tochino Y, Yoshikawa A, Imagawa A, Miyazaki J, Gannon M, Stein R, Shimomura I (2015) Preserving Mafa expression in diabetic islet beta-cells improves glycemic control in vivo. J Biol Chem 290(12):7647–7657. https://doi.org/10.1074/jbc.M114.620739

Maxwell KG, Augsornworawat P, Velazco-Cruz L, Kim MH, Asada R, Hogrebe NJ, Morikawa S, Urano F, Millman JR (2020) Gene-edited human stem cell-derived beta cells from a patient with monogenic diabetes reverse preexisting diabetes in mice. Sci Transl Med 12(540):eaax9106. https://doi.org/10.1126/scitranslmed.aax9106

Maxwell KG, Kim MH, Gale SE, Millman JR (2022) Differential function and maturation of human stem cell-derived islets after transplantation. Stem Cells Transl Med 11(3):322–331. https://doi.org/10.1093/stcltm/szab129

McLellan MA, Rosenthal NA, Pinto AR (2017) Cre-loxP-mediated recombination: general principles and experimental considerations. Curr Protoc Mouse Biol 7(1):1–12. https://doi.org/10.1002/cpmo.21

Meinke G, Bohm A, Hauber J, Pisabarro MT, Buchholz F (2016) Cre recombinase and other tyrosine recombinases. Chem Rev 116(20):12785–12820. https://doi.org/10.1021/acs.chemrev.6b00191

Meran L, Baulies A, Li VSW (2017) Intestinal stem cell niche: the extracellular matrix and cellular components. Stem Cells Int 2017:7970385. https://doi.org/10.1155/2017/7970385

Metzger D, Chambon P (2001) Site- and time-specific gene targeting in the mouse. Methods 24(1):71–80. https://doi.org/10.1006/meth.2001.1163

Miellet S, Maddock M, Napierala JS, Napierala M, Dottori M (2024) Generation of genetically modified Friedreich's ataxia induced pluripotent stem cell lines and isogenic control lines carrying an inducible neurogenin-2 expression cassette. Stem Cell Res 79:103477. https://doi.org/10.1016/j.scr.2024.103477

Mihara Y, Matsuura K, Sakamoto Y, Okano T, Kokudo N, Shimizu T (2017) Production of pancreatic progenitor cells from human induced pluripotent stem cells using a three-dimensional suspension bioreactor system. J Tissue Eng Regen Med 11(11):3193–3201. https://doi.org/10.1002/term.2158

Millman JR, Xie C, Van Dervort A, Gurtler M, Pagliuca FW, Melton DA (2016) Generation of stem cell-derived beta-cells from patients with type 1 diabetes. Nat Commun 7:11463. https://doi.org/10.1038/ncomms11463

Mingozzi F, High KA (2013) Immune responses to AAV vectors: overcoming barriers to successful gene therapy. Blood 122(1):23–36. https://doi.org/10.1182/blood-2013-01-306647

Miyatsuka T, Kaneto H, Kajimoto Y, Hirota S, Arakawa Y, Fujitani Y, Umayahara Y, Watada H, Yamasaki Y, Magnuson MA, Miyazaki J, Hori M (2003) Ectopically expressed PDX-1 in liver initiates endocrine and exocrine pancreas differentiation but causes dysmorphogenesis. Biochem Biophys Res Commun 310(3):1017–1025. https://doi.org/10.1016/j.bbrc.2003.09.044

Moein MM, Javanbakht M, Akbari-Adergani B (2014) Molecularly imprinted polymer cartridges coupled on-line with high performance liquid chromatography for simple and rapid analysis of human insulin in plasma and pharmaceutical formulations. Talanta 121:30–36. https://doi.org/10.1016/j.talanta.2013.12.041

Monia BP, Lesnik EA, Gonzalez C, Lima WF, McGee D, Guinosso CJ, Kawasaki AM, Cook PD, Freier SM (1993) Evaluation of 2′-modified oligonucleotides containing 2′-deoxy gaps as antisense inhibitors of gene expression. J Biol Chem 268(19):14514–14522. https://doi.org/10.1016/S0021-9258(18)83656-6

Moraes-Vieira PM, Yore MM, Dwyer PM, Syed I, Aryal P, Kahn BB (2014) RBP4 activates antigen-presenting cells, leading to adipose tissue inflammation and systemic insulin resistance. Cell Metab 19(3):512–526. https://doi.org/10.1016/j.cmet.2014.01.018

Moriyama H, Abiru N, Paronen J, Sikora K, Liu E, Miao D, Devendra D, Beilke J, Gianani R, Gill RG, Eisenbarth GS (2003) Evidence for a primary islet autoantigen (preproinsulin 1) for insulitis and diabetes in the nonobese diabetic mouse. Proc Natl Acad Sci USA 100(18):10376–10381. https://doi.org/10.1073/pnas.1734309100

Moschetta A, Kliewer SA (2005) Weaving betaKlotho into bile acid metabolism. J Clin Invest 115(8):2075–2077. https://doi.org/10.1172/JCI25627

Motte E, Szepessy E, Suenens K, Stange G, Bomans M, Jacobs-Tulleneers-Thevissen D, Ling Z, Kroon E, Pipeleers D, Beta Cell Therapy Consortium (2014) Composition and function of macroencapsulated human embryonic stem cell-derived implants: comparison with clinical human islet cell grafts. Am J Physiol 307(9):E838–E846. https://doi.org/10.1152/ajpendo.00235.2014

Munday MR (2002) Regulation of mammalian acetyl-CoA carboxylase. Biochem Soc Trans 30(Pt 6):1059–1064. https://doi.org/10.1042/bst0301059

Nagy A (2000) Cre recombinase: the universal reagent for genome tailoring. Genesis 26(2):99–109. https://doi.org/10.1002/(SICI)1526-968X(200002)26:2<99::AID-GENE4>3.0.CO;2-B

Nair GG, Tzanakakis ES, Hebrok M (2020) Emerging routes to the generation of functional beta-cells for diabetes mellitus cell therapy. Nat Rev Endocrinol 16(9):506–518. https://doi.org/10.1038/s41574-020-0379-2

Nimri R, Nir J, Phillip M (2020) Insulin pump therapy. Am J Ther 27(1):e30–e41. https://doi.org/10.1097/MJT.0000000000001084

Nordestgaard BG, Langsted A (2024) Lipoprotein(a) and cardiovascular disease. Lancet 404(10459):1255–1264. https://doi.org/10.1016/S0140-6736(24)00423-5

Norseen J, Hosooka T, Hammarstedt A, Yore MM, Kant S, Aryal P, Kiernan UA, Phillips DA, Maruyama H, Kraus BJ, Usheva A, Davis RJ, Smith U, Kahn BB (2012) Retinol-binding protein 4 inhibits insulin signaling in adipocytes by inducing proinflammatory cytokines in macrophages through a c-Jun N-terminal kinase- and toll-like receptor 4-dependent and retinol-independent mechanism. Mol Cell Biol 32(10):2010–2019. https://doi.org/10.1128/MCB.06591-11

Nostro MC, Sarangi F, Yang C, Holland A, Elefanty AG, Stanley EG, Greiner DL, Keller G (2015) Efficient generation of NKX6-1+ pancreatic progenitors from multiple human pluripotent stem cell lines. Stem Cell Reports 4(4):591–604. https://doi.org/10.1016/j.stemcr.2015.02.021

Noureddin M, Rinella ME, Chalasani NP, Neff GW, Lucas KJ, Rodriguez ME, Rudraraju M, Patil R, Behling C, Burch M, Chan DC, Tillman EJ, Zari A, de Temple B, Shringarpure R, Jain M, Rolph T, Cheng A, Yale K (2025) Efruxifermin in compensated liver cirrhosis caused by MASH. N Engl J Med 392(24):2413–2424. https://doi.org/10.1056/NEJMoa2502301

Nwokolo M, Hovorka R (2023) The artificial pancreas and type 1 diabetes. J Clin Endocrinol Metab 108(7):1614–1623. https://doi.org/10.1210/clinem/dgad067

O'Neil KT, Hoess RH, DeGrado WF (1990) Design of DNA-binding peptides based on the leucine zipper motif. Science 249(4970):774–778. https://doi.org/10.1126/science.2371826

O'Reilly MS, Holmgren L, Shing Y, Chen C, Rosenthal RA, Moses M, Lane WS, Cao Y, Sage EH, Folkman J (1994) Angiostatin: a novel angiogenesis inhibitor that mediates the suppression of metastases by a Lewis lung carcinoma. Cell 79(2):315–328. https://doi.org/10.1016/0092-8674(94)90565-0

Ortega-Azorín C, Sorlí JV, Asensio EM, Coltell O, Martínez-González M, Salas-Salvadó J, Covas MI, Arós F, Lapetra J, Serra-Majem L, Gómez-Gracia E, Fiol M, Sáez-Tormo G, Pintó X, Muñoz MA, Ros E, Ordovás JM, Estruch R, Corella D (2012) Associations of the FTO rs9939609 and the MC4R rs17782313 polymorphisms with type 2 diabetes are modulated by diet, being higher when adherence to the Mediterranean diet pattern is low. Cardiovasc Diabetol 11:137. https://doi.org/10.1186/1475-2840-11-137

Pagliuca FW, Millman JR, Gurtler M, Segel M, Van Dervort A, Ryu JH, Peterson QP, Greiner D, Melton DA (2014) Generation of functional human pancreatic beta cells in vitro. Cell 159(2):428–439. https://doi.org/10.1016/j.cell.2014.09.040

Palomo AB, Lucas M, Dilley RJ, McLenachan S, Chen FK, Requena J, Sal MF, Lucas A, Alvarez I, Jaraquemada D, Edel MJ (2014) The power and the promise of cell reprogramming: personalized autologous body organ and cell transplantation. J Clin Med 3(2):373–387. https://doi.org/10.3390/jcm3020373

Papuc C, Goran GV, Predescu CN, Tudoreanu L, Ştefan G (2022) Plant polyphenols mechanisms of action on insulin resistance and against the loss of pancreatic beta cells. Crit Rev Food Sci Nutr 62(2):325–352. https://doi.org/10.1080/10408398.2020.1821671

Parent AV, Faleo G, Chavez J, Saxton M, Berrios DI, Kerper NR, Tang Q, Hebrok M (2021) Selective deletion of human leukocyte antigens protects stem cell-derived islets from immune rejection. Cell Rep 36(7):109538. https://doi.org/10.1016/j.celrep.2021.109538

Park K, Kim WJ, Cho YH, Lee YI, Lee H, Jeong S, Cho ES, Chang SI, Moon SK, Kang BS, Kim YJ, Cho SH (2008) Cancer gene therapy using adeno-associated virus vectors. Front Biosci 13:2653–2659. https://doi.org/10.2741/2967

Park SJ, Seo SE, Kim KH, Lee SH, Kim J, Ha S, Song HS, Lee SH, Kwon OS (2021) Real-time monitoring of geosmin based on an aptamer-conjugated graphene field-effect transistor. Biosens Bioelectron 174:112804. https://doi.org/10.1016/j.bios.2020.112804

Pechan P, Rubin H, Lukason M, Ardinger J, DuFresne E, Hauswirth WW, Wadsworth SC, Scaria A (2009) Novel anti-VEGF chimeric molecules delivered by AAV vectors for inhibition of retinal neovascularization. Gene Ther 16(1):10–16. https://doi.org/10.1038/gt.2008.165

Peng H, Mu P, Li H, Lin S, Lin C, Lin K, Liu K, Zeng W, Zeng L (2021) Caveolin-1 is essential for the improvement of insulin sensitivity through AKT activation during glargine treatment on diabetic mice. J Diabetes Res 2021:9943344. https://doi.org/10.1155/2021/9943344

Picó C, Serra F, Rodríguez AM, Keijer J, Palou A (2019) Biomarkers of nutrition and health: new tools for new approaches. Nutrients 11(5):1092. https://doi.org/10.3390/nu11051092

Pilkis SJ, Granner DK (1992) Molecular physiology of the regulation of hepatic gluconeogenesis and glycolysis. Annu Rev Physiol 54:885–909. https://doi.org/10.1146/annurev.ph.54.030192.004321

Pinsker JE, Lee JB, Dassau E, Seborg DE, Bradley PK, Gondhalekar R, Bevier WC, Huyett L, Zisser HC, Doyle FJ 3rd (2016) Randomized crossover comparison of personalized MPC and PID control algorithms for the artificial pancreas. Diabetes Care 39(7):1135–1142. https://doi.org/10.2337/dc15-2344

Porsche CE, Delproposto JB, Patrick E, Zamarron BF, Lumeng CN (2020) Adipose tissue dendritic cell signals are required to maintain T cell homeostasis and obesity-induced expansion. Mol Cell Endocrinol 505:110740. https://doi.org/10.1016/j.mce.2020.110740

Prea SM, Chan EC, Dusting GJ, Vingrys AJ, Bui BV, Liu GS (2015) Gene therapy with endogenous inhibitors of angiogenesis for neovascular age-related macular degeneration: beyond anti-VEGF therapy. J Ophthalmol 2015:201726. https://doi.org/10.1155/2015/201726

Price NL, Gomes AP, Ling AJ, Duarte FV, Martin-Montalvo AM, North BJ, Agarwal B, Ye L, Ramadori G, Teodoro JS, Hubbard BP, Varela AT, Davis JG, Varamini B, Hafner A, Moaddel R, Rolo AP, Coppari R, Palmeira CM, de Cabo R, Baur JA, Sinclair DA (2012) SIRT1 is required for AMPK activation and the beneficial effects of resveratrol on mitochondrial function. Cell Metab 15(5):675–690. https://doi.org/10.1016/j.cmet.2012.04.003

Prochorov AV, Tretjak SI, Goranov VA, Glinnik AA, Goltsev MV (2008) Treatment of insulin dependent diabetes mellitus with intravascular transplantation of pancreatic islet cells without immunosuppressive therapy. Adv Med Sci 53(2):240–244. https://doi.org/10.1016/j.advms.2008.04.001

Prudente S, Hribal ML, Flex E, Turchi F, Morini E, De Cosmo S, Bacci S, Tassi V, Cardellini M, Lauro R, Sesti G, Dallapiccola B, Trischitta V (2005) The functional Q84R polymorphism of mammalian Tribbles homolog TRB3 is associated with insulin resistance and related cardiovascular risk in Caucasians from Italy. Diabetes 54(9):2807–2811. https://doi.org/10.2337/diabetes.54.9.2807

Qin L, Fan B, Zhou Y, Zheng J, Diao R, Wang F, Liu J (2025) Targeted gut microbiome therapy: applications and prospects of probiotics, fecal microbiota transplantation and natural products in the management of type 2 diabetes. Pharmacol Res 213:107625. https://doi.org/10.1016/j.phrs.2025.107625

Quemener AM, Bachelot L, Forestier A, Donnou-Fournet E, Gilot D, Galibert MD (2020) The powerful world of antisense oligonucleotides: from bench to bedside. Wiley Interdisc Rev 11(5):e1594. https://doi.org/10.1002/wrna.1594

Rabbani PS, Zhou A, Borab ZM, Frezzo JA, Srivastava N, More HT, Rifkin WJ, David JA, Berens SJ, Chen R, Hameedi S, Junejo MH, Kim C, Sartor RA, Liu CF, Saadeh PB, Montclare JK, Ceradini DJ (2017) Novel lipoproteoplex delivers Keap1 siRNA based gene therapy to accelerate diabetic wound healing. Biomaterials 132:1–15. https://doi.org/10.1016/j.biomaterials.2017.04.004

Rahman SMN, Yung JHM, Volchuk A, Goldenberg NM, Giacca A (2025) Metabolic phenotypes in a Lyz2Cre recombinase mouse model. Front Immunol 16:1499858. https://doi.org/10.3389/fimmu.2025.1499858

Raiten DJ, Bremer AA (2020) Exploring the nutritional ecology of stunting: new approaches to an old problem. Nutrients 12(2):1–21. https://doi.org/10.3390/nu12020456

Rajewsky K, Gu H, Kuhn R, Betz UA, Muller W, Roes J, Schwenk F (1996) Conditional gene targeting. J Clin Invest 98(3):600–603. https://doi.org/10.1172/JCI118875

Ramzy A, Belmonte PJ, Braam MJS, Ida S, Wilts EM, Levings MK, Rezania A, Kieffer TJ (2023) A century-long journey from the discovery of insulin to the implantation of stem cell-derived islets. Endocr Rev 44(2):222–253. https://doi.org/10.1210/endrev/bnac044

Ramzy A, Thompson DM, Ward-Hartstonge KA, Ivison S, Cook L, Garcia RV, Loyal J, Kim PTW, Warnock GL, Levings MK, Kieffer TJ (2021) Implanted pluripotent stem-cell-derived pancreatic endoderm cells secrete glucose-responsive C-peptide in patients with type 1 diabetes. Cell Stem Cell 28(12):2047–2061.e2045. https://doi.org/10.1016/j.stem.2021.09.004

Rasouli M, Allaudin ZN, Omar AR, Ahmad Z (2013) Insulin secreted from genetically engineered intestinal cells reduces blood glucose levels in diabetic mice. Curr Gene Ther 13(4):229–239. https://doi.org/10.2174/1566523211313040006

Ravnskjaer K, Kester H, Liu Y, Zhang X, Lee D, Yates JR 3rd, Montminy M (2007) Cooperative interactions between CBP and TORC2 confer selectivity to CREB target gene expression. EMBO J 26(12):2880–2889. https://doi.org/10.1038/sj.emboj.7601711

Ray MK, Fagan SP, Brunicardi FC (2000) The Cre-loxP system: a versatile tool for targeting genes in a cell- and stage-specific manner. Cell Transplant 9(6):805–815. https://doi.org/10.1177/096368970000900611

Reebye V, Sætrom P, Mintz PJ, Rossi JJ, Kasahara N, Nteliopoulos G, Nicholls J, Haoudi A, Gordon M, Habib NA (2013) A short-activating RNA oligonucleotide targeting the islet β-cell transcriptional factor MafA in CD34(+) cells. Mol Ther Nucleic Acids 2(6):e97. https://doi.org/10.1038/mtna.2013.36

Rezania A, Bruin JE, Arora P, Rubin A, Batushansky I, Asadi A, O'Dwyer S, Quiskamp N, Mojibian M, Albrecht T, Yang YH, Johnson JD, Kieffer TJ (2014) Reversal of diabetes with insulin-producing cells derived in vitro from human pluripotent stem cells. Nat Biotechnol 32(11):1121–1133. https://doi.org/10.1038/nbt.3019

Riedel MJ, Kieffer TJ (2010) Treatment of diabetes with glucagon-like peptide-1 gene therapy. Expert Opin Biol Ther 10(12):1681–1692. https://doi.org/10.1517/14712598.2010.526324

Rist MJ, Wenzel U, Daniel H (2006) Nutrition and food science go genomic. Trends Biotechnol 24(4):172–178. https://doi.org/10.1016/j.tibtech.2006.02.005

Ropper AH, Gorson KC, Gooch CL, Weinberg DH, Pieczek A, Ware JH, Kershen J, Rogers A, Simovic D, Schratzberger P, Kirchmair R, Losordo D (2009) Vascular endothelial growth factor gene transfer for diabetic polyneuropathy: a randomized, double-blinded trial. Ann Neurol 65(4):386–393. https://doi.org/10.1002/ana.21535

Rui L (2014) Energy metabolism in the liver. Compr Physiol 4(1):177–197. https://doi.org/10.1002/cphy.c130024

Russ HA, Parent AV, Ringler JJ, Hennings TG, Nair GG, Shveygert M, Guo T, Puri S, Haataja L, Cirulli V, Blelloch R, Szot GL, Arvan P, Hebrok M (2015) Controlled induction of human pancreatic progenitors produces functional beta-like cells in vitro. EMBO J 34(13):1759–1772. https://doi.org/10.15252/embj.201490490

Ruzzin J, Wagman AS, Jensen J (2005) Glucocorticoid-induced insulin resistance in skeletal muscles: defects in insulin signalling and the effects of a selective glycogen synthase kinase-3 inhibitor. Diabetologia 48(10):2119–2130. https://doi.org/10.1007/s00125-005-1886-x

Rychak JJ, Klibanov AL (2014) Nucleic acid delivery with microbubbles and ultrasound. Adv Drug Deliv Rev 72:82–93. https://doi.org/10.1016/j.addr.2013.11.004

Saarimaki-Vire J, Balboa D, Russell MA, Saarikettu J, Kinnunen M, Keskitalo S, Malhi A, Valensisi C, Andrus C, Eurola S, Grym H, Ustinov J, Wartiovaara K, Hawkins RD, Silvennoinen O, Varjosalo M, Morgan NG, Otonkoski T (2017) An activating STAT3 mutation causes neonatal diabetes through premature induction of pancreatic differentiation. Cell Rep 19(2):281–294. https://doi.org/10.1016/j.celrep.2017.03.069

Saber N, Bruin JE, O'Dwyer S, Schuster H, Rezania A, Kieffer TJ (2018) Sex differences in maturation of human embryonic stem cell-derived beta cells in mice. Endocrinology 159(4):1827–1841. https://doi.org/10.1210/en.2017-00438

Sali S, Azzam L, Jaro T, Ali AAG, Mardini A, Al-Dajani O, Khattak S, Butler AE, Azeez JM, Nandakumar M (2025) A perfect islet: reviewing recent protocol developments and proposing strategies for stem cell derived functional pancreatic islets. Stem Cell Res Ther 16(1):160. https://doi.org/10.1186/s13287-025-03001-4

Samuel VT, Choi CS, Phillips TG, Romanelli AJ, Geisler JG, Bhanot S, McKay R, Monia B, Shutter JR, Lindberg RA, Shulman GI, Veniant MM (2006) Targeting foxo1 in mice using antisense oligonucleotide improves hepatic and peripheral insulin action. Diabetes 55(7):2042–2050. https://doi.org/10.2337/db06-0022

Samuel VT, Liu ZX, Wang A, Beddow SA, Geisler JG, Kahn M, Zhang XM, Monia BP, Bhanot S, Shulman GI (2007) Inhibition of protein kinase Cepsilon prevents hepatic insulin resistance in nonalcoholic fatty liver disease. J Clin Invest 117(3):739–745. https://doi.org/10.1172/JCI30188

Samulski RJ, Muzyczka N (2014) AAV-mediated gene therapy for research and therapeutic purposes. Annu Rev Virol 1(1):427–451. https://doi.org/10.1146/annurev-virology-031413-085355

Sang A, Zhuo S, Bochanis A, Manautou JE, Bahal R, Zhong XB, Rasmussen TP (2024) Mechanisms of action of the US Food and Drug Administration-approved antisense oligonucleotide drugs. BioDrugs 38(4):511–526. https://doi.org/10.1007/s40259-024-00612-8

Sapir T, Shternhall K, Meivar-Levy I, Blumenfeld T, Cohen H, Skutelsky E, Eventov-Friedman S, Barshack I, Goldberg I, Pri-Chen S, Ben-Dor L, Polak-Charcon S, Karasik A, Shimon I, Mor E, Ferber S (2005) Cell-replacement therapy for diabetes: generating functional insulin-producing tissue from adult human liver cells. Proc Natl Acad Sci USA 102(22):7964–7969. https://doi.org/10.1073/pnas.0502885102

Sasaki S, Sun R, Bui HH, Crosby JR, Monia BP, Guo S (2019) Steric inhibition of 5' UTR regulatory elements results in upregulation of human CFTR. Mol Ther 27(10):1749–1757. https://doi.org/10.1016/j.ymthe.2019.07.008

Sauer B (1998) Inducible gene targeting in mice using the Cre/lox system. Methods 14(4):381–392. https://doi.org/10.1006/meth.1997.0511

Sauer B, Henderson N (1988) Site-specific DNA recombination in mammalian cells by the Cre recombinase of bacteriophage P1. Proc Natl Acad Sci USA 85(14):5166–5170. https://doi.org/10.1073/pnas.85.14.5166

Savage DB, Choi CS, Samuel VT, Liu ZX, Zhang D, Wang A, Zhang XM, Cline GW, Yu XX, Geisler JG, Bhanot S, Monia BP, Shulman GI (2006) Reversal of diet-induced hepatic steatosis and hepatic insulin resistance by antisense oligonucleotide inhibitors of acetyl-CoA carboxylases 1 and 2. J Clin Invest 116(3):817–824. https://doi.org/10.1172/JCI26950

Sayed N, Allawadhi P, Khurana A, Singh V, Navik U, Pasumarthi SK, Khurana I, Banothu AK, Weiskirchen R, Bharani KK (2022) Gene therapy: comprehensive overview and therapeutic applications. Life Sci 294:120375. https://doi.org/10.1016/j.lfs.2022.120375

Scalbert A, Morand C, Manach C, Rémésy C (2002) Absorption and metabolism of polyphenols in the gut and impact on health. Biomed Pharmacother 56(6):276–282. https://doi.org/10.1016/S0753-3322(02)00208-5

Scharp DW, Swanson CJ, Olack BJ, Latta PP, Hegre OD, Doherty EJ, Gentile FT, Flavin KS, Ansara MF, Lacy PE (1994) Protection of encapsulated human islets implanted without immunosuppression in patients with type I or type II diabetes and in nondiabetic control subjects. Diabetes 43(9):1167–1170. https://doi.org/10.2337/diab.43.9.1167

Schlein C, Talukdar S, Heine M, Fischer AW, Krott LM, Nilsson SK, Brenner MB, Heeren J, Scheja L (2016) FGF21 lowers plasma triglycerides by accelerating lipoprotein catabolism in white and brown adipose tissues. Cell Metab 23(3):441–453. https://doi.org/10.1016/j.cmet.2016.01.004

Shang L, Hua H, Foo K, Martinez H, Watanabe K, Zimmer M, Kahler DJ, Freeby M, Chung W, LeDuc C, Goland R, Leibel RL, Egli D (2014) Beta-cell dysfunction due to increased ER stress in a stem cell model of Wolfram syndrome. Diabetes 63(3):923–933. https://doi.org/10.2337/db13-1354

Shapiro AMJ, Thompson D, Donner TW, Bellin MD, Hsueh W, Pettus J, Wilensky J, Daniels M, Wang RM, Brandon EP, Jaiman MS, Kroon EJ, D'Amour KA, Foyt HL (2021) Insulin expression and C-peptide in type 1 diabetes subjects implanted with stem cell-derived pancreatic endoderm cells in an encapsulation device. Cell Reports Med 2(12):100466. https://doi.org/10.1016/j.xcrm.2021.100466

Sherr JL, Heinemann L, Fleming GA, Bergenstal RM, Bruttomesso D, Hanaire H, Holl RW, Petrie JR, Peters AL, Evans M (2023) Automated insulin delivery: benefits, challenges, and recommendations. A Consensus Report of the Joint Diabetes Technology Working Group of the European Association for the Study of Diabetes and the American Diabetes Association. Diabetologia 66(1):3–22. https://doi.org/10.1007/s00125-022-05841-3

Sherr JL, Patel NS, Michaud CI, Palau-Collazo MM, Van Name MA, Tamborlane WV, Cengiz E, Carria LR, Tichy EM, Weinzimer SA (2016) Mitigating meal-related glycemic excursions in an insulin-sparing manner during closed-loop insulin delivery: the beneficial effects of adjunctive Pramlintide and Liraglutide. Diabetes Care 39(7):1127–1134. https://doi.org/10.2337/dc16-0240

Shi Y, Yang X, Min J, Kong W, Hu X, Zhang J, Chen L (2024) Advancements in culture technology of adipose-derived stromal/stem cells: implications for diabetes and its complications. Front Endocrinol (Lausanne) 15:1343255. https://doi.org/10.3389/fendo.2024.1343255

Shi ZD, Lee K, Yang D, Amin S, Verma N, Li QV, Zhu Z, Soh CL, Kumar R, Evans T, Chen S, Huangfu D (2017) Genome editing in hPSCs reveals GATA6 haploinsufficiency and a genetic interaction with GATA4 in human pancreatic development. Cell Stem Cell 20(5):675–688 e676. https://doi.org/10.1016/j.stem.2017.03.009

Shifrin AL, Auricchio A, Yu QC, Wilson J, Raper SE (2001) Adenoviral vector-mediated insulin gene transfer in the mouse pancreas corrects streptozotocin-induced hyperglycemia. Gene Ther 8(19):1480–1489. https://doi.org/10.1038/sj.gt.3301556

Simsek S, Zhou T, Robinson CL, Tsai SY, Crespo M, Amin S, Lin X, Hon J, Evans T, Chen S (2016) Modeling cystic fibrosis using pluripotent stem cell-derived human pancreatic ductal epithelial cells. Stem Cells Transl Med 5(5):572–579. https://doi.org/10.5966/sctm.2015-0251

Sintov E, Nikolskiy I, Barrera V, Hyoje-Ryu Kenty J, Atkin AS, Gerace D, Ho Sui SJ, Boulanger K, Melton DA (2022) Whole-genome CRISPR screening identifies genetic manipulations to reduce immune rejection of stem cell-derived islets. Stem Cell Reports 17(9):1976–1990. https://doi.org/10.1016/j.stemcr.2022.07.010

Skovso S, Overby P, Memar-Zadeh J, Lee JTC, Yang JCC, Shanina I, Sidarala V, Levi-D'Ancona E, Zhu J, Soleimanpour SA, Horwitz MS, Johnson JD (2022) Beta-cell Cre expression and reduced Ins1 gene dosage protect mice from type 1 diabetes. Endocrinology 163(11):bnac123. https://doi.org/10.1210/endocr/bnac123

Sloop KW, Cao JX, Siesky AM, Zhang HY, Bodenmiller DM, Cox AL, Jacobs SJ, Moyers JS, Owens RA, Showalter AD, Brenner MB, Raap A, Gromada J, Berridge BR, Monteith DK, Porksen N, McKay RA, Monia BP, Bhanot S, Watts LM, Michael MD (2004) Hepatic and glucagon-like peptide-1-mediated reversal of diabetes by glucagon receptor antisense oligonucleotide inhibitors. J Clin Invest 113(11):1571–1581. https://doi.org/10.1172/JCI20682

Sloop KW, Showalter AD, Cox AL, Cao JX, Siesky AM, Zhang HY, Irizarry AR, Murray SF, Booten SL, Finger EA, McKay RA, Monia BP, Bhanot S, Michael MD (2007) Specific reduction of hepatic glucose 6-phosphate transporter-1 ameliorates diabetes while avoiding complications of glycogen storage disease. J Biol Chem 282(26):19113–19121. https://doi.org/10.1074/jbc.M702850200

Smirnikhina SA, Zaynitdinova MI, Sergeeva VA, Lavrov AV (2022) Improving homology-directed repair in genome editing experiments by influencing the cell cycle. Int J Mol Sci 23(11):6160. https://doi.org/10.3390/ijms23116160

Smith LIF, Hill TG, Bowe JE (2020) Generating Beta-cell-specific transgenic mice using the Cre-lox system. In: King AJF (ed) Animal models of diabetes: methods and protocols. Springer US, New York, pp 181–205. https://doi.org/10.1007/978-1-0716-0852-1_14

So WY, Han W (2024) Gene therapy targeting key beta cell regulators as a potential intervention for diabetes. EMBO Mol Med 16(7):1490–1494. https://doi.org/10.15252/emmm.202517146

So WY, Liao Y, Liu WN, Rutter GA, Han W (2023) Paired box 6 gene delivery preserves beta cells and improves islet transplantation efficacy. EMBO Mol Med 15(12):e17928. 10.15252/emmm.202217928

So WY, Liu WN, Teo AKK, Rutter GA, Han W (2021) Paired box 6 programs essential exocytotic genes in the regulation of glucose-stimulated insulin secretion and glucose homeostasis. Sci Transl Med 13(600):eabh3355. https://doi.org/10.1126/scitranslmed.abh3355

Soltani N, Kumar M, Glinka Y, Prud'homme GJ, Wang Q (2007) In vivo expression of GLP-1/IgG-fc fusion protein enhances beta-cell mass and protects against streptozotocin-induced diabetes. Gene Ther 14(12):981–988. https://doi.org/10.1038/sj.gt.3303017

Song KH, Li T, Owsley E, Strom S, Chiang JY (2009) Bile acids activate fibroblast growth factor 19 signaling in human hepatocytes to inhibit cholesterol 7alpha-hydroxylase gene expression. Hepatology 49(1):297–305. https://doi.org/10.1002/hep.22675

Soon-Shiong P, Heintz RE, Merideth N, Yao QX, Yao Z, Zheng T, Murphy M, Moloney MK, Schmehl M, Harris M et al (1994) Insulin independence in a type 1 diabetic patient after encapsulated islet transplantation. Lancet 343(8903):950–951. https://doi.org/10.1016/S0140-6736(94)92292-1

Stalmans W, Laloux M (1979) Glucocorticoids and hepatic glycogen metabolism. Monogr Endocrinol 12:517–533

Stark R, Guebre-Egziabher F, Zhao X, Feriod C, Dong J, Alves TC, Ioja S, Pongratz RL, Bhanot S, Roden M, Cline GW, Shulman GI, Kibbey RG (2014) A role for mitochondrial phosphoenolpyruvate carboxykinase (PEPCK-M) in the regulation of hepatic gluconeogenesis. J Biol Chem 289(11):7257–7263. https://doi.org/10.1074/jbc.M113.536990

Sternberg N, Austin S, Hamilton D, Yarmolinsky M (1978) Analysis of bacteriophage P1 immunity by using lambda-P1 recombinants constructed in vitro. Proc Natl Acad Sci U S A 75(11):5594–5598. https://doi.org/10.1073/pnas.75.11.5594

Sternberg N, Hamilton D (1981) Bacteriophage P1 site-specific recombination. I. Recombination between loxP sites. J Mol Biol 150(4):467–486. https://doi.org/10.1016/0022-2836(81)90375-2

Sternberg N, Sauer B, Hoess R, Abremski K (1986) Bacteriophage P1 cre gene and its regulatory region. Evidence for multiple promoters and for regulation by DNA methylation. J Mol Biol 187(2):197–212. https://doi.org/10.1016/0022-2836(86)90244-7

Stock AA, Gonzalez GC, Pete SI, De Toni T, Berman DM, Rabassa A, Diaz W, Geary Jr JC, Willman M, Jackson JM, DeHaseth NH, Ziebarth NM, Hogan AR, Ricordi C, Kenyon NS, Tomei AA (2022) Performance of islets of Langerhans conformally coated via an emulsion cross-linking method in diabetic rodents and nonhuman primates. Sci Adv 8(26):eabm3145. https://doi.org/10.1126/sciadv.abm3145

Stroes ESG, Alexander VJ, Karwatowska-Prokopczuk E, Hegele RA, Arca M, Ballantyne CM, Soran H, Prohaska TA, Xia S, Ginsberg HN, Witztum JL, Tsimikas S, Investigators B (2024) Olezarsen, acute pancreatitis, and familial Chylomicronemia syndrome. N Engl J Med 390(19):1781–1792. https://doi.org/10.1056/NEJMoa2309118

Su L, Hong Z, Zhou T, Jian Y, Xu M, Zhang X, Zhu X, Wang J (2022) Health improvements of type 2 diabetic patients through diet and diet plus fecal microbiota transplantation. Sci Rep 12(1):1152. https://doi.org/10.1038/s41598-022-05100-4

Sun P, Liu Z (2019) Overexpressing kringle 1 domain of hepatocyte growth factor with adeno-associated virus inhibits the pathological retinal neovascularization in an oxygen-induced retinopathy mouse model. Biochem Biophys Res Commun 508(1):130–137. https://doi.org/10.1016/j.bbrc.2018.11.086

Swarbrick MM, Havel PJ, Levin AA, Bremer AA, Stanhope KL, Butler M, Booten SL, Graham JL, McKay RA, Murray SF, Watts LM, Monia BP, Bhanot S (2009) Inhibition of protein tyrosine phosphatase-1B with antisense oligonucleotides improves insulin sensitivity and increases adiponectin concentrations in monkeys. Endocrinology 150(4):1670–1679. https://doi.org/10.1210/en.2008-1442

Syed ZQ, Samaraweera S, Wang Z, Krishnan S (2024) Colorimetric nano-biosensor for low-resource settings: insulin as a model biomarker. Sens Diagn 3(10):1659–1671. https://doi.org/10.1515/snd-2024-0163

Takahashi K, Tanabe K, Ohnuki M, Narita M, Ichisaka T, Tomoda K, Yamanaka S (2007) Induction of pluripotent stem cells from adult human fibroblasts by defined factors. Cell 131(5):861–872. https://doi.org/10.1016/j.cell.2007.11.019

Takahashi K, Yamanaka S (2006) Induction of pluripotent stem cells from mouse embryonic and adult fibroblast cultures by defined factors. Cell 126(4):663–676. https://doi.org/10.1016/j.cell.2006.07.024

Takahashi Y, Ohoka N, Hayashi H, Sato R (2008) TRB3 suppresses adipocyte differentiation by negatively regulating PPARgamma transcriptional activity. J Lipid Res 49(4):880–892. https://doi.org/10.1194/jlr.M700480-JLR200

Tan Y, Sun LQ, Kamal MA, Wang X, Seale JP, Qu X (2011) Suppression of retinol-binding protein 4 with RNA oligonucleotide prevents high-fat diet-induced metabolic syndrome and non-alcoholic fatty liver disease in mice. Biochim Biophys Acta 1811(12):1045–1053. https://doi.org/10.1016/j.bbalip.2011.09.004

Tang M, Chen Y, Li B, Sugimoto H, Yang S, Yang C, LeBleu VS, McAndrews KM, Kalluri R (2021) Therapeutic targeting of STAT3 with small interference RNAs and antisense oligonucleotides embedded exosomes in liver fibrosis. FASEB J 35(5):e21557. https://doi.org/10.1096/fj.202001808R

Tavakolidakhrabadi N, Ding WY, Saleem MA, Welsh GI, May C (2024) Gene therapy and kidney diseases. Mol Ther Methods Clin Dev 32(4):101333. https://doi.org/10.1016/j.omtm.2023.101333

Tehrani F, Teymourian H, Wuerstle B, Kavner J, Patel R, Furmidge A, Aghavali R, Hosseini-Toudeshki H, Brown C, Zhang F, Mahato K, Li Z, Barfidokht A, Yin L, Warren P, Huang N, Patel Z, Mercier PP, Wang J (2022) An integrated wearable microneedle array for the continuous monitoring of multiple biomarkers in interstitial fluid. Nat Biomed Eng 6(11):1214–1224. https://doi.org/10.1038/s41551-022-00937-6

Thirawatananond P, Brown ME, Sachs LK, Arnoletti JM, Yeh WI, Posgai AL, Shapiro MR, Chen YG, Brusko TM (2023) Treg-specific CD226 deletion reduces diabetes incidence in NOD mice by improving regulatory T-cell stability. Diabetes 72(11):1629–1640. https://doi.org/10.2337/db23-0133

Thorel F, Nepote V, Avril I, Kohno K, Desgraz R, Chera S, Herrera PL (2010) Conversion of adult pancreatic alpha-cells to beta-cells after extreme beta-cell loss. Nature 464(7292):1149–1154. https://doi.org/10.1038/nature08894

Thorens B, Tarussio D, Maestro MA, Rovira M, Heikkila E, Ferrer J (2015) Ins1(Cre) knock-in mice for beta cell-specific gene recombination. Diabetologia 58(3):558–565. https://doi.org/10.1007/s00125-014-3467-0

Trujillo E, Davis C, Milner J (2006) Nutrigenomics, proteomics, metabolomics, and the practice of dietetics. J Am Diet Assoc 106(3):403–413. https://doi.org/10.1016/j.jada.2005.12.018

Uhl S, Choure A, Rouse B, Loblack A, Reaven P (2024) Effectiveness of continuous glucose monitoring on metrics of glycemic control in type 2 diabetes mellitus: a systematic review and meta-analysis of randomized controlled trials. J Clin Endocrinol Metab 109(4):1119–1131. https://doi.org/10.1210/clinem/dgad821

Uludag H, Ubeda A, Ansari A (2019) At the intersection of biomaterials and gene therapy: Progress in non-viral delivery of nucleic acids. Front Bioeng Biotechnol 7:131. https://doi.org/10.3389/fbioe.2019.00131

Um JE, Park JT, Nam BY, Lee JP, Jung JH, Kim Y, Kim S, Park J, Wu M, Han SH, Yoo TH, Kang SW (2017) Periostin-binding DNA aptamer treatment attenuates renal fibrosis under diabetic conditions. Sci Rep 7(1):8490. https://doi.org/10.1038/s41598-017-08665-2

Vaithilingam V, Bal S, Tuch BE (2017) Encapsulated islet transplantation: where do we stand? Rev Diabetic Stud 14(1):51–78. https://doi.org/10.1900/RDS.2017.14.51

Vazquez-Marroquin G, Ochoa-Précoma R, Porchia LM, Pérez-Fuentes R, Nicolás-Toledo L, Rodríguez-Antolín J, Gonzalez-Mejia ME (2023) The effect of microbiome therapies on waist circumference, a measure of central obesity, in patients with type 2 diabetes: a systematic review and meta-analysis of randomized controlled trials. J Acad Nutr Diet 123(6):933–952.e931. https://doi.org/10.1016/j.jand.2023.01.015

Vegas AJ, Veiseh O, Gurtler M, Millman JR, Pagliuca FW, Bader AR, Doloff JC, Li J, Chen M, Olejnik K, Tam HH, Jhunjhunwala S, Langan E, Aresta-Dasilva S, Gandham S, McGarrigle JJ, Bochenek MA, Hollister-Lock J, Oberholzer J, Greiner DL, Weir GC, Melton DA, Langer R, Anderson DG (2016) Long-term glycemic control using polymer-encapsulated human stem cell-derived beta cells in immune-competent mice. Nat Med 22(3):306–311. https://doi.org/10.1038/nm.4030

Veres A, Faust AL, Bushnell HL, Engquist EN, Kenty JH, Harb G, Poh YC, Sintov E, Gurtler M, Pagliuca FW, Peterson QP, Melton DA (2019) Charting cellular identity during human in vitro beta-cell differentiation. Nature 569(7756):368–373. https://doi.org/10.1038/s41586-019-1168-5

Villa C, Manzoli V, Abreu MM, Verheyen CA, Seskin M, Najjar M, Molano RD, Torrente Y, Ricordi C, Tomei AA (2017) Effects of composition of alginate-polyethylene glycol microcapsules and transplant site on encapsulated islet graft outcomes in mice. Transplantation 101(5):1025–1035. https://doi.org/10.1097/TP.0000000000001576

Wang AY, Ehrhardt A, Xu H, Kay MA (2007) Adenovirus transduction is required for the correction of diabetes using Pdx-1 or Neurogenin-3 in the liver. Mol Ther 15(2):255–263. https://doi.org/10.1038/sj.mt.6300037

Wang C, Abadpour S, Aizenshtadt A, Dalmao-Fernandez A, Hoyem M, Wilhelmsen I, Stokowiec J, Olsen PA, Krauss S, Chera S, Ghila L, Raeder H, Scholz H (2024a) Cell identity dynamics and insight into insulin secretagogues when employing stem cell-derived islets for disease modeling. Front Bioeng Biotechnol 12:1392575. https://doi.org/10.3389/fbioe.2024.1392575

Wang H, Li S, Zhang L, Zhang N (2024b) The role of fecal microbiota transplantation in type 2 diabetes mellitus treatment. Front Endocrinol (Lausanne) 15:1469165. https://doi.org/10.3389/fendo.2024.1469165

Wang L, Liu Y, Yan LS, Nguyen KT, Schroer SA, Suzuki A, Mak TW, Gaisano H, Woo M (2010) Deletion of Pten in pancreatic ss-cells protects against deficient ss-cell mass and function in mouse models of type 2 diabetes. Diabetes 59(12):3117–3126. https://doi.org/10.2337/db10-0570

Wang LH, Ernst AU, Flanders JA, Liu W, Wang X, Datta AK, Epel B, Kotecha M, Papas KK, Ma M (2021) An inverse-breathing encapsulation system for cell delivery. Sci Adv 7(20). https://doi.org/10.1126/sciadv.abf0343

Wang Q, Donelan W, Ye H, Jin Y, Lin Y, Wu X, Wang Y, Xi Y (2019a) Real-time observation of pancreatic beta cell differentiation from human induced pluripotent stem cells. Am J Transl Res 11(6):3490–3504

Wang W, Yang J, Liu H, Lu D, Chen X, Zenonos Z, Campos LS, Rad R, Guo G, Zhang S, Bradley A, Liu P (2011) Rapid and efficient reprogramming of somatic cells to induced pluripotent stem cells by retinoic acid receptor gamma and liver receptor homolog 1. Proc Natl Acad Sci U S A 108(45):18283–18288. https://doi.org/10.1073/pnas.1112140108

Wang X, Lai S, Ye Y, Hu Y, Pan D, Bai X, Shen J (2020) Conditional knockout of pyruvate dehydrogenase in mouse pancreatic beta-cells causes morphological and functional changes. Mol Med Rep 21(4):1717–1726. https://doi.org/10.3892/mmr.2020.10977

Wang X, Sterr M, Ansarullah BI, Bottcher A, Beckenbauer J, Siehler J, Meitinger T, Haring HU, Staiger H, Cernilogar FM, Schotta G, Irmler M, Beckers J, Wright CVE, Bakhti M, Lickert H (2019b) Point mutations in the PDX1 transactivation domain impair human beta-cell development and function. Mol Metab 24:80–97. https://doi.org/10.1016/j.molmet.2019.02.007

Wardani NI, Kangkamano T, Wannapob R, Kanatharana P, Thavarungkul P, Limbut W (2023) Electrochemical sensor based on molecularly imprinted polymer cryogel and multiwalled carbon nanotubes for direct insulin detection. Talanta 254:124137. https://doi.org/10.1016/j.talanta.2023.124137

Warren L, Manos PD, Ahfeldt T, Loh YH, Li H, Lau F, Ebina W, Mandal PK, Smith ZD, Meissner A, Daley GQ, Brack AS, Collins JJ, Cowan C, Schlaeger TM, Rossi DJ (2010) Highly efficient reprogramming to pluripotency and directed differentiation of human cells with synthetic modified mRNA. Cell Stem Cell 7(5):618–630. https://doi.org/10.1016/j.stem.2010.08.012

Watts LM, Manchem VP, Leedom TA, Rivard AL, McKay RA, Bao D, Neroladakis T, Monia BP, Bodenmiller DM, Cao JX, Zhang HY, Cox AL, Jacobs SJ, Michael MD, Sloop KW, Bhanot S (2005) Reduction of hepatic and adipose tissue glucocorticoid receptor expression with antisense oligonucleotides improves hyperglycemia and hyperlipidemia in diabetic rodents without causing systemic glucocorticoid antagonism. Diabetes 54(6):1846–1853. https://doi.org/10.2337/diabetes.54.6.1846

Weinzimer SA, Sherr JL, Cengiz E, Kim G, Ruiz JL, Carria L, Voskanyan G, Roy A, Tamborlane WV (2012) Effect of pramlintide on prandial glycemic excursions during closed-loop control in adolescents and young adults with type 1 diabetes. Diabetes Care 35(10):1994–1999. https://doi.org/10.2337/dc12-0171

Weismann D, Erion DM, Ignatova-Todorava I, Nagai Y, Stark R, Hsiao JJ, Flannery C, Birkenfeld AL, May T, Kahn M, Zhang D, Yu XX, Murray SF, Bhanot S, Monia BP, Cline GW, Shulman GI, Samuel VT (2011) Knockdown of the gene encoding drosophila tribbles homologue 3 (Trib3) improves insulin sensitivity through peroxisome proliferator-activated receptor-gamma (PPAR-gamma) activation in a rat model of insulin resistance. Diabetologia 54(4):935–944. https://doi.org/10.1007/s00125-010-2027-6

Wiggins SC, Abuid NJ, Gattas-Asfura KM, Kar S, Stabler CL (2020) Nanotechnology approaches to modulate immune responses to cell-based therapies for type 1 diabetes. J Diabetes Sci Technol 14(2):212–225. https://doi.org/10.1177/1932296819878952

Wilson JT, Cui W, Chaikof EL (2008) Layer-by-layer assembly of a conformal nanothin PEG coating for intraportal islet transplantation. Nano Lett 8(7):1940–1948. https://doi.org/10.1021/nl080924j

Witztum JL, Gaudet D, Freedman SD, Alexander VJ, Digenio A, Williams KR, Yang Q, Hughes SG, Geary RS, Arca M, Stroes ESG, Bergeron J, Soran H, Civeira F, Hemphill L, Tsimikas S, Blom DJ, O'Dea L, Bruckert E (2019) Volanesorsen and triglyceride levels in familial Chylomicronemia syndrome. N Engl J Med 381(6):531–542. https://doi.org/10.1056/NEJMoa1815037

Woltjen K, Michael IP, Mohseni P, Desai R, Mileikovsky M, Hamalainen R, Cowling R, Wang W, Liu P, Gertsenstein M, Kaji K, Sung HK, Nagy A (2009) piggyBac transposition reprograms fibroblasts to induced pluripotent stem cells. Nature 458(7239):766–770. https://doi.org/10.1038/nature07863

Xiao X, Guo P, Shiota C, Zhang T, Coudriet GM, Fischbach S, Prasadan K, Fusco J, Ramachandran S, Witkowski P, Piganelli JD, Gittes GK (2018) Endogenous reprogramming of alpha cells into beta cells, induced by viral gene therapy, reverses autoimmune diabetes. Cell Stem Cell 22(1):78–90.e4. https://doi.org/10.1016/j.stem.2017.10.014

Xiao Z, Wei T, Ge R, Li Q, Liu B, Ji Z, Chen L, Zhu J, Shen J, Liu Z, Huang Y, Yang Y, Chen Q (2022) Microfluidic production of zwitterion coating microcapsules with low foreign body reactions for improved islet transplantation. Small 18(29):e2202596. https://doi.org/10.1002/smll.202202596

Yabe SG, Nishida J, Fukuda S, Takeda F, Nashiro K, Ibuki M, Okochi H (2019) Definitive endoderm differentiation is promoted in suspension cultured human iPS-derived spheroids more than in adherent cells. Int J Dev Biol 63(6–7):271–280. https://doi.org/10.1387/ijdb.190132sy

Yang J, Luo S, Zhang J, Yu T, Fu Z, Zheng Y, Xu X, Liu C, Fan M, Zhang Z (2021a) Exosome-mediated delivery of antisense oligonucleotides targeting alpha-synuclein ameliorates the pathology in a mouse model of Parkinson's disease. Neurobiol Dis 148:105218. https://doi.org/10.1016/j.nbd.2020.105218

Yang J, Wang W, Ooi J, Campos LS, Lu L, Liu P (2015) Signalling through retinoic acid receptors is required for reprogramming of both mouse embryonic fibroblast cells and epiblast stem cells to induced pluripotent stem cells. Stem Cells 33(5):1390–1404. https://doi.org/10.1002/stem.1940

Yang K, O'Cearbhaill ED, Liu SS, Zhou A, Chitnis GD, Hamilos AE, Xu J, Verma MKS, Giraldo JA, Kudo Y, Lee EA, Lee Y, Pop R, Langer R, Melton DA, Greiner DL, Karp JM (2021b) A therapeutic convection-enhanced macroencapsulation device for enhancing beta cell viability and insulin secretion. Proc Natl Acad Sci U S A 118(37):e2106538118. https://doi.org/10.1073/pnas.2106538118

Yang Q, Graham TE, Mody N, Preitner F, Peroni OD, Zabolotny JM, Kotani K, Quadro L, Kahn BB (2005) Serum retinol binding protein 4 contributes to insulin resistance in obesity and type 2 diabetes. Nature 436(7049):356–362. https://doi.org/10.1038/nature03711

Yin X, Mead BE, Safaee H, Langer R, Karp JM, Levy O (2016) Engineering stem cell organoids. Cell Stem Cell 18(1):25–38. https://doi.org/10.1016/j.stem.2015.12.005

Yoshida W, Mochizuki E, Takase M, Hasegawa H, Morita Y, Yamazaki H, Sode K, Ikebukuro K (2009) Selection of DNA aptamers against insulin and construction of an aptameric enzyme subunit for insulin sensing. Biosens Bioelectron 24(5):1116–1120. https://doi.org/10.1016/j.bios.2008.09.004

Yoshihara E, O'Connor C, Gasser E, Wei Z, Oh TG, Tseng TW, Wang D, Cayabyab F, Dai Y, Yu RT, Liddle C, Atkins AR, Downes M, Evans RM (2020) Immune-evasive human islet-like organoids ameliorate diabetes. Nature 586(7830):606–611. https://doi.org/10.1038/s41586-020-2784-4

Yu J, Vodyanik MA, Smuga-Otto K, Antosiewicz-Bourget J, Frane JL, Tian S, Nie J, Jonsdottir GA, Ruotti V, Stewart R, Slukvin II, Thomson JA (2007) Induced pluripotent stem cell lines derived from human somatic cells. Science 318(5858):1917–1920. https://doi.org/10.1126/science.1151526

Yu XX, Murray SF, Pandey SK, Booten SL, Bao D, Song XZ, Kelly S, Chen S, McKay R, Monia BP, Bhanot S (2005) Antisense oligonucleotide reduction of DGAT2 expression improves hepatic steatosis and hyperlipidemia in obese mice. Hepatology 42(2):362–371. https://doi.org/10.1002/hep.20707

Yue Z, Liu X, Coates PT, Wallace GG (2016) Advances in printing biomaterials and living cells: implications for islet cell transplantation. Curr Opin Organ Transplant 21(5):467–475. https://doi.org/10.1097/MOT.0000000000000318

Zanardi TA, Han SC, Jeong EJ, Rime S, Yu RZ, Chakravarty K, Henry SP (2012) Pharmacodynamics and subchronic toxicity in mice and monkeys of ISIS 388626, a second-generation antisense oligonucleotide that targets human sodium glucose cotransporter 2. J Pharmacol Exp Ther 343(2):489–496. https://doi.org/10.1124/jpet.112.196293

Zanotti G, Berni R (2004) Plasma retinol-binding protein: structure and interactions with retinol, retinoids, and transthyretin. Vitam Horm 69:271–295. https://doi.org/10.1016/S0083-6729(04)69009-5

Zemany L, Bhanot S, Peroni OD, Murray SF, Moraes-Vieira PM, Castoldi A, Manchem P, Guo S, Monia BP, Kahn BB (2015) Transthyretin antisense oligonucleotides lower circulating RBP4 levels and improve insulin sensitivity in obese mice. Diabetes 64(5):1603–1614. https://doi.org/10.2337/db14-1255

Zeng H, Guo M, Zhou T, Tan L, Chong CN, Zhang T, Dong X, Xiang JZ, Yu AS, Yue L, Qi Q, Evans T, Graumann J, Chen S (2016) An isogenic human ESC platform for functional evaluation of genome-wide-association-study-identified diabetes genes and drug discovery. Cell Stem Cell 19(3):326–340. https://doi.org/10.1016/j.stem.2016.06.007

Zhang MM, Bahal R, Rasmussen TP, Manautou JE, Zhong XB (2021a) The growth of siRNA-based therapeutics: updated clinical studies. Biochem Pharmacol 189:114432. https://doi.org/10.1016/j.bcp.2021.114432

Zhang W, Mehta A, Tong Z, Esser L, Voelcker NH (2021b) Development of polymeric nanoparticles for blood-brain barrier transfer—strategies and challenges. Adv Sc 8(10):2003937. https://doi.org/10.1002/advs.202003937

Zhang X, Das SK, Passi SF, Uehara H, Bohner A, Chen M, Tiem M, Archer B, Ambati BK (2015) AAV2 delivery of Flt23k intraceptors inhibits murine choroidal neovascularization. Mol Ther 23(2):226–234. https://doi.org/10.1038/mt.2014.219

Zhang Y, Lei T, Huang JF, Wang SB, Zhou LL, Yang ZQ, Chen XD (2011) The link between fibroblast growth factor 21 and sterol regulatory element binding protein 1c during lipogenesis in hepatocytes. Mol Cell Endocrinol 342(1–2):41–47. https://doi.org/10.1016/j.mce.2011.01.015

Zhang Z, Gallagher T, Scherer PE, Beutler B (2020) Tissue-specific disruption of Kbtbd2 uncovers adipocyte-intrinsic and -extrinsic features of the teeny lipodystrophy syndrome. Proc Natl Acad Sci U S A 117(21):11829–11835. https://doi.org/10.1073/pnas.1922201117

Zhang Z, Turer E, Li X, Zhan X, Choi M, Tang M, Press A, Smith SR, Divoux A, Moresco EM, Beutler B (2016) Insulin resistance and diabetes caused by genetic or diet-induced KBTBD2 deficiency in mice. Proc Natl Acad Sci U S A 113(42):E6418–E6426. https://doi.org/10.1073/pnas.1611235113

Zhou Q, Brown J, Kanarek A, Rajagopal J, Melton DA (2008) In vivo reprogramming of adult pancreatic exocrine cells to beta-cells. Nature 455(7213):627–632. https://doi.org/10.1038/nature07214

Zhou X, Zheng W, Kong W, Zhang J, Liao Y, Min J, Zeng T (2025) Glucose parameters, inflammation markers, and gut microbiota changes of gut microbiome-targeted therapies in type 2 diabetes mellitus: a systematic review and meta-analysis of randomized controlled trials. J Clin Endocrinol Metab 110:2980

Zinker BA, Rondinone CM, Trevillyan JM, Gum RJ, Clampit JE, Waring JF, Xie N, Wilcox D, Jacobson P, Frost L, Kroeger PE, Reilly RM, Koterski S, Opgenorth TJ, Ulrich RG, Crosby S, Butler M, Murray SF, McKay RA, Bhanot S, Monia BP, Jirousek MR (2002) PTP1B antisense oligonucleotide lowers PTP1B protein, normalizes blood glucose, and improves insulin sensitivity in diabetic mice. Proc Natl Acad Sci USA 99(17):11357–11362. https://doi.org/10.1073/pnas.172405899

Åvall K, Ali Y, Leibiger IB, Leibiger B, Moede T, Paschen M, Dicker A, Daré E, Köhler M, Ilegems E, Abdulreda MH, Graham M, Crooke RM, Tay VS, Refai E, Nilsson SK, Jacob S, Selander L, Berggren PO, Juntti-Berggren L (2015) Apolipoprotein CIII links islet insulin resistance to β-cell failure in diabetes. Proc Natl Acad Sci USA 112(20):E2611–E2619. https://doi.org/10.1073/pnas.1503536112

Nutrigenomics: Unraveling the Gene–Diet Interactions for Personalized Nutrition and Optimizing Health Outcomes

Rownock Afruza

Abstract This book chapter provides an in-depth exploration of nutrigenomics, an emerging field that explores the complex relationship between genetics and nutrition. By investigating how dietary components influence gene expression and how genetic variations shape individual responses to nutrients, nutrigenomics lays the foundation for personalized nutrition. It emphasizes the potential to tailor dietary recommendations based on an individual's genetic makeup, aiming to optimize overall well-being. The chapter highlights the significant role of the gut microbiome in modulating gene-diet interactions, which influence nutrient metabolism, immune function, and physiological responses to dietary intake. Additionally, the integration of metabolomics with nutrigenomics offers deeper insights into metabolic pathways, facilitating the impact of precise nutritional interventions tailored to an individual's genetic and biochemical profile. Key topics covered include the impact of nutrigenomics on gut health, metabolic disorders, as well as its relevance in cancer prevention and neurodegenerative diseases. The chapter also addresses the potential challenges in nutrigenomics, such as the complexity of gene-diet interactions, ethical considerations, and the need for large-scale validation studies. Despite these challenges, the chapter concludes with a forward-looking perspective, emphasizing how advancements in nutrigenomics, integrated with genomics, metabolomics, and microbiome data, will likely revolutionize personalized nutrition and contribute to improved health outcome.

Keywords Gene–diet interactions · Gut-dysbiosis · Nutrigenomics · Microbiome · Personalized medicine · Personalized nutrition

R. Afruza (✉)
Translational Hepatology Section, National Institute of Diabetes and Digestive and Kidney Diseases, National Institutes of Health, Bethesda, MD, USA
e-mail: rownock.afruza@nih.gov

A. H. M. N. Nabi, H. Uddin Shekhar (eds.), *Biotechnological Advances in Healthomics*, Interdisciplinary Biotechnological Advances,
https://doi.org/10.1007/978-981-95-3612-2_10

Abbreviations

AKI	Acute kidney injury
AMY1	Amylase 1
APOE	Apolipoprotein E
BDNF	Brain-derived neurotrophic factor
BMI	Body mass index
CNVs	Copy number variations
CYP1A2	Cytochrome P450 1A2
DC12	Dodecanedioic acid
DC8	Octanedioic acid
DNMTs	DNA methyltransferases
DTC	Direct-to-consumer
EGCG	Epigallocatechin gallate
FADS1	Fatty acid desaturase 1
FMT	Fecal microbiota transplantation
FUT2	Alpha-1,2-L-fucosyltransferase
GDPR	General Data Protection Regulation
GSTs	Glutathione S-transferases
HATs	Histone acetyltransferases
HDACs	Histone deacetylases
HIPAA	Health Insurance Portability and Accountability Act
iHMP	Integrative Human Microbiome Project
LCT	Lactase
LDL	Low-density lipoprotein
LPS	Lipopolysaccharide
MFTO gene	Fat mass and obesity-associated gene
miRNA	MicroRNA
mRNAs	Messenger RNAs
MTHFR	Methylenetetrahydrofolate reductase
NGS	Next-generation sequencing
NMR	Nuclear magnetic resonance
NOD2	Nucleotide-binding oligomerization domain-containing protein 2
NTDs	Neural tube defects
OGG1	8-oxoguanine DNA glycosylase
PD	Parkinson's disease
PRRs	Pattern recognition receptors
PRS	Polygenic risk scores
PUFAs	Polyunsaturated fatty acids
SAM	S-adenosylmethionine
SCFAs	Short-chain fatty acids
SNPs	Single nucleotide polymorphisms
TCF7L2	Transcription factor 7-like 2
TMAO	Trimethylamine-N-oxide
VDR	Vitamin D receptor
XRCC1	X-ray repair cross complementing 1

1 Introduction to Nutrigenomics

1.1 Definition and Scope of Nutrigenomics

Nutrigenomics is an emerging field within nutritional science that investigates the interaction between nutrients and genes at the molecular level. It encompasses the study of how individual genetic differences affect nutrient metabolism and dietary responses, as well as how nutrients influence gene expression (Ferguson et al. 2016). At its core, nutrigenomics seeks to understand the bidirectional relationship between diet and the genome, aiming to inform dietary recommendations that are tailored to genetic profiles. This dynamic and interdisciplinary field integrates genomics, transcriptomics, proteomics, and metabolomics to uncover the molecular mechanisms underpinning nutrition-related health outcomes (Corella and Ordovás 2015). The purpose of this chapter is to explore the basic science of nutrigenomics and its role in shaping personalized nutrition through gene–diet interactions, epigenetic regulation, and microbiome dynamics. The central research question addressed is: *How can insights from nutrigenomics and multi-omics be applied to develop precise, evidence-based dietary interventions tailored to individual genetic and metabolic profiles?*

1.2 Historical Evolution and Technological Advances

The conceptual foundations of nutrigenomics can be traced back to the broader scientific recognition of the gene–environment interaction. However, it was not until the completion of the Human Genome Project in 2003 that nutrigenomics began to gain traction as a distinct discipline (Afman and Müller 2006). This landmark achievement paved the way for high-throughput genomic technologies and bioinformatics tools that could be applied to nutrition research. The emergence of omics platforms such as transcriptomics (gene expression profiling), proteomics (protein analysis), and metabolomics (metabolite profiling) has significantly enhanced the ability to study diet-gene interactions at a systems biology level (Müller and Kersten 2003).

During the early 2000s, researchers began identifying single nucleotide polymorphisms (SNPs) that influence nutrient metabolism (Ordovas and Mooser 2004). These discoveries marked the beginning of genotype-informed dietary guidance. Advances in microarray technologies and next-generation sequencing (NGS) have since expanded the capacity to analyze genetic and epigenetic changes in response to dietary interventions across diverse populations.

Technological advances facilitate the integration of epigenomics into nutrigenomic research, revealing how dietary factors can influence DNA methylation, histone modification, and noncoding RNA expression (Mathers 2017). This has important implications for developmental programming and transgenerational inheritance of disease risks. For example, maternal nutrition during pregnancy can influence the epigenome of the fetus, affecting lifelong health outcomes. Moreover, computational biology and machine learning approaches are increasingly being utilized to handle the vast datasets generated in nutrigenomic studies, enabling more refined modeling of gene–diet interactions.

1.3 Relationship to Personalized Nutrition and Precision Medicine

Personalized nutrition and precision medicine both focus on tailoring health interventions but differ in scope. Personalized nutrition focuses on dietary recommendations based on individual characteristics, while precision medicine is a broader approach encompassing various treatments and interventions, including nutritional ones, tailored to an individual's specific needs. Traditional dietary guidelines have typically been based on population averages, which may not account for individual genetic variability. In contrast, personalized nutrition leverages genetic, phenotypic, and lifestyle information to tailor dietary recommendations that optimize health and minimize disease risk on an individual basis (de Toro-Martín et al. 2017). Nutrigenomic insights allow practitioners to design dietary interventions that consider specific genetic markers, thereby enhancing efficacy and compliance. For instance, individuals carrying the fat mass and obesity-associated (FTO) gene variant, which is linked to increased risk for obesity, may benefit more from higher-protein diets or increased physical activity than those without the variant (Phillips 2013). Similarly, carriers of certain variants in the Cytochrome P450 1A2 (CYP1A2) gene metabolize caffeine more slowly and may have a higher risk of hypertension when consuming high amounts of coffee (Cornelis et al. 2006).

Precision medicine—an approach that customizes healthcare based on individual variability in genes, environment, and lifestyle—has increasingly embraced the contributions of nutrigenomics. The integration of genetic data into clinical nutrition allows for targeted prevention and treatment strategies, especially for metabolic and lifestyle-related diseases. This holistic approach is exemplified by precision nutrition programs that incorporate genetic testing, microbiome analysis, metabolic biomarkers, and wearable technologies to deliver real-time, actionable health insights.

2 Genetic Variations and Nutrient Interactions

SNPs and copy number variations (CNVs) are the most studied genetic variations in nutrigenomics. SNPs are single-base-pair substitutions in the DNA sequence and represent the most common type of genetic variation in the human genome. They occur approximately once every 300 base pairs, meaning that the human genome may contain over 10 million SNPs (Collins et al. 2003). SNPs may be in coding regions, affecting amino acid sequences, or in noncoding regions, where they can influence gene regulation, splicing, or messenger ribonucleic acid (mRNA) stability. In the context of nutrigenomics, SNPs are particularly relevant because they can alter the metabolism of key nutrients, affect taste preferences, modulate nutrient transport, and modify disease susceptibility based on dietary intake. For instance, the Methylenetetrahydrofolate reductase (MTHFR) C677T polymorphism leads to reduced activity of methylenetetrahydrofolate reductase, affecting folate metabolism and homocysteine levels (Bailey and Gregory III 1999). Similarly, SNPs in the Apolipoprotein E (APOE) gene are associated with differential lipid responses to dietary fat intake and cardiovascular disease (CVD) risk (Ordovas and Mooser 2004). CNVs refer to structural variations in the genome that result in the duplication or deletion of large DNA segments, typically encompassing one or more genes. These variations can significantly influence gene dosage and expression, with potential impacts on nutrient metabolism and disease risk. An example of CNVs in nutrition is the amylase gene (AMY1), which encodes salivary amylase responsible for starch digestion. Individuals with a higher copy number of AMY1 genes tend to produce more salivary amylase and exhibit better glycemic control when consuming starch-rich diets, suggesting evolutionary adaptation to dietary habits (Perry et al. 2007). CNVs have also been implicated in taste receptor genes, influencing dietary behaviors and preferences.

Several well-characterized gene–nutrient interactions demonstrate the functional consequences of genetic variation on nutrient metabolism and associated clinical implications (Fig. 1). These interactions form the basis of genotype-informed dietary recommendations and highlight the relevance of integrating genetic data into nutritional practice.

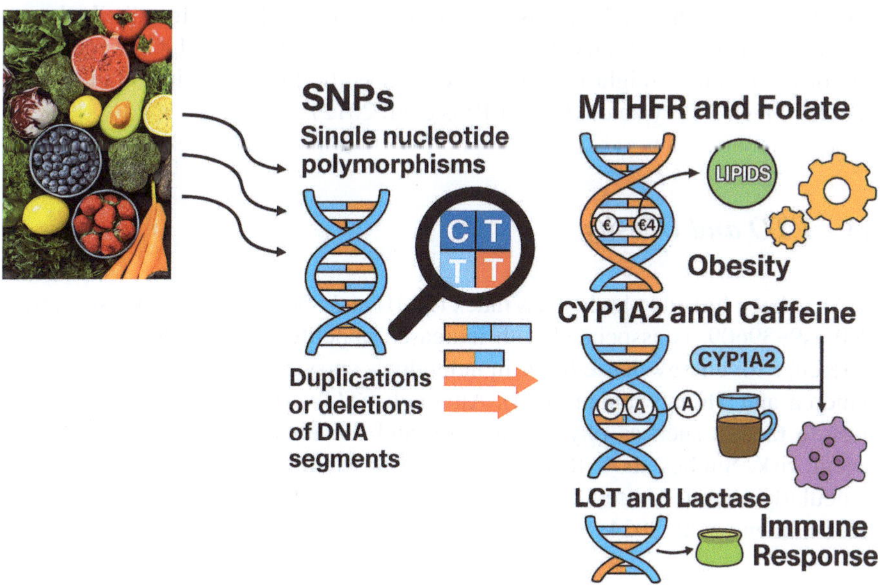

Fig. 1 Schematic representation of gene–diet interactions in nutrigenomics

2.1 MTHFR and Folate

The MTHFR gene encodes an enzyme crucial for converting 5,10-methylenetetrahydrofolate to 5-methyltetrahydrofolate, the bioactive form of folate involved in homocysteine remethylation to methionine. The most studied SNP in this gene is C677T (rs1801133) (Santilli et al. 2016; Froese et al. 2019). Individuals homozygous for the T allele (TT genotype) exhibit reduced enzyme activity, leading to elevated plasma homocysteine and lower folate levels, especially under conditions of low dietary folate intake (Bailey and Gregory III 1999). Clinically, this genotype has been associated with increased risk for neural tube defects (NTDs), cardiovascular disease, and possibly colorectal cancer. Ensuring sufficient folate intake, particularly in women of childbearing age, can mitigate these risks. Some evidence also suggests that riboflavin supplementation may help normalize blood pressure in individuals with the TT genotype by enhancing enzyme function (Rooney et al. 2022; McNulty et al. 2017).

2.2 APOE and Lipids

The Apolipoprotein E (APOE) gene encodes apolipoprotein E, a key player in lipid metabolism and transport. There are three common alleles: ε2, ε3, and ε4, resulting in six genotypes. The ε4 allele is associated with increased low-density lipoprotein (LDL) cholesterol levels and greater sensitivity to dietary saturated fat, whereas the ε2 allele is often linked with lower LDL but higher triglycerides (Minihane et al. 2007). Individuals with the ε4 allele are at elevated risk for cardiovascular disease and late-onset Alzheimer's disease. Evidence indicates that such individuals may particularly benefit from diets low in saturated fat and cholesterol. Conversely, carriers of the ε2 allele might respond better to specific lipid-lowering strategies that include omega-3 fatty acids (Garcia-Rios et al. 2012).

2.3 FTO and Obesity

FTO has been linked to body mass index (BMI) and obesity risk. The most common SNP, rs9939609, is associated with increased appetite, higher caloric intake, and preference for energy-dense foods in individuals carrying the risk allele (AA or AT) (Nindrea and Thongwichian 2024). Although FTO variants are linked to obesity, lifestyle factors such as physical activity and dietary composition can attenuate genetic risk. Studies show that high-protein or Mediterranean-style diets may be particularly effective in reducing weight in FTO risk allele carriers (Gholami 2024; Pérez-Gimeno et al. 2024; Seral-Cortes et al. 2022).

Dietary nutrients influence genetic expression through variations such as SNPs and DNA segment duplications or deletions. These genetic differences affect individual responses to nutrients and metabolic processes. The figure illustrates the core concept of nutrigenomics, highlighting how dietary components interact with genetic variations—specifically single nucleotide polymorphisms (SNPs) and structural variations (such as duplications or deletions)—to influence key physiological outcomes.

2.4 CYP1A2 and Caffeine Metabolism

CYP1A2 encodes a liver enzyme responsible for metabolizing caffeine. The rs762551 SNP affects enzyme activity, categorizing individuals as either "fast" or "slow" caffeine metabolizers. Slow metabolizers have been associated with increased risk of hypertension and myocardial infarction with high caffeine intake (Cornelis et al. 2006; Virgili et al. 2023). This gene–nutrient interaction highlights the importance of moderation in caffeine consumption for individuals with the slow metabolizer genotype and exemplifies how simple genetic testing can inform daily dietary habits.

2.5 LCT and Lactose Intolerance

The Lactase (LCT) gene encodes lactase, the enzyme responsible for lactose digestion. A common SNP (rs4988235) upstream of LCT determines lactase persistence or nonpersistence. Individuals with the CC genotype typically exhibit lactose intolerance, whereas those with CT or TT genotypes continue to express lactase into adulthood (Enattah et al. 2002; Smith et al. 2009). For lactase-nonpersistent individuals, personalized nutrition strategies might include lactose-free diets, fermented dairy products, or lactase supplementation to avoid gastrointestinal symptoms and maintain calcium intake.

3 Epigenetics and Nutritional Regulation of Gene Expression

Expanding on the foundational insights into how genetic variations such as SNPs and CNVs affect nutrient metabolism and dietary responses, it is equally important to explore how nutrition can regulate gene expression through epigenetics. Epigenetics refers to heritable changes in gene expression that occur without alterations in the underlying DNA sequence. These modifications play a crucial role in development, cellular differentiation, and disease susceptibility. Nutrients and dietary bioactives can influence the epigenome dynamically, thereby modulating gene expression in health and disease. Nutritional epigenomics investigates how specific dietary components modify epigenetic marks such as DNA methylation, histone modifications, and microRNA (miRNA) expression, with implications for lifelong health trajectories.

3.1 DNA Methylation, Histone Modification, and miRNA Pathways

3.1.1 DNA Methylation

DNA methylation involves the addition of a methyl group to the 5′ carbon of cytosine residues, primarily in CpG dinucleotides. This modification is catalyzed by DNA methyltransferases (DNMTs) and typically represses gene transcription when located in gene promoter regions (Moore et al. 2013). Methylation patterns are established during early development but can be modified throughout life by environmental factors, including diet. Aberrant DNA methylation is associated with various diseases, including cancer, metabolic syndrome, and neurological disorders. Hypomethylation can lead to genomic instability, while hypermethylation of tumor suppressor genes contributes to carcinogenesis (Sheaffer et al. 2016). Nutritional inputs play a critical role in maintaining proper methylation patterns by supplying methyl donors and cofactors for the one-carbon metabolism pathway.

3.1.2 Histone Modification

Histones are proteins around which DNA winds to form chromatin. Posttranslational modifications of histone tails—such as acetylation, methylation, phosphorylation, and ubiquitination—affect chromatin structure and gene accessibility. For instance, histone acetylation by histone acetyltransferases (HATs) generally promotes transcription, while deacetylation by histone deacetylases (HDACs) represses it (Kouzarides 2007). Histone modifications can be influenced by dietary components that modulate the activity of HATs and HDACs. For example, short-chain fatty acids (SCFAs) like butyrate, produced from dietary fiber fermentation in the colon, act as HDAC inhibitors and promote histone acetylation, thus enhancing gene expression linked to anti-inflammatory responses (Davie 2003).

3.1.3 miRNAs

miRNAs are small, noncoding RNAs (approximately 22 nucleotides) that regulate gene expression posttranscriptionally by binding to target mRNAs, leading to their degradation or translational repression. Each miRNA can target multiple messenger RNAs (mRNAs), and their expression is tightly regulated during development and in response to environmental stimuli. Diet influences miRNA expression directly through bioactive compounds and indirectly by altering the cellular environment. For instance, polyphenols like epigallocatechin gallate (EGCG) from green tea can modulate miRNA profiles associated with cell cycle regulation and apoptosis (Mokhtari et al. 2020; Wang et al. 2023; Dharshini and Mandal 2024). Dysregulated miRNA expression is implicated in cancer, obesity, and cardiovascular disease.

3.2 *Role of Nutrients in Epigenetic Regulation*

Nutritional epigenetics focuses on how micronutrients and phytochemicals affect epigenetic processes. Several classes of nutrients—such as B vitamins, polyphenols, and fatty acids—serve as substrates or cofactors in pathways that establish or modify epigenetic marks.

3.2.1 B Vitamins and One-Carbon Metabolism

B vitamins, particularly folate (B9), vitamin B12, vitamin B6, and riboflavin (B2), are essential for one-carbon metabolism, which provides methyl groups for DNA and histone methylation. Folate is converted into 5-methyltetrahydrofolate, the primary methyl donor in the remethylation of homocysteine to methionine, which is subsequently used to generate S-adenosylmethionine (SAM), the universal methyl donor (Stover 2009). Deficiencies in these vitamins can result in hypomethylation and aberrant gene expression. For example, maternal folate deficiency during pregnancy is associated with impaired fetal DNA methylation and increased risk of neural tube defects and other developmental anomalies (Crider et al. 2011). Moreover, polymorphisms in the MTHFR gene can further modulate the impact of folate on the epigenome.

3.2.2 Polyphenols and Phytochemicals

Polyphenols, such as curcumin, resveratrol, and EGCG, exhibit epigenetic activity by interacting with enzymes that modify DNA and histones. These compounds can inhibit DNMTs and HDACs, restore expression of silenced tumor suppressor genes, and modulate inflammation-related gene networks (Link et al. 2010). For instance, curcumin, a compound from turmeric, has been shown to reduce the expression of DNMT1 and HDACs in cancer cell lines, leading to reactivation of pro-apoptotic genes (Sharma et al. 2005). Resveratrol, found in red grapes, enhances histone acetylation and miRNA expression that regulate lipid metabolism and aging-related genes (Howitz et al. 2003).

3.2.3 Fatty Acids

Certain fatty acids, particularly omega-3 polyunsaturated fatty acids (PUFAs), exert epigenetic effects by influencing histone acetylation and DNA methylation. Omega-3 PUFAs can alter the activity of transcription factors and chromatin-modifying enzymes involved in inflammatory and metabolic pathways (Niculescu and Lupu 2011). Moreover, maternal intake of omega-3 fatty acids during pregnancy has been shown to influence the epigenetic programming of metabolic genes in offspring, potentially reducing the risk of obesity and insulin resistance later in life.

4 The Gut Microbiome: A Mediator of Gene–Diet Interactions

The gut microbiome, a vast community of microorganisms residing in the gastrointestinal tract, plays a central role in human health through its interactions with diet and host genetics. With over 100 trillion microbial cells and thousands of species, the microbiome acts as an intermediary between dietary inputs and host physiological responses (Lynch and Pedersen 2016). Recent advances in high-throughput sequencing and metagenomics have uncovered how the gut microbiome contributes to nutrient metabolism, immune system development, and gene regulation. This interplay forms a crucial axis—microbiome–diet–genome—that influences disease susceptibility and therapeutic responses.

4.1 Influence of Diet on Microbiome Composition

Diet is one of the most potent and modifiable factors shaping the gut microbiome. Long-term dietary habits, as well as short-term changes, can significantly alter the abundance and diversity of microbial taxa (David et al. 2014). The gut microbiota adapts dynamically to macronutrient composition, fiber intake, polyphenols, and food additives, ultimately influencing metabolic and immunological outcomes.

4.1.1 Macronutrient Composition

High-fat, high-sugar Western diets are associated with a reduced microbial diversity and an increased Firmicutes–to–Bacteroidetes ratio, a pattern often linked to obesity and metabolic inflammation (Turnbaugh et al. 2006). In contrast, plant-based and high-fiber diets promote the proliferation of beneficial bacteria such as *Bifidobacterium* and *Lactobacillus*, which produce short-chain fatty acids (SCFAs) and support gut barrier function (Flint et al. 2012). Protein intake also influences microbiota composition. Animal-derived proteins favor bile-tolerant microorganisms like *Alistipes* and *Bilophila*, which are associated with inflammatory responses. Meanwhile, plant-derived proteins tend to support more diverse and anti-inflammatory microbial populations (Singh et al. 2017).

4.1.2 Dietary Fiber and Polyphenols

Dietary fibers, especially nondigestible oligosaccharides, serve as prebiotics that selectively enhance the growth of beneficial microbes. These fibers are fermented by gut bacteria into SCFAs such as acetate, propionate, and butyrate, which contribute to energy homeostasis, anti-inflammatory effects, and epigenetic regulation (Koh et al. 2016). Polyphenols, abundant in fruits, vegetables, tea, and wine, also exert prebiotic effects. Although poorly absorbed in the small intestine, polyphenols are metabolized by gut bacteria into bioactive compounds with antioxidant and anti-inflammatory properties. These compounds, in turn, modulate microbial diversity and can inhibit pathogenic bacteria (Cardona et al. 2013).

4.1.3 Food Additives and Supplements

Modern diets often contain emulsifiers, artificial sweeteners, and preservatives that have an impact on the gut microbiome (Gerasimidis et al. 2020). Emulsifiers such as carboxymethylcellulose have been shown to disrupt the mucosal barrier and promote inflammatory microbiota configurations, contributing to colitis and metabolic dysregulation (Chassaing et al. 2015). Recent studies have demonstrated that dietary supplementation with dicarboxylic acids, such as octanedioic acid (DC8) and dodecanedioic acid (DC12), offers protective effects against acute kidney injury (AKI) and obesity-related metabolic disorders in mice. DC8 enhances peroxisomal fatty acid oxidation, preserving kidney function during AKI, while DC12 increases metabolic rate, reduces body and liver fat, and improves glucose tolerance without being stored in adipose tissue, indicating potential therapeutic applications for these compounds in renal and metabolic health (Barbosa et al. 2024; Goetzman et al. 2024).

4.2 Gut Microbiota in Nutrient Metabolism and Immune Modulation

The gut microbiome serves as an additional organ system with metabolic and immunological capabilities. It enables the host to extract energy from otherwise indigestible polysaccharides, synthesize essential vitamins, regulate lipid metabolism, and educate the immune system (Chen et al. 2015). Gut microbes play a vital role in fermenting complex carbohydrates and producing SCFAs, which serve as energy sources for colonocytes and systemic metabolic signals (Wang et al. 2019). Butyrate has been shown to regulate gene expression through histone deacetylase inhibition, thereby influencing inflammation and tumor suppression (Canani et al. 2011). Furthermore, gut bacteria are involved in the synthesis of vitamins such as vitamin K, biotin, and folate. These microbial-derived vitamins contribute to host metabolic needs and participate in key processes like one-carbon metabolism and blood clotting (LeBlanc et al. 2013). Microbiota also modulate bile acid metabolism, transforming primary bile acids into secondary bile acids that influence lipid absorption and act as signaling molecules in glucose and energy homeostasis via the Farnesoid X receptor (a nuclear receptor primarily expressed in the liver and intestine that functions as a key regulator of bile acid homeostasis) and Takeda G protein-coupled receptor 5 (also known as GPBAR1, is a bile acid-activated receptor expressed in the intestine, liver, and immune cells and plays a key role in energy metabolism, glucose homeostasis, inflammation, and bile acid signaling) pathways (Wahlström et al. 2016). Ali et al. used longitudinal multi-omics profiling to demonstrate that progression of hepatitis C virus infection and cirrhosis is characterized by alterations in bile acid metabolism, especially the accumulation of taurine-conjugated bile acids (Ali et al. 2023). These metabolites were linked to hepatic immune activation, lipid remodeling, and fibrogenesis.

The gut microbiome is crucial for the development and calibration of the host immune system. Germ-free animal studies have demonstrated that the absence of microbes results in impaired immune responses, including reduced numbers of regulatory T cells and defective mucosal immunity (Belkaid and Hand 2014). Microbial metabolites such as SCFAs and tryptophan derivatives shape immune cell function by influencing cytokine production, dendritic cell maturation, and T cell differentiation. For instance, butyrate enhances the induction of regulatory T cells, which suppress autoimmunity and maintain immune tolerance (Furusawa et al. 2013). Pathobionts—normally harmless microbes that become pathogenic under stress—can trigger inflammatory diseases when microbial balance is disrupted. Dysbiosis, or the loss of microbial diversity, has been associated with inflammatory bowel disease, asthma, allergies, and autoimmune conditions (Haag and Siegmund 2015).

4.3 Microbiome in Disease Pathogenesis

The disruption of microbiome-host interactions has been increasingly recognized as a critical factor in the pathogenesis of a wide range of diseases. In metabolic disorders like obesity and type 2 diabetes, microbial-derived metabolites such as imidazole propionate impair insulin signaling, highlighting a mechanistic link between dysbiosis and metabolic dysfunction (Koh et al. 2018). Similarly, in cardiovascular disease, gut microbial metabolism of dietary choline and carnitine into trimethylamine-N-oxide (TMAO) promotes atherosclerosis, establishing the microbiome as a key player in systemic disease (Tang et al. 2013).

Recent research has substantially advanced our understanding of the gut–liver axis, a critical conduit linking the microbiome to liver and systemic diseases (Fig. 2). Taurine-conjugated bile acids have been shown to engage hepatic sphingosine-1-phosphate receptor 2 signaling, promoting inflammation and fibrosis, thereby highlighting the role of microbial metabolites in liver disease pathogenesis (Ali et al. 2024). Evidence from treated HIV patients indicates that persistent immune activation is more closely associated with specific microbial community profiles than with total microbial load, suggesting that microbial quality critically shapes systemic and hepatic immune responses (Nganou-Makamdop et al. 2021). Hepatic microbial DNA has emerged as a marker of liver inflammation and fibrosis, providing compelling evidence that microbial translocation contributes to chronic liver injury.

Beyond the presence of microbes, both host genetic factors and microbial functional capacities play pivotal roles in determining disease susceptibility and clinical outcomes. An enrichment of pathogenic taxa such as *Veillonella* and *Enterococcus* has been observed in cirrhotic patients, correlating with systemic inflammation and altered bile acid profiles (Quinn et al. 2023). Loss of butyrate-producing bacteria, coupled with increased microbial DNA translocation and activation of toll-like receptor pathways, perpetuates a pro-inflammatory milieu in the liver (Tripathi et al. 2018). The concept of a "blood microbiome" has gained traction with the detection of circulating microbial DNA encoding genes for lipopolysaccharide (LPS) biosynthesis and bile acid modification, suggesting that microbial influences extend beyond the liver to affect systemic physiology (Castillo et al. 2019).

Microbial functionality, rather than taxonomy alone, appears to be a more reliable predictor of disease outcomes. In models of metabolic syndrome and

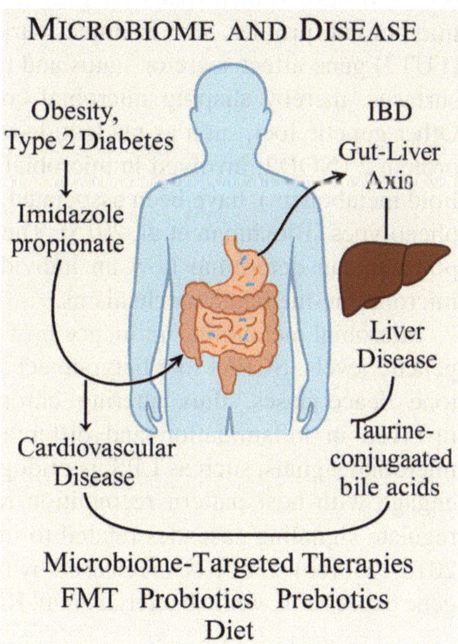

Fig. 2 Microbiome and disease: pathways and therapeutic interventions. This diagram illustrates the interplay between the gut microbiome and chronic diseases. It highlights how microbial metabolites like imidazole propionate and TMAO are linked to obesity, type 2 diabetes, and cardiovascular disease. It also depicts the gut–liver axis in inflammatory bowel disease (IBD) and liver disease, influenced by bile acid metabolism. The bottom section presents microbiome-targeted therapies, including FMT, probiotics, prebiotics, and diet as strategies to modulate disease risk

inflammatory bowel disease, disrupted microbial functions—such as impaired SCFA production—have been linked to adverse metabolic and immune profiles (Thaiss et al. 2014). From a clinical standpoint, microbial signatures in portal vein blood have shown strong correlations with liver disease severity, supporting their potential use as biomarkers for diagnosis and prognosis (Quinn et al. 2023; Ali et al. 2024). Moreover, hepatic microbial DNA has been proposed as a stratification tool for assessing fibrosis risk (Schierwagen et al. 2023).

Microbiome-based therapies are gaining importance across diverse clinical domains. Strategies including probiotics, fecal microbiota transplantation (FMT), and diet-based interventions are being actively explored for conditions such as liver disease, inflammatory bowel disease, and even cancer (Kao et al. 2017). Restoring microbial balance through these interventions holds promise for modulating disease trajectories and enhancing therapeutic efficacy. Additionally, pharmacomicrobiomics—the study of microbiome–drug interactions—is emerging as a key component of precision medicine, offering new pathways for personalized therapeutic approaches.

4.4 Microbiome-Gene Interactions

Interactions between the host genome and the microbiome are bidirectional and complex. Host genetic variations can influence the composition of the gut microbiota, while microbial metabolites can regulate host gene expression through epigenetic and receptor-mediated pathways. Twin and genome-wide association studies (GWAS) have shown that host genetics contribute to interindividual variation in microbiome profiles. For example, variants in the alpha-1,2-L-fucosyltransferase (FUT2) gene affect secretor status and the presence of certain glycans on mucosal surfaces, thereby shaping microbial colonization patterns (Rausch et al. 2011). Other genetic loci, such as nucleotide-binding oligomerization domain-containing protein 2 (NOD2; involved in microbial sensing) and apolipoprotein A5 (linked to lipid metabolism), have been associated with specific microbial taxa and metabolic phenotypes (Blekhman et al. 2015). These interactions suggest that genetic predisposition can determine how an individual responds to dietary interventions via microbiome-mediated mechanisms.

Microbial metabolites influence host gene expression at transcriptional and epigenetic levels. SCFAs like butyrate act as epigenetic modulators by inhibiting histone deacetylases, thus altering chromatin accessibility and regulating genes involved in inflammation and differentiation (Krautkramer et al. 2016). Other microbial signals, such as LPS, peptidoglycan fragments, and secondary bile acids, engage with host pattern recognition receptors (PRRs) and nuclear receptors to regulate signaling cascades related to immunity and metabolism (Nicholson et al. 2012). Moreover, miRNAs released by host epithelial cells can influence microbial gene expression, while bacterial small RNAs may impact host mRNA stability. This bidirectional RNA-based crosstalk is an emerging area of microbiome-genome interaction research.

5 Integration of Metabolomics with Nutrigenomics

Nutrigenomics is increasingly being enhanced by the complementary field of metabolomics, which systematically identifies and quantifies small-molecule metabolites in biological samples. Metabolomics provides a dynamic biochemical snapshot of physiological and pathological states, thereby serving as a vital bridge between genotype and phenotype. This integration supports a comprehensive, systems-level understanding of nutrient-gene interactions, offering novel insights into personalized nutrition, disease mechanisms, and predictive health outcomes. Metabolomics involves the thorough analysis of metabolites—the end products of cellular metabolism—within diverse biological matrices such as blood, urine, saliva, and tissue extracts, reflecting responses to environmental factors including diet, physical activity, stress, and toxins. The primary analytical platforms used in metabolomics are nuclear magnetic resonance (NMR) spectroscopy and mass spectrometry, frequently combined with liquid chromatography or gas chromatography to enhance detection and separation capabilities (Patti et al. 2012). Metabolomics can be either targeted, focusing on specific metabolites associated with known pathways, or untargeted, offering a holistic view to uncover unexpected metabolic changes. Tools such as principal component analysis, partial least squares discriminant analysis, and machine learning are widely employed to classify sample groups and identify discriminative metabolites (Saccenti et al. 2014). Public databases like the Human Metabolome Database and the Kyoto Encyclopedia of Genes and Genomes are essential for metabolite annotation and mapping metabolic pathways. Integrating metabolomics with genetic data provides a powerful approach to deciphering the complex interactions between genes and the environment, particularly in the context of diet-related phenotypes. One important method is the identification of metabolite quantitative trait loci, which are genomic regions associated with variations in metabolite concentrations. These loci help to explain interindividual variability in metabolism and dietary response, revealing genetic influences on lipid metabolism, amino acid synthesis, and detoxification pathways (Illig et al. 2010). For instance, SNPs in the fatty acid desaturase 1 (FADS1) gene, encoding fatty acid desaturase, are linked to variations in blood levels of PUFAs, which influence inflammation and cardiovascular risk (Lemaitre et al. 2011). Such information allows for genotype-guided dietary advice, such as tailoring PUFA intake based on FADS1 genotype. A further advancement is the identification of nutrient–gene–metabolite triads—three-way interactions where a genetic variant alters a nutrient's metabolic impact. For example, individuals with the MTHFR C677T polymorphism show impaired folate metabolism, evident by elevated homocysteine levels—a metabolomic marker of disrupted one-carbon metabolism (Donze et al. 2019). Addressing these disruptions through targeted nutritional interventions, such as methylated folate supplementation, can normalize metabolic profiles and reduce disease risks. This integrative approach underscores the transformative potential of combining genomics and metabolomics to refine dietary recommendations and optimize individual health outcomes.

6 Systems Biology and Multi-omics Approaches in Nutrition

Integrating metabolomics with genomics, transcriptomics, proteomics, and microbiomics within a systems biology framework enables a holistic understanding of nutrition-related health and disease. This multi-omics approach is transforming the field of nutritional science from reductionism to complexity, allowing researchers to model biological networks and interactions. Multi-omics integration involves the horizontal (same data type across conditions) and vertical (different data types in the same samples) fusion of datasets. Studies such as the Personalized Nutrition Project and Integrative Human Microbiome Project (iHMP) have successfully used multi-omics to map personalized responses to diet and identify metabolic signatures predictive of glucose tolerance, lipid metabolism, and inflammation (Zeevi et al. 2015). A study by Krug et al. used transcriptomics and metabolomics to investigate the metabolic impact of a high-fat meal in humans, revealing coordinated changes in lipid metabolites and inflammation-related gene expression (Krug et al. 2012). Another investigation linked gut microbial composition with host transcriptomic and metabolomic profiles, illustrating how fiber intake modulates SCFA production and immune-related gene expression (Henriquez et al. 2020).

Beyond traditional omics integration, recent advances have been driven by artificial intelligence (AI) and deep learning, which offer powerful tools for handling high-dimensional multi-omics datasets and uncovering nonlinear gene–nutrient–disease relationships. Machine learning models are increasingly used to predict individual responses to dietary interventions, identify hidden patterns in complex biological networks, and develop personalized nutrition algorithms. Moreover, large-scale longitudinal studies like the UK Biobank provide vast, diverse, and well-phenotyped cohorts essential for validating multi-omics findings, training robust predictive models, and enabling trans-ethnic analyses that enhance generalizability and reduce bias (Sudlow et al. 2015; Topol 2019).

These technological and infrastructural advances underscore the shifting landscape of modern nutrigenomics, where computational integration of omics data and population-scale resources are becoming central to the development of precise, individualized dietary recommendations.

7 Clinical Applications of Nutrigenomics

The integration of genetic insights into nutritional science has started a new era of precision health, where dietary recommendations are tailored to individual genetic profiles. Nutrigenomics plays a pivotal role in preventing and managing chronic diseases by identifying gene–diet interactions that influence susceptibility, progression, and treatment response. From metabolic disorders to cancer and neurodegeneration, understanding how genes and nutrients interact enables targeted interventions to optimize health outcomes (Table 1).

7.1 Obesity

Obesity is a multifactorial condition influenced by genetic, environmental, and behavioral factors. Numerous gene–diet interactions have been identified in obesity risk, particularly involving variants in FTO, melanocortin 4 receptor (MC4R), and transmembrane protein 18 (TMEM18). The FTO rs9939609 polymorphism is one of the most studied obesity-related variants. Individuals with the A allele have higher susceptibility to weight gain, partly due to increased appetite and caloric intake. However, studies have shown that physical activity and dietary interventions, particularly low-fat diets, can attenuate the obesogenic effect of this variant (Livingstone et al. 2016). Additionally, macronutrient composition can differentially impact weight management based on genotype. For example, high-protein diets may be more effective for individuals with certain transcription factor 7-like 2 (TCF7L2) variants, while others may benefit from lower carbohydrate intake (Hansen et al. 2021).

7.2 Type 2 Diabetes Mellitus (T2DM)

Type 2 diabetes is strongly influenced by both genetic predisposition and modifiable lifestyle factors. Key genes implicated in T2DM risk include TCF7L2, peroxisome proliferator-activated receptor gamma (PPARG), solute carrier family 30 (zinc transporter), member 8 (SLC30A8), and potassium inwardly rectifying channel subfamily J member 11 (KCNJ11). The TCF7L2 rs7903146 variant has been consistently associated with increased risk due to its role in insulin secretion and glucose homeostasis. Dietary fiber intake appears to mitigate T2DM risk in individuals with high-risk TCF7L2 genotypes, suggesting that complex carbohydrates and resistant starches may provide protective effects through modulation of gut microbiota and postprandial glucose response (Corella and Ordovás 2015). Moreover, nutrigenomic profiling can help predict individual glycemic responses to foods. For example, microbiome and genetic data could accurately forecast postprandial

Table 1 Gene–nutrient interactions and personalized nutritional strategies across disease areas

Disease area	Gene/ polymorphism	Nutrient(s) involved	Genotype effect/ clinical implication	Nutritional strategy
Obesity	FTO (rs9939609)	Total energy intake	A allele associated with a higher obesity risk	High-protein, Mediterranean or low-fat diet; physical activity; appetite regulation
	MC4R	Macronutrient balance	Affects energy balance and appetite	Same as above
Type 2 diabetes	TCF7L2 (rs7903146)	Fiber, complex carbohydrates	Risk allele affects insulin secretion and glucose metabolism	Low-GI diet; high fiber; modulation of gut microbiota
Cardiovascular disease	APOE (ε2/ε3/ε4)	Saturated fat, cholesterol, omega-3 s	ε4 is associated with higher LDL and CVD risk	Low-saturated-fat diet; increased omega-3 intake
	MTHFR (rs1801133, C677T)	Folate, B-vitamins	TT genotype reduces enzyme activity, increasing homocysteine and CVD risk	Folate and B12 supplementation
Cancer (detoxification)	GSTM1, GSTT1, GSTP1	Isothiocyanates (cruciferous vegetables)	Variants affect detoxification enzyme activity	High cruciferous vegetable intake; phase II enzyme support
Cancer (DNA repair)	XRCC1, OGG1	Vitamins C, E, and carotenoids	Affects DNA repair and oxidative stress response	Antioxidant-rich diet; reduction of oxidative stress
Neurodegeneration	BDNF (Val66Met), VDR variants	Omega-3 s, vitamin D, polyphenols	Variants influence neuroplasticity and vitamin D metabolism	Omega-3 supplementation; vitamin D optimization; polyphenol-rich foods
Caffeine sensitivity	CYP1A2 (rs762551)	Caffeine	A allele = slow metabolizer; risk of hypertension and heart disease with high caffeine intake	Limit caffeine intake in slow metabolizers
Lactose intolerance	LCT (rs4988235)	Lactose	CC genotype = lactase non-persistence (lactose intolerance)	Avoid dairy or use lactose-free alternatives
Starch metabolism	AMY1 (copy number variation)	Starch	High copy number = increased amylase activity; better starch digestion	Starch-rich diets may be better tolerated in individuals with high AMY1 copy number

glucose levels, facilitating the development of personalized dietary interventions to prevent or manage diabetes (Zeevi et al. 2015).

7.3 Cardiovascular Disease

Cardiovascular disease (CVD) encompasses a broad spectrum of disorders, including coronary artery disease, hypertension, and dyslipidemia. Nutrigenomic research has identified key gene–diet interactions influencing lipid metabolism, inflammation, and endothelial function. The APOE gene, which encodes apolipoprotein E, is a central player in lipid transport and cholesterol homeostasis. The three common isoforms—APOE2, APOE3, and APOE4—have distinct impacts on lipid profiles and CVD risk. Individuals carrying the APOE4 allele tend to have elevated LDL cholesterol levels and heightened sensitivity to dietary saturated fat intake. Clinical studies suggest that reducing saturated fat and increasing monounsaturated and polyunsaturated fatty acids can significantly improve lipid profiles in APOE4 carriers (Minihane et al. 2007). Conversely, APOE2 carriers may exhibit a favorable lipid profile but have an increased risk of developing type III hyperlipoproteinemia under certain dietary conditions. Nutrigenomics thus provides actionable insights for tailoring lipid-lowering diets based on APOE genotype. The MTHFR C677T polymorphism affects folate metabolism and homocysteine levels, both of which are linked to cardiovascular risk. Individuals with the TT genotype often have reduced enzyme activity and elevated homocysteine levels, contributing to endothelial dysfunction and atherogenesis (Luo et al. 2018). Folate-rich diets and supplementation with methylated folate or B-vitamins (B6, B12) have been shown to normalize homocysteine levels and may reduce cardiovascular risk in genetically susceptible individuals.

7.4 Cancer

Cancer is a complex disease involving genetic mutations, epigenetic alterations, and environmental exposures. Dietary components can modulate carcinogenesis through multiple pathways, including DNA repair, oxidative stress, inflammation, and detoxification. Glutathione S-transferases (GSTs) are enzymes that detoxify carcinogens and protect cells from oxidative stress. Polymorphisms in glutathione S-transferase Mu 1 (GSTM1), Glutathione S-transferase T1 (GSTT1), and Glutathione S-transferase P1 (GSTP1) affect enzyme activity and influence cancer risk. Consumption of cruciferous vegetables (e.g., broccoli, brussels sprouts, and cabbage) provides glucosinolates, which are converted into isothiocyanates—potent inducers of detoxifying enzymes, including GSTs. Individuals with null variants of GSTM1 or GSTT1 may derive greater benefit from cruciferous vegetable intake due to compensatory upregulation of other detoxification pathways (Abbaoui et al.

2018). Clinical and epidemiological studies suggest that high cruciferous vegetable intake is associated with reduced risk of lung, breast, and colorectal cancers, particularly in individuals with specific GST genotypes (Lampe et al. 2000). Variants in X-Ray Repair Cross Complementing 1 (XRCC1), 8-oxoguanine DNA glycosylase (OGG1), and other DNA repair genes influence cellular resilience to oxidative DNA damage. Diets rich in antioxidants such as vitamins C and E, carotenoids, and polyphenols may reduce cancer risk in individuals with suboptimal DNA repair capacity (Yeh et al. 2005). Nutrigenomic approaches enable the identification of gene–nutrient combinations that offer maximal protective effects against carcinogenesis and provide personalized strategies for cancer prevention.

7.5 Neurodegenerative Diseases

Neurodegenerative disorders such as Alzheimer's disease, Parkinson's disease (PD), and cognitive decline are increasingly linked to gene–diet interactions that affect brain structure, function, and plasticity. Brain-derived neurotrophic factor (BDNF) is a key regulator of synaptic plasticity, neurogenesis, and cognitive function. The BDNF Val66Met polymorphism affects activity-dependent BDNF release and is associated with memory deficits and increased risk of Alzheimer's disease. Dietary and lifestyle factors can modulate BDNF expression. Omega-3 fatty acids, polyphenols (e.g., from berries and green tea), and caloric restriction have been shown to upregulate BDNF and support neuroprotection (Gómez-Pinilla 2008). Individuals with the Met allele may benefit from targeted interventions to enhance BDNF signaling and mitigate cognitive decline. Vitamin D exerts neuroprotective effects through its roles in calcium homeostasis, inflammation, and immune function. The vitamin D receptor (VDR) gene mediates these effects, and polymorphisms such as FokI, BsmI, and TaqI have been associated with altered cognitive performance and neurodegenerative disease risk. Vitamin D deficiency is prevalent in older adults and has been linked to increased risk of dementia and PD. Genotype-specific responses to vitamin D supplementation highlight the importance of personalized dosing to optimize cognitive outcomes (Annweiler et al. 2009).

Despite the promising clinical potential of nutrigenomics, several limitations hinder its widespread application and generalizability. Many genome-wide association studies (GWAS) and twin studies suffer from small sample sizes, limiting statistical power and the ability to detect modest but meaningful gene–diet interactions (Ordovas and Mooser 2004; Manolio et al. 2009). Additionally, population stratification—genetic differences among subpopulations—can confound results, leading to false associations or reduced reproducibility across diverse cohorts (Ioannidis et al. 2001). A further concern is the prevalent reliance on observational associations rather than establishing causality, making it challenging to translate findings into actionable clinical recommendations (Corella and Ordovás 2015). These issues underscore the need for larger, multi-ethnic cohorts, rigorous validation studies, and mechanistic research to strengthen the scientific foundation of nutrigenomics and support its integration into precision nutrition frameworks.

8 Designing Gene-Based Dietary Recommendations

Personalized nutrition stands at the nexus of genomics, lifestyle, and environmental interactions, aiming to optimize dietary interventions based on individual genetic profiles. The central premise is that individuals respond differently to the same diet, and genetic variations play a pivotal role in modulating these responses (Ferguson et al. 2016). Designing gene-based dietary plans requires a multi-step approach, as shown in Fig. 3. First, genetic screening identifies polymorphisms associated with nutrient metabolism, absorption, or physiological outcomes. Notable examples include SNPs in genes like FTO, MTHFR, transcription factor 7-like 2 (TCF7L2), APOE, and CYP1A2, each linked to responses to macronutrients, vitamins, or

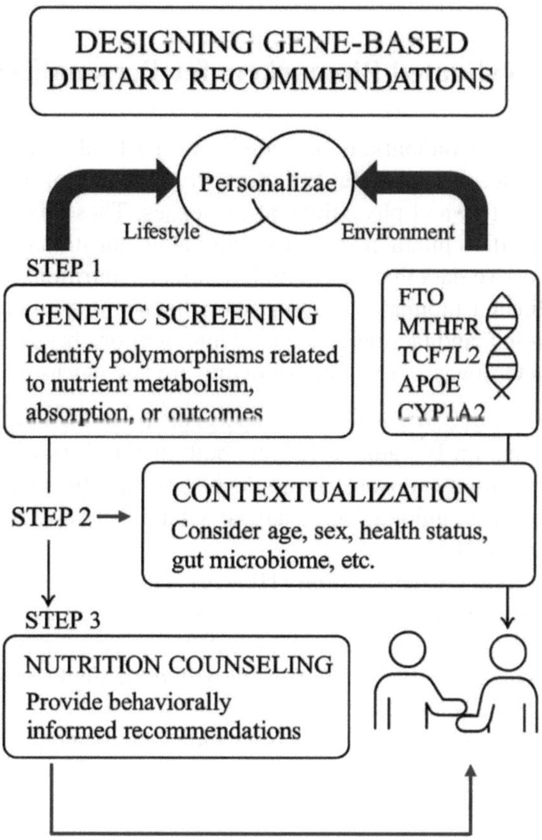

Fig. 3 Designing gene-based dietary recommendations: a stepwise approach to personalized nutrition. The current flowchart illustrates the three core phases of gene-based dietary planning. It begins with genetic screening to identify relevant polymorphisms, integrates contextual factors such as age, microbiome, and lifestyle, and concludes with behaviorally informed nutrition counseling. The diagram emphasizes the cyclical and interconnected nature of these stages in developing effective, personalized dietary strategies

stimulants like caffeine (Ordovas and Taib 2008). For instance, individuals carrying specific variants of the CYP1A2 gene metabolize caffeine more slowly, which may elevate the risk of hypertension or myocardial infarction with high coffee consumption (Cornelis et al. 2006). Once the genetic data is obtained, it must be contextualized within a broader matrix that includes age, sex, health status, gut microbiome profile, and lifestyle behaviors. Tools such as polygenic risk scores (PRS, under active validation) offer a quantitative metric for susceptibility to conditions like obesity or type 2 diabetes and help in framing dietary guidelines that counteract these risks (Livingstone et al. 2021). A PRS indicating a high predisposition to insulin resistance might prompt personalized advice to reduce refined carbohydrate intake and increase dietary fiber. The third phase involves behaviorally informed nutrition counseling. Even the most precise gene-based recommendations are ineffective without adherence. Behavioral interventions such as motivational interviewing, goal setting, and digital health apps tailored to genetic risk profiles improve compliance and long-term health outcomes (Celis-Morales et al. 2017). Integration with mobile health platforms also allows dynamic updates based on biometric data, further enhancing personalization.

9 Validated Biomarkers for Personalized Nutrition

Validated biomarkers are crucial in the field of personalized nutrition, providing reliable, reproducible, and clinically meaningful measures of dietary intake, nutrient status, and physiological responses. These biomarkers help translate complex gene–diet interactions into actionable nutritional recommendations by offering objective data that reflect an individual's nutritional state or risk profile. The use of validated biomarkers strengthens dietary assessments, supports clinical decision-making, and facilitates targeted interventions based on biological evidence (Corella and Ordovás 2015). The following are key nutritional biomarkers with validation:

1. Serum 25-hydroxyvitamin D: This biomarker is the gold standard for assessing vitamin D status. It reflects both dietary intake and endogenous synthesis from sunlight exposure. Serum levels are used to evaluate deficiency, sufficiency, and excess, guiding personalized vitamin D supplementation strategies (Del Valle et al. 2011).
2. Glycated hemoglobin (HbA1c): HbA1c reflects average blood glucose levels over the previous 2–3 months and is widely used to monitor the effectiveness of dietary and pharmacologic interventions in individuals with diabetes or prediabetes. It helps in tailoring carbohydrate intake and monitoring metabolic control (Jones et al. 2012).
3. Serum ferritin: Ferritin is a validated indicator of iron stores and is used to assess iron deficiency and overload. It allows for personalized iron supplementation or dietary adjustments based on individual iron status. However, as an acute-phase

reactant, it must be interpreted carefully in inflammatory conditions (Dragsted et al. 2018).

4. Plasma carotenoids (e.g., β-carotene, lutein, lycopene): These biomarkers reflect fruit and vegetable intake and are commonly used in dietary validation studies. Plasma carotenoid levels correlate with antioxidant status and overall dietary quality (Al-Delaimy et al. 2005).
5. Serum omega-3 index (EPA + DHA in RBC membranes): The omega-3 index reflects long-term intake of omega-3 fatty acids and is associated with cardiovascular risk. A higher index is linked to improved cardiovascular outcomes, and this biomarker is used to guide fish oil supplementation and dietary fat modification (Von Schacky 2014).
6. Urinary sodium and potassium: Twenty-four-hour urinary sodium and potassium excretion are considered gold standards for evaluating salt and potassium intake. These biomarkers help tailor dietary recommendations for blood pressure and cardiovascular risk reduction (Brown et al. 2009).
7. Serum homocysteine: Elevated plasma homocysteine is a marker of B-vitamin deficiencies, particularly folate, vitamin B6, and B12. It is used in personalized nutrition to monitor methylation-related gene–nutrient interactions, especially in individuals with MTHFR polymorphisms (Corella and Ordovás 2015).

Although these biomarkers are validated, they are not without limitations. Many are influenced by nondietary factors such as inflammation (e.g., ferritin), hydration status (e.g., urinary sodium), or genetic variability (e.g., homocysteine). Proper contextual interpretation is critical to avoid misclassification or inappropriate recommendations (Dragsted et al. 2018). Additionally, validated biomarkers must meet criteria such as sensitivity, specificity, reproducibility, and population-level relevance. The Biomarkers of Nutrition for Development (BOND) program and other initiatives continue to evaluate and promote global standards for biomarker validation (King et al. 2016).

10 Engaging the Public in Personalized Nutrition

The successful implementation of personalized nutrition depends not only on scientific innovation but also on public awareness, acceptance, and engagement. Nutrigenomics—a complex field that blends genetics, nutrition, and individualized health strategies—must be communicated effectively to the public to gain broader societal support. Ensuring that individuals understand the benefits, limitations, and ethical considerations of personalized nutrition is crucial for maximizing its potential and fostering trust in its application.

10.1 Importance of Public Engagement

1. Improving health literacy: As personalized nutrition relies on genomic, metabolic, and behavioral data, individuals must have a baseline understanding of genetic variation and its impact on diet and disease. Health literacy enables informed decision-making and increases the likelihood of adherence to personalized dietary recommendations (De Roos 2013).
2. Building trust and transparency: Public concerns regarding data privacy, genetic discrimination, and the commercial use of personal health information can hinder participation. Transparent communication about data handling, consent, and the intended use of nutrigenomic findings is essential for building trust (Nielsen and El-Sohemy 2014).
3. Empowering individuals: When individuals are engaged in their own health journeys, they are more likely to make sustainable lifestyle changes. Providing actionable insights derived from their genetic and metabolic profiles empowers them to take control of their nutrition and health (Mathers 2017).

10.2 Strategies for Public Education and Engagement

To promote acceptance and participation in nutrigenomic-based initiatives, a combination of communication strategies and educational frameworks is needed:

1. Simplifying scientific communication

   Complex genomic concepts should be translated into clear, relatable messages. Using analogies, infographics, interactive tools, and storytelling helps demystify personalized nutrition and increases comprehension among diverse populations (Treweek et al. 2013).
2. Incorporating personalized nutrition into school and community programs

   Integrating genetics and nutrition education into school curricula, public health campaigns, and community workshops can foster early awareness. Programs that emphasize the role of genes in shaping dietary needs may help normalize genetic testing and personalized dietary planning (Aruoma et al. 2019).
3. Leveraging digital platforms

   Mobile health apps, social media, and personalized nutrition websites offer scalable ways to reach the public. These platforms can deliver tailored information, track user engagement, and provide interactive features that enhance motivation and behavior change.
4. Promoting participatory research and citizen science

   Encouraging public participation in nutrigenomic research—such as through biobanking, dietary tracking, and microbiome studies—can increase ownership of scientific discovery. Citizen science initiatives empower individuals to contribute data and benefit from findings, fostering a collaborative model of health innovation.

5. Engaging health-care providers and nutrition professionals
   As trusted intermediaries, clinicians and dietitians play a pivotal role in educating patients about the implications of nutrigenomic testing. Training professionals to communicate genomic concepts in an accessible language and interpret genetic data responsibly ensures that personalized nutrition is integrated effectively into health care (Grimaldi et al. 2017).

10.3 Concerns and Future Directions

Despite the promise of personalized nutrition, public engagement with nutrigenomics faces significant barriers such as limited access to genetic testing due to high costs or inadequate infrastructure, concerns over genetic privacy and the potential misuse of health data, and cultural skepticism toward biotechnology and individualized medicine. To overcome these challenges, it is crucial to uphold ethical principles of autonomy, justice, and beneficence, ensuring that individuals are fully informed and their rights protected. Public education campaigns should not only communicate the benefits of personalized nutrition but also emphasize the safeguards in place to ensure privacy and data security (Bloss et al. 2011). Looking ahead, greater investment in public outreach and science communication is essential to build trust and promote informed participation. Collaborative efforts among governments, educational institutions, and healthcare organizations are needed to develop inclusive strategies that engage vulnerable and underrepresented populations. Co-designing interventions with communities and stakeholders fosters cultural sensitivity and enhances the relevance of personalized approaches. Furthermore, establishing transparent data governance frameworks, ensuring equitable access to genomic services, and implementing clear regulations for commercial genetic tests will be critical to strengthening public confidence and maximizing the societal benefits of nutrigenomic innovations.

11 Ethical, Legal, and Social Implications

As nutrigenomics continues to evolve and personalized nutrition becomes more embedded in healthcare systems, it raises complex ethical, legal, and social implications (ELSI). Grounding these concerns in foundational bioethical principles—autonomy, beneficence, justice, and nonmaleficence—is critical for ensuring that innovation progresses responsibly and equitably.

One of the most pressing concerns is the protection of genetic privacy and data security. With the proliferation of direct-to-consumer (DTC) and clinical genetic testing platforms, individuals increasingly share their DNA without fully understanding the implications. Genetic information is unique and immutable—once compromised, it cannot be changed, raising serious questions about autonomy and informed consent. Studies have shown that some commercial companies share consumer data with third parties for marketing or research, often without adequate transparency or consent mechanisms (Tandy-Connor et al. 2018). Additionally, advances in data analytics have made it possible to re-identify anonymized genetic datasets, undermining promises of privacy (Gymrek et al. 2013).

Although frameworks like the General Data Protection Regulation (GDPR) in the European Union and the Health Insurance Portability and Accountability Act (HIPAA) in the United States offer some protection, global oversight remains fragmented and inconsistent (Phillips 2016). To uphold the principle of autonomy, policy reforms should include mandatory transparent consent agreements from all DTC platforms and require user-friendly disclosures about data use, storage, and sharing.

The ethical principle of justice demands equitable representation in genomic research and personalized nutrition initiatives. However, most existing genomic datasets are disproportionately composed of individuals of European ancestry, limiting the applicability of nutrigenomic findings to diverse populations (Lambert 2019; Sirugo et al. 2019). This lack of inclusivity risks exacerbating existing health disparities by offering less effective or irrelevant dietary recommendations to underrepresented groups.

To address this, research initiatives should be incentivized to include ethnically diverse populations, and funding bodies could prioritize proposals that demonstrate inclusive sampling strategies. Moreover, public health frameworks must work toward integrating personalized nutrition without marginalizing communities lacking access to genetic testing or follow-up care.

The rapid growth of DTC nutrigenomic testing has outpaced regulatory mechanisms. Many companies claim to provide highly personalized diet plans based on genetic profiles, yet the clinical validity and utility of such services remain variable and largely unverified. Discrepancies in test interpretation between providers can result in confusion, unnecessary dietary restrictions, or misplaced reassurance (Hogarth and Saukko 2017), contravening the principle of nonmaleficence.

While agencies like the U.S. Food and Drug Administration (FDA) and the European Food Safety Authority (EFSA) are developing clearer guidelines, enforcement remains limited, especially for international or online services. Policy recommendations include the implementation of standardized labeling or certification systems for DTC nutrigenomic services to ensure quality, accuracy, and consumer protection. Genetic literacy programs should also be expanded to help consumers interpret their results critically.

Despite promising advances, many gene–diet interactions remain scientifically uncertain. Traits such as obesity, diabetes, and metabolic syndrome are polygenic and environmentally influenced, which limits the predictive power of single nucleotide polymorphisms (SNPs) (Wang and Hu 2018). This raises concerns under the principle of beneficence, as premature or oversimplified recommendations may not truly serve individuals' health interests.

To strengthen the scientific foundation of nutrigenomics, rigorous study designs—such as longitudinal analyses and randomized controlled trials—are essential (Livingstone et al. 2016). Additionally, polygenic risk scoring models need refinement, and data from diverse populations should be integrated into algorithm development. Global collaborations like the Global Alliance for Genomics and Health are helping facilitate standardized and ethical data sharing (Knoppers 2014).

The integration of artificial intelligence (AI), machine learning, and digital health tools offers exciting possibilities for real-time, personalized dietary guidance. Wearables, mobile health apps, and blockchain systems may enhance individual empowerment, security, and engagement (Zeevi et al. 2015; Topol 2019). However, these innovations bring new ethical considerations. Algorithmic bias poses a significant risk if tools are trained on nonrepresentative datasets, potentially perpetuating existing health inequities. To align with the principle of justice, AI-driven dietary tools must be reviewed to ensure fairness, inclusivity, and transparency in their design and application. Oversight committees, auditing protocols, and regulatory guidelines will be essential to monitor and mitigate algorithmic bias.

The future of nutrigenomics depends not only on scientific breakthroughs but also on robust ethical governance. By grounding policy and innovation in bioethical principles and implementing concrete regulatory measures—such as transparent consent policies, inclusive research funding, and certification standards for DTC services—the field can advance in ways that are both scientifically credible and socially responsible.

12 Conclusion

In summary, nutrigenomics bridges diet, genes, and the microbiome, revealing how individual genetic makeup and gut microbes influence responses to nutrients and overall health. The gut microbiome acts as a key mediator between diet and gene expression by producing metabolites that regulate immune function, metabolism, and epigenetic modifications. Gene–nutrient interactions are highly individualized, and common genetic variants can affect metabolism, disease risk, and optimal dietary needs. Personalized nutrition is the future of healthcare, using genetic and microbiome data to tailor dietary recommendations that improve disease prevention and treatment outcomes.

References

Abbaoui B, Lucas CR, Riedl KM, Clinton SK, Mortazavi A (2018) Cruciferous vegetables, isothiocyanates, and bladder cancer prevention. Mol Nutr Food Res 62:1800079. https://doi.org/10.1002/mnfr.201800079

Afman L, Müller M (2006) Nutrigenomics: from molecular nutrition to prevention of disease. J Am Diet Assoc 106:569–576. https://doi.org/10.1016/j.jada.2006.02.017

Al-Delaimy W, Ferrari P, Slimani N, Pala V, Johansson I, Nilsson S, Mattisson I, Wirfalt E, Galasso R, Palli D (2005) Plasma carotenoids as biomarkers of intake of fruits and vegetables: individual-level correlations in the European Prospective Investigation into Cancer and Nutrition (EPIC). Eur J Clin Nutr 59:1387–1396. https://doi.org/10.1038/sj.ejcn.1602242

Ali RO, Quinn GM, Umarova R, Haddad JA, Zhang GY, Townsend EC, Scheuing L, Hill KL, Gewirtz M, Rampertaap S (2023) Longitudinal multi-omics analyses of the gut–liver axis reveals metabolic dysregulation in hepatitis C infection and cirrhosis. Nat Microbiol 8:12–27. https://doi.org/10.1038/s41564-022-01234-5

Ali RO, Haddad JA, Quinn GM, Zhang GY, Townsend E, Scheuing L, Hill KL, Menkart M, Oringher JL, Umarova R (2024) Taurine-conjugated bile acids and their link to hepatic S1PR2 play a significant role in hepatitis C–related liver disease. Hepatol Commun 8:e0478. https://doi.org/10.1002/hep4.478

Annweiler C, Allali G, Allain P, Bridenbaugh S, Schott AM, Kressig RW, Beauchet O (2009) Vitamin D and cognitive performance in adults: a systematic review. Eur J Neurol 16:1083–1089. https://doi.org/10.1111/j.1468-1331.2009.02703.x

Aruoma OI, Hausman-Cohen S, Pizano J, Schmidt MA, Minich DM, Joffe Y, Brandhorst S, Evans SJ, Brady DM (2019) Personalized nutrition: translating the science of nutrigenomics into practice: proceedings from the 2018 American College of Nutrition Meeting. J Am Coll Nutr 38:287–301. https://doi.org/10.1080/07315724.2019.1580120

Bailey LB, Gregory JF III (1999) Folate metabolism and requirements. J Nutr 129:779–782. https://doi.org/10.1093/jn/129.5.779

Barbosa AC, Pfister KE, Chiba T, Bons J, Rose JP, Burton JB, King CD, O'Broin A, Young V, Zhang B (2024) Dicarboxylic acid dietary supplementation protects against acute kidney injury. J Am Soc Nephrol 10:1681. https://doi.org/10.1681/ASN.20240168

Belkaid Y, Hand TW (2014) Role of the microbiota in immunity and inflammation. Cell 157:121–141. https://doi.org/10.1016/j.cell.2014.03.011

Blekham R, Goodrich JK, Huang K, Sun Q, Bukowski R, Bell JT, Spector TD, Keinan A, Ley RE, Gevers D (2015) Host genetic variation impacts microbiome composition across human body sites. Genome Biol 16:1–12. https://doi.org/10.1186/s13059-015-0703-0

Bloss CS, Schork NJ, Topol EJ (2011) Effect of direct-to-consumer genomewide profiling to assess disease risk. N Engl J Med 364:524–534. https://doi.org/10.1056/NEJMoa1011893

Brown IJ, Tzoulaki I, Candeias V, Elliott P (2009) Salt intakes around the world: implications for public health. Int J Epidemiol 38:791–813. https://doi.org/10.1093/ije/dyp139

Canani RB, Di Costanzo M, Leone L, Pedata M, Meli R, Calignano A (2011) Potential beneficial effects of butyrate in intestinal and extraintestinal diseases. World J Gastroenterol 17:1519. https://doi.org/10.3748/wjg.v17.i12.1519

Cardona F, Andrés-Lacueva C, Tulipani S, Tinahones FJ, Queipo-Ortuño MI (2013) Benefits of polyphenols on gut microbiota and implications in human health. J Nutr Biochem 24:1415–1422. https://doi.org/10.1016/j.jnutbio.2013.05.001

Castillo DJ, Rifkin RF, Cowan DA. Potgieter M(2019) The healthy human blood microbiome: fact or fiction? Front Cell Infect Microbiol 9:148. https://doi.org/10.3389/fcimb.2019.00148

Celis-Morales C, Livingstone KM, Marsaux CF, Macready AL, Fallaize R, O'Donovan CB, Woolhead C, Forster H, Walsh MC, Navas-Carretero S (2017) Effect of personalized nutrition on health-related behaviour change: evidence from the Food4Me European randomized controlled trial. Int J Epidemiol 46:578–588. https://doi.org/10.1093/ije/dyw186

Chassaing B, Koren O, Goodrich JK, Poole AC, Srinivasan S, Ley RE, Gewirtz AT (2015) Dietary emulsifiers impact the mouse gut microbiota promoting colitis and metabolic syndrome. Nature 519:92–96. https://doi.org/10.1038/nature14232

Chen J, Li Y, Tian Y, Huang C, Li D, Zhong Q, Ma X (2015) Interaction between microbes and host intestinal health: modulation by dietary nutrients and gut-brain-endocrine-immune axis. Curr Protein Pept Sci 16:592–603. https://doi.org/10.2174/1389203716666150320104323

Collins FS, Green ED, Guttmacher AE, Guyer MS, Institute UNHGR (2003) A vision for the future of genomics research. Nature 422:835–847. https://doi.org/10.1038/nature01626

Corella D, Ordovás JM (2015) Biomarkers: background, classification and guidelines for applications in nutritional epidemiology. Nutr Hosp 31:177–188. https://doi.org/10.3305/nh.2015.31.1.7630

Cornelis MC, El-Sohemy A, Kabagambe EK, Campos H (2006) Coffee, CYP1A2 genotype, and risk of myocardial infarction. JAMA 295:1135–1141. https://doi.org/10.1001/jama.295.10.1135

Crider KS, Bailey LB, Berry RJ (2011) Folic acid food fortification—its history, effect, concerns, and future directions. Nutrients 3:370–384. https://doi.org/10.3390/nu3040370

David LA, Maurice CF, Carmody RN, Gootenberg DB, Button JE, Wolfe BE, Ling AV, Devlin AS, Varma Y, Fischbach MA (2014) Diet rapidly and reproducibly alters the human gut microbiome. Nature 505:559–563. https://doi.org/10.1038/nature12820

Davie JR (2003) Inhibition of histone deacetylase activity by butyrate. J Nutr 133:2485S–2493S. https://doi.org/10.1093/jn/133.7.2485S

De Roos B (2013) Personalised nutrition: ready for practice? Proc Nutr Soc 72:48–52. https://doi.org/10.1017/S0029665112002895

De Toro-Martín J, Arsenault BJ, Després JP, Vohl MC (2017) Precision nutrition: a review of personalized nutritional approaches for the prevention and management of metabolic syndrome. Nutrients 9:913. https://doi.org/10.3390/nu9090913

Del Valle HB, Yaktine AL, Taylor CL, Ross AC (2011) Dietary reference intakes for calcium and vitamin D. https://doi.org/10.17226/13050

Dharshini LCP, Mandal AKA (2024) Regulation of gene expression by modulating microRNAs through Epigallocatechin-3-gallate in cancer. Mol Biol Rep 51:230. https://doi.org/10.1007/s11033-024-08555-3

Donze SH, De Weerd AW, Van den Bossche RA, Joosten KF, Hokken-Koelega AC (2019) Sleep-related breathing disorders in young adults with Prader-Willi syndrome: a placebo-controlled, crossover GH trial. J Clin Endocrinol Metabol 104:3931–3938. https://doi.org/10.1210/jc.2019-00412

Dragsted LO, Gao Q, Scalbert A, Vergères G, Kolehmainen M, Manach C, Brennan L, Afman LA, Wishart DS, Andres Lacueva C (2018) Validation of biomarkers of food intake—critical assessment of candidate biomarkers. Genes Nutr 13:14. https://doi.org/10.1186/s12263-018-0605-1

Enattah NS, Sahi T, Savilahti E, Terwilliger JD, Peltonen L, Järvelä I (2002) Identification of a variant associated with adult-type hypolactasia. Nat Genet 30:233–237. https://doi.org/10.1038/ng826

Ferguson LR, De Caterina R, Görman U, Allayee H, Kohlmeier M, Prasad C, Choi MS, Curi R, De Luis DA, Gil Á (2016) Guide and position of the international society of nutrigenetics/nutrigenomics on personalised nutrition: part 1-fields of precision nutrition. Lifestyle Genomics 9:12–27. https://doi.org/10.1159/000441022

Flint HJ, Scott KP, Duncan SH, Louis P, Forano E (2012) Microbial degradation of complex carbohydrates in the gut. Gut Microbes 3:289–306. https://doi.org/10.4161/gmic.19897

Froese DS, Fowler B, Baumgartner MR (2019) Vitamin B12, folate, and the methionine remethylation cycle—biochemistry, pathways, and regulation. J Inherit Metab Dis 42:673–685. https://doi.org/10.1002/jimd.12026

Furusawa Y, Obata Y, Fukuda S, Endo TA, Nakato G, Takahashi D, Nakanishi Y, Uetake C, Kato K, Kato T (2013) Commensal microbe-derived butyrate induces the differentiation of colonic regulatory T cells. Nature 504:446–450. https://doi.org/10.1038/nature12721

Garcia-Rios A, Perez-Martinez P, Delgado-Lista J, Lopez-Miranda J, Perez-Jimenez F (2012) Nutrigenetics of the lipoprotein metabolism. Mol Nutr Food Res 56:171–183. https://doi.org/10.1002/mnfr.201100496

Gerasimidis K, Bryden K, Chen X, Papachristou E, Verney A, Roig M, Hansen R, Nichols B, Papadopoulou R, Parrett A (2020) The impact of food additives, artificial sweeteners and domestic hygiene products on the human gut microbiome and its fibre fermentation capacity. Eur J Nutr 59:3213–3230. https://doi.org/10.1007/s00394-020-02298-3

Gholami M (2024) FTO is a major genetic link between breast cancer, obesity, and diabetes. Breast Cancer Res Treat 204:159–169. https://doi.org/10.1007/s10549-024-07398-1

Goetzman ES, Zhang BB, Zhang Y, Bharathi SS, Bons J, Rose J, Shah S, Solo KJ, Schmidt AV, Richert AC (2024) Dietary dicarboxylic acids provide a nonstorable alternative fat source that protects mice against obesity. J Clin Invest 134. https://doi.org/10.1172/JCI169349

Gómez-Pinilla F (2008) Brain foods: the effects of nutrients on brain function. Nat Rev Neurosci 9:568–578. https://doi.org/10.1038/nrn2421

Grimaldi KA, Van Ommen B, Ordovas JM, Parnell LD, Mathers JC, Bendik I, Brennan L, Celis-Morales C, Cirillo E, Daniel H (2017) Proposed guidelines to evaluate scientific validity and evidence for genotype-based dietary advice. Genes Nutr 12:35. https://doi.org/10.1186/s12263-017-0575-8

Gymrek M, McGuire AL, Golan D, Halperin E, Erlich Y (2013) Identifying personal genomes by surname inference. Science 339:321–324. https://doi.org/10.1126/science.1229566

Haag LM, Siegmund B (2015) Intestinal microbiota and the innate immune system–a cross-talk in Crohn's disease pathogenesis. Front Immunol 6:489. https://doi.org/10.3389/fimmu.2015.00489

Hansen TT, Astrup A, Sjödin A (2021) Are dietary proteins the key to successful body weight management? A systematic review and meta-analysis of studies assessing body weight outcomes after interventions with increased dietary protein. Nutrients 13:3193. https://doi.org/10.3390/nu13103193

Henriquez T, Lenzi A, Baldi A, Marvasi M (2020) Frontiers in plant breeding: perspectives for the selection of vegetables less susceptible to enteric pathogens. Front Microbiol 11:1087. https://doi.org/10.3389/fmicb.2020.01087

Hogarth S, Saukko P (2017) A market in the making: the past, present and future of direct-to-consumer genomics. Taylor & Francis. https://doi.org/10.4324/9781315185738

Howitz KT, Bitterman KJ, Cohen HY, Lamming DW, Lavu S, Wood JG, Zipkin RE, Chung P, Kisielewski A, Zhang LL (2003) Small molecule activators of sirtuins extend Saccharomyces cerevisiae lifespan. Nature 425:191–196. https://doi.org/10.1038/nature01960

Illig T, Gieger C, Zhai G, Römisch-Margl W, Wang-Sattler R, Prehn C, Altmaier E, Kastenmüller G, Kato BS, Mewes HW (2010) A genome-wide perspective of genetic variation in human metabolism. Nat Genet 42:137–141. https://doi.org/10.1038/ng.507

Ioannidis J, Ntzani EE, Trikalinos TA, Contopoulos-Ioannidis DG (2001) Replication validity of genetic association studies. Nat Genet 29:306–309. https://doi.org/10.1038/ng1001-306

Jones DP, Park Y, Ziegler TR (2012) Nutritional metabolomics: progress in addressing complexity in diet and health. Annu Rev Nutr 32:183–202. https://doi.org/10.1146/annurev-nutr-071811-150608

Kao D, Roach B, Silva M, Beck P, Rioux K, Kaplan GG, Chang HJ, Coward S, Goodman KJ, Xu H (2017) Effect of oral capsule–vs colonoscopy-delivered fecal microbiota transplantation on recurrent Clostridium difficile infection: a randomized clinical trial. JAMA 318:1985–1993. https://doi.org/10.1001/jama.2017.15391

King JC, Brown KH, Gibson RS, Krebs NF, Lowe NM, Siekmann JH, Raiten DJ (2016) Biomarkers of nutrition for development (BOND)—zinc review. J Nutr 146:858S–885S. https://doi.org/10.3945/jn.115.220079

Knoppers BM (2014) Framework for responsible sharing of genomic and health-related data. HUGO J 8:3. https://doi.org/10.1007/s11568-014-9160-0

Koh A, De Vadder F, Kovatcheva-Datchary P, Bäckhed F (2016) From dietary fiber to host physiology: short-chain fatty acids as key bacterial metabolites. Cell 165:1332–1345. https://doi.org/10.1016/j.cell.2016.05.041

Koh A, Molinaro A, Ståhlman M, Khan MT, Schmidt C, Mannerås-Holm L, Wu H, Carreras A, Jeong H, Olofsson LE (2018) Microbially produced imidazole propionate impairs insulin signaling through mTORC1. Cell 175(947–961):e17. https://doi.org/10.1016/j.cell.2018.09.055

Kouzarides T (2007) Chromatin modifications and their function. Cell 128:693–705. https://doi.org/10.1016/j.cell.2007.02.005

Krautkramer KA, Kreznar JH, Romano KA, Vivas EI, Barrett-Wilt GA, Rabaglia ME, Keller MP, Attie AD, Rey FE, Denu JM (2016) Diet-microbiota interactions mediate global epigenetic programming in multiple host tissues. Mol Cell 64:982–992. https://doi.org/10.1016/j.molcel.2016.10.025

Krug S, Kastenmüller G, Stückler F, Rist MJ, Skurk T, Sailer M, Raffler J, Römisch-Margl W, Adamski J, Prehn C (2012) The dynamic range of the human metabolome revealed by challenges. FASEB J 26:2607–2619. https://doi.org/10.1096/fj.11-192133

Lambert J (2019) Huge US government study will offer genetic counselling. Nature Publishing Group, London. https://doi.org/10.1038/d41586-019-02952-7

Lampe JW, King IB, Li S, Grate MT, Barale KV, Chen C, Feng Z, Potter JD (2000) Brassica vegetables increase and apiaceous vegetables decrease cytochrome P450 1A2 activity in humans: changes in caffeine metabolite ratios in response to controlled vegetable diets. Carcinogenesis 21:1157–1162. https://doi.org/10.1093/carcin/21.6.1157

LeBlanc JG, Milani C, De Giori GS, Sesma F, Van Sinderen D, Ventura M (2013) Bacteria as vitamin suppliers to their host: a gut microbiota perspective. Curr Opin Biotechnol 24:160–168. https://doi.org/10.1016/j.copbio.2012.08.005

Lemaitre RN, Tanaka T, Tang W, Manichaikul A, Foy M, Kabagambe EK, Nettleton JA, King IB, Weng LC, Bhattacharya S (2011) Genetic loci associated with plasma phospholipid n-3 fatty acids: a meta-analysis of genome-wide association studies from the CHARGE Consortium. PLoS Genet 7:e1002193. https://doi.org/10.1371/journal.pgen.1002193

Link A, Balaguer F, Goel A (2010) Cancer chemoprevention by dietary polyphenols: promising role for epigenetics. Biochem Pharmacol 80:1771–1792. https://doi.org/10.1016/j.bcp.2010.06.036

Livingstone KM, Celis-Morales C, Papandonatos GD, Erar B, Florez JC, Jablonski KA, Razquin C, Marti A, Heianza Y, Huang T (2016) FTO genotype and weight loss: systematic review and meta-analysis of 9563 individual participant data from eight randomised controlled trials. BMJ 354. https://doi.org/10.1136/bmj.i4707

Livingstone KM, Celis-Morales C, Navas-Carretero S, San-Cristobal R, Forster H, Woolhead C, O'Donovan CB, Moschonis G, Manios Y, Traczyk I (2021) Personalised nutrition advice reduces intake of discretionary foods and beverages: findings from the Food4Me randomised controlled trial. Int J Behav Nutr Phys Act 18:1–12. https://doi.org/10.1186/s12966-021-01146-0

Luo Z, Lu Z, Muhammad I, Chen Y, Chen Q, Zhang J, Song Y (2018) Associations of the MTHFR rs1801133 polymorphism with coronary artery disease and lipid levels: a systematic review and updated meta-analysis. Lipids Health Dis 17:1–15. https://doi.org/10.1186/s12944-018-0784-7

Lynch SV, Pedersen O (2016) The human intestinal microbiome in health and disease. N Engl J Med 375:2369–2379. https://doi.org/10.1056/NEJMra1600266

Manolio TA, Collins FS, Cox NJ, Goldstein DB, Hindorff LA, Hunter DJ, McCarthy MI, Ramos EM, Cardon LR, Chakravarti A (2009) Finding the missing heritability of complex diseases. Nature 461:747–753. https://doi.org/10.1038/nature08494

Mathers JC (2017) Nutrigenomics in the modern era. Proc Nutr Soc 76:265–275. https://doi.org/10.1017/S0029665116000667

McNulty H, Strain J, Hughes CF, Ward M (2017) Riboflavin, MTHFR genotype and blood pressure: a personalized approach to prevention and treatment of hypertension. Mol Asp Med 53:2–9. https://doi.org/10.1016/j.mam.2016.08.001

Minihane A, Jofre-Monseny L, Olano-Martin E, Rimbach G (2007) ApoE genotype, cardiovascular risk and responsiveness to dietary fat manipulation: symposium on 'molecular basis for diseases'. Proc Nutr Soc 66:183–197. https://doi.org/10.1017/S0029665107005510

Mokhtari H, Yaghmaei B, Sirati-Sabet M, Jafari N, Mardomi A, Abediankenari S, Mahrooz A (2020) Epigallocatechin-3-gallate enhances the efficacy of MicroRNA-34a mimic and MiR-93 inhibitor co-transfection in prostate cancer cell line. Iran J Allergy Asthma Immunol 19:1–10. https://doi.org/10.18502/ijaai.v19i1.4116

Moore LD, Le T, Fan G (2013) DNA methylation and its basic function. Neuropsychopharmacology 38:23–38. https://doi.org/10.1038/npp.2012.112

Müller M, Kersten S (2003) Nutrigenomics: goals and strategies. Nat Rev Genet 4:315–322. https://doi.org/10.1038/nrg1048

Nganou-Makamdop K, Talla A, Sharma AA, Darko S, Ransier A, Laboune F, Chipman JG, Beilman GJ, Hoskuldsson T, Fourati S (2021) Translocated microbiome composition determines immunological outcome in treated HIV infection. Cell 184: 3899–3914 e16. https://doi.org/10.1016/j.cell.2021.05.019

Nicholson JK, Holmes E, Kinross J, Burcelin R, Gibson G, Jia W, Pettersson S (2012) Host-gut microbiota metabolic interactions. Science 336:1262–1267. https://doi.org/10.1126/science.1223813

Niculescu MD, Lupu DS (2011) Nutritional influence on epigenetics and effects on longevity. Curr Opin Clin Nutr Metab Care 14:35–40. https://doi.org/10.1097/MCO.0b013e3283426d2a

Nielsen DE, El-Sohemy A (2014) Disclosure of genetic information and change in dietary intake: a randomized controlled trial. PLoS One 9:e112665. https://doi.org/10.1371/journal.pone.0112665

Nindrea RD, Thongwichian P (2024) FTO gene polymorphisms rs9939609 and the risk of obesity among adults in Western and Asian countries: a systematic review and meta-analysis. Clin Epidemiol Glob Health:101621. https://doi.org/10.1016/j.cegh.2024.101621

Ordovas JM, Mooser V (2004) Nutrigenomics and nutrigenetics. Curr Opin Lipidol 15:101–108. https://doi.org/10.1097/00041433-200404000-00003

Ordovas J, Taib E (2008) Curr Opin Lipidol 19:1–2. https://doi.org/10.1097/MOL.0b013e3282f0c14a

Patti GJ, Yanes O, Siuzdak G (2012) Metabolomics: the apogee of the omics trilogy. Nat Rev Mol Cell Biol 13:263–269. https://doi.org/10.1038/nrm3314

Pérez-Gimeno G, Seral-Cortes M, Sabroso-Lasa S, Esteban LM, Widhalm K, Gottrand F, Stehle P, Meirhaeghe A, Muntaner M, Kafatos A (2024) Interplay of the Mediterranean diet and genetic hypertension risk on blood pressure in European adolescents: findings from the HELENA study. Eur J Pediatr 183:2101–2110. https://doi.org/10.1007/s00431-024-05072-x

Perry GH, Dominy NJ, Claw KG, Lee AS, Fiegler H, Redon R, Werner J, Villanea FA, Mountain JL, Misra R (2007) Diet and the evolution of human amylase gene copy number variation. Nat Genet 39:1256–1260. https://doi.org/10.1038/ng2123

Phillips CM (2013) Nutrigenetics and metabolic disease: current status and implications for personalised nutrition. Nutrients 5:32–57. https://doi.org/10.3390/nu5010032

Phillips AM (2016) Only a click away—DTC genetics for ancestry, health, love… and more: a view of the business and regulatory landscape. Appl Transl Genomics 8:16–22. https://doi.org/10.1016/j.atg.2016.04.001

Quinn G, Ali RO, Zhang GY, Hill K, Townsend E, Umarova R, Chakraborty M, Ahmad MF, Gewirtz M, Haddad J (2023) Non-selective dampening of the host immune response after hepatitis C clearance and its association with circulating chemokine and endotoxin levels. Liver Int 43:2701–2712. https://doi.org/10.1111/liv.15746

Rausch P, Rehman A, Künzel S, Häsler R, Ott SJ, Schreiber S, Rosenstiel P, Franke A, Baines JF (2011) Colonic mucosa-associated microbiota is influenced by an interaction of Crohn disease and FUT2 (Secretor) genotype. Proc Natl Acad Sci 108: 19030–19035. https://doi.org/10.1073/pnas.1108694108

Rooney M, Hughes CF, Strain J, Clements M, McNulty H, Ward M (2022) Impact of the MTHFR C677T polymorphism on blood pressure and related central haemodynamic parameters in healthy adults. J Hum Nutr Diet 35:689–700. https://doi.org/10.1111/jhn.12988

Saccenti E, Hoefsloot HC, Smilde AK, Westerhuis JA, Hendriks MM (2014) Reflections on univariate and multivariate analysis of metabolomics data. Metabolomics 10:361–374. https://doi.org/10.1007/s11306-013-0577-6

Santilli F, Davi G, Patrono C (2016) Homocysteine, methylenetetrahydrofolate reductase, folate status and atherothrombosis: a mechanistic and clinical perspective. Vasc Pharmacol 78:1–9. https://doi.org/10.1016/j.vph.2015.12.005

Schierwagen R, Gu W, Brieger A, Brüne B, Ciesek S, Đikić I, Dimmeler S, Geisslinger G, Greten FR, Hermann E (2023) Pathogenetic mechanisms and therapeutic approaches of acute-to-chronic liver failure. Am J Phys Cell Phys 325:C129–C140. https://doi.org/10.1152/ajpcell.00315.2022

Seral-Cortes M, Larruy-García A, De Miguel-Etayo P, Labayen I, Moreno LA (2022) Mediterranean diet and genetic determinants of obesity and metabolic syndrome in European children and adolescents. Genes 13:420. https://doi.org/10.3390/genes13030420

Sharma RA, Gescher AJ, Steward WP (2005) Curcumin: the story so far. Eur J Cancer 41:1955–1968. https://doi.org/10.1016/j.ejca.2005.05.009

Sheaffer KL, Elliott EN, Kaestner KH (2016) DNA hypomethylation contributes to genomic instability and intestinal cancer initiation. Cancer Prev Res 9:534–546. https://doi.org/10.1158/1940-6207.CAPR-15-0349

Singh RK, Chang HW, Yan D, Lee KM, Ucmak D, Wong K, Abrouk M, Farahnik B, Nakamura M, Zhu TH (2017) Influence of diet on the gut microbiome and implications for human health. J Transl Med 15:1–17. https://doi.org/10.1186/s12967-017-1175-y

Sirugo G, Williams SM, Tishkoff SA (2019) The missing diversity in human genetic studies. Cell 177:26–31. https://doi.org/10.1016/j.cell.2019.02.048

Smith GD, Lawlor DA, Timpson NJ, Baban J, Kiessling M, Day IN, Ebrahim S (2009) Lactase persistence-related genetic variant: population substructure and health outcomes. Eur J Hum Genet 17:357–367. https://doi.org/10.1038/ejhg.2008.178

Stover PJ (2009) One-carbon metabolism–genome interactions in folate-associated pathologies. J Nutr 139:2402–2405. https://doi.org/10.3945/jn.109.108287

Sudlow C, Gallacher J, Allen N, Beral V, Burton P, Danesh J, Downey P, Elliott P, Green J, Landray M (2015) UK biobank: an open access resource for identifying the causes of a wide range of complex diseases of middle and old age. PLoS Med 12:e1001779. https://doi.org/10.1371/journal.pmed.1001779

Tandy-Connor S, Guiltinan J, Krempely K, LaDuca H, Reineke P, Gutierrez S, Gray P, Davis BT (2018) False-positive results released by direct-to-consumer genetic tests highlight the importance of clinical confirmation testing for appropriate patient care. Genet Med 20:1515–1521. https://doi.org/10.1038/gim.2018.38

Tang WW, Wang Z, Levison BS, Koeth RA, Britt EB, Fu X, Wu Y, Hazen SL (2013) Intestinal microbial metabolism of phosphatidylcholine and cardiovascular risk. N Engl J Med 368:1575–1584. https://doi.org/10.1056/NEJMoa1109400

Thaiss CA, Zeevi D, Levy M, Zilberman-Schapira G, Suez J, Tengeler AC, Abramson L, Katz MN, Korem T, Zmora N (2014) Transkingdom control of microbiota diurnal oscillations promotes metabolic homeostasis. Cell 159:514–529. https://doi.org/10.1016/j.cell.2014.09.048

Topol EJ (2019) High-performance medicine: the convergence of human and artificial intelligence. Nat Med 25:44–56. https://doi.org/10.1038/s41591-018-0300-7

Treweek S, Lockhart P, Pitkethly M, Cook JA, Kjeldstrøm M, Johansen M, Taskila TK, Sullivan FM, Wilson S, Jackson C (2013) Methods to improve recruitment to randomised controlled trials: Cochrane systematic review and meta-analysis. BMJ Open 3:e002360. https://doi.org/10.1136/bmjopen-2012-002360

Tripathi A, Debelius J, Brenner DA, Karin M, Loomba R, Schnabl B, Knight R (2018) The gut–liver axis and the intersection with the microbiome. Nat Rev Gastroenterol Hepatol 15:397–411. https://doi.org/10.1038/nrgastro.2018.32

Turnbaugh PJ, Ley RE, Mahowald MA, Magrini V, Mardis ER, Gordon JI (2006) An obesity-associated gut microbiome with increased capacity for energy harvest. Nature 444:1027–1031. https://doi.org/10.1038/nature05414

Virgili J, Motitis P, Julal G, Mavrommatis Y, Pilic L (2023) The impact of genetic variability on the relationship between caffeine and cardiometabolic outcomes: a systematic review. Nutr Bull 48:28–42. https://doi.org/10.1111/nbu.12571

Von Schacky C (2014) Omega-3 index and cardiovascular health. Nutrients 6:799–814. https://doi.org/10.3390/nu6020799

Wahlström A, Sayin SI, Marschall H-U, Bäckhed F (2016) Intestinal crosstalk between bile acids and microbiota and its impact on host metabolism. Cell Metab 24:41–50. https://doi.org/10.1016/j.cmet.2016.05.005

Wang DD, Hu FB (2018) Precision nutrition for prevention and management of type 2 diabetes. Lancet Diabetes Endocrinol 6:416–426. https://doi.org/10.1016/S2213-8587(18)30053-2

Wang G, Yu Y, Wang YZ, Wang JJ, Guan R, Sun Y, Shi F, Gao J, Fu XL (2019) Role of SCFAs in gut microbiome and glycolysis for colorectal cancer therapy. J Cell Physiol 234:17023–17049. https://doi.org/10.1002/jcp.27801

Wang C, Bai M, Sun Z, Yao N, Zhang A, Guo S, Asemi Z (2023) Epigallocatechin-3-gallate and cancer: focus on the role of microRNAs. Cancer Cell Int 23:241. https://doi.org/10.1186/s12935-023-03015-7

Yeh CC, Hsieh LL, Tang R, Chang-Chieh CR, Sung FC (2005) MS-920: DNA repair gene polymorphisms, diet and colorectal cancer risk in Taiwan. Cancer Lett 224:279–288. https://doi.org/10.1016/j.canlet.2004.10.012

Zeevi D, Korem T, Zmora N, Israeli D, Rothschild D, Weinberger A, Ben-Yacov O, Lador D, Avnit-Sagi T, Lotan-Pompan M (2015) Personalized nutrition by prediction of glycemic responses. Cell 163:1079–1094. https://doi.org/10.1016/j.cell.2015.11.001

MT-CYB and *MT-ATP6* Polymorphisms: Gateway to Understanding COVID-19 Susceptibility

Gazi Nurun Nahar Sultana, Arindita Das, and Hossain Uddin Shekhar

Abstract The coronavirus disease 2019 (COVID-19) pandemic, caused by the severe acute respiratory syndrome coronavirus 2 (SARS-CoV-2), has exposed striking interindividual and interpopulation differences in disease susceptibility, severity, and outcomes. While demographic factors and preexisting comorbidities explain part of this variability, it has become increasingly evident that genetic determinants play a crucial role in shaping host responses to SARS-CoV-2 infection. Among the genetic factors under investigation, mitochondrial DNA (mtDNA) variations have emerged as significant contributors to the complex landscape of COVID-19 pathogenesis and severity. Mitochondria are vital for energy metabolism, immune responses, and antiviral signaling. Consequently, mitochondrial dysfunction or genetic variations can influence viral replication, immune responses, and inflammation, potentially altering disease outcomes. Among 13 protein-coding mitochondrial genes, *MT-CYB* and *MT-ATP6* encode essential proteins for the electron transport chain and ATP synthesis. Thus, polymorphisms in these genes can disrupt mitochondrial function. This chapter aims to elucidate the emerging links between polymorphisms in these two key mitochondrial genes and COVID-19 severity. It will explore the biological functions of these mitochondrial genes, discuss how their variants may influence mitochondrial physiology and immune mechanisms, and summarize current evidence and omics-based approaches connecting these polymorphisms to COVID-19 outcomes within the broader context of health-omics.

G. N. N. Sultana (✉)
Centre for Advanced Research in Sciences (CARS), University of Dhaka, Dhaka, Bangladesh
e-mail: nngazi@du.ac.bd

A. Das
Department of Genetic Engineering and Biotechnology, University of Dhaka, Dhaka, Bangladesh

H. U. Shekhar
Department of Biochemistry and Molecular Biology, University of Dhaka, Dhaka, Bangladesh

A. H. M. N. Nabi, H. Uddin Shekhar (eds.), *Biotechnological Advances in Healthomics*, Interdisciplinary Biotechnological Advances,
https://doi.org/10.1007/978-981-95-3612-2_11

Keywords Mitochondrial DNA variations · *MT-ATP6* · *MT-CYB* · Mitochondrial dysfunction · SARS-CoV-2 · COVID-19 · Disease susceptibility · Omics-based approaches

Abbreviations

ATP	Adenosine Triphosphate
CYB	Cytochrome B
DNA	Deoxyribonucleic Acid
NAC	N-acetylcysteine
NMN	nicotinamide mononucleotide
NR	nicotinamide riboside
ns	non-synonymous
ORF	Open Reading Frame
PCR	Polymerase Chain Reaction
RAAS	Renin-Angiotensin-Aldosterone System
RNA	Ribonucleic Acid
ROS	Reactive Oxygen Species
RSA	Relative Surface Accessibility
SARS	Severe Acute Respiratory Syndrome
SNP	Single Nucleotide Polymorphism

1 Introduction

1.1 Fundamentals of Mitochondria

Mitochondria are subcellular, double-membraned organelles often referred to as the "powerhouses" of the cell, as they generate energy in the form of adenosine triphosphate (ATP) through oxidative phosphorylation. Thought to have evolved approximately 2 billion years ago through an endosymbiotic event in which an aerobic bacterium integrated into a primitive eukaryotic cell, mitochondria are unique among organelles in that they possess their own DNA and can replicate independently.

A defining characteristic of mitochondria is the presence of their own mitochondrial DNA (mtDNA), distinct from the nuclear genome. The human mitochondrial genome comprises 16,569 base pairs arranged in a circular structure. It encodes 13 essential proteins required for the formation of oxidative phosphorylation complexes, as well as 22 transfer RNAs and 2 ribosomal RNAs necessary for mitochondrial RNA (mtRNA) translation (Duchen 2004). While the mitochondrial genome encodes some critical components, the nuclear genome provides the majority of proteins essential for mitochondrial structure and function.

Mitochondria are highly dynamic organelles, capable of rapid adaptation to cellular demands through the processes of fusion and fission. This dynamic behavior is vital for regulating cellular metabolism, calcium homeostasis, reactive oxygen species (ROS) production, and mitochondrial quality control (Hopper et al. 2006). However, mitochondria are particularly vulnerable to physiological and environmental stresses, including bacterial, fungal, and viral infections. Among these, viral infections are especially disruptive, impairing mitochondrial structure and function, and consequently affecting cellular metabolism and immune signaling (Schon et al. 2012).

Recent studies have underscored the central role of mitochondria in orchestrating both innate and adaptive immune responses, as key immune signaling pathways converge within these organelles (Salimi and Hamlyn 2020). Viruses exploit various strategies to manipulate host mitochondria to support their replication and survival, weakening the host's immune defenses and promoting cell death. This emerging interplay between mitochondrial function and viral hijacking of host immune mechanisms represents a promising avenue for research, offering novel insights for the clinical management and control of viral infections.

1.2 Mitochondria in Innate and Adaptive Immune Responses

Innate immunity represents the body's initial defense against harmful foreign substances. It acts as the first line of defense, engaging cells such as monocytes/macrophages, neutrophils, dendritic cells, plasma proteins, endothelial cells, and natural killer T-cells (NK-T cells). Unlike adaptive immunity, innate immunity is inherently present, responds rapidly, and does not require prior exposure to be activated. Recent studies highlight the critical role of mitochondria in innate immune pathways, where they function as signaling hubs and contribute to immune responses (Valdes-Aguayo et al. 2021). Beyond regulating antiviral signaling, mitochondria enhance antimicrobial immunity by generating reactive oxygen species, which induce cellular damage and stress, key components of the innate immune response (Arnoult et al. 2011).

Viral open reading frames (ORFs) can manipulate host mitochondria, leading to the release of mtDNA into the cytoplasm, which triggers mtDNA-induced inflammasome activation and suppresses both innate and adaptive immunity. Evidence suggests that the distinct localization of viral RNA and proteins within mitochondria plays a crucial role in syndrome coronavirus 2 (SARS-CoV-2) pathogenesis (Bhowal et al. 2023). Mitochondria also regulate adaptive immunity, particularly through T cell activation. T cells, vital components of adaptive immunity, respond to antigens during infection. Upon encountering an antigen, naïve T cells (Tn) proliferate rapidly into effector T cells (T_e). Most T_e cells undergo apoptosis after the infection resolves, leaving behind a small pool of long-lived memory T cells (Tm).

These memory T cells can quickly reactivate and expand into T_e cells if the same pathogen reemerges, enabling a swift immune response. Research has established that mitochondrial localization to the immune synapse is essential for T cell activation (Gill and Levine 2013). Thus, exploring the mechanisms by which viruses interact with host mitochondria could offer critical insights into the pathology of COVID-19 and related viral diseases.

1.3 Mitochondrial Subunit Genes **MT-CYB** *and* **MT-ATP6**

The mitochondrial genome, though small, encodes proteins fundamental to cellular respiration and energy production. Among these, the *mitochondrial cytochrome b* (*MT-CYB*) and *mitochondrially encoded ATP synthase membrane subunit 6* (*MT-ATP6*) genes are of particular interest due to their critical roles within the mitochondrial electron transport chain (ETC) and ATP synthesis machinery. The *mitochondrial cytochrome b* (*MT-CYB*) gene encodes Cytochrome b, a critical subunit of respiratory Complex III, also known as the cytochrome bc1 complex or ubiquinol-cytochrome c oxidoreductase. This complex is a crucial component of the mitochondrial ETC, embedded within the inner mitochondrial membrane. Complex III facilitates the transfer of electrons from ubiquinol (reduced Coenzyme Q10) to cytochrome c. This electron transfer process is coupled with the active pumping of protons across the inner mitochondrial membrane, thereby contributing significantly to the proton motive force. This electrochemical gradient is the driving force for ATP synthesis by Complex V (de Las Heras et al. 2020). Cytochrome b plays a pivotal role in this process, being directly involved in the Q-cycle, a sophisticated mechanism that facilitates the efficient transfer of electrons and contributes to the generation of the proton gradient.

Similarly, the *MT-ATP6* gene, officially termed *mitochondrially encoded ATP synthase membrane subunit 6*, encodes the protein ATP synthase F_O subunit '6' or subunit 'a' (Das et al. 2023). This is a subunit protein of Complex V, commonly referred to as ATP synthase. Complex V harnesses the proton motive force, which is established by the preceding Complexes I, III, and IV, to catalyze the synthesis of ATP from adenosine diphosphate (ADP) and inorganic phosphate. Thus, this subunit is an integral component of the proton channel and plays a vital role in oxidative phosphorylation (Yeagle 2016).

Mutations in *MT-CYB* and *MT-ATP6* are not merely isolated genetic defects; they directly impair fundamental processes of cellular energy production, rendering cells more vulnerable to metabolic stress. The proteins encoded by these genes are implicated in various cellular processes, including inducing inflammatory responses in the lungs and hepatocytes, apoptosis, renal fibrosis, and chronic kidney disease (CKD) (Das et al. 2023). Complex III and Complex V are sequential and interdependent steps within the oxidative phosphorylation (OXPHOS) pathway. A defect in Complex III, caused by a mutation in *MT-CYB*, would directly reduce the proton gradient across the inner mitochondrial membrane, thereby significantly impacting

Complex V's ability to synthesize ATP. Conversely, a defect in Complex V, resulting from a mutation in *MT-ATP6*, would directly diminish ATP production even if the proton gradient were to be maintained. This interdependence means that both genes are critical bottlenecks for efficient energy production. Impairment in either leads to a systemic energy deficit, compromising numerous cellular functions that require high ATP turnover. This diminished cellular energy reserve makes cells less resilient to various stressors, including viral infection, which inherently impose significant metabolic demands and can further exacerbate mitochondrial dysfunction.

1.4 Structure of SARS-CoV-2 and Its Multi-organ Impact Through Mitochondrial Dysfunction

SARS-CoV-2, the virus that causes COVID-19, is an enveloped single-stranded positive-sense RNA virus that encodes 4 structural and 25 nonstructural proteins (Madsen et al. 2024). The spike protein (S), nucleocapsid protein (N), membrane protein (M), and envelope protein (E) make up the structural proteins of SARS-CoV-2. While the N protein bound to the viral genomic RNA is packed inside the virion, the structural proteins S, E, and M are incorporated into the virion membrane (Tebha et al. 2024). The S protein, assembled as a trimer, gives the appearance of a crown (corona), mediates major entry steps, including receptor binding and membrane fusion. The S protein uses the S1 domain to bind to the host cell's angiotensin-converting enzyme 2 (ACE2) receptor. Consequently, the S protein can be cleaved by the transmembrane protease, serine 2 (TMPRSS2) at the cell surface, activating the S2 domain for membrane fusion (Hoffmann et al. 2020). As a result of activated S2 fusing the host and viral lipid bilayers, the virus's positive-sense, single-stranded RNA genome is deposited into the host cell. The E and M proteins contribute to virus assembly and budding through interactions with other viral proteins.

The primary site of host infection includes the nasal cavity and lung epithelium, but viral particles from SARS-CoV-2 infection have been observed in multiple tissues, for example, the heart, kidney, and gastrointestinal tract, combined with hyperinflammation, indicating that COVID-19 may result in multi-organ pathogenesis (Madsen et al. 2024). The etiology of COVID-19 is complex, and there are multiple factors involved; decreased organ function has been reported with mitochondrial dysfunction for both the acute and post-acute phases of COVID-19. For instance, SARS-CoV-2 infection can cause long-lasting damage to energy production by mitochondria in many body organs, and almost 70% of people showed reduced organ function even after a few months of COVID-19 diagnosis (Madsen et al. 2024). Such reduced organ function may be linked with an increased risk of the development of noncommunicable diseases (NCDs) such as heart failure, kidney dysfunction, and neurological disorders. Furthermore, preexisting variations or mutations in an individual's mitochondrial DNA can influence their susceptibility to infection and the trajectory of the disease, highlighting the importance of mitochondrial genetics in personalized medicine approaches to COVID-19 (Bľandová et al. 2024).

Given that SARS-CoV-2 infection has been linked to mtDNA mutations and mitochondrial dysfunction, these factors may contribute to the pathogenesis observed in clinical symptoms (Rurek 2024). However, there is limited information on this topic. In this chapter, the roles of mtDNA variations, especially the *MT-CYB* and *MT-ATP6* gene variations, in COVID-19 susceptibility and severity are discussed. This chapter aims to provide insights into how mitochondrial genetics can inform our understanding of COVID-19.

2 Key Findings

A case-control study was used to investigate the association between mutations in the *MT-CYB* and *MT-ATP6* genes and COVID-19 disease outcome. After DNA isolation from the blood samples, the *MT-CYB* and *MT-ATP6* genes were amplified by polymerase chain reaction (PCR). PCR products were purified and sequenced. Using Geneious Prime software V202, the mtDNA sequences were aligned and compared with the revised Cambridge Reference Sequence (rCRS) (NCBI Reference Sequence: NC_012920.1), and genetic variations were found. Genomics-driven analyses through in-silico tools and statistical analysis were performed to detect the impacts of identified mtDNA mutations on the structure and stability of these two gene products. The consequences were analyzed for their effects on energy metabolism and mitochondrial dysfunction contributing to COVID-19 pathogenesis. Findings from this study, along with evidence from other studies, are discussed in this section.

2.1 Demographic Characteristics of the Study Cohort

The average age of COVID-19-infected patients was 52.72 ± 9.11 years, while the average age of healthy participants was 56.19 ± 8.46 years. Statistical analysis revealed no significant difference in the age distribution between the two groups, mitigating the potential confounding effects of age on the study outcomes. Among the patients, 62.8% were male and 37.2% were female, whereas 46.61% were male and 52.38% were female among healthy individuals. The mean calculated body mass indices (BMIs) of patients and healthy participants were similar, suggesting that obesity, another known risk factor for severe COVID-19, was not a significant differentiating factor between the two cohorts. This careful demographic matching strengthens the validity of the subsequent genetic analyses.

2.2 Mutations Found Within the *MT-CYB* and *MT-ATP6* Gene Regions

Analysis of the *MT-CYB* gene region revealed nine distinct single nucleotide polymorphisms (SNPs) (Table 11.1). Notably, five of these mutations were found exclusively in the COVID-19 patient group, suggesting a potential link to disease susceptibility. The remaining four variations were present in both patients and healthy controls. Among the variations observed in the control group, three (m.14942A>C, m.15301G>A, and m.15315C>T) were identified as nonsynonymous single nucleotide polymorphisms (nsSNPs), meaning they result in a change in the amino acid sequence of the cytochrome b protein. In Table 11.1, all other nucleotide changes leading to identical amino acids (e.g., Thr to Thr, Gly to Gly, or Phe to Phe) are considered synonymous.

A closer look at the frequencies of these *MT-CYB* variants (Table 11.1) highlighted a particularly striking difference for the m.14942A>C mutation. This specific variant was found in 20% of the patients but was completely absent (0%) in the control group, yielding a statistically significant *p*-value of 0.01165. This strong association hinted at a potential role for this specific *MT-CYB* mutation in influencing COVID-19 susceptibility.

The investigation extended to the *MT-ATP6* gene, where eight different SNPs were identified (Table 11.2). In contrast to *MT-CYB*, only one variation in *MT-ATP6* was found to be specific to the patient group. The remaining seven mutations were present in both cases and controls. Within the *MT-ATP6* variants, four (m.8701A>G, m.8744T>G, m.8794C>T, and m.8860A>G) were classified as nsSNPs, potentially altering the amino acid sequence of the ATP6 protein.

Frequency analysis of the *MT-ATP6* variants (Table 11.2) revealed significant differences for two specific mutations. The m.8701A>G variant was observed in 70% of patients compared to 41.25% of controls (*p*-value = 0.0064). Similarly, the m.8744T>G variant showed a higher frequency in patients (23%) compared to controls (1.25%), with a p-value of 0.009. These statistically significant differences suggested a potential association of these *MT-ATP6* mutations with COVID-19.

Table 11.1 Variants detected in the *MT-CYB* gene of the study participants with their frequency (Sultana et al. 2025)

mtDNA position	Nucleotide change	Amino acid change	Frequency in sample (*n* = 30)	Frequency in controls (*n* = 80)	*p*-value
14942	A → C	Ile66Leu	20% (6/30)	0% (0/80)	0.01165
14956	T → C	Thr70Thr	3.33% (1/30)	0% (0/80)	0.325582
15043	G → A	Gly99Gly	80% (24/30)	63% (50/80)	0.062096
15049	C → T	Gly101Gly	3.33% (1/30)	0% (0/80)	0.325582
15295	C → T	Phe183Phe	3.33% (1/30)	0% (0/80)	0.325582
15301	G → A	Leu185Leu	70% (21/30)	52.85% (42/80)	0.091667
15315	C → T	Ala190Val	3.33% (1/30)	0% (0/80)	0.325582
15326	A → G	Thr194Ala	100% (29/30)	100% (80/80)	0.325582
15346	G → A	Leu200Leu	3.33% (1/30)	0% (0/80)	0.325582

Table 11.2 Variants detected in the *MT-ATP6* gene of the study participants with their frequency (Sultana et al. 2025)

mtDNA position	Nucleotide change	Amino acid change	Frequency in sample (n = 30)	Frequency in controls (n = 80)	p-value
8701	A → G	Thr59Ala	70% (21/30)	41.25% (33/80)	0.0064
8744	T → G	Val73Gly	23% (7/30)	1.25% (1/80)	0.009
8794	C → T	His90Tyr	3.33% (1/30)	5% (4/80)	0.6885
8860	A → G	Thr112Ala	100% (30/30)	100% (80/80)	–
8911	T → C	Leu129Leu	6.66% (2/30)	0% (0/80)	0.1607
8928	T → C	Pro134Pro	3.33% (1/30)	1.25% (1/80)	0.5619
9003	C → T	Arg159Arg	13.33% (4/30)	6.25% (5/80)	0.3091
9004	C → T	Leu160Leu	13.33% (4/30)	5% (4/80)	0.1693

Table 11.3 Frequency of the variants found in the patients, MITOMAP, and gnomAD3.1

Gene	Variations	Patients (%)	MITOMAP (%)	gnomAD 3.1 (%)
MT-CYB	m.14942A>C	20	0	0
MT-CYB	m.15315C>T	3.33	0.34	0.04
MT-CYB	m.15326A>G	100	98.605	99.342
MT-ATP6	m.8701A>G	70	31.225	30.319
MT-ATP6	m.8744T>G	23	0.002	0.004
MT-ATP6	m.8794C>T	3.33	2.859	4.75
MT-ATP6	m.8860A>G	100	98.522	99.381

To contextualize the observed frequencies within the research population, the global frequencies of all nsSNPs found in *MT-CYB* and *MT-ATP6* genes were determined from the MITOMAP and gnomAD3.1 databases. Patterns of frequency distribution with respect to the identified variants have been presented in Table 11.3. A particularly noteworthy finding was that the m.14942A>C mutation in the *MT-CYB* gene, which was significantly enriched in the patient group and absent in local controls, was also unreported in both global databases. This classification as a novel variant further strengthens its specific association with COVID-19 susceptibility. The mutation frequency m.8744T>G in the *MT-ATP6 gene* is also significantly higher in patients compared to that in global databases MITOMAP and gnomAD 3.1. However, only one nsSNP m.15326A>G (causing amino acid alteration Thr194Ala) was found in 100% (n = 30) patients and 100% (n = 80) healthy controls.

2.3 *Association of Mitochondrial Variants with COVID-19 Susceptibility and Disease Outcome*

The disease outcome was studied for all variations found within the *MT-CYB* and *MT-ATP6* genes using SNPStats. Fig. 11.1a shows that in the case of the *MT-CYB* gene, m.14942A > C had a significant correlation (p = 0.012 < 0.05) with the

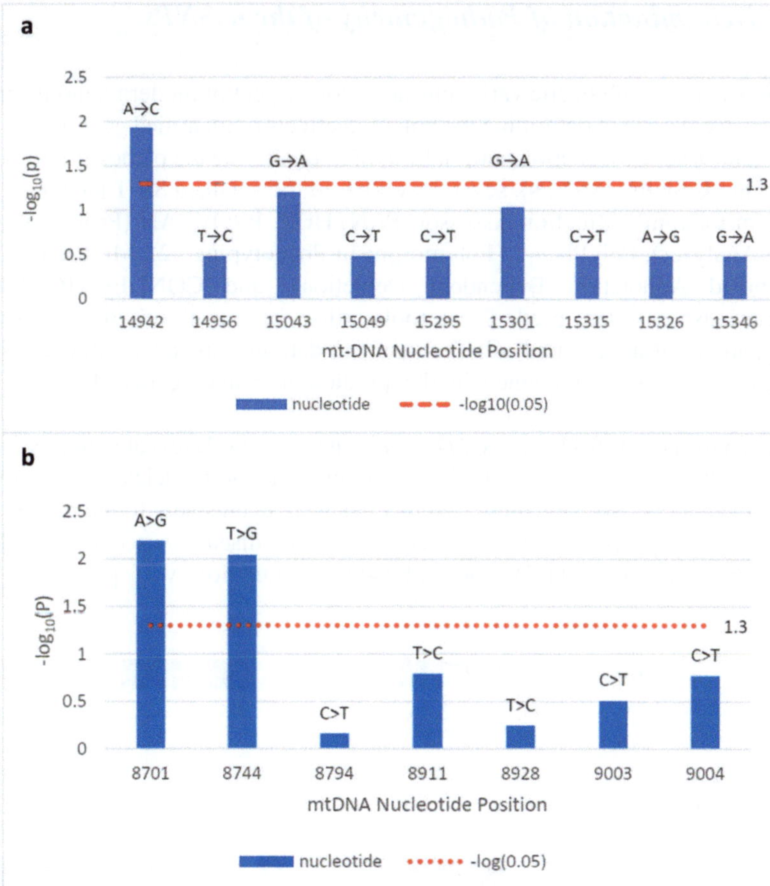

Fig. 11.1 Bar plots illustrating the significance of the association between disease outcome and mutations within mitochondrial genes. (**a**) *MT-CYB* (14747–15887 bp) and (**b**) *MT-ATP6* (8527–9207 bp). The y axis indicates the $-\log_{10}$ of p values, while the x axis indicates nucleotide locations. The red dotted line denotes the significance threshold, i.e., $-\log_{10}(0.05)$. Any points that are greater than or equal to the line are designated as statistically significant (p-value ≤ 0.05)

likelihood of COVID-19 infectivity, exhibiting an odds ratio of 19.75. This substantial odds ratio indicates a nearly 20-fold increased likelihood of COVID-19 in individuals carrying this specific mitochondrial variant. Similarly, within the *MT-ATP6* gene, both the m.8701A>G and m.8744T>G mutations showed significant associations with COVID-19 infectivity, as represented by Fig. 11.1b. The m.8701A>G variant had an odds ratio of 3.36 ($p = 0.006 < 0.05$). The m.8744T>G variant displayed a remarkably high odds ratio of 24.049 ($p = 0.009 < 0.05$), suggesting a strong positive association with COVID-19 susceptibility.

2.4 Determination of Pathogenicity of the nsSNPs

In-silico analyses of genetic variations are a core aspect of modern genomics; these analyses were done to determine the potential effects of identified genetic variations on protein structure, function, and stability. For predicting the pathogenicity of non-synonymous SNPs (nsSNPs), several web-based tools named SIFT (sorting intolerant from tolerant), Mutation Assessor, PANTHER, PROVEAN (Protein Variation Effect Analyzer), PolyPhen-2 (Polymorphism Phenotyping v2), MutPred2, CADD (Combined Annotation Dependent Depletion), and CONDEL (CONsensus DELeteriousness) were used. These tools utilize various algorithms to assess the pathogenicity of amino acid substitutions and datasets, so consistency across several tools increases confidence in the prediction, reducing the chance of false positives.

The analysis of nsSNPs in the *MT-CYB* gene (Fig. 11.2) revealed that two mutations, m.14942A>C and m.15315C>T, were predicted to be deleterious by five out of the seven tools. Notably, these two nsSNPs were exclusively found in the patient group, further supporting their potential role in disease susceptibility. In the *MT-ATP6* gene (Fig. 11.2), the m.8744T>G mutation was predicted to be

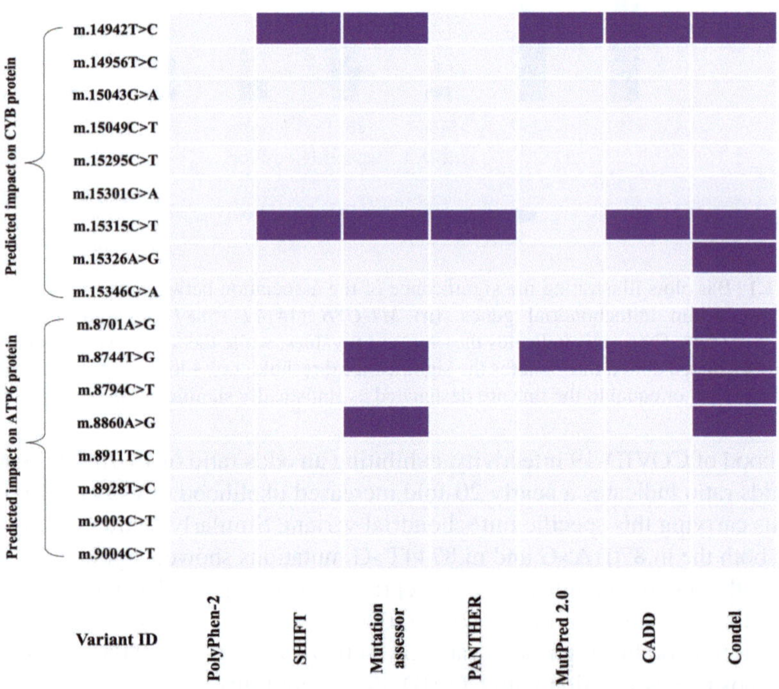

Fig. 11.2 Heatmap of the predicted deleterious impact on CYB and ATP6 proteins due to nsSNPs using seven pathogenicity prediction tools (Polyphen-2, SIFT, Mutation assessor, PANTHER, Mutpred 2.0, CADD, and CONDEL). The purple color indicates a deleterious impact on protein

deleterious by four of the seven tools. Interestingly, this variant was also observed at a significantly higher frequency in the patient group compared to the healthy controls. A "deleterious" prediction suggests that the amino acid change is likely to impair protein function, folding, or stability. This computational validation moves beyond mere statistical association to suggest a plausible biological mechanism: by directly compromising the function of critical mitochondrial proteins involved in energy production, these mutations could contribute to disease.

2.5 Effects of nsSNPs on Protein Stability

In addition to assessing pathogenicity, this study investigated how the identified nonsynonymous SNPs (nsSNPs) might affect the stability of the resulting proteins. Protein stability plays a vital role in ensuring proper folding, functionality, and molecular interactions. To predict the change in unfolding free energy ($\Delta\Delta G$) caused by amino acid substitutions, five computational tools—DUET, mCSM, EnCOM, INPS-MD, and MuPro—were used. A negative $\Delta\Delta G$ value indicates a reduction in protein stability, whereas a positive value signifies enhanced stability.

Analysis of the *MT-CYB* nsSNPs (Fig. 11.3b) showed that the m.14942A>C mutation was consistently predicted to be destabilizing by all five tools. This finding aligns with the observed significant association of this novel variant with COVID-19 and its predicted deleterious impact on protein function. In contrast, the predictions for m.15315C>T and m.15326A>G were less consistent, with some tools predicting destabilization and others suggesting stabilization (Sultana et al. 2025).

For the *MT-ATP6* nsSNPs (Fig. 11.3a), the m.8744T>G and m.8860A>G mutations were predominantly predicted to be destabilizing by all tools. However, m.8701A>G was destabilizing by four out of the five tools (mCSM, EnCOM, INPS, and MuPro), while DUET predicted a stabilizing effect (Sultana et al. 2025). The consistent destabilizing predictions for m.8744T>G, which also showed a strong

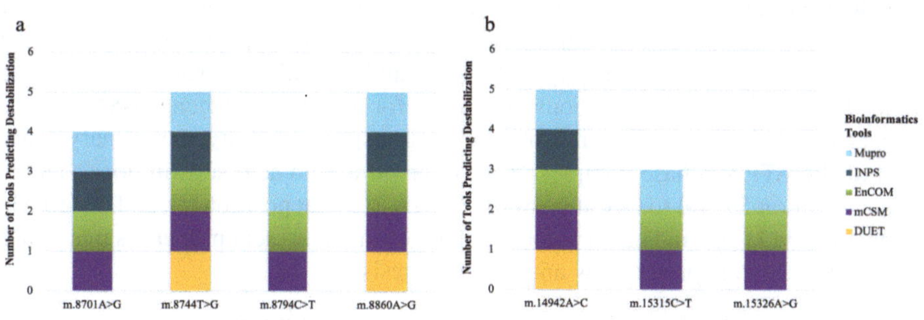

Fig. 11.3 Stacked bar plot presenting the destabilizing effect on (**a**) ATP6 protein and (**b**) CYB protein due to nsSNPs predicted by five different tools (Mupro, INPS, EnCOM, mCSM, and DUET)

association with COVID-19 and predicted deleterious impact, further highlight its potential functional relevance.

2.6 Surface Accessibility of nsSNP Sites

To detect the potential mechanisms by which these nsSNPs might exert their effects, the surface accessibility of the affected amino acid residues was investigated using the NetSurfP 3.0 tool. The relative surface accessibility (RSA) value determines whether an amino acid is exposed (RSA $> 25\%$) or buried (RSA $\leq 25\%$) within the protein structure. Surface-exposed residues are often involved in protein–protein interactions or interactions with other molecules, making mutations at these sites potentially impactful on protein function.

In the *MT-CYB* protein, the m.14942A>C mutation was found to occur at a residue that is not exposed to the protein surface. Conversely, the amino acid changes resulting from m.15315C>T and m.15326A>G were located on the protein surface. In the *MT-ATP6* protein, all nsSNPs, except for m.8860A>G, were predicted to occur at surface-exposed residues. This suggests that the amino acid alterations resulting from these mutations could potentially affect the protein's interactions with other components of the ATP synthase complex or with other cellular factors.

2.7 Impacts of nsSNPs on the Functions of CYB and ATP6 Proteins

To understand the potential functional consequences of the identified nsSNPs, the study utilized the HOPE tool to analyze the physicochemical properties of the wild-type and mutant amino acids (Table 11.4). For the *MT-CYB* gene, it was anticipated that nsSNPs could affect crucial activities such as electron transfer and ubiquinol cytochrome-c reductase activity, given the role of cytochrome b in Complex III of the electron transport chain. The analysis of the amino acid changes (Ile to Leu at position 66, Ala to Val at position 190, and Thr to Ala at position 194) revealed potential alterations in hydrophobicity, size, and overall properties, which could indeed impact these vital functions. Similarly, for the *MT-ATP6* gene, nsSNPs were expected to influence proton transmembrane transporter activity, the primary function of the ATP synthase complex. The observed amino acid substitutions (Thr to Ala at position 59, Val to Gly at position 73, His to Tyr at position 90, and Thr to Ala at position 112) also indicated potential changes in amino acid properties that could disrupt the delicate proton flow and ATP synthesis.

Table 11.4 Prediction of structural effect in the CYB and ATP6 proteins due to nsSNPs using the HOPE webtool

Mutation	Altered size	Altered hydrophobicity	Significant impacts
m.14942A>C	No Data	No Data	Located in a domain that is important for the activity and in contact with another domain that is also important for the activity.
m.15315C>T	Bigger	Increased	Could affect the local stability and can affect contacts with the lipid membrane.
m.15326A>G	Smaller	Increased	Causes a space in the core of the protein, leading to disturbance in correct folding.
m.8701A>G	Smaller	Increased	Located in a domain that is important for the activity and in contact with another important domain.
m.8744T>G	Smaller	Increased	This mutation could affect multimer contacts. Glycine is very flexible and can disturb the required rigidity of the protein at this position.
m.8794C>T	Bigger	Increased	The mutation will cause a loss of hydrogen bonds in the core of the protein and disturb correct folding.
m.8860A>G	Smaller	Increased	The mutation will cause a loss of hydrogen bonds in the core of the protein and disturb correct folding.

2.8 *Alteration of Protein Structure Due to m.14942A>C Mutation in the* MT-CYB *Gene*

Given its significant association with disease outcome, predicted deleterious impact, and consistent destabilizing effect on protein stability, the m.14942A>C (Ile66Leu) mutation in *MT-CYB* was further investigated for its structural consequences. The protein sequence was retrieved from UniProtKB (P00156), and the three-dimensional structures of the wild-type and mutant cytochrome b protein were retrieved from the Dynamut webserver (Fig. 11.4).

The analysis revealed that the Ile to Leu substitution at position 66 leads to notable alterations in the protein's interactions with its neighboring amino acids. Specifically, the mutation resulted in the loss of a hydrogen bond with a preceding residue and the formation of new hydrogen bonds. These changes in interatomic interactions can have significant repercussions for the protein's local structure, potentially affecting its overall stability, folding, and interaction with other components of Complex III. The proximity of the mutated residue to a domain critical for protein function further suggests that these structural alterations could disrupt the normal electron transfer process. This suggests a cascade: the mutation (Ile66Leu) leads to subtle hydrogen bond changes, which cause local structural perturbations, propagate instability within the protein, and ultimately impair electron transfer through Complex III. This directly links the genetic variant to bioenergetic dysfunction, explaining increased susceptibility.

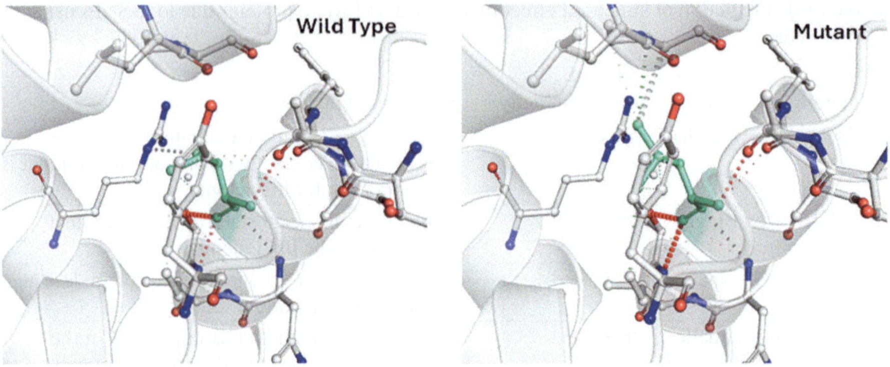

Fig. 11.4 Interatomic interaction variations due to the m.14942A>C mutation in the *MT-CYB* gene (Sultana et al. 2025). Wild-type and mutant residues are colored light green and are also represented as sticks alongside the surrounding residues, which are involved in any type of interaction

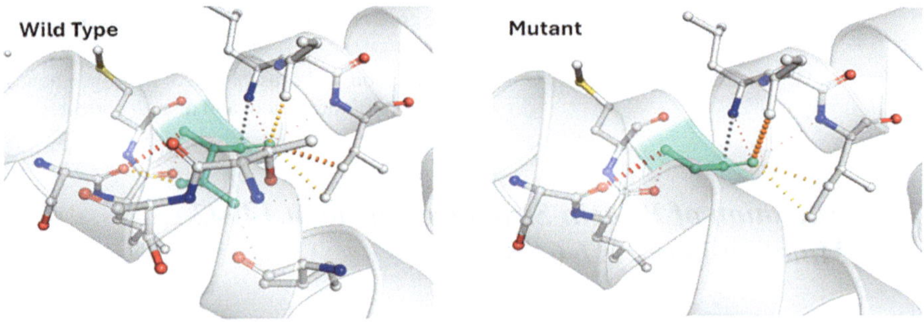

Fig. 11.5 Interatomic interaction variations due to the m.8744T>G mutation in the gene *MT-ATP6* (Sultana et al. 2025). Wild-type and mutant residues are colored light green and are also represented as sticks alongside the surrounding residues, which are involved in any type of interaction

2.9 Alterations of Protein Structure Due to m.8744T>G Mutation in the MT-ATP6 *Gene*

The m.8744T>G (Val73Gly) mutation in *MT-ATP6* also exhibited a strong association with COVID-19, predicted deleterious impact, and a tendency toward destabilizing the protein. Consequently, to examine its structural consequences, the protein sequence was retrieved from UniProtKB (P00846), and the protein's wild-type and mutant three-dimensional structure was retrieved using the Dynamut webserver (Fig. 11.5). The substitution of valine, a relatively bulky and hydrophobic amino acid, with glycine, the smallest and most flexible amino acid, at position 73 appears to have a substantial impact on the local protein environment. The analysis indicated a considerable reduction in contacts with surrounding amino acids in the mutant protein compared to the wild-type. Furthermore, the mutation led to the loss of a hydrogen bond and the formation of new hydrogen bonds. ATP synthase is a large, multi-subunit complex whose function relies on precise structural rigidity and specific subunit interfaces for efficient proton translocation and rotational catalysis. The introduction of a highly flexible glycine residue at this position could compromise the structural rigidity required for proper multimer interactions within the ATP synthase complex. The change to a less hydrophobic residue might also disrupt crucial hydrophobic interactions necessary for protein stability and function.

The location of this mutation within a domain critical for ATP6 function suggests that its structural impact is highly relevant. The extreme flexibility of glycine replacing a bulkier valine at a critical surface-exposed site in *MT-ATP6* likely disrupts the precise protein–protein interactions necessary for the assembly and function of the ATP synthase complex, leading to impaired ATP synthesis. This provides a clear structural and functional pathway: the mutation (Val73Gly) leads to increased local flexibility and reduced contacts, which compromise the structural rigidity of ATP6, impair multimer interactions within ATP synthase, and ultimately reduce proton flow and ATP synthesis. This directly links the genetic variant to the core bioenergetic deficit, explaining its strong association with COVID-19 susceptibility.

3 Healthomics Perspective: Mitochondrial Genetics in COVID-19

This study significantly contributes to a healthomics understanding of COVID-19 by integrating mitochondrial genetic data with the complex pathogenesis of the disease. Healthomics, with its holistic approach, seeks to combine various omics technologies to provide a comprehensive view of health and disease. Our findings, which highlight the association of specific mtDNA mutations in *MT-CYB* and *MT-ATP6* genes with COVID-19 susceptibility, align with the core principles of healthomics by linking genetic variations to disease outcomes. The identification of the novel m.14942A>C mutation in *MT-CYB* and the significant association of the m.8744T>G mutation in *MT-ATP6* with COVID-19 open avenues for personalized medicine.

Mitochondrial dysfunction has emerged as a critical factor in the pathogenesis of COVID-19, as SARS-CoV-2 has been shown to impair host mitochondrial function (Singh et al. 2020). An intricate link exists between impaired mitochondrial function and the hyper-inflammatory state characteristic of severe COVID-19. Mitochondrial dysfunction leads to an increased production of ROS, which are crucial mediators of cellular damage and inflammation (Green and Kroemer 2004). This excessive ROS production triggers the release and amplification of pro-inflammatory cytokines, including TNF-α, IL-1β, IL-6, and IL-18, a phenomenon often referred to as a "cytokine storm syndrome" (Valdes-Aguayo et al. 2021). Inflammatory cytokines themselves, such as TNF-α, IL-6, and IFN-γ, can further increase mitochondrial ROS production and impede OXPHOS, creating a detrimental vicious cycle (Valdes-Aguayo et al. 2021). This cascade can ultimately lead to mitochondrial membrane permeabilization, the release of mtDNA as mitochondrial damage-associated molecular patterns into the cytoplasm, and altered mitochondrial dynamics, potentially culminating in cell death. The identified mtDNA mutations, by predisposing individuals to mitochondrial dysfunction, could lower the threshold for SARS-CoV-2-induced oxidative stress and cytokine storm, thereby increasing the risk of severe outcomes. If baseline mitochondrial function is already compromised due to mutations in *MT-CYB* or *MT-ATP6*, cells are less efficient at producing ATP and potentially more prone to ROS leakage from a dysfunctional ETC (Dirican et al. 2022). When SARS-CoV-2 actively targets mitochondria and induces further ROS, these "pre-damaged" mitochondria are overwhelmed more quickly, leading to an earlier or more severe activation of inflammatory pathways, such as inflammasome activation triggered by mtDNA release (Valdes-Aguayo et al. 2021). This suggests a "two-hit" mechanism: individuals with preexisting mtDNA mutations have a diminished mitochondrial reserve capacity, and when infected with SARS-CoV-2, their cells rapidly cross a critical threshold into severe dysfunction, leading to uncontrolled inflammation and tissue damage. This offers a powerful explanation for differential disease severity. Our research supports this notion by demonstrating that mutations in mitochondrial genes, particularly those encoding key respiratory chain components, are associated with increased susceptibility to the disease. This insight underscores the importance of considering mitochondrial health in the management of COVID-19 and the development of therapeutic strategies aimed at preserving or restoring mitochondrial function.

It is important to note that the involvement of mitochondrial DNA in COVID-19 is not limited to the *MT-CYB* and *MT-ATP6* genes. Other studies have identified additional mtDNA mutations uniquely associated with COVID-19 severity, including 15 mutations in *MT-ND5*, *MT-ND4*, *MT-ND2*, and *MT-COI* genes, which have been shown to affect the secondary structure of proteins in COVID-19-positive subjects (Kumari et al. 2023). Furthermore, haplogroup analysis suggests that specific mtDNA haplogroups, such as M3d1a and W3a1b, might be potentially associated with COVID-19 pathophysiology (Kumari et al. 2023). This broader perspective indicates that the findings on *MT-CYB* and *MT-ATP6* are part of a larger pattern of mitochondrial genetic involvement in COVID-19, highlighting the complex interplay of host genetics and viral pathogenesis. Mitochondrial genetic profiling has the potential to identify individuals at an elevated risk of developing severe COVID-19. This stratification could facilitate targeted interventions, such as early administration of antiviral therapies or immunomodulatory treatments, potentially improving patient outcomes and alleviating the burden on healthcare systems.

The implications of mtDNA mutations extend beyond the acute phase of COVID-19. Mitochondrial dysfunction has been associated with decreased organ function in both the acute and post-acute phases of COVID-19. Mitochondrial dysfunction contributes significantly to the diverse and persistent symptoms observed in Long COVID. Chronic fatigue, a primary symptom, is directly linked to deficits in mitochondrial ATP production (Molnar et al. 2024). When mitochondrial function is compromised, reduced ATP production leads to energy deficits that manifest as profound and persistent fatigue, impacting muscle function, cognitive processes, and overall physical capacity. Similarly, cognitive disturbances and muscle weakness are also linked to compromised mitochondrial energy production (Molnar et al. 2024). Increased oxidative stress and inflammation, stemming from mitochondrial impairment, contribute to widespread tissue damage across various organs. Neurologically, SARS-CoV-2 exhibits neuroinvasive capabilities, and mitochondrial dysfunction in brain microglia plays a critical role in neuroinflammation, apoptosis, and neurocognitive impairments (Pliss et al. 2022). The impact extends to endothelial dysfunction and impaired blood flow, contributing to symptoms such as blood pressure variations and an increased risk of thrombosis.

Hence, it is evident that mitochondria are ubiquitous and essential for all energy-demanding processes. If their function is compromised, it can manifest differently depending on the tissue's metabolic demands and sensitivity. Tissues with high energy requirements, such as the brain (neurons and microglia), muscles, endothelium, and gut, are particularly vulnerable (Shenoy 2020). This suggests that genetic predispositions to mitochondrial impairment could influence both acute severity and the development or persistence of Long COVID symptoms. Therefore, understanding the role of mitochondrial genetics in COVID-19 could also provide insights into the long-term health consequences of the disease.

4 Future Research Directions and Therapeutic Implications

The compelling evidence linking mitochondrial gene mutations and broader mitochondrial dysfunction to COVID-19 pathogenesis and pathophysiology opens numerous promising avenues for future research and therapeutic development.

4.1 Understanding Disease Etiology

To move beyond mere association to a causal understanding, rigorous functional validation is critically needed. This includes conducting larger, multiethnic cohort studies to validate the findings on m.14942A>C and m.8744T>G and to identify other potentially significant mtDNA variants associated with COVID-19 susceptibility and severity across diverse populations. Experimental validation, utilizing advanced models such as cell lines, animal models, or patient-derived induced pluripotent stem cells, is essential to confirm the predicted deleterious effects of these mutations on protein structure, stability, and function (e.g., electron transfer and ATP synthesis) (Dirican et al. 2022). Such studies should assess their impact on mitochondrial respiration, membrane potential, and ROS production. This step is fundamental for translating findings into clinical utility; if these mutations can be functionally validated, they become stronger candidates for diagnostic markers or targets for precision medicine, moving beyond population-level associations to individual patient risk assessment.

Further research should investigate how these mtDNA mutations interact with environmental factors, such as comorbidities like obesity, diabetes, and hypertension, which are linked to Renin–Angiotensin–Aldosterone System (RAAS) overactivation and increased ROS production (de Las Heras et al. 2020; Valdes-Aguayo et al. 2021). Understanding these gene-environment interactions, as well as interactions with other host genetic factors (e.g., nuclear genome variations), is crucial for a comprehensive understanding of disease outcome. Additionally, longitudinal studies are advocated to track individuals with identified mtDNA mutations to understand their long-term health trajectories post-COVID-19, particularly in relation to the development and persistence of Long COVID (Molnar et al. 2024).

4.2 Biomarker Identification

The identification of specific pathogenic mtDNA mutations presents a significant opportunity for developing personalized medicine approaches in COVID-19. This includes the potential for developing mitochondrial biomarkers, such as specific mtDNA variants, mitochondrial function parameters, or circulating mitochondrial damage products like cell-free mtDNA, for early detection, prognosis, and monitoring of COVID-19 severity and Long COVID (Molnar et al. 2024). If these mutations are strong predictors of susceptibility or severity, screening for them could identify individuals at high risk. This could inform tailored vaccination strategies, early monitoring protocols, or even prophylactic treatments. This moves the research from basic science to practical application, shifting from reactive treatment to preventative or early intervention strategies.

4.3 Mitochondrial-Targeted Therapies

The convergence of genetic predisposition and viral assault on mitochondria points to mitochondrial health as a central therapeutic target. Several pharmacological interventions show promise. Antioxidants, such as Coenzyme Q10, MitoQ, N-acetylcysteine (NAC), resveratrol, and alpha-lipoic acid, are suggested to reduce oxidative stress within mitochondria, thereby improving their function (Polyak et al. 2018; Oskuye et al. 2024). L-carnitine is another compound that facilitates fatty acid transport into mitochondria for energy production and could therefore be beneficial (Virmani and Cirulli 2022). Furthermore, nicotinamide adenine dinucleotide (NAD+) boosters, including nicotinamide riboside (NR) and nicotinamide mononucleotide (NMN), can enhance mitochondrial function by increasing NAD+ levels, a critical coenzyme for cellular energy production and repair. These supplements may help counteract mitochondrial dysfunction and improve energy metabolism (Amjad et al. 2021). It is noteworthy that some of these compounds, including CoQ10, MitoQ, alpha-lipoic acid, nicotinamide riboside, and resveratrol, are currently being tested in clinical trials for Long COVID (Livieratos et al. 2024). Additionally, Vitamin D has been shown to play a role in normalizing mitochondrial dynamics, acting as a powerful antioxidant, and reducing RAAS activation, thereby potentially improving the prognosis of SARS-CoV-2 infection (de Las Heras et al. 2020).

Beyond pharmacological approaches, lifestyle modifications offer accessible avenues for supporting mitochondrial health. Tailored exercise and physical rehabilitation programs for long COVID patients may enhance mitochondrial biogenesis, promoting the growth and division of existing mitochondria (Jia and Su 2024). Nutritional support through dietary interventions, emphasizing a diet rich in nutrients that support mitochondrial health, such as omega-3 fatty acids, B vitamins, and minerals like magnesium, could further support recovery (Lepretti et al. 2018).

The rationale for developing and testing mitochondria-targeted therapies is compelling. Instead of solely treating symptoms, these interventions aim at a fundamental cellular pathology. This could lead to novel, more effective treatments for both acute COVID-19 severity and the debilitating symptoms of Long COVID, potentially benefiting individuals with or without these specific mtDNA mutations.

4.4 Integrative Approaches for COVID-19 Management

An integrative approach, combining genetic, molecular, and clinical data, is crucial for developing personalized medicine strategies for COVID-19. Understanding the intricate interaction between mitochondrial dysfunction and other pathophysiological processes, such as immune dysregulation, endothelial dysfunction, and intestinal issues, could lead to the development of combination therapies that address multiple aspects of the disease simultaneously (Molnar et al. 2024). This holistic perspective is essential for comprehensive disease management.

5 Conclusion

This chapter provides compelling evidence for the potential involvement of specific mitochondrial DNA variations in influencing susceptibility and severity of COVID-19. The identification of the novel *MT-CYB* mutation m.14942A>C, its strong association with the disease, predicted deleterious impact, destabilizing effect on protein stability, and the observed structural alterations highlight its potential functional relevance. Similarly, the *MT-ATP6* mutation m.8744T>G demonstrated a significant association with COVID-19 and was predicted to be deleterious and destabilizing, with notable structural consequences. The evidence suggests that these genetic predispositions, combined with the direct and multifaceted assault by SARS-CoV-2 on mitochondria, contribute to a cascade of cellular dysfunction. This includes heightened oxidative stress, the amplification of a cytokine storm, detrimental metabolic reprogramming, and ultimately, the multi-systemic dysfunction observed in severe COVID-19 and the lingering symptoms of Long COVID.

These findings underscore the importance of considering mitochondrial genetics in understanding the host response to viral infections. Further research, including functional studies to validate the impact of these mutations on mitochondrial respiratory chain activity and ATP production, is warranted to fully elucidate the mechanisms by which these mtDNA variants might contribute to COVID-19 susceptibility and severity. The insights gained from such studies could potentially pave the way for novel diagnostic or therapeutic strategies targeting mitochondrial function in the context of viral infections. This includes the identification of novel biomarkers for early detection and prognosis, as well as the development of targeted mitochondrial therapies aimed at restoring cellular energy balance and mitigating inflammation. A holistic approach that integrates genetic insights with molecular mechanisms and clinical observations promises to manage the ongoing and evolving impact of SARS-CoV-2 on global health.

References

Amjad S, Nisar S, Bhat AA et al (2021) Role of NAD+ in regulating cellular and metabolic signaling pathways. Mol Metab 49:101195. https://doi.org/10.1016/j.molmet.2021.101195

Arnoult D, Soares F, Tattoli I, Girardin SE (2011) Mitochondria in innate immunity. EMBO Rep 12:901–910. https://doi.org/10.1038/embor.2011.157

Bhowal C, Ghosh S, Ghatak D, De R (2023) Pathophysiological involvement of host mitochondria in SARS-CoV-2 infection that causes COVID-19: a comprehensive evidential insight. Mol Cell Biochem 478:1325–1343. https://doi.org/10.1007/s11010-022-04593-z

Bľandová G, Janoštiaková N, Kodada D et al (2024) Mitochondrial DNA variability and Covid-19 in the Slovak population. Mitochondrion 75:101827. https://doi.org/10.1016/j.mito.2023.101827

Das A, Ahsan N, Begum R, Sultana GNN (2023) Association of MT-ATP6 gene mutations with the pathogenesis of chronic kidney disease in human. Hum Gene Ther 36:201155. https://doi.org/10.1016/j.humgen.2023.201155

de Las Heras N, Martín Giménez VM, Ferder L et al (2020) Implications of oxidative stress and potential role of mitochondrial dysfunction in COVID-19: therapeutic effects of vitamin D. Antioxidants 9:897. https://doi.org/10.3390/antiox9090897

Dirican E, Savrun ŞT, Aydın İE et al (2022) Analysis of mitochondrial DNA cytochrome-b (CYB) and ATPase-6 gene mutations in COVID-19 patients. J Med Virol 94:3138–3146. https://doi.org/10.1002/jmv.27704

Duchen MR (2004) Mitochondria in health and disease: perspectives on a new mitochondrial biology. Mol Asp Med 25:365–451. https://doi.org/10.1016/j.mam.2004.03.001

Gill T, Levine AD (2013) Mitochondria-derived hydrogen peroxide selectively enhances T cell receptor-initiated signal transduction. J Biol Chem 288:26246–26255. https://doi.org/10.1074/jbc.M113.476895

Green DR, Kroemer G (2004) The pathophysiology of mitochondrial cell death. Science (1979) 305:626–629. https://doi.org/10.1126/science.1099320

Hoffmann M, Kleine-Weber H, Schroeder S et al (2020) SARS-CoV-2 cell entry depends on ACE2 and TMPRSS2 and is blocked by a clinically proven protease inhibitor. Cell 181:271–280. https://doi.org/10.1016/j.cell.2020.02.052

Hopper RK, Carroll S, Aponte AM et al (2006) Mitochondrial matrix phosphoproteome: effect of extra mitochondrial calcium. Biochemistry 45:2524–2536. https://doi.org/10.1021/bi052475e

Jia G, Su C-H (2024) Tailored physical activity interventions for long COVID: current approaches and benefits—a narrative review. Healthcare 12:1539. https://doi.org/10.3390/healthcare12151539

Kumari D, Singh Y, Singh S et al (2023) Mitochondrial pathogenic mutations and metabolic alterations associated with COVID-19 disease severity. J Med Virol 95:e28553. https://doi.org/10.1002/jmv.28553

Lepretti M, Martucciello S, Burgos Aceves MA et al (2018) Omega-3 fatty acids and insulin resistance: focus on the regulation of mitochondria and endoplasmic reticulum stress. Nutrients 10:350. https://doi.org/10.3390/nu10030350

Livieratos A, Gogos C, Akinosoglou K (2024) Beyond antivirals: alternative therapies for long COVID. Viruses 16:1795. https://doi.org/10.3390/v16111795

Madsen HB, Durhuus JA, Andersen O et al (2024) Mitochondrial dysfunction in acute and post-acute phases of COVID-19 and risk of non-communicable diseases. npj Metab Health Dis 2:36. https://doi.org/10.1038/s44324-024-00038-x

Molnar T, Lehoczki A, Fekete M et al (2024) Mitochondrial dysfunction in long COVID: mechanisms, consequences, and potential therapeutic approaches. Geroscience 46:5267–5286. https://doi.org/10.1007/s11357-024-01165-5

Oskuye ZZ, Mehri K, Mokhtari B et al (2024) Cardioprotective effect of antioxidant combination therapy: a highlight on MitoQ plus alpha-lipoic acid beneficial impact on myocardial ischemia-reperfusion injury in aged rats. Heliyon 10:e28158. https://doi.org/10.1016/j.heliyon.2024.e28158

Pliss A, Kuzmin AN, Prasad PN, Mahajan SD (2022) Mitochondrial dysfunction: a prelude to neuropathogenesis of SARS-CoV-2. ACS Chem Neurosci 13:308–312. https://doi.org/10.1021/acschemneuro.1c00675

Polyak E, Ostrovsky J, Peng M et al (2018) N-acetylcysteine and vitamin E rescue animal longevity and cellular oxidative stress in pre-clinical models of mitochondrial complex I disease. Mol Genet Metab 123:449–462. https://doi.org/10.1016/j.ymgme.2018.02.013

Rurek M (2024) Mitochondria in COVID-19: from cellular and molecular perspective. Front Physiol 15:1406635. https://doi.org/10.3389/fphys.2024.1406635

Salimi S, Hamlyn JM (2020) COVID-19 and crosstalk with the hallmarks of aging. J Gerontol 75:e34–e41. https://doi.org/10.1093/gerona/glaa149

Schon EA, DiMauro S, Hirano M (2012) Human mitochondrial DNA: roles of inherited and somatic mutations. Nat Rev Genet 13:878–890. https://doi.org/10.1038/nrg3275

Shenoy S (2020) Coronavirus (Covid-19) sepsis: revisiting mitochondrial dysfunction in pathogenesis, aging, inflammation, and mortality. Inflamm Res 69:1077–1085. https://doi.org/10.1007/s00011-020-01389-z

Singh KK, Chaubey G, Chen JY, Suravajhala P (2020) Decoding SARS-CoV-2 hijacking of host mitochondria in COVID-19 pathogenesis. Am J Phys Cell Phys. https://doi.org/10.1152/ajpcell.00224.2020

Sultana GNN, Hasan Z, Das A, Sarker R, Shekhar HU, Begum R (2025) Evaluating the MT-CYB and MT-ATP6 variations in COVID-19 patients: A case-control study PLOS One 20(8):e0329866. https://doi.org/10.1371/journal.pone.0329866

Tebha SS, Tameezuddin A, Bajpai S, Zaidi AK (2024) Chapter One - SARS-CoV-2—virus structure and life cycle. In: Zaidi AK (ed) Progress in molecular biology and translational science. Academic Press, pp 1–23

Valdes-Aguayo JJ, Garza-Veloz I, Badillo-Almaraz JI et al (2021) Mitochondria and mitochondrial DNA: key elements in the pathogenesis and exacerbation of the inflammatory state caused by COVID-19. Medicina (B Aires) 57:928. https://doi.org/10.3390/medicina57090928

Virmani MA, Cirulli M (2022) The role of l-carnitine in mitochondria, prevention of metabolic inflexibility and disease initiation. Int J Mol Sci 23:2717. https://doi.org/10.3390/ijms23052717

Yeagle PL (2016) The membranes of cells. Academic Press

High-Resolution Structural Approaches for Drug Discovery: Targeting Terminal Oxidase in Cellular Respiration

Atsuhiro Shimada

Abstract Understanding how drugs interact with their target proteins is fundamental to the development of effective therapeutics. Most drugs act by binding to specific proteins involved in diseases or key biological pathways, thereby modulating their activity. When high-resolution structural information of a protein–drug complex is available, it enables detailed elucidation of the drug's mechanism of action. Such structural insights also facilitate rational design and optimization of molecules that more effectively bind to the target protein to either enhance or inhibit its function. This chapter focuses on cytochrome *c* oxidase (CcO), the terminal oxidase of the mitochondrial respiratory chain, as a model system to illustrate structure-based drug discovery. CcO plays a central role in cellular energy production, and its activity is tightly regulated in response to metabolic demands. By examining recent advances in high-resolution structural studies of CcO in complex with small molecules, we highlight how structural biology can inform and accelerate the development of new therapeutic strategies.

Keywords Respiratory chain protein · Oxidative phosphorylation · Metalloprotein · Membrane protein · High-resolution crystallography · Catalytic intermediates · Proton pump protein

Abbreviations

ADP	Adenosine diphosphate
ATP	Adenosine triphosphate
CcO	Cytochrome c oxidase
CHCHD2	Coiled-coil-helix-coiled-coil-helix domain-containing protein 2

A. Shimada (✉)
Biomolecular chemistry group, Department of Applied Life Science, Faculty of Applied Biological Sciences, Gifu University, Gifu, Japan
e-mail: shimada.atsuhiro.s2@f.gifu-u.ac.jp

A. H. M. N. Nabi, H. Uddin Shekhar (eds.), *Biotechnological Advances in Healthomics*, Interdisciplinary Biotechnological Advances,
https://doi.org/10.1007/978-981-95-3612-2_12

CN^-	Cyanide
CO	Carbon monoxide
ETC	Electron transport chain
Higd1	Hypoxia-inducible gene domain 1
MNRR1	Mitochondria Nuclear Retrograde Regulator 1
mtDNA	Mitochondrial DNA
N_3^-	Azide
NO	Nitric oxide
N-side	Negative side
OXPHOS	Oxidative phosphorylation
P-side	Positive side
Rcf1	Respiratory supercomplex factor 1
XFEL	X-ray free electron laser

1 Introduction

The majority of energy required to sustain life in aerobic organisms is produced via oxidative phosphorylation (OXPHOS). This process is driven by a series of protein complexes constituting the mitochondrial electron transport chain (ETC) and adenosine triphosphate (ATP) synthase. Cytochrome *c* oxidase (CcO), also known as complex IV, functions as the terminal oxidase of the ETC and is responsible for the four-electron reduction of molecular oxygen (triplet O_2) to water (H_2O) (Kaila et al. 2010; Yoshikawa et al. 2011). During this catalytic reaction, the electrons necessary for the reduction of O_2 are supplied from the positive side (P-side) of the mitochondrial inner membrane (or the plasma membrane in prokaryotes), while the protons required for water formation are taken up from the negative side (N-side). This asymmetric proton movement contributes to the generation of an electrochemical membrane potential. In addition, CcO actively pumps protons from the N-side to the P-side utilizing the free energy released by the O_2 reduction reaction, thereby creating a proton gradient across the membrane (Fig. 1) (Kaila et al. 2010; Yoshikawa

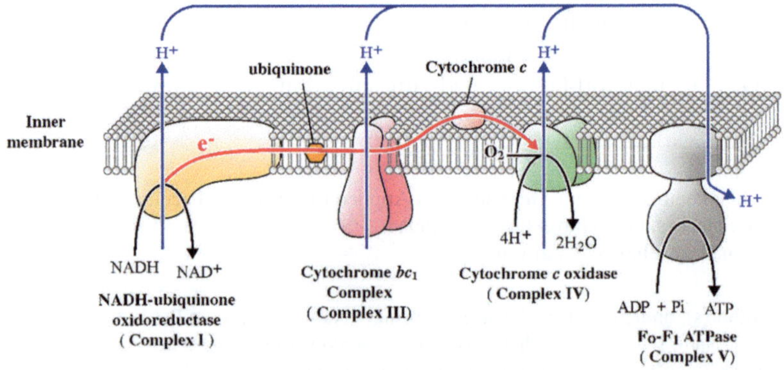

Fig. 1 Electron transfer chain complexes and ATP synthase

et al. 2011; Yoshikawa and Shimada 2015). The resulting proton motive force (PMF)—composed of the membrane potential and the proton gradient—drives ATP synthesis by ATP synthase, the molecular machine responsible for cellular energy currency production.

Given its central role in cellular bioenergetics, CcO also participates in the regulation of diverse physiological functions (Sebastián et al. 2017). In eukaryotes, CcO dysfunction impairs oxidative phosphorylation, leading to mitochondrial metabolic disorders (Diaz 2010; Srinivasan and Avadhani 2012; Fernandez-Vizarra and Zeviani 2021). Genetic or acquired deficiencies in CcO activity have been associated with a spectrum of diseases, particularly neurodegenerative disorders caused by mitochondrial dysfunction (Cottrell et al. 2002; Fukui et al. 2007; Bi et al. 2018; Danyu et al. 2019; Zong et al. 2024). Conversely, elevated CcO activity has been linked to excessive ATP production, which may contribute to the development of metabolic syndromes such as obesity and type 2 diabetes (Čapková et al. 2002; Van der Schueren et al. 2015; Aharon-Hananel et al. 2022; LaMoia et al. 2022; Zong et al. 2024). These dual aspects of CcO function have drawn increasing attention to the development of pharmacological agents capable of either positively or negatively modulating its activity. Moreover, the identification of pathogen-specific features in microbial CcOs has opened new avenues for designing selective inhibitors as potential antimicrobial agents (Nishida et al. 2022). High-resolution X-ray crystallographic structures of CcO, particularly those derived from bovine heart mitochondria, have provided critical insights into its functional mechanisms and enabled structure-based drug design strategies (Tsukihara et al. 1995; Yano et al. 2016; Shimada et al. 2018, 2020, 2021).

In this chapter, we first review the historical development of structural analyses of CcO, with an emphasis on the mechanistic insights derived from its three-dimensional architecture. We then discuss the current status and future prospects of drug discovery efforts targeting CcO in the context of human diseases and infectious pathogens.

2 Disorders Involving Cytochrome c Oxidase (CcO)

The association between deficiencies in CcO subunits and diseases or symptoms is summarized in Table 1.

2.1 Neurodegenerative Diseases Caused by Abnormal CcO Activity

A decrease in cytochrome *c* oxidase (CcO) activity can result from congenital genetic abnormalities or the accumulation of damage to genes due to oxidative stress (Srinivasan and Avadhani 2012; Fernandez-Vizarra and Zeviani 2021; Zong

Table 1 List of mutations or defects in CcO subunits and associated diseases

Subunit name	Mutation or defect	Associated diseases or symptoms
COX1	Point mutation, heteroplasmy	Myopathy (Valente et al. 2009), Leber hereditary optic neuropathy (LHON) and hearing loss, myoglobinuria (Nishigaki et al. 2010)
COX2	Missense mutation	Encephalomyopathy (Clark et al. 1999), LHON (Abu-Amero and Bosley 2006), myopathy (Rahman et al. 1999)
COX3	Stop gain, microdeletion	Myopathy (Horváth et al. 2002), LHON (Pereira et al. 2011), myoglobinuria (Figueroa-Martínez et al. 2011), severe encephalopathy (Hanna et al. 1998)
COX6B1	Deletion/point mutation	Infantile encephalopathy (Massa et al. 2008)
COX6A1	Point mutation	Charcot-Marie-Tooth type X (Tamiya et al. 2014)
COX7B	Point mutation	Microphthalmia with skin lesions (Indrieri et al. 2012)
COX4I2 (COX6A1, COX6A2, COX7A1, COX7A2)	Point mutation	Exocrine pancreatic insufficiency, dyserythropoietic anemia, calvarial hyperostosis (EPIDACH) (Vondrackova et al. 2012)

Adapted with modifications from Table 1 in Rak et al. (2016)

et al. 2024). This dysfunction primarily induces disorders of the nervous and muscular systems. Representative conditions include mitochondrial encephalomyopathies (such as mitochondrial encephalomyopathy, lactic acidosis, and stroke-like episodes [MELAS], including Leigh syndrome) and mitochondrial myopathies (Diaz 2010; Srinivasan and Avadhani 2012; Fernandez-Vizarra and Zeviani 2021; Zong et al. 2024). Alzheimer's disease and Parkinson's disease have also been reported to involve, albeit indirectly, reduced CcO activity (Cottrell et al. 2002; Fukui et al. 2007; Bi et al. 2018; Danyu et al. 2019).

Mitochondrial encephalomyopathies are a group of progressive disorders caused by mitochondrial dysfunction, affecting both the nervous and muscular systems, with prominent impairments in energy metabolism, particularly in the central nervous system and skeletal muscles (Robinson 2000). These disorders are caused by mutations in mitochondrial DNA (mtDNA) or nuclear DNA, and abnormalities in respiratory chain enzymes—especially CcO—play a central role in their pathogenesis (Antonicka et al. 2003). Depending on the subtype, they are classified as MELAS syndrome, myoclonic epilepsy with ragged red fibers (MERRF) syndrome, Kearns–Sayre syndrome, and others. Leigh syndrome is also considered a subtype of MELAS. It is a progressive neurodegenerative disease that typically manifests during infancy and is believed to result from impaired mitochondrial energy metabolism, leading to energy deficiency and central nervous system dysfunction (Robinson 2000; Dimauro and Schon 2003). Many patients with Leigh syndrome carry mutations in CcO subunits such as COX10 or genes involved in CcO assembly, such as SURF1 (Péquignot et al. 2001; Antonicka et al. 2003).

Mitochondrial myopathy refers to a group of disorders characterized by chronic muscle weakness and exercise intolerance caused by impaired mitochondrial function in muscle tissue. The underlying cause is often dysfunction of enzymes in the mitochondrial respiratory chain, especially CcO, leading to a significant decrease in energy production in muscle cells (Ronchi et al. 2023; Aich et al. 2024). These diseases are progressive and generally have a poor prognosis, with many patients succumbing to respiratory or cardiac failure within a few years after onset (DiMauro et al. 1990). Despite the severity of these conditions, there is currently no curative treatment available, and the standard management consists of supportive care including nutritional supplementation and administration of cofactors to alleviate symptoms and slow disease progression (Schon al. 1988; DiMauro et al. 1990).

In Alzheimer's and Parkinson's diseases, reduced CcO activity is also implicated in their pathogenesis (Fernandez-Vizarra and Zeviani 2021). Decreased CcO activity leads to a reduction in ATP production via oxidative phosphorylation, while simultaneously promoting the accumulation of reactive oxygen species (ROS) (Fukui et al. 2007; Srinivasan and Avadhani 2012). These changes accelerate the deposition of amyloid-β, a hallmark of Alzheimer's disease. In Parkinson's disease, a marked decrease in CcO activity has been observed in the substantia nigra pars compacta, increasing oxidative stress and contributing to the degeneration of dopaminergic neurons (Robinson 2000; Zong et al. 2024). Furthermore, one of the known causative genes for Parkinson's disease, mitochondrial nuclear retrograde regulator 1 (MNRR1) (also known as coiled-coil-helix-coiled-coil-helix domain-containing protein 2 [CHCHD2]), has been shown to activate CcO. Dysfunction of MNRR1 and the resulting decline in CcO activity are strongly implicated in the pathogenesis of Parkinson's disease (Meng et al. 2017; Kee et al. 2021).

2.2 *Metabolic Syndrome Caused by Abnormal CcO Activity*

Abnormal CcO activity is not only linked to neurodegenerative diseases, but also to metabolic conditions such as obesity and metabolic syndrome (Robinson 2000; Fernandez-Vizarra and Zeviani 2021). When energy intake becomes excessive, for example, due to overeating, the balance of oxidative phosphorylation in the respiratory chain is disrupted, leading to excessive ATP production and fat accumulation, which contributes to obesity (Čapková et al. 2002; Sebastián et al. 2017). Moreover, decreased CcO activity in brown adipose tissue reduces energy expenditure, further exacerbating obesity (Alcalá et al. 2017; Wang et al. 2021).

In addition, reduced CcO activity in adipose tissue and skeletal muscle leads to a decrease in ATP production efficiency, which in turn impairs insulin signaling and contributes to the development of insulin resistance (Nelson et al. 2014). In the liver, decreased CcO activity results in lower ATP levels, which inhibits activation of AMPK. This prevents suppression of gluconeogenesis and ultimately contributes to the onset of type 2 diabetes (Van der Schueren et al. 2015).

While certain medications, such as metformin, which indirectly activates CcO activity, are effective in treating metabolic syndrome, lifestyle modifications including dietary restriction and moderate physical activity are also essential for prevention (Apostolova et al. 2020; LaMoia et al. 2022). However, many working adults have limited time and are frequently exposed to diets high in fat, making prevention difficult in practice. Therefore, the development of drugs that specifically regulate CcO activity is an urgent need in the fight against metabolic syndrome.

3 The Structure of CcOs

Drug discovery efforts targeting the regulation of CcO activity fundamentally rely on obtaining high-resolution structural information. A detailed understanding of its reaction mechanism, based on this three-dimensional structure, is crucial for the rational design of therapeutic agents. Cytochrome *c* oxidase is classified into several families based on its structural characteristics, but its fundamental function remains conserved across these families (Yoshikawa and Shimada 2015). In this section, we describe the X-ray crystallographic structure of bovine heart CcO, which belongs to the A-family and is the most extensively studied, and also discuss the structures of CcOs from other families.

3.1 High-Resolution Crystal Structures of Bovine Heart CcO

Even long before the determination of the X-ray crystallographic structure, the existence of 13 distinct subunits in purified bovine CcO had been reported based on sodium dodecyl sulfate–polyacrylamide gel electrophoresis (SDS-PAGE) analysis (Kadenbach and Merle 1981; Kadenbach and Hüttemann 2015). The subunit composition suggested by these biochemical analyses was strongly supported by the crystal structure resolved in 1995 (Tsukihara et al. 1995). Although the physiological roles of all 13 subunits have yet to be fully elucidated, several subunits—particularly subunits I and II, which contain redox-active metal centers, as well as subunit III, which includes the oxygen transport pathway—have been demonstrated to be essential for the enzymatic activity of CcO. These three subunits, constituting the functional core of CcO, are referred to as core subunits and are encoded by the mitochondrial genome. In contrast, the remaining ten subunits are encoded by the nuclear genome. While the core subunits are highly conserved across both prokaryotes and eukaryotes, the ten additional subunits surrounding the core are unique to eukaryotes and are thus considered to have been acquired during the course of evolution (Grossman and Lomax 1997; Kadenbach and Hüttemann 2015). The physiological functions of these nuclear-encoded subunits remain a topic of ongoing debate and are not yet clearly understood. However, given that bovine CcO has yielded higher-resolution crystal structures compared to its bacterial counterparts, it is strongly suggested that one of the roles of these additional subunits is to stabilize the overall structure of the enzyme.

3.1.1 Metal Centers in CcO

Bovine heart CcO contains four redox-active metal centers designated Cu_A, heme *a* (Fe_a), Cu_B, and heme a_3 (Fe_{a_3}), each of which can accept one electron equivalent (Yoshikawa and Shimada 2015). The two iron ions are coordinated with porphyrins to form heme *a*, as illustrated in Fig. 2. One of these hemes, referred to as heme *a*, is fixed within the protein environment and exists in a low-spin, six-coordinate state. The other heme, known as heme a_3, is in a five-coordinate, high-spin state in the fully reduced enzyme, lacking a sixth ligand. In this fully reduced state, all four redox-active metal centers are in the one-electron reduced form. As shown in Fig. 1, both hemes are oriented essentially perpendicular to the plane of the mitochondrial membrane. The fifth ligands (histidines) of both hemes originate from residues 376 and 378 on helix X (unless otherwise noted, residue numbers refer to the bovine sequence throughout this review), indicating that the two hemes are in very close proximity. The propionate groups of the C-rings of heme *a* and heme a_3 are ionically bridged with the guanidino groups of R439 and R438, respectively (as shown in Fig. 2). Among these, only heme a_3 serves as the oxygen-binding site in the reduced state.

One of the copper centers, Cu_A, is a dinuclear copper site with a cupredoxin-like geometry bridged by two cysteine residues ($Cu^{2+}/Cu^{1+} + e^- \rightleftharpoons Cu^{1+}/Cu^{1+}$) (Tsukihara et al. 1995). In this chapter, we refer to the oxidized and reduced forms of this center as Cu_A^{2+} and Cu_A^{1+}, respectively. Cu_A is located in subunit II, the third-largest subunit of the enzyme, positioned near the P-side of the mitochondrial membrane. Structural studies of the CcO–cytochrome *c* complex, as well as physiological analyses, have demonstrated that Cu_A acts as the initial electron acceptor from cytochrome *c* (Shimada et al. 2017b). One of the Cu_A ligands—an imidazole ring from H204 in subunit II—forms a hydrogen bond with the carbonyl group of the peptide backbone, which, in turn, is hydrogen-bonded via the NH group of the peptide chain of R438 to one of the propionate groups of heme *a* (Tsukihara et al. 1996). This configuration suggests a viable electron transfer pathway from Cu_A to heme *a*. As there is no direct experimental evidence supporting electron transfer from Cu_A to heme a_3, it is generally believed that electrons donated from cytochrome *c* are transferred sequentially via the pathway: $Cu_A \rightarrow$ heme $a \rightarrow$ heme a_3.

The second copper site, Cu_B, adopts a trigonal planar geometry in the fully reduced state, indicating that Cu_B functions both as a weak ligand acceptor and a weak electron donor. This trigonal geometry is essentially coplanar with the plane of heme a_3, and together they constitute the O_2 reduction site. Although heme a_3 is the direct O_2-binding center, the reduction of both Fe_{a_3} and Cu_B is required for effective oxygen binding to Fe_{a_3} (Dodson et al. 1996).

In addition to the redox-active centers, bovine heart CcO contains three redox-inactive metal ions—Zn^{2+}, Mg^{2+}, and Na^+—all of which are detectable in the X-ray

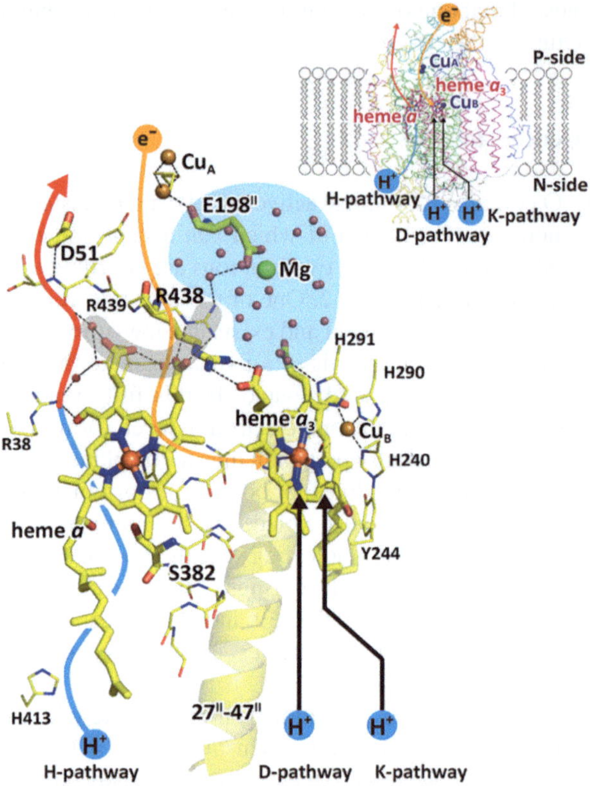

Fig. 2 Metal sites are indicated by brown, beige, and green spheres for iron, copper, and magnesium ions, respectively. Porphyrins of heme *a* and heme a_3 and amino acid residues showing redox-coupled conformational changes are represented by the thick stick models as labeled. Within the stick models, dark blue, red and yellow portions are nitrogen, oxygen and carbon, respectively. A beige arrow indicates the location of the electron transfer passage, while two black arrows indicate those for protons involved in producing water molecules. The hydrogen-bond network and the water channel of the H-pathway are indicated by the red and blue portions of the leftmost curved arrow, respectively. The Mg/H_2O cluster (the blue area) is attached to the hydrogen-bond network of the H-pathway via a short hydrogen-bond network (the gray area). Small beige spheres mark the water molecule positions. The formyl group and one of the propionate groups of heme *a* are hydrogen-bonded with Arg^{38} and a fixed water molecule in the hydrogen-bond network of the H-pathway, respectively. The location of the residues 27–47 of one of the transmembrane helices of subunit II is shown by a yellow ribbon model as labeled. The inset shows the overall locations of the redox-active metal sites and pathways for transportation of electrons and protons within the CcO structure, indicated by Cα-backbone traces. (Adapted from Shimada et al. (2023a, b), licensed under CC BY 4.0)

crystallographic structure (Yoshikawa et al. 1998). The zinc site is located in subunit Vb, which is encoded by nuclear DNA, and is coordinated by four cysteine residues. The physiological function of this Zn^{2+} site remains unclear. In contrast, Mg^{2+} is situated within a large internal water cluster in CcO and has been suggested to participate in the transient storage of protons during pumping (Yano et al. 2016).

Furthermore, the Na^+ ion is coordinated by residues within the proton transfer pathway and is proposed to play a role in facilitating proton translocation (Yoshikawa et al. 1998; Kamiya et al. 2007).

3.1.2 Proton Transfer Pathway

It has been proposed that internal hydrogen-bonding networks and mobile water molecules within the protein matrix facilitate proton translocation in CcO (Yoshikawa and Shimada 2015; Shimada et al. 2023a, b). High-resolution X-ray crystallography of bovine heart CcO has revealed three potential proton-conducting pathways: the D-, K-, and H-pathways. The D- and K-pathways connect the N-side (matrix side) surface of the protein to the oxygen reduction site, as illustrated in Fig. 3.

The D-pathway is characterized by the presence of conserved residues D91, N98, M71, and E242, along with several ordered water molecules situated between

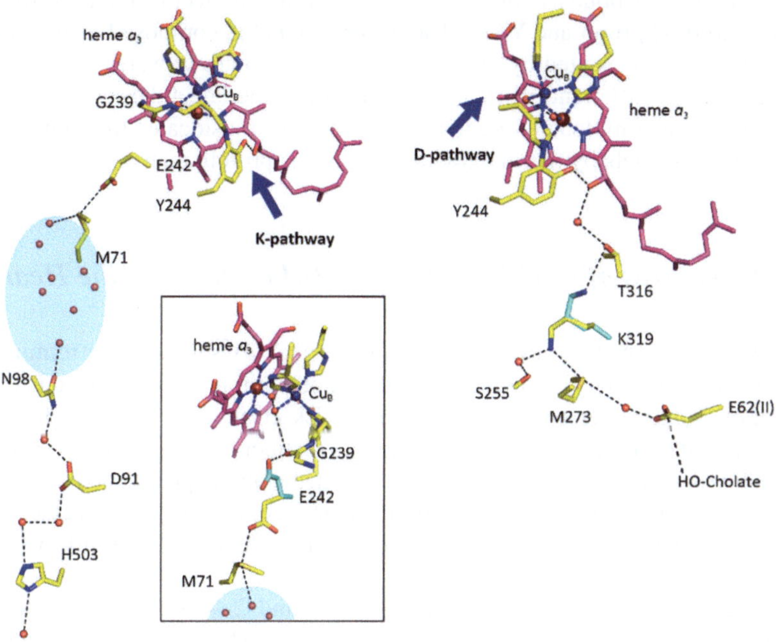

Fig. 3 Possible structural changes in K and D-pathways, which could trigger proton transfers from the N-side to the O_2 reduction site. The red and dark blue spheres denote the locations of Fe_{a_3} and Cu_B, respectively. The small red spheres are placed at the sites of fixed water or its derivatives (H_3O^+, OH^-, or O^{2-}) in the pathways. The dark blue arrows denote the locations of the O_2 reduction site ends of the two pathways. The dotted lines are the hydrogen bonds in the pathways. The left panel shows the atomic model of the D-pathway with an inset showing the possible structural changes in Glu^{242} in light blue. The right panel includes possible structural changes in Lys^{319} in light blue. (Adapted from Shimada et al. (2021), licensed under CC BY 4.0)

these residues. Although the direct structural connection between E242 and the oxygen reduction site—composed of heme a_3 and Cu_B—remains unclear, the proposed movement of E242 and G239, in concert with the insertion of OH^- or oxide (O^{2-}) at the oxygen reduction site (as shown in the inset), could provide a viable and efficient pathway for proton delivery from E242 to the reduction center.

The K-pathway includes residues K319 and T316, the hydroxyl group of the hydroxyfarnesylethyl substituent on heme a_3, and Y244, which is covalently linked to one of the histidine ligands coordinating Cu_B.

The third pathway, the H-pathway, spans the N-phase to the P-phase of the membrane and consists of a hydrogen-bonded network and a water channel (Yano et al. 2016). This hydrogen-bonding network connects residue D51 near the P-side with R38, which is hydrogen-bonded to the formyl group of heme *a*. The network comprises peptide bonds, a water molecule fixed to the propionate group on the D-ring of heme *a*, and residue Y371. The hydrogen bond between R38 and the heme *a* formyl group is structurally connected to the N-side surface of the protein via a water channel that allows access for water molecules from the N-phase to this interaction site. This hydrogen-bonding network interacts strongly with heme *a* by forming two hydrogen bonds between R38 and a water molecule fixed between the peptide carbonyl group and Y371. The water channel is composed of five discrete water cavities, represented by ellipses, connected by narrow water conduits. Each water cavity has sufficient internal volume to accommodate at least one water molecule. These water molecules are hypothesized to be translocated through the water cavities and channels by thermal fluctuations of the protein matrix.

3.2 Structures of CcO from Sources Other than Bovine Heart

In addition to bovine heart cytochrome *c* oxidase (CcO), several other terminal oxidases, which couple oxygen reduction with proton pumping, are involved in cellular respiration (Ostermeier et al. 1997; Svensson-Ek et al. 2002). Although some of these terminal oxidases utilize electron donors distinct from cytochrome *c*, they all contain an oxygen reduction site composed of heme and copper centers. Therefore, these enzymes are collectively referred to as heme–copper oxygen reductases (commonly abbreviated as HCOs; however, to avoid confusion, the term CcO will be used throughout this chapter).

CcOs are classified into three families—A, B, and C—based on the architecture of their proton transfer pathways (Pereira et al. 2008). The A-family, which includes bovine CcO, possesses both D- and K-pathways. The D-pathway is responsible for transporting both pumped and chemical protons, while the K-pathway has been traditionally associated only with the delivery of chemical protons (Konstantinov et al. 1997). The A-family can be further subdivided into A1 and A2 subfamilies. In the A2-type CcOs, the glutamate residue (E242) in the D-pathway of A1-type enzymes is replaced by a tyrosine residue located in a turn structure adjacent to the helix that contains E242 in the A1-type. The hydroxyl group of this tyrosine occupies a spatial position analogous to that of the carboxyl group in the A1-type (Pereira et al. 2006).

CcOs in the B-family contain only a single proton transfer pathway, analogous to the K-pathway in A-family enzymes, which is responsible for transporting both pumped and chemical protons (Pereira et al. 2008). Notably, CcOs in this family preserve a covalent linkage between one of the three imidazole ligands bound to the copper center and a conserved tyrosine residue (Y244). The C-family also lacks the D-pathway, and in these enzymes, the conserved Y244 is substituted by a tyrosine residue located on an adjacent helix (Pereira et al. 2008). In other words, Y244 is not conserved in C-family HCOs, although the histidine–tyrosine covalent cross-link is retained.

In summary, while the structural features of the oxygen reduction site—including the imidazole–tyrosine covalent bond—are highly conserved among all families, the architecture of the proton transfer pathways is not.

3.3 O_2 Reduction Mechanism

3.3.1 Investigation of CcO Intermediates by Spectroscopic and Electrochemical Analyses

Although O_2 resists one-electron reduction, it is more susceptible to two-electron reduction (Caughey and Rabinowitz 1976). This property is considered a key factor in the stability of oxyhemoglobin and oxymyoglobin (OxyMb) (Hodel et al. 1992). In these globins, the Fe^{2+}–O_2 adduct is isolated within the protein matrix, preventing the efficient delivery of a second electron. In contrast, the O_2 reduction site of cytochrome *c* oxidase (CcO) contains a copper ion (Cu_B) that can act as a second electron donor. Consequently, the oxy form typical of hemoglobin is predicted to be highly unstable in CcO. Instead, the first intermediate is expected to be a peroxide-bound species [$Fe_{a_3}^{3+}$–O–O^{2-}–Cu_B^{2+}] (Hirata et al. 2014).

Surprisingly, however, resonance Raman spectroscopy has shown that the first intermediate is in fact an oxy-type species, and the peroxide intermediate is too unstable to be detected (Takahashi et al. 1993). Rather, the second intermediate contains fully reduced oxygen (O^{2-}), corresponding to a four-electron reduced state. These have been designated as intermediate A ([$Fe_{a_3}^{2+}$–O_2]) and intermediate P ([$Fe_{a_3}^{4+} = O^{2-}$, Cu_B^{2+}–OH^-]), respectively (Yoshikawa and Shimada 2015). A total of four electrons are transferred from cytochrome *c* via Cu_A and heme *a* to intermediate P to regenerate the fully reduced intermediate R. During this process, intermediate F, O, and E are sequentially formed after each one-electron transfer to P (Yoshikawa and Shimada 2015). Each of the P → F, F → O, O → E, and E → R transitions is coupled to the translocation of a single proton (Shimada et al. 2023b).

3.3.2 X-Ray Crystallographic Structures of Reaction Intermediates

In multistep reactions such as O_2 reduction, synchronizing intermediate formation is inherently difficult, leading to mixtures of intermediates in crystalline state. A method was developed to resolve individual intermediate structures from such mixtures by analyzing temperature factors (Shimada et al. 2021, 2023a, b). This method has allowed the determination of all spectroscopically identified intermediates at 1.8 Å resolution by X-ray crystallography (Yano et al. 2016). Crystals of intermediates P and F were prepared via H_2O_2 treatment, while other intermediate crystals were obtained by varying the O_2 exposure time of fully reduced CcO crystals (Shimada et al. 2020, 2021), as shown in Fig. 4.

3.3.3 Intermediates R and A

The ligand-bound geometry of intermediate A closely resembles that of oxymyoglobin (Fig. 4b). The distance between O_2 coordinated to Fe_{a_3} and Cu_B is unusually long for a coordination bond, suggesting extremely weak interaction. This implies that simultaneous two-electron reduction of the O_2 bound to Fe_{a_3} is hindered, making the O_2 moiety sufficiently stable for Raman detection. This structural insight corroborates the resonance Raman findings reported nearly three decades earlier (Ogura et al. 1993).

3.3.4 Intermediates P and F

The crystal structure of a stoichiometric mixture of intermediates P and F was determined; however, structural differences between the two were not resolved due to limited resolution (Fig. 4c). The Fe_{a_3}–oxygen distance (1.70 Å) in the P/F crystal is

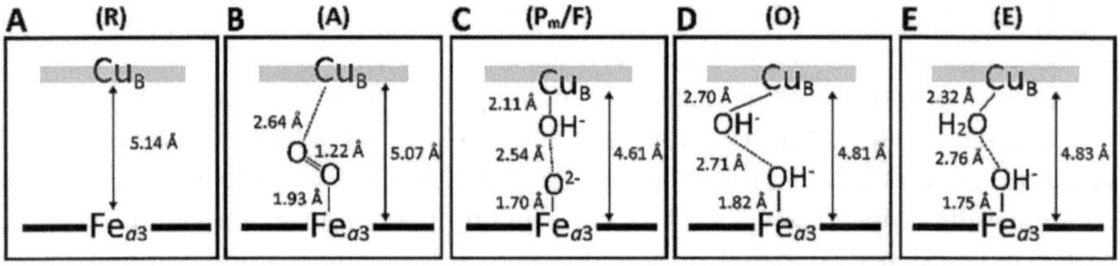

Fig. 4 Schematic representations of the X-ray structures of the intermediate forms. Panels **a**, **b**, **c**, **d**, and **e** show those of the R-, A-, Pm/F-, O- and E-forms, respectively. These structures were obtained from the final atomic models determined by X-ray structural analyses of fully reduced bovine heart CcO crystals exposed to excess O_2 for different periods. The structural analyses were performed taking the possibility of multiple structures into account. The accuracy of the determined structures depends on the occupancy of each intermediate form. Because of the low occupancy of the A-form (~20%), the reliability of these structural results is significantly lower than that of the other forms. However, the difference in the atomic distances between Cu_B–O and Fe_{a_3}–O in panel **b** is significant. The structure given in panel **c** was determined from the crystals of a 1:1 Pm/F mixture. The structural difference between the two forms is too small to detect at the resolution of the X-ray diffraction experiments. (Reprinted with permission from Shimada et al. (2023b))

shorter than that in intermediate O (1.82 Å), consistent with a ferryl-type $Fe_{a_3}^{4+} = O^{2-}$ species, supporting resonance Raman data (Ogura et al. 1993; Shimada et al. 2021). The distance between OH^- coordinated to Cu_B^{2+} and O^{2-} bound to $Fe_{a_3}^{4+}$ (2.54 Å) is shorter than that of a typical hydrogen bond, indicating the presence of a low-barrier hydrogen bond. The $Fe_{a_3}^{4+} = O^{2-}$ bond is not perpendicular to the heme plane; the N(imidazole)–$Fe_{a_3} = O$ angle is 175°, as shown in Fig. 4, confirming the nonlinearity inferred from Raman data (Shimada et al. 2023b). A structural water molecule capable of forming hydrogen bonds with Tyr244 and oxygen atoms was also detected in intermediates O and E. Additionally, a water channel with an opening to the membrane surface was identified, distinct from the O_2 delivery pathway and characterized by polar but uncharged residues, making it hydrophilic and impermeable to O_2. This spatial separation of substrate and product pathways is maintained in all other intermediates.

3.3.5 Intermediates O and E

In intermediate O, the Cu_B^{2+}–OH^- distance is unusually long (2.70 Å) (Fig. 4d), suggesting a high electron affinity of Cu_B^{2+}. In intermediate E, this distance shortens to 2.32 Å (Fig. 4e), though the heme a_3 geometry and O-ligand distances remain similar to those in O. Combined with minimal spectral changes during the O → E transition, this implies that the single electron transferred is accepted exclusively by Cu_B, forming Cu_B^+. This electron transfer shortens the Cu_B–O distance. If OH^- were retained near Cu_B^+, the resulting drop in electron affinity could cause partial reduction of Fe_{a_3}. Thus, Cu_B^+ is likely coordinated by H_2O.

3.3.6 Mechanism of O_2 Reduction

The combined Raman and crystallographic data suggest the O_2 reduction mechanism shown in Fig. 5. O_2 binds to Fe_{a_3} of intermediate R to form A (Fig. 5a). Four electrons are then transferred to generate intermediate P (Fig. 5c). During this step, Tyr244 captures an internal water molecule and, as indicated by the red arrows, establishes a low-barrier hydrogen-bond network enabling proton transfer from the D-pathway to Tyr244. The nonlinear geometry of N–$Fe_{a_3}^{4+} = O^{2-}$ in P decreases proton affinity and facilitates efficient proton transport. Electron transfer from heme *a* reduces the Tyr244 radical, and a proton is transferred via the D-pathway to yield intermediate F (Fig. 5d). A second proton from the D-pathway, coupled with electron transfer, forms intermediate O (Fig. 5d). Protons for subsequent E and R formation are provided via the K-pathway (Fig. 5e, f), aided by internal water molecules.

High energy conversion efficiency requires effective irreversibility in key steps (Shimada et al. 2023b). Electron transfer driving the P → F transition corresponds to Tyr244 radical reduction, and F → O involves ferryl reduction—both essentially irreversible processes (Shimada et al. 2020). The long Cu_B^{2+}–OH^- distance (2.70 Å)

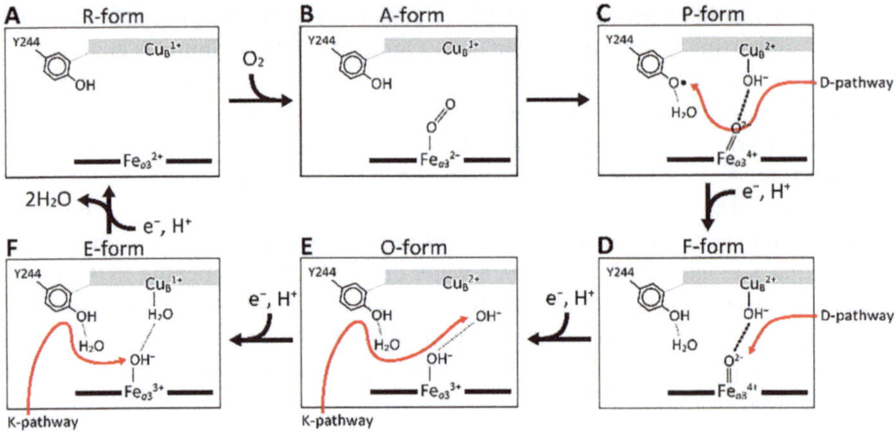

Fig. 5 A schematic representation of the catalytic cycle of bovine CcO based on the X-ray structures of the catalytic intermediate forms of bovine CcO. The O_2 reduction site includes Fe_{a_3}, Cu_B, and Tyr244OH group. When the R-form is attained (**a**), the $Fe_{a_3}^{2+}$ receives O_2 to yield the A-form providing a weak interaction between Cu_B^{1+} and the bound O_2, which is displayed by the absence of any line between Cu_B^{1+} and the O_2 (**b**). The present OxyMb-type structure supports the weak interaction between O_2 and Cu_B^{1+} in the A-form which has been proposed by resonance Raman and X-ray structural analyses (1, 12, 27). The A-form is relaxed into the P-form in which the bound O_2 has been completely reduced, giving $Fe_{a_3}^{4+} = O^{2-}$, Cu_B^{2+}–OH^-, and a Tyr244 radical with a hydrogen-bonded water (the interstitial water) (**c**). During the P → F transition, the Tyr244 radical is transformed to Tyr244OH by a proton-coupled electron transfer. The red arrow in (**d**) marks the substrate proton transfer pathway from D-pathway to the Tyr244 radical through the OH^- group at the Cu_B^{2+}, the O^{2-} at the $Fe_{a_3}^{4+}$, and the interstitial water. The bent structure of HisN–$Fe_{a_3}^{4+} = O^{2-}$ of the P-form (24) is illustrated schematically in (**c**). The second proton-coupled electron transfer to $Fe_{a_3}^{4+} = O^{2-}$ of the F-form from the D-pathway via the OH^- group, as illustrated by a red arrow in (**d**), forms the O-form (**e**). The low-barrier hydrogen bonds between $Fe_{a_3}^{4+} = O^{2-}$ and Cu_B^{2+}–OH^- in both the P- and F-forms are shown by thick dotted lines in (**c**) and (**d**). The unusually long distance between Cu_B^{2+} and OH^- is illustrated by deleting any lines between them (**e**). In the O to E transition (**e** to **f**), the substrate protons are transferred from K-pathway to the OH^- group near the Cu_B^{2+} through the interstitial water and the OH^- group at the $Fe_{a_3}^{3+}$ (**e**), giving Cu_B^{1+}–H_2O (**f**) as illustrated by red arrows (**e**). In the O- and E-forms, normal hydrogen bonds are detectable between the two oxygen atoms as illustrated by thin dotted lines (**e** and **f**). The fourth proton-coupled electron transfer through the K-pathway and interstitial water, illustrated by a red arrow (**f**), regenerates the R-form (**a**), releasing two water molecules. The interstitial water molecules, hydrogen-bonded to Tyr244 in (**c**), (**d**), (**e**), and (**f**), could be reversibly transferred from a storage site near the O_2 reduction site, not from the bulk water phase. (Adapted from Shimada et al. (2021), licensed under CC BY 4.0).

in intermediate O suggests high electron affinity. In E, Cu_B remains highly electron-affine while Fe_{a_3} remains nearly fully oxidized ($Fe_{a_3}^{3+}$–OH^-), which enhances Fe_{a_3} electron affinity (Shimada et al. 2021). These findings indicate that both O → E and E → R transitions are effectively irreversible.

3.4 Proton Pumping Mechanism

Two competing hypotheses have been proposed for the pathway by which pumped protons are transported in CcO: one involving the H-pathway and the other the D-pathway (Kaila et al. 2010; Yoshikawa and Shimada 2015; Shimada et al. 2023a, b). The debate remains unresolved. In this section, we present representative experimental findings that support each hypothesis and discuss the proposed reaction mechanisms in detail.

3.4.1 Structural Changes of Asp51 and the Water Channel in the H-Pathway

Figure 6 shows the X-ray crystallographic structure and schematic diagram of the H-pathway. Asp51, located at the exit on the P-side, is connected via a hydrogen-bonded network to a water channel. This residue is also connected through hydrogen bonds with the propionate and formyl (–CHO) groups of the heme *a* side chain (Shimada et al. 2023b). Given that heme *a* is a six-coordinate, low-spin species and thus uncharged in its reduced state, the observed structural arrangement suggests that upon oxidation, the resulting positive charge on heme *a* becomes delocalized. This delocalization may induce electrostatic repulsion with the pumped proton located on the hydrogen-bonded network, thereby driving proton translocation.

It is noteworthy that this hydrogen-bonded network includes peptide bonds. Proton transport across peptide bonds is facilitated via an imidic acid intermediate (–C(OH) = $N^{+}H$–), as previously demonstrated (Kamiya et al. 2007). Computational studies also support the feasibility of proton transport through this peptide-bonded hydrogen network in CcO. As illustrated in Fig. 6a, Asp51 is exposed to the molecular surface (P-side) in the R intermediate (cyan), whereas in other intermediates (magenta), it is buried within the protein, suggesting that the –COOH group is dissociated only in the R intermediate.

The water channel includes several internal cavities of sufficient size to permit free movement of water molecules (H_2O or H_3O^+) (Fig. 6b). In these regions, the mobility of water is not restricted by the protein environment, which likely facilitates rapid proton transport. X-ray crystallography has revealed that a cavity near the top of the water channel observed in the R intermediate (cyan circle in Fig. 6b) disappears in other intermediates, suggesting that proton transport via the water channel is markedly suppressed in all but the R intermediate (Shimada et al. 2020, 2021). These findings imply that four equivalents of pumped protons are initially loaded into the Mg^{2+}-containing water cluster in the R intermediate. Following O_2 binding (formation of intermediate A), the water channel is closed, and proton pumping is driven four times by the oxidation of heme *a*, as described above.

Additionally, X-ray crystallographic studies have demonstrated that formation of the CcO–N_3^- complex opens the water channel, allowing the removal of pumped protons (Shimada et al. 2018). This removal enhances the electron affinity of heme

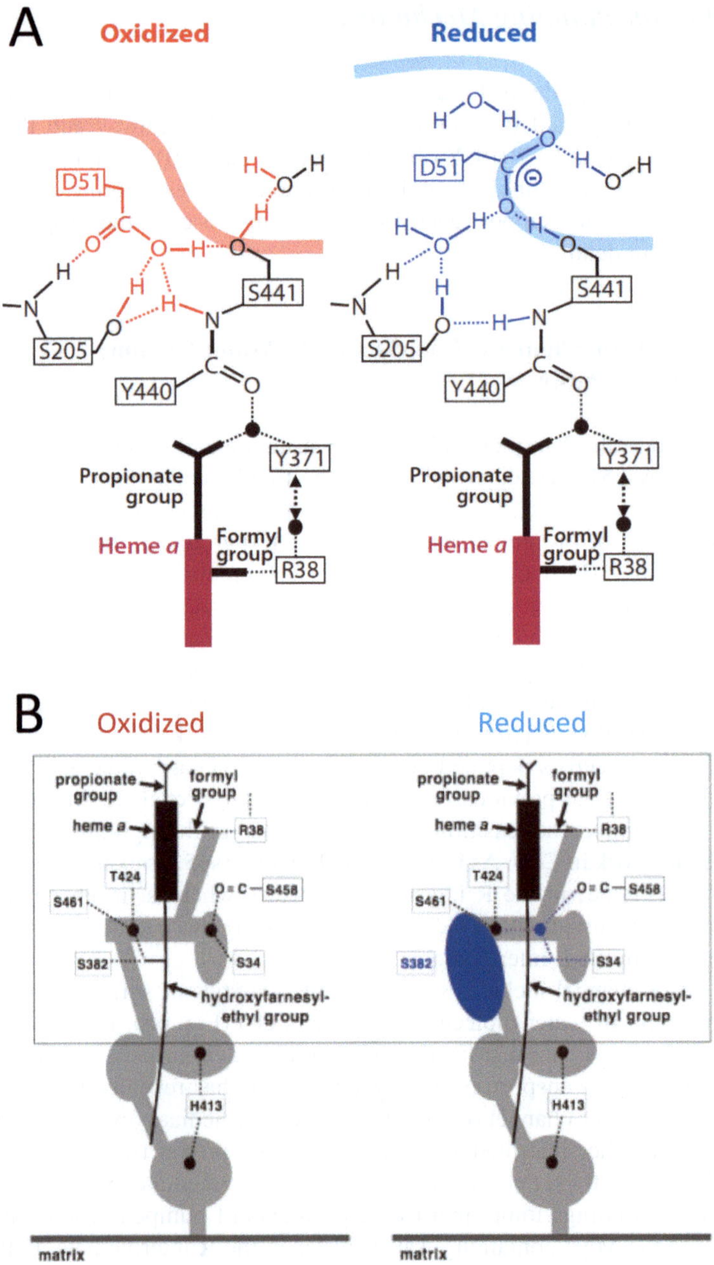

Fig. 6 Previously reported redox-coupled X-ray structural changes. (**a**) Schematic representation of Asp51 in the fully oxidized form structure (left) and the fully reduced form structure (right). The smooth thick curves denote the molecular surface accessible to the water molecules in the P-side phase. (**b**) Schematic representation of the redox-coupled structural changes in the water channel detectable in the residues 380–385 region including Ser382. The water cavity colored in blue in the fully reduced form (right) is eliminated upon complete oxidation to the fully oxidized form (left). (Adapted from Shimada et al. (2023a), licensed under CC BY 4.0)

a, strongly suggesting the existence of electrostatic interactions between heme *a* and the pumped protons. On the other hand, it has been experimentally confirmed that with each release of a pumped proton to the P-side, a new pumped proton is taken up from the N-side (Belevich et al. 2006). This implies that CcO incorporates pumped protons from a distinct proton-loading site—separate from the water cluster—concurrently with their release. When the R intermediate is formed and the water channel opens, the pumped protons are translocated from the proton-loading site to the water cluster. Experimental evidence supporting the existence of such a proton-loading site has been reported (Björck and Brzezinski 2018; Shimada et al. 2023b).

Time-resolved X-ray crystallography, performed at SPring-8 Angstrom Compact Free Electron Laser (SACLA) using carbon monoxide (CO) as a model ligand for O_2, has proposed a water-channel closure mechanism (Shimada et al. 2017a). One of the ligands of Cu_B, His291, forms a hydrogen bond with a water molecule fixed within the Mg^{2+}-containing water cluster (Fig. 7a). The saturation of the water cluster with protons is sensed by Cu_B via this fixed water and His291, triggering the acceptance of CO (or O_2) (Fig. 7b). Structural rearrangements involving the vinyl group of heme a_3 and the neighboring residues L381 and S382 on the same helix then induce closure of the water channel (Fig. 7c). This closure ensures that the water channel is sealed before O_2 binds to heme a_3 and the reduction reaction (and proton pumping) commences. Subsequently, the oxidation of heme *a* generates an electrostatic repulsion (double arrows) that drives proton pumping (Fig. 7d).

3.4.2 Mutational Analysis of the H-Pathway in Bovine Heart CcO

To evaluate the role of the H-pathway in proton pumping, a system was developed in HeLa cells for expressing bovine/human hybrid CcO that includes all residues of the H-pathway in subunit I (Shimokata et al. 2007). Three mutant variants—Asp51Asn, Tyr440Pro, and Val386Leu/Met390Trp—were found to lack proton pumping activity while retaining oxygen reduction capability. These findings strongly suggest distinct roles for Asp51 in proton translocation, for the peptide bond in forming imidic acid intermediates, and for the water channel in water molecule transport (Kamiya et al. 2007).

3.4.3 D-Pathway Mechanism

Based on mutagenesis studies in bacterial CcO, it has been proposed that the D-pathway (Fig. 3) serves not only for the delivery of chemical protons for water formation but also as a route for pumped protons (Kaila et al. 2010). The D-pathway extends from Asp91 at the N-side entry to Glu242 at the terminal end near the O_2 reduction site. X-ray crystallographic data suggest that Glu242 receives protons from Asp91 and delivers them to the O_2 reduction center (Fig. 3). The fact that the Glu242Gln mutation inactivates CcO indicates that Glu242 undergoes cycles of

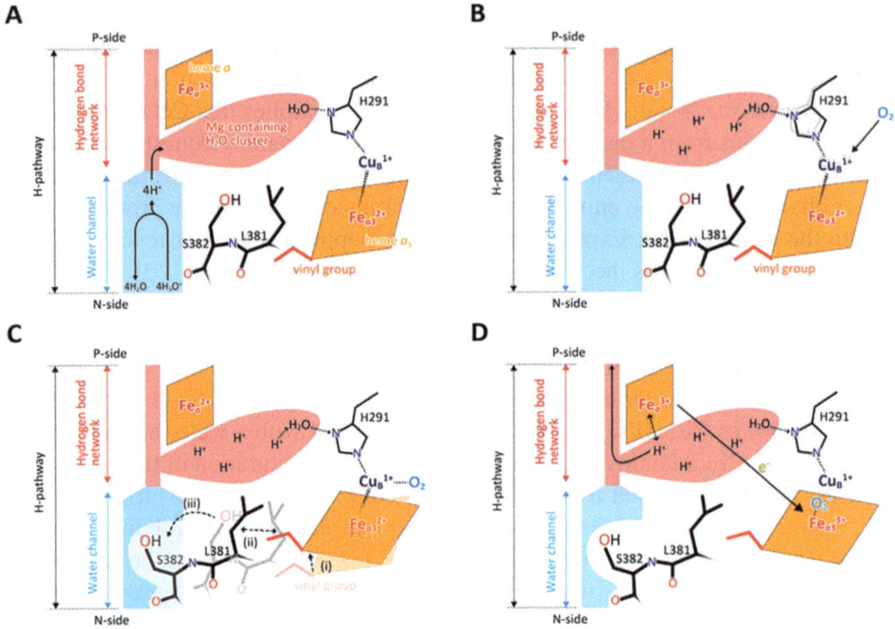

Fig. 7 A schematic representation of the water channel closure mechanism. (**a**) The water channel closure system in which the O_2 reduction site composed of Cu_B and Fe_{a_3} is in the fully reduced state under turnover conditions. The Mg-containing H_2O cluster is the H^+ storage site which connects with Cu_B via H291 and a fixed water molecule (HOH10 in Fig. 1) in the storage site. The heme a_3 vinyl group is in van der Waals contact with L381. Protons are transferred by H_3O^+ through the open water channel. (**b**) Increase in the O_2 affinity of Cu_B^{1+} by distortion of the regular trigonal coordination of Cu_B^{1+} by protonation of the fixed water upon full protonation of the Mg-containing H_2O cluster. (**c**) Closure of the water channel upon O_2 binding to Cu_B as follows: (i) migration of the plane of heme a_3, (ii) structural change of L381, and (iii) formation of the bulge at S382 to close the channel. (**d**) Proton pump after the channel closure. After migration of O_2 to Fe_{a_3}, Fe_a^{2+} is oxidized to pump protons. (Adapted from Shimada et al. (2017a), licensed under CC BY 4.0).

protonation and deprotonation to transport protons (Adelroth et al. 1997; Konstantinov et al. 1997).

Furthermore, the Asn98Asp mutation, which is adjacent to Asp91 within the D-pathway, abolishes proton pumping activity without significantly impairing O_2 reduction (Brändén et al. 2006). Based on these results, it has been proposed that the D-pathway also transports pumped protons as follows: Glu242 translocates a pumped proton to a loading site in concert with electron transfer from heme *a* to heme a_3. It then delivers a chemical proton to the O_2 reduction site, and the resulting electrostatic repulsion expels the loaded proton to the P-side. However, because no direct pathway has been observed for transporting protons from the D-pathway to the Mg^{2+}-containing water cluster, it is unlikely that this cluster serves as the loading site. Despite various mutational analyses, the identity of the loading site remains undetermined (Popović and Stuchebrukhov 2005; Cai et al. 2018).

While Asn98 is part of the D-pathway, the loss of pumping activity in the Asn98Asp mutant does not necessarily confirm the D-pathway as the pumping route (Pawate et al. 2002; Han et al. 2005). Structural conversion from Asn to Asp might be expected to facilitate, rather than hinder, proton transport. Therefore, the observed loss of pumping activity may result from secondary structural perturbations of the functional center involved in pumping. However, the data do not permit the conclusion that such secondary effects are limited solely to the D-pathway. Conversely, since the proton pumping function is presumed to be evolutionarily conserved, it seems unlikely that a residue such as Asp51—conserved only in mammals—would play a central role in such a fundamental function. Thus, despite the aforementioned issue, the D-pathway, being conserved even in bacteria, was long considered the likely route for pumped protons. However, the discovery of CcO variants lacking a conserved D-pathway has refuted this evolutionary argument, and the nonconservation of the proton pumping mechanism is now widely accepted.

4 Previously Identified Regulators of CcO Activity

A wide range of molecules capable of positively or negatively regulating the catalytic activity of CcO have been reported to date (Yoshikawa and Shimada 2015). While some of these molecules may be directly applicable as therapeutic agents for modulating CcO activity, many of them are not immediately suitable for practical use. This is often due to limitations such as chemical instability or unfavorable pharmacokinetic properties that hinder their administration in vivo.

Nevertheless, structural analyses of CcO in complex with these regulatory molecules—even those based on computational predictions such as molecular docking simulations—provide invaluable insights that can significantly contribute to the rational design of more stable and practical modulators of CcO activity. These structure-based approaches are expected to play a critical role in drug development targeting mitochondrial bioenergetics.

An equally important consideration in the development of CcO-targeted modulators is tissue specificity (LaMoia et al. 2022). That is, the effectiveness of a given regulatory molecule may depend on whether it is capable of modulating the CcO isoform expressed in the target tissue. Notably, all structurally characterized mammalian CcO complexes to date have been derived from cardiac tissue. In contrast, no high-resolution structures of CcO isolated from other pharmacologically relevant tissues, such as the liver or nervous system, have yet been reported. Previous studies have demonstrated that the composition of nuclear-encoded subunit isoforms in CcO varies depending on the tissue of expression. Therefore, elucidating the full structures of CcO derived from noncardiac tissues remains an urgent objective in the field. Nonetheless, the catalytic core subunits essential for enzymatic activity are conserved across tissues (Yoshikawa and Shimada 2015). Consequently, structural and biochemical analyses of CcO from cardiac tissue—including investigations of its interactions with regulatory molecules—remain highly valuable and should not be disregarded when considering therapeutic applications.

4.1 Inhibitors of CcO

4.1.1 Oxygen Analogs

A group of small molecules—including carbon monoxide (CO), azide (N_3^-), cyanide (CN^-), nitric oxide (NO), and adenosine triphosphate (ATP)—are known to act as inhibitors of CcO activity (Yoshikawa and Shimada 2015). Among them, CO, NO, CN^-, and N_3^- function as oxygen analogs by competitively binding to the oxygen reduction site of CcO, thereby blocking its catalytic function (Yoshikawa and Caughey 1990, 1992; Muramoto et al. 2010). As described previously, CcO-mediated proton pumping is tightly coupled to the electron transfer from cytochrome *c* to the oxygen reduction center. Therefore, when these oxygen analogs occupy the site in place of molecular oxygen, proton translocation is also abrogated.

CO, which shares similar chemical properties and size with O_2, binds to the O_2 reduction center in an almost identical manner (Muramoto et al. 2010). CO binding induces a slight displacement of heme a_3 toward the helix X region, located near the H-pathway. This conformational shift results in the collapse of a major water cavity that constitutes part of the H-pathway's water channel, effectively sealing the pathway and preventing proton uptake and back leakage (Shimada et al. 2017a, b). As a consequence, the proton pump activity is abolished. Nitric oxide (NO) inhibits CcO via a similar mechanism, also binding competitively to the oxygen reduction site and causing a structural rearrangement that impairs H-pathway function (Muramoto et al. 2010).

In contrast, azide (N_3^-) and cyanide (CN^-) elicit distinct structural effects upon binding. In the case of azide, two N_3^- molecules simultaneously bind to the oxygen reduction site (Yoshikawa and Caughey 1992; Shimada et al. 2018). The electrostatic and steric repulsion between these anions displaces heme a_3 away from helix X—that is, in the opposite direction of the CO- or NO-induced shift. CN^- binding induces a similar displacement but follows a different mechanism (Shimada et al. 2023a). To access the oxygen reduction site, CN^- must first enter the CcO interior in its neutral form as hydrogen cyanide (HCN). Once inside, HCN undergoes deprotonation, and the liberated proton must be accommodated by proton-accepting functional groups in the vicinity of the binuclear center. High-resolution X-ray crystallographic studies have revealed that this proton is accepted by His290, one of the three histidine ligands coordinating Cu_B. As a result, His290 becomes protonated and detaches from Cu_B, allowing CN^- to bridge between Cu_B and Fe_{a_3}. This bridging configuration causes heme a_3 to shift in the same direction as that observed with N_3^- binding.

Interestingly, while CO and NO binding lead to closure of the H-pathway by collapsing water cavities, CN^- and N_3^- binding expand the water channel. At first glance, this structural expansion may seem inconsistent with the inhibitory effect. However, persistent opening of the H-pathway permits backflow of protons from the hydrogen-bonded network to the N-side, resulting in a collapse of the proton gradient and functional inhibition of CcO. Thus, even structurally divergent modes of inhibition ultimately converge on the same outcome—loss of proton pumping and catalytic activity.

Among all known inhibitors of CcO, oxygen analogs represent the most extensively characterized group. While they effectively abolish CcO activity, their practical application as regulatory agents is severely limited by their high affinity for a broad range of heme-containing proteins, including hemoglobin and myoglobin (Schuresko and Webb 1978; Barends et al. 2015). Therefore, their therapeutic utility as CcO-selective inhibitors is highly constrained. Nevertheless, structural studies of CcO in complex with these inhibitors have significantly advanced our understanding of the catalytic mechanism. In particular, they have elucidated critical aspects of the structural interplay between the oxygen reduction center and the H-pathway, underscoring the profound importance of these analogs as mechanistic probes rather than direct therapeutic agents (Muramoto et al. 2010; Shimada et al. 2018, 2023a).

4.1.2 Small-Molecule Inhibitors of CcO

Among small-molecule inhibitors of CcO, adenosine triphosphate (ATP) is one of the most well-characterized (Napiwotzki et al. 1997; Bender and Kadenbach 2000; Ramzan et al. 2022). Given that CcO plays a central role in oxidative phosphorylation and thereby in the synthesis of cellular ATP, it is physiologically reasonable that ATP itself acts as an allosteric inhibitor under conditions of elevated intracellular ATP concentrations. ATP-binding sites are primarily located on nuclear-encoded subunits of CcO, with subunit IV being particularly important for mediating ATP-dependent inhibition. Binding of ATP to this subunit has been shown to reduce proton pumping efficiency by approximately 50% ($H^+/e^- = 0.5$).

This ATP-mediated regulatory mechanism is of significant interest not only for understanding feedback inhibition in bioenergetics, but also for elucidating the functional roles of noncore subunits. However, the structural basis of this inhibition remains poorly understood. No high-resolution structure of ATP-bound CcO has been reported to date, and the manner in which ATP binding to subunit IV influences the enzyme's active site—comprising the oxygen reduction center and proton translocation pathways—remains entirely unknown (Bender and Kadenbach 2000; Kadenbach and Hüttemann 2015). In existing CcO crystal structures, the putative ATP-binding sites are occupied by cholate, a detergent used during protein purification, which may compete with ATP and prevent its binding (Ramzan et al. 2021).

Hormonal inhibitors such as thyroid hormones and steroid hormones have also been reported to bind directly to CcO and inhibit its enzymatic activity (Oleynikov et al. 2020, 2022). These molecules are predicted to bind at lipid- or cholate-binding regions near the entrance of the K-pathway, a proton uptake channel located within the core subunits. Biochemical analyses have demonstrated that such binding interferes with the oxygen reduction reaction catalyzed by CcO. Because the binding sites for these hormones are located within highly conserved regions of the core subunits, it is strongly suggested that such inhibition constitutes a physiologically relevant regulatory mechanism that is conserved across species.

Guanidine- and biguanide-based drugs—including metformin, galegine, and phenformin—are widely known for their use in the treatment of type 2 diabetes (LaMoia et al. 2022). Due to severe side effects, only metformin remains in clinical use today. These compounds were originally believed to act primarily on complex I of the mitochondrial respiratory chain (Vial et al. 2019). However, more recent studies have indicated that they may also directly interact with CcO (Bridges et al. 2014; LaMoia et al. 2022). In particular, their inhibitory effects are enhanced at slightly alkaline pH values (~7.9), which suggests that their activity may be modulated by the protonation state of certain residues within or near the binding site. Nevertheless, the precise binding site(s) remain controversial. Given their molecular size, it is unlikely that these drugs bind directly to the deeply embedded oxygen reduction center; instead, they are presumed to associate with surface-exposed regions of the core or accessory subunits.

Because ATP, guanidine–/biguanide-class drugs, and endogenous hormones inhibit CcO activity in a moderate and potentially reversible manner, they represent attractive candidates for the development of CcO activity modulators. To facilitate rational drug design and better understand the underlying regulatory mechanisms, structural elucidation of CcO in complex with these inhibitors will be essential.

4.2 Activators of CcO

4.2.1 ADP

The activation of cytochrome *c* oxidase (CcO) by adenosine diphosphate (ADP) mirrors, in many respects, the inhibitory effect exerted by ATP (Napiwotzki et al. 1997). Specifically, under conditions of energy depletion—characterized by a significantly reduced intracellular ATP/ADP ratio—ADP enhances CcO activity to stimulate oxidative phosphorylation and thereby promote ATP synthesis (Napiwotzki and Kadenbach 1998; Beauvoit and Rigoulet 2001). Due to structural similarity between ADP and ATP, multiple ADP-binding sites are presumed to exist on CcO, particularly on the nuclear-encoded subunits. Among these, binding to subunit IV has been shown to play a critical role in modulating enzymatic activity.

As with ATP, the three-dimensional structure of an ADP-bound CcO complex remains unresolved. Nonetheless, the fact that the presence or absence of a single phosphate group can result in diametrically opposing regulatory effects on enzyme activity underscores the mechanistic sophistication of this regulation. This binary control mechanism—whereby ADP activates and ATP inhibits CcO—offers a compelling model for energy-responsive modulation of mitochondrial respiration and may present a valuable framework for the development of novel therapeutic agents targeting mitochondrial bioenergetics.

4.2.2 Higd1a

Proteins belonging to the hypoxia-inducible gene domain 1 (Higd1) family are a class of hypoxia-inducible factors that are evolutionarily conserved across species from bacteria to eukaryotes (Hayashi et al. 2015). Under hypoxic conditions, oxidative phosphorylation is downregulated due to diminished oxygen availability, resulting in a drop in intracellular ATP levels. Because CcO is the only terminal oxidase in the electron transport chain that directly reduces molecular oxygen, a decline in its activity under hypoxia has a profound impact on the efficiency of oxidative phosphorylation.

Members of the Higd1 family are upregulated in response to hypoxia and are also known by other names such as "respiratory supercomplex factor 1" (Rcf1). In yeast, both Rcf1 and a yeast-specific homolog Rcf2 have been identified; both proteins are predicted to span the mitochondrial inner membrane twice (Hoang et al. 2019; Strogolova et al. 2019). These proteins interact directly with subunit III of CcO, promoting the assembly of CcO subunits and contributing to the structural stability of the enzyme complex. Importantly, CcO undergoes a phenomenon known as "suicide inactivation" after repeated catalytic cycles, whereby its activity gradually decreases. Rcf1 and Rcf2 are thought to suppress this inactivation, thereby preventing the backflow of pumped protons through inactivated CcO and helping to maintain the mitochondrial membrane potential (Hoang et al. 2019; Timón-Gómez et al. 2020).

In addition to Rcf1, the human homolog Higd1a has been identified and is also predicted to contain two transmembrane domains. Like Rcf1 and Rcf2, Higd1a has been reported to interact directly with CcO and to enhance its proton pumping activity (Hayashi et al. 2015). While this direct interaction has been demonstrated spectroscopically, the precise binding site remains unknown. Molecular docking simulations suggest that Higd1a may interact with helix IX of subunit I and helix I of subunit II of CcO (Shimada et al. 2023a, b). These interactions are predicted to influence the structure of the oxygen reduction center and the H-pathway, potentially widening the water channel within the H-pathway. However, this remains speculative, as it is based solely on in silico modeling, and the detailed mechanism by which Higd1a activates CcO is yet to be fully elucidated.

Recent studies have revealed that the expression of Higd1 family proteins is not limited to hypoxic conditions but can also be induced under low glucose availability and oxidative stress, suggesting that these proteins may function as versatile regulators of CcO activity in response to various cellular stresses (Nagao et al. 2020). Structural elucidation of Higd1 family proteins and their complexes with CcO is expected to significantly advance our understanding of the positive regulation of CcO activity and may provide new opportunities for the development of therapeutic agents targeting mitochondrial dysfunction.

4.2.3 MNRR1 (CHCHD2)

Mitochondrial nuclear retrograde regulator 1 (MNRR1), also known as coiled-coil-helix-coiled-coil-helix domain-containing protein 2 (CHCHD2), is a soluble protein encoded by the nuclear genome and is well known as one of the genes implicated in the pathogenesis of Parkinson's disease (Meng et al. 2017; Kee et al. 2021). MNRR1 contains four cysteine residues that form two intramolecular disulfide bonds. Aside from the domain containing these disulfide linkages, the remainder of the protein is considered to be intrinsically disordered, as no stable tertiary structure has been predicted by current structural prediction algorithms. Consequently, the complete three-dimensional structure of MNRR1 remains unresolved.

MNRR1 has been shown to physically associate with cytochrome *c* oxidase (CcO) through co-immunoprecipitation assays, suggesting a direct interaction with the enzyme (Aras et al. 2015; Yanagisawa et al. 2024). Functional analyses have demonstrated that MNRR1 acts as a positive regulator of CcO activity. However, due to the absence of structural data for the MNRR1–CcO complex, the precise binding site of MNRR1 on CcO remains unknown, and the molecular mechanism by which it activates CcO has yet to be elucidated.

Given that MNRR1 is a soluble protein, it is evident that its binding mode to CcO differs from that of the HigD1 family proteins, which are membrane-spanning. While the HigD1 family proteins are expected to bind primarily through hydrophobic interactions with the transmembrane regions of CcO, MNRR1 is presumed to interact with the hydrophilic surface of CcO, particularly within the intermembrane space. This distinction strongly suggests that MNRR1 and HigD1 proteins activate CcO through independent mechanisms.

Accordingly, small-molecule activators designed based on the MNRR1–CcO interaction are expected to differ in both structure and binding mode from those derived from HigD1-based mechanisms. The existence of multiple regulatory mechanisms for CcO activation highlights the potential for developing a diverse array of CcO activators. Such diversity enables tailored drug discovery strategies depending on the desired regulatory pathway and thus holds significant promise for the development of targeted therapeutics for mitochondrial dysfunction.

5 Structure-Based Drug Discovery Targeting High-Resolution Structures of CcO

High-resolution structural data of CcO, particularly from bovine heart mitochondria, have been extensively reported (Shinzawa-Itoh et al. 2021; Shimada et al. 2023a). However, despite the availability of these structural insights, crystal structures of CcO in complex with physiological regulatory ligands remain scarce, with the notable exception of oxygen analogs. One plausible reason for this scarcity is that many regulatory molecules bind to surface-accessible regions of CcO with relatively low affinity. Given the central role of CcO in oxidative phosphorylation, it is reasonable to speculate that physiological regulators are evolutionarily tuned to bind weakly, in order to avoid irreversible perturbation of mitochondrial energy metabolism.

Another contributing factor is the interference of detergents commonly used in purification protocols. These detergents may occupy potential binding sites for allosteric modulators, thereby hindering ligand association (Shinzawa-Itoh et al. 2007). To overcome this, attempts have been made to replace bound detergents with regulatory compounds via dialysis. While such approaches have yielded biochemical evidence of ligand binding and activity modulation, structural characterization of the resulting complexes remains elusive.

A noteworthy example of progress in this area is the work by Nishida, Shintani, and colleagues, who systematically investigated allosteric regulation of bacterial CcO (Fig. 8) (Nishida et al. 2022). They identified a highly conserved allosteric binding site among bacterial heme–copper oxidases. Interestingly, although this allosteric pocket is conserved across species, it is structurally occluded in eukaryotic CcOs due to additional helical segments. Capitalizing on this structural divergence, they designed synthetic inhibitors that selectively bind to the bacterial allosteric site without affecting mammalian CcO. Their work demonstrates a rational approach to developing pathogen-specific antibiotics based on subtle differences within otherwise conserved enzyme architectures.

This strategy highlights a promising avenue in drug discovery: targeting evolutionarily variable allosteric sites to achieve species selectivity and minimize host toxicity. Because allosteric regions are less evolutionarily constrained than catalytic cores, they offer attractive targets for the development of highly selective antimicrobials. Furthermore, this principle may be extended to other protein targets, paving the way for novel antibiotics with unique mechanisms of action and potential positive allosteric modulators for treating loss-of-function diseases.

6 Future Perspectives

Although drug discovery targeting CcO is still in its early stages and tangible outcomes are limited, a wealth of high-resolution structural data is now available. These include multiple crystal structures of CcO in various redox states and in complexes with small molecules, as well as recent advances in time-resolved structural studies using X-ray free-electron lasers (XFEL). Such developments are expected to significantly deepen our understanding of the CcO catalytic mechanism.

In structure-based drug design, understanding not only the reaction mechanism but also the conformational dynamics that underlie enzymatic activity is critical. With increasing structural information on CcO, rational development of activity-modulating compounds—either inhibitors or activators—will become increasingly feasible.

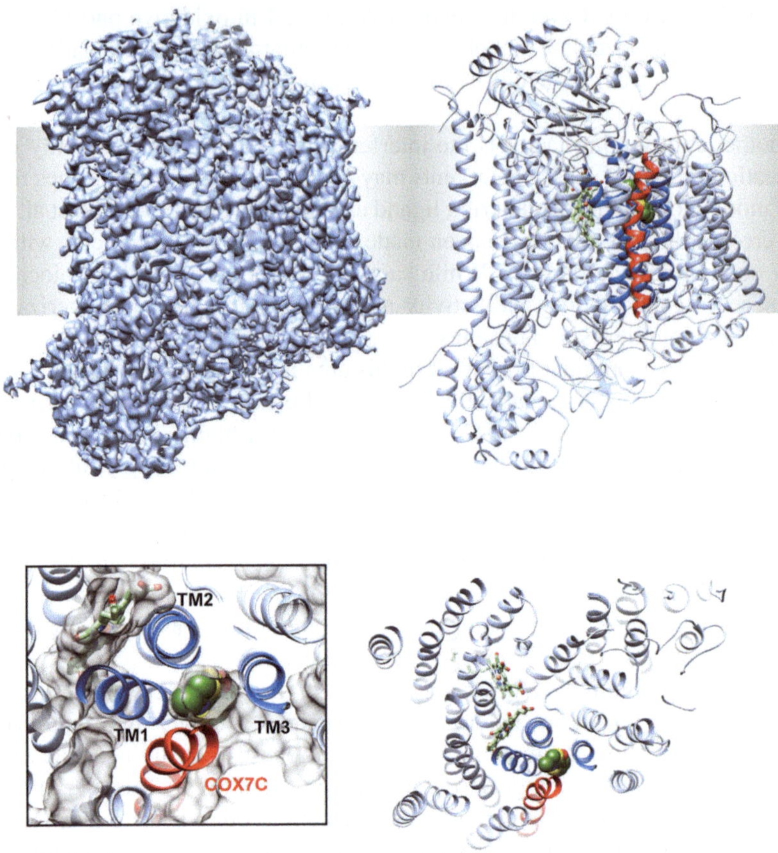

Fig. 8 The allosteric site for T113 is buried inside eukaryotic mtCcO. The structure of mtCcO with T113 was determined by X-ray crystallography. The electron density map (2Fo–Fc), contoured at 1σ, is shown in the upper left panel. The ribbon model of mtCcO with T113 in the sphere is shown on the upper right panel. T113 is covered by COX7C (red), hidden from the surface (lower right panel). T113 is surrounded by four transmembrane helices (TM1–3, COX7C) and buried from the surface, viewed from the intermembrane space. Protein molecular surface is shown as gray in the close-up view. Three helices of subunit I surrounding the allosteric site are shown in dark blue, the other helices of subunit I in pale blue, and subunit COX7C in mtCcO is shown in red (lower left panel). (Adapted from Nishida et al. (2022), licensed under CC BY 4.0)

However, caution is essential: Since CcO plays a central role in oxidative phosphorylation, even modest perturbations in its activity may lead to severe physiological consequences. Therefore, future drug design efforts must prioritize not only efficacy and selectivity but also precise modulation of enzyme activity to avoid systemic toxicity.

References

Abu-Amero KK, Bosley TM (2006) Mitochondrial abnormalities in patients with LHON-like optic neuropathies. Invest Ophthalmol Vis Sci 47(10):4211–4220. https://doi.org/10.1167/iovs.06-0295

Adelroth P, Svensson Ek M, Mitchell D, Gennis R, Brzezinski P(1997) Glutamate 286 in cytochrome aa3 from Rhodobacter sphaeroides is involved in proton uptake during the reaction of the fully-reduced enzyme with dioxygen. Biochemistry 36(45):13824–13829. https://doi.org/10.1021/bi9629079

Aharon-Hananel G, Romero-Afrima L, Saada A, Mantzur C, Raz I, Weksler-Zangen S (2022) Cytochrome c oxidase activity as a metabolic regulator in pancreatic beta-cells. Cells 11(6):929. https://doi.org/10.3390/cells11060929

Aich A, Boshnakovska A, Witte S, Gall T, Untham-Fechner K, Yousefi R, Chowdhury A, Dahal D, Methi A, Kaufmann S, Sibern I, Prochazka J, Nichtova Z, Palkova M, Raishbrook M, Koubkova G, sedlacek R, Troder S, Zevnik B, Riedel D, Michanski W, Strobel P, Luchtenborg C, Giavalisco P, Urlaub H, Fischer A, Brugger B, Jakobs S, Rehling P (2024) Defective mitochondrial COX1 translation due to loss of COX14 function triggers ROS-induced inflammation in mouse liver. Nat Commun 15(1):6914. https://doi.org/10.1038/s41467-024-51109-y

Alcalá M, Calderon-Dominguez M, Bustos E, Ramos P, Casals N, Serra D, Viana M, Herrero L (2017) Increased inflammation, oxidative stress and mitochondrial respiration in brown adipose tissue from obese mice. Sci Rep 7(1):16082. https://doi.org/10.1038/s41598-017-16463-6

Antonicka H, Leary SC, Guercin GH, Agar JN, Horvath R, Kennaway NG, Shoubridge EA (2003) Mutations in COX10 result in a defect in mitochondrial heme A biosynthesis and account for multiple, early-onset clinical phenotypes associated with isolated COX deficiency. Hum Mol Genet 12(20):2693–2702. https://doi.org/10.1093/hmg/ddg284

Apostolova N, Iannantuoni F, Gruevska A, Muntane J, Rocha M, Victor V (2020) Mechanisms of action of metformin in type 2 diabetes: effects on mitochondria and leukocyte-endothelium interactions. Redox Biol 34:101517. https://doi.org/10.1016/J.REDOX.2020.101517

Aras S, Bai M, Lee I, Springett R, Hüttemann M, Grossman L (2015) MNRR1 (formerly CHCHD2) is a bi-organellar regulator of mitochondrial metabolism. Mitochondrion 20:43–51. https://doi.org/10.1016/J.MITO.2014.10.003

Barends TRM, Foucar L, Ardevo A, Nass K, Aquila A, Botha S, Doak RB, Fakahati K, Hartmann E, Hilpert M, Heinz M, Hoffmann MC, Kofinger J, Koglin JE, Kovacsova G, Liang M, Milathianaki D, Lemke HT, Reinstein J, Roome CM, Shoeman RL, Boutet S, Schlichting I(2015) Direct observation of ultrafast collective motions in CO myoglobin upon ligand dissociation. Science (New York, N.Y.) 350(6259):445–450. https://doi.org/10.1126/science.aac5492

Beauvoit B, Rigoulet M (2001) Regulation of cytochrome c oxidase by adenylic nucleotides. Is oxidative phosphorylation feedback regulated by its end-products? IUBMB Life 52(3–5):143–152. https://doi.org/10.1080/152165401317316545

Belevich I, Verkhovsky MI, Wikström M (2006) Proton-coupled electron transfer drives the proton pump of cytochrome c oxidase. Nature 440(7085):829–832. https://doi.org/10.1038/nature04619

Bender E, Kadenbach B (2000) The allosteric ATP-inhibition of cytochrome c oxidase activity is reversibly switched on by cAMP-dependent phosphorylation. FEBS Lett 466(1):130–134. https://doi.org/10.1016/S0014-5793(99)01773-1

Bi R, Zhang W, Zhang D, Xu M, Fan Y, Hu Q, Jiang H, Tan L, Li T, Fang Y, Zhag C, Yao Y (2018) Genetic association of the cytochrome c oxidase-related genes with Alzheimer's disease in Han Chinese. Neuropsychopharmacology 43(11):2264–2276. https://doi.org/10.1038/s41386-018-0144-3

Björck ML, Brzezinski P (2018) Control of transmembrane charge transfer in cytochrome c oxidase by the membrane potential. Nat Commun 9(1):3187. https://doi.org/10.1038/s41467-018-05615-5

Brändén G, Pawate SA, Gennis BR, Brzezinski P (2006) Controlled uncoupling and recoupling of proton pumping in cytochrome c oxidase. Proc Natl Acad Sci USA 103(2):317–322. https://doi.org/10.1073/pnas.0507734103

Bridges HR, Jones AJ, Pollak MN, Hirst J (2014) Effects of metformin and other biguanides on oxidative phosphorylation in mitochondria. Biochem J 462(3):475–487. https://doi.org/10.1042/BJ20140620

Cai X, Haider K, Lu J, Radic S, Son YC, Cui Q, Gunner MR (2018) Network analysis of a proposed exit pathway for protons to the P-side of cytochrome c oxidase. Biochim Biophys Acta Bioenerg 1859(10):997–1005. https://doi.org/10.1016/j.bbabio.2018.05.010

Čapková M, Houstek J, Hansikova H, Hainer V, Kunesova M, Zeman J (2002) Activities of cytochrome c oxidase and citrate synthase in lymphocytes of obese and normal-weight subjects. Int J Obes 26(8):1110–1117. https://doi.org/10.1038/sj.ijo.0802055

Caughey SW, Wallace JW, Volpe AJ, Yoshikawa S (1976) Cytochrome c oxidase. In: Boyer PD (ed.) The enzymes 3rd ed. 13th edn. Academic Press, New York, pp 299–344

Clark KM, Taylor RW, Johnson MA, Chinnery PF, Chrzanowska-Lightowlers ZMA, Andrews RM, Nelson IP, Wood NW, Lamont PJ, Hanna MG, Lightowlers RN, Turnbull DM (1999) An mtDNA mutation in the initiation codon of the cytochrome C oxidase subunit II gene results in lower levels of the protein and a mitochondrial encephalomyopathy. Am J Hum Genet 64(5):1330–1339. https://doi.org/10.1086/302361

Cottrell DA, Borthwick GM, Johnson MA, Ince PG, Turnbull DM (2002) The role of cytochrome c oxidase deficient hippocampal neurones in Alzheimer's disease. Neuropathol Appl Neurobiol 28(5):390–396. https://doi.org/10.1046/j.1365-2990.2002.00414.x

Danyu L, Yanran L, Xiuna J, Ying C, Sudan P, Tianen Z, Zhifen Z, Dezhi Z, Kaixun H, Yingyu X, Enxiang T (2019) α-Synuclein induced mitochondrial dysfunction via cytochrome c oxidase subunit 2 in SH-SY5Y cells. Exp Cell Res 378(1):57–65. https://doi.org/10.1016/J.YEXCR.2019.02.006

Diaz F (2010) Cytochrome c oxidase deficiency: patients and animal models. Biochim Biophys Acta 1802(1):100–110. https://doi.org/10.1016/J.BBADIS.2009.07.013

Dimauro S, Schon EA (2003) Mechanisms of disease. Mitochondrial Respiratory-Chain Diseases. Available at: www.nejm.org.

DiMauro S, Lombes A, Nakase H, Mita S, Fabrizi GM, Bonilla E, Miranda A, DeVivo DC, Schon EA (1990) Cytochrome c oxidase deficiency. Pediatr Res 28(5):536–541. https://doi.org/10.1203/00006450-199011000-00025

Dodson ED, Zhao XJ, Caughey WS, Elliott CM (1996) Redox dependent interactions of the metal sites in carbon monoxide-bound cytochrome c oxidase monitored by infrared and UV/visible spectroelectrochemical methods. Biochemistry 35(2):444–452. https://doi.org/10.1021/bi951313n

Fernandez-Vizarra E, Zeviani M (2021) Mitochondrial disorders of the OXPHOS system. FEBS Lett 595(8):1062–1106. https://doi.org/10.1002/1873-3468.13995

Figueroa-Martínez F, Vazquez-Acevedo M, Cortes-Hernandez P, Garcia-Trejo JJ, Davidson E, King MP, Gonzalez-Halphen D (2011) What limits the allotopic expression of nucleus-encoded mitochondrial genes? The case of the chimeric Cox3 and Atp6 genes. Mitochondrion 11(1):147–154. https://doi.org/10.1016/J.MITO.2010.09.003

Fukui H, Diaz F, Garcia S, Moraes CT, (2007) Cytochrome c oxidase deficiency in neurons decreases both oxidative stress and amyloid formation in a mouse model of Alzheimer's disease. Proc Natl Acad Sci 104(35):14163–14168. https://doi.org/10.1073/pnas.0705738104

Grossman LI, Lomax MI (1997) Nuclear genes for cytochrome c oxidase. Biochim Biophys Acta 1352(2):174–192. https://doi.org/10.1016/S0167-4781(97)00025-0

Han D, Morgan JE, Gennis RB (2005) G204D, a mutation that blocks the proton-conducting D-channel of the aa3-type cytochrome c oxidase from Rhodobacter sphaeroides. Biochemistry 44(38):12767–12774. https://doi.org/10.1021/bi051141m

Hanna MG, Nelson I, Rahman S, Lane RJM, Land J, Heales S, Cooper MJ, Schapira AHV, Morgan-Hughes JA, Wood NW (1998) Cytochrome c oxidase deficiency associated with the first stop-codon point mutation in human mtDNA. Am J Hum Genet 63(1):29–36. https://doi.org/10.1086/301910

Hayashi T, Asano Y, Shintani Y, Aoyama H, Kioka H, Hikit M, Shinzawa-Itoh K, Takafuji K, Higo S, Kato H, Yamazaki S, Matsuoka K, Nakan A, Asanuma H, Asakr M, Minamino T, Got Y, Ogura T, Kitakaze M, Komuro I, Sakat Y, Tsukihar T, Yoshikwa S, Takashima S (2015) Higd1a is a positive regulator of cytochrome c oxidase. Proc Natl Acad Sci USA 112(5):1553–1558. https://doi.org/10.1073/pnas.1419767112

Hirata K, Shinzawa-Itoh K, Yano N, Takemura S, Kat K, Hatank M, Muramoto K, Kawahara T, Tsukihara T, Yamashita E, Tono K, Ueno G, Hikima T, Murakami H, Inubushi Y, Yabashi M, Ishikawa T, Yamamoto M, Ogura T, Sugimoto H, Shen J, Yoshikawa S, Ago H (2014) Determination of damage-free crystal structure of an X-ray-sensitive protein using an XFEL. Nat Methods 11(7):734–736. https://doi.org/10.1038/nmeth.2962

Hoang NH, Strogolova V, Mosley JJ, Stuart RA, Hosler J (2019) Hypoxia-inducible gene domain 1 proteins in yeast mitochondria protect against proton leak through complex IV. J Biol Chem 294(46):17669–17677. https://doi.org/10.1074/JBC.RA119.010317

Hodel AT, Kim SH, Brünger AT (1992) Model bias in macromolecular crystal structures. Acta Crystallogr Sect A 48(6):851–858. https://doi.org/10.1107/S0108767392006044

Horváth R, Scharfe C, Hoeltzenbein M, Do BH, Schroder C, Warzok R, Vogelgesang S, Lochmuller H, Muller-Hocker J, Gerbitz KD, Oefner PJ Jaksch M (2002) Childhood onset mitochondrial myopathy and lactic acidosis caused by a stop mutation in the mitochondrial cytochrome c oxidase III gene. J Med Genet 39(11):812–816. https://doi.org/10.1136/jmg.39.11.812

Indrieri A, Alexandra van Rahden VA, Tiranti V, Morleo M, Iaconis D, Tammaro R, D'Amato I, Conte I, Maystadt I, Demuth S, Zvulunov A, Kutsche K, Zeviani M, Franco B (2012) Mutations in COX7B cause microphthalmia with linear skin lesions, an unconventional mitochondrial disease. Am J Hum Genet 91(5):942–949. https://doi.org/10.1016/J.AJHG.2012.09.016

Kadenbach B, Hüttemann M (2015) The subunit composition and function of mammalian cytochrome c oxidase. Mitochondrion. https://doi.org/10.1016/j.mito.2015.07.002

Kadenbach B, Merle P (1981) On the function of multiple subunits of cytochrome c oxidase from higher eukaryotes. FEBS Lett 135(1):1–11. https://doi.org/10.1016/0014-5793(81)80932-5

Kaila VRI, Verkhovsky MI, Wikström M (2010) Proton-coupled electron transfer in cytochrome oxidase. Chem Rev 110(12):7062–7081. https://doi.org/10.1021/cr1002003

Kamiya K, Boero M, Tateno M, Shiraishi K, Oshiyama A (2007) Possible mechanism of proton transfer through peptide groups in the H-pathway of the bovine cytochrome c oxidase. J Am Chem Soc 129(31):9663–9673. https://doi.org/10.1021/ja070464y

Kee TR, Gonzalez PE, Wehinger JL, Bukhari MZ, Ermekbaeva A, Sista A, Kotsiviras P, Liu T, Kang DE, Woo JAA (2021) Mitochondrial CHCHD2: disease-associated mutations, physiological functions, and current animal models. Front Aging Neurosci 13(0):660843. https://doi.org/10.3389/fnagi.2021.660843

Konstantinov AA, Siletsky S, Mitchell D, Kaulen A, Gennis RB (1997) The roles of the two proton input channels in cytochrome c oxidase from Rhodobacter sphaeroides probed by the effects of site-directed mutations on time-resolved electrogenic intraprotein proton transfer. Proc Natl Acad Sci U S A 94(17):9085–9090. Available at: http://www.pubmedcentral.nih.gov/articlerender.fcgi?artid=23042&tool=pmcentrez&rendertype=abstract. Accessed 8 Mar 2014

LaMoia TE, Butrico GM, Kalpage HA, Goedeke L, Hubbard BT, Vatner DF, Gaspar RC, Zhang X, Cline GW, Nakahara K, Woo S, Shimada A, Huttemann M, Shulman GI (2022) Metformin, phenformin, and galegine inhibit complex IV activity and reduce glycerol-derived gluconeogenesis. Proc Natl Acad Sci USA 119(10):e2122287119. https://doi.org/10.1073/pnas.2122287119

Massa V, Fernandez-Vizarra E, Alshahwan S, Bakhsh E, Goffrini P, Ferrero I, Mereghetti P, D'Adamo P, Gasparini P, Zeviani M (2008) Severe infantile encephalomyopathy caused by a mutation in COX6B1, a nucleus-encoded subunit of cytochrome c oxidase. Am J Hum Genet 82(6):1281–1289. https://doi.org/10.1016/J.AJHG.2008.05.002

Meng H, Yamashita C, Shiba-Fukushima K, Inoshita T, Funayama M, Sato S, Hatta T, Natsume T, Umitsu M, Takagi J, Umai Y, Hattori N (2017) Loss of Parkinson's disease-associated protein CHCHD2 affects mitochondrial crista structure and destabilizes cytochrome c. Nat Commun 8(1):15500. https://doi.org/10.1038/ncomms15500
Muramoto K, Ohta K, Shinzawa-Itoh K, Kanda K, Taniguchi M, Nabekura H, Yamashita E, Tsukihara T, Yoshikawa S (2010) Bovine cytochrome c oxidase structures enable O2 reduction with minimization of reactive oxygens and provide a proton-pumping gate. Proc Natl Acad Sci USA 107(17):7740–7745. https://doi.org/10.1073/pnas.0910410107
Nagao T, Shintani Y, Hayashi T, Kioka H,Kato H, Nishida Y, Yamazaki S, Tsukamoto O, Yashirogi S, Yazawa I, Asano Y, Shinzawa-Itoh K, Imamura H, Suzuki T, Suzuki T, Goto Y, Takashima S (2020) Higd1a improves respiratory function in the models of mitochondrial disorder. FASEB J 34(1):1859–1871. https://doi.org/10.1096/fj.201800389R
Napiwotzki J, Kadenbach B (1998) Extramitochondrial ATP/ADP-ratios regulate cytochrome c oxidase activity via binding to the cytosolic domain of subunit IV. Biol Chem 379(3):335–340. https://doi.org/10.1515/bchm.1998.379.3.335
Napiwotzki J, Shinzawa-Itoh K, Yoshikawa S, Kadenbach B (1997) ATP and ADP bind to cytochrome c oxidase and regulate its activity. Biol Chem 378(9):1013–1022. https://doi.org/10.1515/bchm.1997.378.9.1013
Nelson VLB, Jiang Y, Dickman KG, Ballou LM, Lin RZ (2014) Adipose tissue insulin resistance due to loss of PI3K p110α leads to decreased energy expenditure and obesity. Am J Physiol 306(10):E1205–E1216. https://doi.org/10.1152/ajpendo.00625.2013
Nishida Y, Yanagisawa S, Morita R, Shigematsu H, Yuki H, Ogasawara S, Shimuta K, Iwamoto T, Nakabayashi C, Matsumura W, Kato H, Gopalasingam C, Nagao T, Qaqorh T, Takahashi Y, Yamazaki S, Kamiya K, Harada R, Mizuno N, Takahashi H, Akeda Y, Ohnishi M, Ishii M, Kumasaka T, Murata T, Muramoto K, Tosha T, Shiro Y, Honma T, Shigeta Y, Kubo M,Takashima S, Shintani Y (2022) Identifying antibiotics based on structural differences in the conserved allostery from mitochondrial heme-copper oxidases. Nat Commun 13(1):7591. https://doi.org/10.1038/s41467-022-34771-y
Nishigaki Y, Ueno H, Coku J, Koga Y, Fujii T, Sahashi K, Nakano K, Yoneda M, Nonaka M, Tang L, Liou C, Paquis-Flucklinger V, Harigaya Y, Ibi T, Goto Y, Hosoya H DiMauro S, Hirano M, Tanaka M (2010) Extensive screening system using suspension array technology to detect mitochondrial DNA point mutations. Mitochondrion 10(3):300–308. https://doi.org/10.1016/J.MITO.2010.01.003
Ogura T, Takahashi S, Hirota S, Shinzawa-Itoh K, Yoshikawa S, Appleman EH, Kitagawa T (1993) Time-resolved resonance Raman elucidation of the pathway for dioxygen reduction by cytochrome c oxidase. J Am Chem Soc 115(19):8527–8536. https://doi.org/10.1021/ja00072a002
Oleynikov IP, Azarkina NV, Vygodina TV, Konstantinov AA (2020) Interaction of cytochrome c oxidase with steroid hormones. Cells 9(10):2211. https://doi.org/10.3390/cells9102211
Oleynikov IP, Sudakov RV, Azarkina NV, Vygodina T (2022) Direct interaction of mitochondrial cytochrome c oxidase with thyroid hormones: evidence for two binding sites. Cells 11(5):908. https://doi.org/10.3390/cells11050908
Ostermeier C, Harrenga A, Ermler U, Michel H (1997) Structure at 2.7 Å resolution of the Paracoccus denitrificans two-subunit cytochrome c oxidase complexed with an antibody FV fragment. Proc Natl Acad Sci USA 94(20):10547–10553
Pawate AS, Morgan J, Namslaue A, Mills D, Brzezinski P, Ferguson-Miller S, Gennis RB (2002) A mutation in subunit I of cytochrome oxidase from Rhodobacter sphaeroides results in an increase in steady-state activity but completely eliminates proton pumping. Biochemistry 41(45):13417–13423. https://doi.org/10.1021/bi026582
Péquignot MO, Dey R, Zeviani M, Tiranti V, Godinot C, Poyau A, Sue C, DiMauro S, Abitbol M, Marsac C (2001) Mutations in the SURF1 gene associated with Leigh syndrome and cytochrome c oxidase deficiency. Hum Mutat 17(5):374–381. https://doi.org/10.1002/humu.1112

Pereira MM, Sousa FL, Teixeira M, Nyquist RM, Heberle J (2006) A tyrosine residue deprotonates during oxygen reduction by the caa3 reductase from Rhodothermus marinus. FEBS Lett 580(5):1350–1354. https://doi.org/10.1016/j.febslet.2006.01.055

Pereira MM, Sousa FL, Verissimo AF, Teixeira M (2008) Looking for the minimum common denominator in heme-copper oxygen reductases: towards a unified catalytic mechanism. Biochim Biophys Acta 1777(7–8):929–934. https://doi.org/10.1016/j.bbabio.2008.05.441

Pereira L, Soares P, Radivojac P, Li B, Samuels D (2011) Comparing phylogeny and the predicted pathogenicity of protein variations reveals equal purifying selection across the global human mtDNA diversity. Am J Hum Genet 88(4):433–439. https://doi.org/10.1016/J.AJHG.2011.03.006

Popović DM, Stuchebrukhov AA (2005) Proton exit channels in bovine cytochrome c oxidase. J Phys Chem B 109(5):1999–2006. https://doi.org/10.1021/jp0464371

Rahman S, Taanman J, Cooper JM, Nelson I, Hargreaves I, Meunier B, Hanna MG, Garcia JJ, Capaldi RA, Lake BD, Leonard JV, Schapira AHV (1999) A missense mutation of cytochrome oxidase subunit II causes defective assembly and myopathy. Am J Hum Genet 65(4):1030–1039. https://doi.org/10.1086/302590

Rak M, Benit P, Chretien D, Bouchereau J, Schiff M, ElKhooury R, Tzagoloff A, Rustin P (2016) Mitochondrial cytochrome c oxidase deficiency. Clin Sci 130(6):393–407. https://doi.org/10.1042/CS20150707

Ramzan R, Kadenbach B, Vogt S (2021) Multiple mechanisms regulate eukaryotic cytochrome c oxidase. Cells 10(3):514. https://doi.org/10.3390/cells10030514

Ramzan R, Dolga AM, Michels S, Weber P, Culmsee C, Rastan AJ, Vogt S (2022) Cytochrome c oxidase inhibition by ATP decreases mitochondrial ROS production. Cells 11(6):992. https://doi.org/10.3390/cells11060992

Robinson BH (2000) Human cytochrome oxidase deficiency. Pediatr Res 48(5):581–585. https://doi.org/10.1203/00006450-200011000-00004

Ronchi D, Mosca F, Corti S, Iascone M, Comi PG (2023) A biallelic variant in COX18 causes isolated Complex IV deficiency associated with neonatal encephalo-cardio-myopathy and axonal sensory neuropathy. Eur J Hum Genet 31(12):1414–1420. https://doi.org/10.1038/s41431-023-01433-6

Schon EA, Bonilla E, Lombes A, Moraes CT, Nakase H, Rizzuto R, Zeviani M, DiMauro S (1988) Clinical and biochemical studies on cytochrome oxidase deficiencies. Ann N Y Acad Sci 550(1):348–359. https://doi.org/10.1111/j.1749-6632.1988.tb35349.x

Schuresko DD, Webb WW (1978) Carboxylation kinetics of hemoglobin and myoglobin: linear transient response to step perturbation by laser photolysis. Biophys J 24(1):382–383. https://doi.org/10.1016/S0006-3495(78)85390-9

Sebastián D, Palacín M, Zorzano A (2017) Mitochondrial dynamics: coupling mitochondrial fitness with healthy aging. Trends Mol Med 23(3):201–215. https://doi.org/10.1016/J.MOLMED.2017.01.003

Shimada A, Kubo M, Baba S, Yamashita K, Hirata K, Ueno G, Nomura T, Kimura T, Shinzawa-Itoh K, Baba J, Hatano K, Eto Y, Miyamoto A, Murakami H, Kumasaka T, Owada S, Tono K, Yabashi M, Yamaguchi Y, Yanagisawa S, Sakaguchi M, Ogura T, Komiya R, Yan J, Yamashita E, Yamamoto M, Ago H, Yoshikawa S, Tsukihara T (2017a) A nanosecond time-resolved XFEL analysis of structural changes associated with CO release from cytochrome c oxidase. Sci Adv 3(7):e1603042. https://doi.org/10.1126/sciadv.1603042

Shimada S, Shinzawa-Itoh K, Baba J, Aoe S, Shimada A, Yamashita E, Kang J, Tateno M, Yoshikawa S, Tsukihara T (2017b) Complex structure of cytochrome c–cytochrome c oxidase reveals a novel protein–protein interaction mode. EMBO J 36(3):291-300. https://doi.org/10.15252/embj.201695021

Shimada A, Hatano K, Tadehara H, Yano N, Shinzawa-Itoh K, Yamashita E, Muramoto K, Tsukihara T, Yoshikawa S (2018) X-ray structural analyses of azide-bound cytochrome c oxidases reveal that the H-pathway is critically important for the proton-pumping activity. J Biol Chem 293(38):14868–14879. https://doi.org/10.1074/jbc.RA118.003123

Shimada A, Etoh Y, Kitoh-Fujisawa R, Sasaki A, Shinzawa-Itoh K, Hiromoto T, Yamashita E, Muramoto K, Tsukihara T, Yoshikawa S (2020) X-ray structures of catalytic intermediates of cytochrome c oxidase provide insights into its O2 activation and unidirectional proton-pump mechanisms. J Biol Chem 295(17):5818–5833. https://doi.org/10.1074/JBC.RA119.009596

Shimada A, Hara F, Shinzawa-Itoh K, Kanehisa N, Yamashita E, Muramoto K, Tsukihara T, Yoshikawa S (2021) Critical roles of the CuB site in efficient proton pumping as revealed by crystal structures of mammalian cytochrome c oxidase catalytic intermediates. J Biol Chem 297(3):100967. https://doi.org/10.1016/J.JBC.2021.100967

Shimada A, Baba J, Nagao S, Shinzawa-Itoh K, Yamshita E, Muramoto K, Tsukihara T, Yoshikawa S (2023a) Crystallographic cyanide-probing for cytochrome c oxidase reveals structural bases suggesting that a putative proton transfer H-pathway pumps protons. J Biol Chem 299(11):105277. https://doi.org/10.1016/J.JBC.2023.105277

Shimada A, Tsukihara T, Yoshikawa S (2023b) Recent progress in experimental studies on the catalytic mechanism of cytochrome c oxidase. Front Chem 11:1108190. https://doi.org/10.3389/fchem.2023.1108190

Shimokata K, Katayama Y, Murayama H, Suematsu M, Tsukihara T, Muramoto K, Aoyama H, Yoshikawa S, Shimada H (2007) The proton pumping pathway of bovine heart cytochrome c oxidase. Proc Natl Acad Sci USA 104(10):4200–4205. https://doi.org/10.1073/pnas.0611627104

Shinzawa-Itoh K, Aoyama H, Muramoto K, Terada H, Kurauchi T, Tadehara Y, Yamasaki A, Sugimura T, Kurono S, Tsujimoto K, Mizushima T, Yamashita E, Tsukihara T, Yoshikawa S (2007) Structures and physiological roles of 13 integral lipids of bovine heart cytochrome c oxidase. EMBO J 26(6):1713–1725. https://doi.org/10.1038/sj.emboj.7601618

Shinzawa-Itoh K, Hatanaka M, Fujita K, Yano N, Ogasawara Y, Iwata J, Yamashita E, Tsukihara T, Yoshikawa S, Muramoto K (2021) The 1.3-Å resolution structure of bovine cytochrome c oxidase suggests a dimerization mechanism. BBA Adv 1:100009. https://doi.org/10.1016/J.BBADVA.2021.100009

Srinivasan S, Avadhani NG (2012) Cytochrome c oxidase dysfunction in oxidative stress. Free Radic Biol Med 53(6):1252–1263. https://doi.org/10.1016/J.FREERADBIOMED.2012.07.021

Strogolova V, Hoang NH, Nosler J, Stuart RA (2019) The yeast mitochondrial proteins Rcf1 and Rcf2 support the enzymology of the cytochrome c oxidase complex and generation of the proton motive force. J Biol Chem 294(13):4867–4877. https://doi.org/10.1074/JBC.RA118.006888

Svensson-Ek M, Abramson J, Larsson G, Törnroth S, Brzezinski P, Iwata S (2002) The X-ray crystal structures of wild-type and EQ(I-286) mutant cytochrome c oxidases from Rhodobacter sphaeroides. J Mol Biol 321(2):329–339. https://doi.org/10.1016/s0022-2836(02)00619-8

Takahashi S, Ogura T, Shinzawa-Itoh K, Yoshikawa S, Kitagawa T (1993) Resonance Raman studies on the CuA site of cytochrome c oxidase using a multichannel scanning Raman spectrometer with a CCD detector. Biochemistry 32(14):3664–3670. https://doi.org/10.1021/bi00065a020

Tamiya G, Makino S, Hayashi M, Abe A, Numakura C, Ueki M, Tanaka A, Ito C, Toshimori K, Ogawa N, Terashima T, Maegawa H, Yanagisawa D, Toyama I, Tada M, Onodera O, Hayasaka K (2014) A mutation of COX6A1 causes a recessive axonal or mixed form of Charcot-Marie-tooth disease. Am J Hum Genet 95(3):294–300. https://doi.org/10.1016/J.AJHG.2014.07.013

Timón-Gómez A, Bartley-Dier EL, Fontanesi F, Barrientos A (2020) HIGD-driven regulation of cytochrome c oxidase biogenesis and function. Cells 9(12):2620. https://doi.org/10.3390/cells9122620

Tsukihara T, Aoyama H, Yamashita E, Tomizaki T, Yamaguchi H, Shinzawa-Itoh K, Nakashima R, Yaono R, Yoshikawa S (1995) Structures of metal sites of oxidized bovine heart cytochrome c oxidase at 2.8 Å. Science 269(5227):1069–1074. https://doi.org/10.1126/science.7652554

Tsukihara T, Aoyama H, Yamashita E, Tomizaki T, Yamaguchi H, Shinzawa-Itoh K, Nakashima R, Yaono R, Yoshikawa S (1996) The whole structure of the 13-subunit oxidized cytochrome c oxidase at 2.8 Å. Science 272(5265):1136–1144. https://doi.org/10.1126/science.272.5265.1136

Valente L, Piga D, Lamantea E, Carrara F, Uziel G, Cudia P, Zani A, Farina L, Morandi L, Mora M, Spinazzola A, Zeviani M, Tiranti V (2009) Identification of novel mutations in five patients with mitochondrial encephalomyopathy. Biochim Biophys Acta 1787(5):491–501. https://doi.org/10.1016/J.BBABIO.2008.10.001

Van der Schueren B, Vangoitsenhoven R, Geeraert B, DeKeyzer D, Hulsmans M, Lannoo M, Huber HJ, Mathier C, Holvoet P (2015) Low cytochrome oxidase 4I1 links mitochondrial dysfunction to obesity and type 2 diabetes in humans and mice. Int J Obes 39(8):1254–1263. https://doi.org/10.1038/ijo.2015.58

Vial G, Detaille D, Guigas B (2019) Role of mitochondria in the mechanism(s) of action of metformin. Front Endocrinol 10:442419. https://doi.org/10.3389/fendo.2019.00294

Vondrackova A, Vesela K, Hansikova H, Docekalova ZD, Rozsypalova E, Zeman J, Tesarova M (2012) High-resolution melting analysis of 15 genes in 60 patients with cytochrome-c oxidase deficiency. J Hum Genet 57(7):442–448. https://doi.org/10.1038/jhg.2012.49

Wang B, Tsakiridis EE, Zhang S, Llanos A, Desjardins EM, Yabut JM, Green AE,Day EA, Smith BK, Lally JSV, Wu J, Raphenya AR, Srinivasan KA, McArthur AG, Kajimura S, Patel JS, Wade MG, Morrison KM, Holloway AC, Steinberg GR (2021) The pesticide chlorpyrifos promotes obesity by inhibiting diet-induced thermogenesis in brown adipose tissue. Nat Commun 12(1):5163. https://doi.org/10.1038/s41467-021-25384-y

Yanagisawa S, Kamei T, Shimada A, Gladyck S, Aras S, Huttemann M, Grossman LI, Kubo M (2024) Resonance Raman spectral analysis of the heme site structure of cytochrome c oxidase with its positive regulator CHCHD2. J Inorg Biochem 260:112673. https://doi.org/10.1016/J.JINORGBIO.2024.112673

Yano N, Muramoto K, Shimada A, Takemura S, Baba J, Fujisawa H, Mochizuki M, Shinzawa-Itoh K, Yamashita E, Tsukihara T, Yoshikawa S (2016) The Mg2+-containing water cluster of mammalian cytochrome c oxidase collects four pumping proton equivalents in each catalytic cycle. J Biol Chem 291(46):23882–23894. https://doi.org/10.1074/jbc.M115.711770

Yoshikawa S, Caughey WS (1990) Infrared evidence of cyanide binding to iron and copper sites in bovine heart cytochrome c oxidase. Implications regarding oxygen reduction. J Biol Chem 265(14):7945–7958. https://doi.org/10.1016/S0021-9258(19)39023-4

Yoshikawa S, Caughey WS (1992) Infrared evidence of azide binding to iron, copper, and non-metal sites in heart cytochrome c oxidase. J Biol Chem 267(14):9757–9766. Available at: http://www.ncbi.nlm.nih.gov/pubmed/1315769

Yoshikawa S, Shimada A (2015) Reaction mechanism of cytochrome c oxidase. Chem Rev 115(4):1936–1989. https://doi.org/10.1021/cr500266a

Yoshikawa S, Shinzawa-Itoh K, Nakashima R, Yaono R, Yamashita E, Inoue N, Yao M, Fei MJ, Libeu CP, Mizushima T, Yamaguchi H, Tomizaki T, Tsukihara T (1998) Redox-coupled crystal structural changes in bovine heart cytochrome c oxidase. Science 280(5370):1723–1729. Available at: http://www.ncbi.nlm.nih.gov/pubmed/9624044

Yoshikawa S, Muramoto K, Shinzawa-Itoh K (2011) Proton-pumping mechanism of cytochrome C oxidase. Annu Rev Biophys 40:205–223. https://doi.org/10.1146/annurev-biophys-042910-155341

Zong Y, Li H, Liao P, Chen L, Pan Y, Zheng Y, Zhang C, Liu D, Zheng M, Gao J (2024) Mitochondrial dysfunction: mechanisms and advances in therapy. Signal Transduct Target Ther 9(1):124. https://doi.org/10.1038/s41392-024-01839-8

Protein Folding and Stabilisation Through Non-covalent Stapling

Ivan Maslov, Hongkang Wu, Isabelle Riches, Diana Vidovic, and Mohammed Akhter Hossain

Abstract Proteins are incredible yet complex biomolecules with unique shapes. The primary protein structure is a polypeptide made of amino acids, the simplest protein building blocks, which form long chains when linked together. However, the resulting linear structures must be folded into specific shapes. Proteins are reshaped in native environments, where spontaneous protein folding and unfolding can occur. This dynamic process is influenced by covalent peptide and disulphide bonds that interconnect distant cysteine residues, as well as weaker non-covalent interactions. Initiating from a simple polypeptide chain, protein folding is facilitated by various non-covalent forces. These forces influence stability and flexibility, ultimately defining the protein function. Various cross-linking tools can facilitate correct folding in synthetic peptides, including covalent stapling and non-covalent stabilisation methods. In synthetic polypeptides, non-covalent structural stabilisation involves ionic bridging, aromatic π-stacking and metal coordination, where weak interactions can improve fold integrity without imposing significant structural limitations. Both *in silico* techniques and strategic chemical synthesis have influenced medicinal chemistry, transforming unstable polypeptide chains into robust therapeutics and allowing for the development of protease-resistant drugs with improved *in vivo* characteristics, including increased potency and circulation stability.

Keywords Peptides · Stapling · Proteins · Folding · Non-covalent interactions · α-helix · β-strands

I. Maslov (✉) · H. Wu · I. Riches · D. Vidovic · M. A. Hossain
Florey Institute of Neuroscience and Mental Health, University of Melbourne, Parkville, VIC, Australia

Department of Biochemistry and Pharmacology, University of Melbourne, Parkville, VIC, Australia
e-mail: ivan.maslov@florey.edu.au

A. H. M. N. Nabi, H. Uddin Shekhar (eds.), *Biotechnological Advances in Healthomics*, Interdisciplinary Biotechnological Advances,
https://doi.org/10.1007/978-981-95-3612-2_13

Abbreviations

Aib	α-aminoisobutyric acid
BCAA	Branched-chain amino acid
CD	Circular dichroism
cryo-EM	Cryogenic electron microscopy
GFP	Green fluorescent protein
H-bond	Hydrogen bond
HSP	Heat shock protein
MD	Molecular dynamics
MS	Mass spectrometry
NCI	Non-covalent interaction
NMR	Nuclear magnetic resonance
PCNA	Proliferating cell nuclear antigen
PrP	Prion protein
PTM	Post-translational modification
PDB	Protein data bank
SPPS	Solid-phase peptide synthesis
tRNA	Transfer ribonucleic acid

1 Introduction

The central dogma of molecular biology states that the genetic code determines the foundation of biological systems, where genes encode proteins and proteins define 'life' (Crick 1970). Proteins are made from amino acids, the building blocks that can be sequentially linked to form long polypeptide chains. There are 20 types of natural amino acids, providing us with an unlimited number of sequence variations of protein chains. The linear chain conformation of protein is referred to as the primary structure. Linear primary structures can be folded into complex three-dimensional shapes through a delicate protein folding process or disassembled back into amino acid monomers during the process known as proteolytic degradation (Alberts 2017).

Each amino acid is unique, bearing a distinctive side chain with specific physicochemical properties. For example, the side chain can be hydrophobic, which repels water, and hydrophilic, which attracts it. As a result, the shape and arrangement of different side chains can influence how the linear primary sequence folds into compact three-dimensional structures. Moreover, the spatial contacts, usually weak non-covalent interactions, among the nearby side chains in the folded conformation stabilise the folded conformation and the functions of the proteins. Proteins are highly diverse in terms of their size, structure and functions. For example, some proteins may be made up of just a few amino acids, as observed in short signalling peptides, whereas other proteins may contain several thousand amino acid residues, as seen in human connection, which contains 27,000 amino acid residues (Wang et al. 1979). However, most proteins begin as small linear fragments, where the growing peptide chain is organised into more complex shapes (which may appear planar, helical or globular) through a process called co-translational folding (Waudby et al. 2019). This differential and/or spatial organisation is mainly driven by the side chains of nearby amino acids. Hydrophobic side chains tend to avoid interaction with water molecules and are therefore buried within the hydrophobic core of folded proteins. Conversely, hydrophilic side chains prefer interactions with each other and surrounding water molecules and remain on the surface of the growing polypeptide. These delicate side-chain arrangements within the protein's core and on its surface enable proteins to adopt a wide range of stable conformations and functional three-dimensional structures (Chandler 2005). Natural protein folding is a complex yet rapid and relatively straightforward process. Proteins can be isolated and purified from natural sources, which is often tedious, time-consuming and expensive, not to mention the extremely low yields. Alternatively, proteins can be produced chemically or through the use of recombinant DNA technology, followed by protein expression in genetically modified organisms, such as bacteria or yeast (Tripathi and Shrivastava 2019). This approach makes targeted protein isolation and purification simpler, faster, more cost-effective and scalable, ranging from bench-top experimentation to industrial applications (Rosano and Ceccarelli 2014). Recent advancements in site-specific incorporation of non-natural amino acids during ribosomal translation have significantly expanded the toolkit of protein therapeutics. This is achieved through reassigning nonsense codons or rare sense codons, as well as by utilising engineered orthogonal tRNA-synthase systems. This area of research has revolutionised structural biology, enabling the development of customised protein folds (Chin 2014; Lang and Chin 2014; Koch et al. 2021). However, proteins obtained from bacterial and yeast expression systems often lack the necessary post-translational modifications (PTMs) (Cain et al. 2014), and classical synthetic chemistry offers a robust yet low-scale method for producing high-purity proteins with unlimited potential for structural modifications. Chemical synthesis can be automated or semi-automated, often using peptide synthesisers for solid-phase peptide synthesis (SPPS), which rapidly assembles monomers through chemical coupling (Behrendt et al. 2016). While SPPS is a highly reliable technique for creating long polypeptide chains, it does not include chaperones, which are often crucial for correct protein folding in living organisms. Therefore, ensuring correct folding after chemical synthesis remains a considerable challenge for maintaining the biological functions of proteins.

2 Protein Folding

All primary linear protein sequences tend to adopt compact three-dimensional conformations with minimal free energy in natural environments. This is achieved by folding of the primary amino acid sequence into secondary structure components,

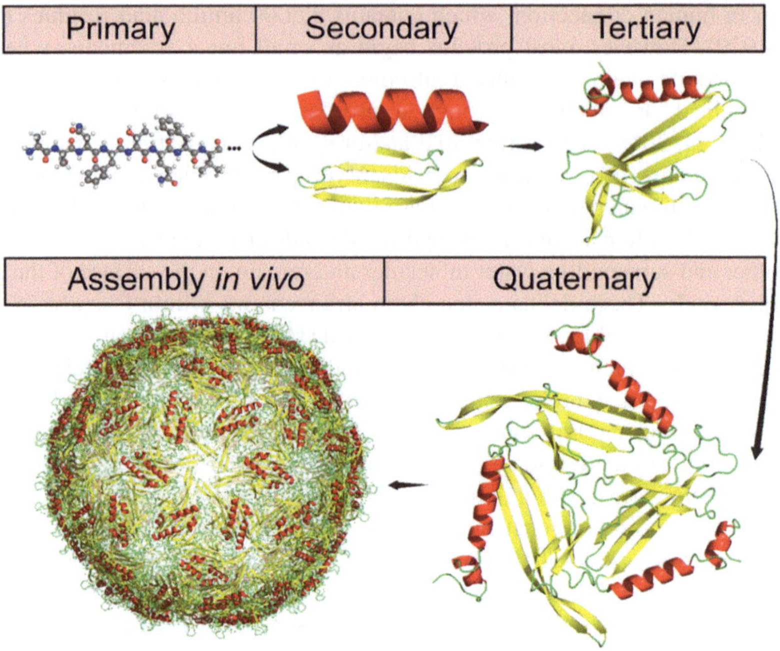

Fig. 1 Protein structure hierarchy, starting from the linear polypeptide chain of amino acids that can be folded into various secondary structure elements, such as helices (shown in red), strands (yellow) and connecting elements (green). In bacteriophage MS2 (*Emesvirus zinderi*), these secondary structure elements form the tertiary structure of a coat protein, whereas three individual foldamers can be subsequently organised into a quaternary fold. The functional state can be achieved when 180 copies of a coat protein assemble into a highly symmetrical three-dimensional shell (Golmohammadi et al. 1993). The figure was generated using the 2MS2 in Protein Data Bank (PDB) (Golmohammadi et al. 1993) using PyMOL (The PyMOL Molecular Graphics System, version 3.0 Schrödinger, LLC.)

which fold back onto themselves, resulting in complex three-dimensional structures with minimal Gibbs free energy (Fig. 1). During and/or after secondary and tertiary structure formation, several non-covalent interactions develop among side chains and backbone atoms, stabilising the protein folds. Such non-covalent tethering among amino acids may not only be the primary driving force behind the protein folding process; they also give rise to stable protein fold architectures. Remarkably, these individual non-covalent interactions are generally weak; yet, they can sometimes create extremely stable structures, such as viral capsids (Katen and Zlotnick 2009; Gertsman et al. 2010). Half a century ago, Christian B. Anfinsen demonstrated in an elegant experiment involving ribonuclease A that protein folding occurs swiftly and spontaneously: after being denatured to its primary polypeptide sequence, the enzyme could spontaneously refold into its active state (Anfinsen 1973). When returned to its native three-dimensional conformation, the enzyme restored its enzymatic activity. This experiment highlights that protein folding is also reversible: tertiary and quaternary proteins with specific functions can revert to their non-functional primary polypeptide sequence and then naturally regain their functional three-dimensional fold.

Even though proteins can adopt multiple potential conformations, natural refolding (i.e. search for the *correct* conformation with the lowest Gibbs free energy) happens rapidly in the natural environment (Englander and Mayne 2014). It suggests that the minimum Gibbs free energy conformation is imprinted within the primary polypeptide sequence, and all structural transitions should occur swiftly (Zwanzig et al. 1992). Interestingly, most proteins can still access several low-energy states, although some conformations are not always favourable, since only one conformation has an actual minimum in Gibbs free energy. Consequently, proteins can adopt altered folding, and these situations are referred to as kinetic traps (Shin et al. 2012).

2.1 Proteins Reshaping Proteins

Molecular chaperones are companion proteins that assist in rescuing misfolded proteins; they are found in all living organisms (e.g. prokaryotes, eukaryotes and viruses), and are essential for protein folding. Chaperones do not determine the proper steric arrangement of each polypeptide chain (which is influenced by the primary sequence itself); they only assist in the correct folding by preventing proteins from becoming trapped in misfolded states (Ellis 1990). Specific chaperones are expressed in response to cellular stress; for example, some chaperones are called heat shock proteins (HSPs) because they protect against heat-induced protein damage (Hu et al. 2022). However, chaperones cannot always restore misfolded proteins to their native states, resulting in disordered proteins with altered functions (Iakoucheva et al. 2002). Disordered regions are harmless as long as misfolded segments do not aggregate and can be processed by enzymatic systems (van der Velden and Hulsmann 1999). Short peptides, peptide hormones, trafficking signalling sequences and unstructured interdomain loops are typically removed by various enzymes during proteolytic degradation. When this cannot be done, unstructured proteins aggregate and accumulate, often leading to diseases (Rufo et al. 2014).

Certain disordered proteins can convert correctly folded protein molecules into misfolded forms, altering their normal function. This avalanching process creates multiple copies of abnormal protein folds, which can accumulate and form stable aggregates. This chain reaction resembles a pathogenic process. The misfolded protein structure is known as a prion (from protein infection) and ultimately leads to severe neurodegenerative diseases. The accumulation of pathogenic prion aggregates is schematically illustrated in Fig. 2, where human prion protein (PrP) loses its correct helical structure and adopts a misfolded planar state, enabling it to reshape other prion proteins and initiate protein aggregation (Calzolai et al. 2000; Wang et al. 2020), leading to neurodegenerative diseases. These aggregations include PrP and other proteins, such as Tau, α-synuclein, huntingtin, c-Myc and p53. Their misfolding and aggregation also resemble prion-like behaviour that ultimately leads to

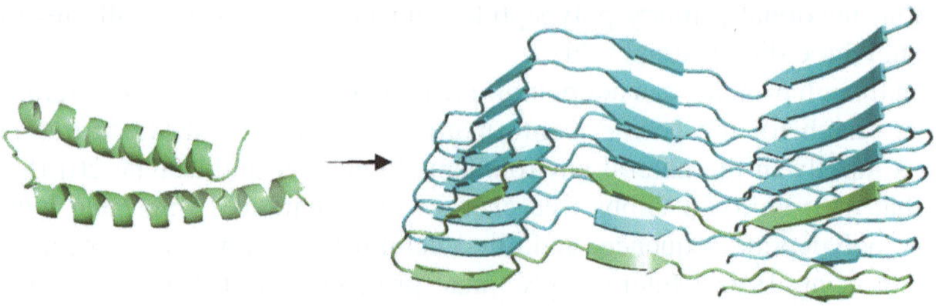

Fig. 2 Prion-mediated conformational changes in human prion protein (PrP) lead to amyloid fibril formation. A C-terminal helical fragment of human PrP, shown in green (left), illustrates the normal protein state. In disease, PrP can undergo spontaneous misfolding to adopt an infectious planar state, also shown in green (right), which converts other copies of PrP, thus leading to protein aggregates (cyan) and subsequent neurodegenerative diseases. The figure was generated in PyMOL using 1E1G (Calzolai et al. 2000) and 6LN1 (Wang et al. 2020) PDB models

Alzheimer's disease (Irvine et al. 2008), Parkinson's disease (Stefanis 2012), Huntington's disease (Arrasate and Finkbeiner 2012) and cancer (Miller et al. 2012; Soragni et al. 2016), respectively. Some therapeutics act as pharmacological 'chaperones' that can either prevent protein aggregation or promote their clearance (Soto and Pritzkow 2018). Nonetheless, in terms of protein folding, prions and prion-like structures are extremely interesting: similar to living organisms that store and replicate their genetic information, these disordered proteins can transmit their misfolded state to other proteins, reshaping them under various pathological conditions. Therefore, the understanding of the barrier between '*molecules*' and '*organisms*' must also be reshaped. And to achieve this, the fundamental processes of protein folding will be further discussed in this chapter.

3 Non-covalent Interactions (NCIs)

Non-covalent interactions (NCIs) within proteins and peptides are weak and reversible forces that play an essential role in molecular folding, stability and recognition. These interactions generally occur between two amino acid residues separated in the primary polypeptide sequence. Among these NCIs, hydrogen bonds (H-bonds) are one of the main bonds that stabilise the secondary structure in proteins, as they can form between the backbone amide and carbonyl groups (-NH ••• O=C-) of two different amino acids. Furthermore, side-chain hydroxyls (found in Ser or Thr) can also engage in H-bonds with backbone carbonyls, further strengthening the spatial organisation in proteins (-OH ••• O=C-). Moreover, the polar groups of hydrophilic side chains form H-bonds with surrounding water molecules, helping to make the protein folds stable, soluble and functional.

The other non-covalent interactions include charge–charge interactions and salt bridges among charged side chains ($-NH_3^+$ and $^-OOC-$), π-interactions and π-π stacking among aromatic side chains and several types of hydrophobic interactions such as van der Waals interactions and London dispersion forces among non-polar side chains. All these non-covalent forces collectively drive and stabilise the folded structures. The forces mentioned above are the main non-covalent tethers relevant to proteins and will be further discussed in this chapter in the context of non-covalent stapling methods. There are also some less common NCIs that can help facilitate the formation of secondary structures, such as helices, strands and loops, from linear polypeptide chains in aqueous environments (Jena et al. 2022; Kojasoy and Tantillo 2022; Adhav and Saikrishnan 2023). Combined with canonical NCIs, they maintain the structural integrity of proteins and prevent their unfolding.

4 Secondary Structure Elements in Proteins: Helices

Helix is the most common secondary structure element in proteins (Sun et al. 2004). A primary polypeptide chain can adopt a helical structure in aqueous, polar solutions because polar amino acids generally appear on the water-exposed surface, while non-polar residues are typically tucked inside. This dual hydrophilic and hydrophobic behaviour explains why many peptides and protein fragments are forced to adopt a helical shape. As a result, the structural elements display hydrophilic and hydrophobic surfaces that help them interact with other structural components (Gimenez-Andres et al. 2018).

Among all helices, the α-helix is the most prevalent. It is right-handedly twisted and consists of 3.6 amino acid residues per turn (Sun et al. 2004) (Fig. 3a). H-bonds are the main tethering forces linking the main chain carbonyl oxygens and amide hydrogens of two amino acids that are four residues apart ($i \rightarrow i + 4$, where i refers to any position within the helical segment). Additionally, salt bridges (e.g. when Arg/Lys and Asp/Glu residues are four positions apart on the hydrophilic face of the helix), π-interactions (e.g. between closely oriented side chains of Trp and Phe) and van der Waals forces also play important roles to further stabilise the α-helical structure. Introducing a single non-canonical amino acid can affect local helicity in synthetic peptides, as non-natural amino acids can form helices with entirely different characteristics (Eddinger and Gellman 2018). For example, a polypeptide composed of *D*-amino acid monomers forms a left-handed α-helix, which is the mirror image of the typical right-handed structure (Fig. 3b). This left-twisting helix is often not recognised by proteolytic enzymes in living organisms, especially eukaryotes, which are adapted for *L*-amino acids and right-twisting helices (Feng and Xu 2016). This makes synthetic *D*-polypeptides resistant to degradation, a feature also observed in some bacteria that produce polypeptides containing both L- and D-amino acids (Sieradzki and Tomasz 1996; Bellais et al. 2006).

Various factors, such as pH, ionic strength and temperature, can either stabilise or disrupt helical fragments (Forood et al. 1993; Li and Deber 1993). A balance of polar and non-polar (often branched-chain amino acids [BCAAs]) residues favours α-helix formation, whereas Gly and Pro residues generally mark the α-helix end. Gly disrupts protein helicity as it has a small achiral structure (Aravinda et al. 2008),

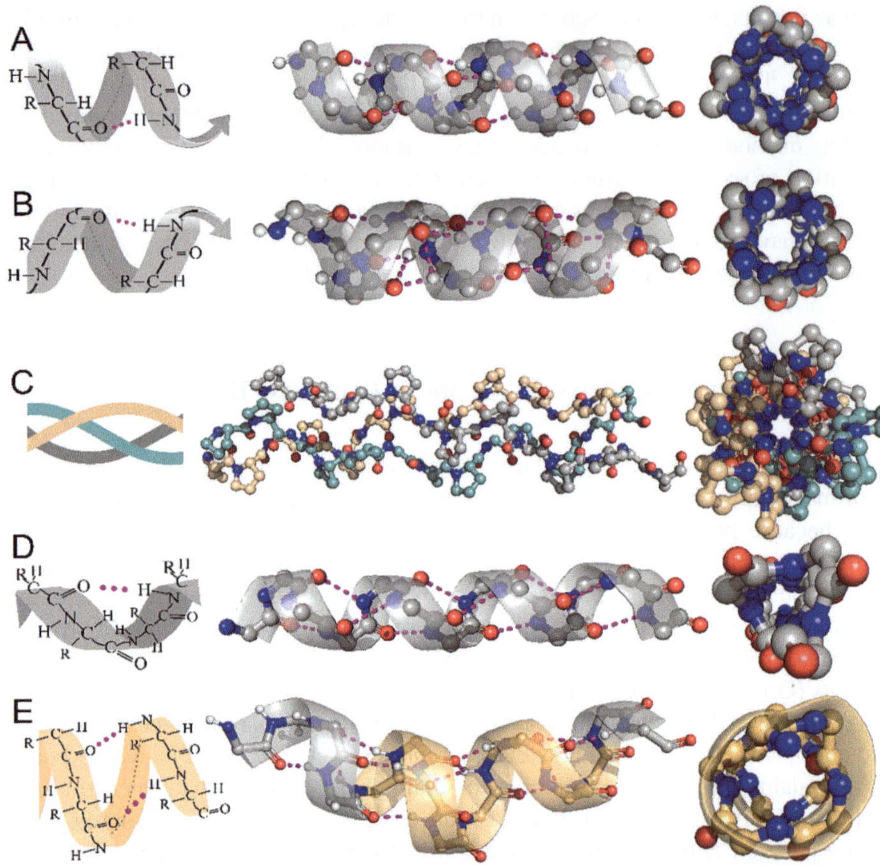

Fig. 3 Different helical elements of protein secondary structure. (**a**) schematic (left), side (middle) and top (right) view images of the α-helix formed by L-amino acids, illustrating its right-handedness. Main-chain carbons (grey), oxygens (red), nitrogens (blue) and amide hydrogens (white) are shown as spheres; side chains are not shown. H-bonds between carbonyl oxygens and amide hydrogens are shown in magenta within the main chain. (**b**) schematic (left), side (middle) and top (right) view images of the α-helix formed by D-amino acids illustrating its left-handedness. (**c**) schematic (left), side (middle) and top (right) view images of the collagen triple helix (the individual chains are shown in grey, teal and light-orange), illustrating the absence of H-bonds. The figure shown is a structural fragment of collagen-like polypeptides adapted from 1K6F in PDB (Berisio et al. 2002). (**d**) schematic (left), side (middle) and top (right) view images of the left-handed 3_{10}-helix adapted from 7QDI in PDB (Kumar et al. 2022). H-bonds between carbonyl oxygens and amide hydrogens (although H atoms could not be visualised as they are missing in the original model) within the main chain are shown as dashes (magenta). (**e**) schematic (left), side (middle) and top (right) view images of the α-helix (shown in grey) with a π-helical insertion element (shown in light-orange). H-bonds between carbonyl oxygens and amide hydrogens are shown as dashes (magenta) within the main chain. The shown structure was adapted from 2A65 in PDB (Yamashita et al. 2005). All structure visualisations were created using PyMOL

while Pro is the only cyclic natural amino acid that cannot form classic intrahelical H-bonds and so bends the chain, restricting its movement. This leads to noticeable structural changes in Pro-rich polypeptides, resulting in unique helices that differ from standard α-helices. Such Pro-rich helices do not contain H-bonds, thus being drastically different from all other helical structures. In collagen, three consecutive Pro residues form a triple helix by coiling around each other (Berisio et al. 2002; Fidler et al. 2018) (Fig. 3c).

The 3_{10}-helix is a less common secondary structure in proteins. It differs from the traditional α-helix in H-bonding patterns, structural configuration and functions (Vieira-Pires and Morais-Cabral 2010; Ceccacci et al. 2013) (Fig. 3d). Although the 3_{10}-helix is also right-handed, it is not as wide as a typical α-helix. The 3_{10}-helix includes three amino acid residues per turn, with H-bonds between residues three positions apart within the polypeptide chain ($i \rightarrow i + 3$). The 3_{10}-helix can be easily recognised by its characteristic triangular shape when viewing the polypeptide backbone (Kumar et al. 2022). Compared to the canonical α-helix, 3_{10}-helical segments are more elongated and more prone to enzymatic degradation. Also, these helices rarely form tertiary or quaternary structures (Kumar et al. 2022) and are often found near canonical α-helices in proteins. Another secondary structure variation, the π-helix (Fig. 3e), contains 4.4 residues per turn, forming H-bonds between residues five positions apart ($i \rightarrow i + 5$). While less favoured than the α-helix, the π-helix enhances protein functionality and is often observed as an insertional element in metalloproteins and transmembrane channels (Cooley et al. 2010; Kumar and Bansal 2015).

5 Secondary Structure Elements in Proteins: β-Strands and β-Sheets

β-strands are the most common non-helical elements of protein secondary structure. Unlike the coiled helical pattern, β-strands have an extended polypeptide backbone with a characteristic zigzag shape. Two or more adjacent β-strands align laterally via H-bonds, forming parallel or anti-parallel β-sheets. In parallel β-sheets, strands align in the same direction, causing the carbonyl oxygens of neighbouring strands to face each other (Fig. 4a). This results in angled interstrand H-bonds that are comparatively weaker due to the offset position of the carbonyl and amide groups (Nowick 2008). In contrast, anti-parallel β-sheets have adjacent β-strands oriented in opposite directions (Fig. 4b), allowing their carbonyl and amide groups to align for connections that are close to perpendicular. This is facilitated by H-bonds perpendicular to the strand axis. Although anti-parallel β-sheets are planar, they often can twist or curl to make packing more efficient.

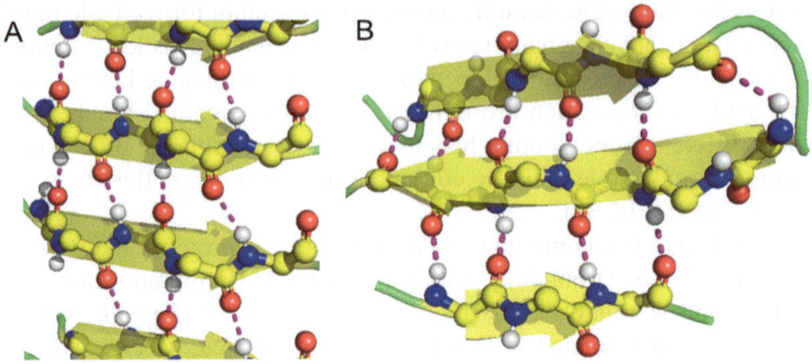

Fig. 4 β-sheets as secondary structure elements in proteins (shown in yellow), where individual β-strands are connected with turn segments (shown in green). Parallel β-sheets (**a**) and anti-parallel β-sheets (**b**) featuring different spatial organisation of neighbouring β-strands, resulting in two variants of H-bonding patterns (dashed lines shown in magenta). The structure visualisations were created in PyMOL, using the three-dimensional model of the human insulin receptor obtained from 7YQ6 in PDB (Kim et al. 2022)

6 Secondary Structure Elements in Proteins: Turns and Loops

Apart from previously discussed helical elements and β-sheets, there are some 'unstructured' elements in the secondary structural organisation of proteins. These are linear oligopeptide fragments that connect helices and β-strands, namely β-turns, γ-turns and Ω-loops (Fetrow 1995; Dhar and Chakrabarti 2015) (shown in green in Figs. 4 and 5). They are important because they help to change the direction of the polypeptide chain and optimise the fold. β-turns are four-residue motifs linking strands in anti-parallel β-sheets, stabilised by an H-bond between the first and last amino acids ($i \rightarrow i + 3$). γ-turns are similar elements and consist of three amino acid residues forming a bend and linked with H-bonds two amino acids apart ($i \rightarrow i + 2$). Lastly, Ω-loops are unique disordered loops that lack strict H-bonding patterns. These elements are often longer than β-turns and γ-turns and are frequently observed in more complex supersecondary structures, where they help resolve space constraints and provide additional flexibility. Sometimes, these fragments can be cleaved by enzymes without affecting the overall functionality of the protein (Fetrow 1995).

7 Higher-Order Protein Structures

Just as amino acids act as building blocks to form the linear polypeptide chain, the secondary structure elements mentioned above serve as structural units to produce higher-order protein folds, resulting in supersecondary, tertiary and quaternary

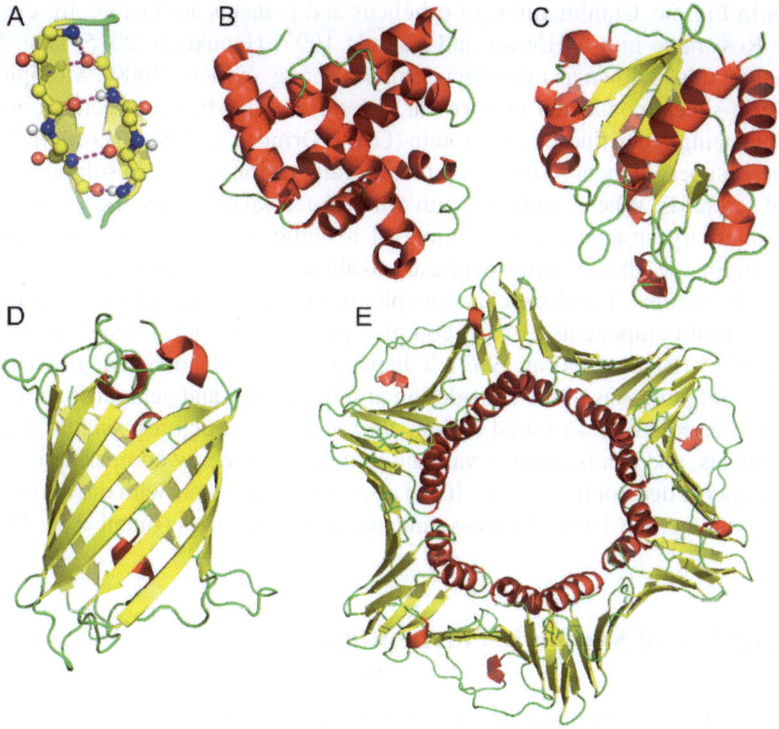

Fig. 5 Higher-order protein structures containing α-helices (shown in red), β-strands (yellow) and connecting elements (green) as parts of supersecondary, tertiary and quaternary folds. (**a**) β-hairpin formed by H-bonding (dashed lines shown in magenta) in human insulin receptor (Kim et al. 2022) obtained from 7YQ6 in PDB. (**b**) Sperm whale myoglobin illustrating an α-helical bundle formed by several supersecondary 'helix–loop–helix' domains in the structural model obtained from 1MBN (Watson and Kendrew 1961) in PDB. (**c**) *Escherichia coli* flavodoxin containing the Rossmann motifs (Hoover and Ludwig 1997) in the structural model obtained from 1AHN in PDB. (**d**) *Aequorea victoria* GFP, an 11-stranded beta-barrel with short coaxial and peripheral helical fragments (Ormo et al. 1996) in the structural model obtained from 1EMA in PDB. (**e**) Symmetrical quaternary organisation (a trimeric ring) of human proliferating cell nuclear antigen (PCNA) complexed with a short peptide derived from the C-terminal region of p21 cell-cycle checkpoint protein to form a central hole necessary for DNA interaction (Gulbis et al. 1996) in the structural model obtained from 1AXC in PDB. All structure visualisations were created using PyMOL

structures. In supersecondary structures, combinations of helices or β-strands are connected by turns and loops, as well as parallel or anti-parallel β-sheets linked by α-helices. A simple example here is the β-hairpin, a configuration of two β-strands forming an anti-parallel β-sheet connected by a turn (DuPai et al. 2021), as shown in Fig. 5a. In the 'helix–loop–helix' and 'helix–turn–helix' motifs, two α-helices are linked together through a connecting element to improve intermolecular recognition. Occasionally, such motifs can form helical bunches, which can be observed in the tertiary structure of myoglobin (Watson and Kendrew 1961) (1MBN.PDB), shown in Fig. 5b. Combinations of α-helices and β-sheets are visible, for example, in the Rossmann motif (Hoover and Ludwig 1997; Hanukoglu 2015) (Fig. 5c) or Greek key (also known as meander) domain (Zhang and Kim 2000). Multiple anti-parallel β-sheets connected by turns in Ω-loops form β-barrel structures, with an example being green fluorescent protein (GFP) (Ormo et al. 1996), shown in Fig. 5d.

These supersecondary structures incorporate various NCIs, resulting in stable spatial scaffolds. When combined with other structural elements, they form tertiary and quaternary structures (an example of a higher-order protein organisation is shown in Fig. 5e) that are often necessary to allow proteins to mediate diverse physiological functions. The Ramachandran plot principles can be utilised to understand the structural components of each complex protein. The plot is essential in structural biology (Ramachandran 1963). It displays all possible torsion angles and polypeptide configurations while considering atomic clashes and geometric limitations. The plot can highlight favoured regions for secondary structures, such as α-helices and β-sheets, and can be used to validate protein structure by comparing it to theoretically permitted configurations. It can also reveal fragments with uncommon conformations (e.g. left-handed helices) and strained geometries (Imani et al. 2023).

8 The Use of Synthetic Polypeptides

In living organisms, peptides are swiftly released and rapidly broken down through proteolysis. Similarly, in large proteins, shorter elements can become disordered fragments that cannot mediate any biological functions, which can often pose a problem in disease contexts. Synthetic peptides and larger proteins can be engineered to function as more stable or potent alternatives, as they often include non-natural amino acid side chains or other synthetic groups that resist proteolytic breakdown (Iegre et al. 2018). In other words, short proteins hold great therapeutic potential: they are simple enough for rational drug design yet complex enough for advanced biological functions.

In nature, folded and unfolded protein forms exist in a balance, and minor structural adjustments can shift this balance toward the stability of the folded state, preventing spontaneous unfolding. In medicinal chemistry, cross-linking is a powerful technique that restricts the entropic freedom of unfolded states. Although various stabilisation methods exist, we broadly categorise them into two main groups: covalent and non-covalent stabilisation techniques. Natural or synthetic cross-linking imposes molecular constraints that can be either permanent (through covalent bonding) or reversible (via NCIs).

To illustrate the difference, imagine travelling from point A to point B under two conditions. First, we are on a flight with flexible seatbelt rules: fastening is only required during take-off and landing, allowing us to unbuckle when in the air. We can stretch our legs, switch seats or even lie down. The second scenario involves cabin crew restraining us and fastening our seatbelts permanently, with no option to unbuckle (fine motor skills are allowed)—no loitering, no reclining, no lying down. While this situation might improve our chances of arriving safely at point B in one piece, the seatbelt becomes an inescapable part of our lives. Jokes aside, the first scenario is an oversimplified non-covalent stabilisation. It is characterised by dynamic and reversible NCIs that temporarily reduce conformational flexibility, whereas overall adaptability remains intact. The second one illustrates covalent stabilisation, where permanent cross links immobilise the structure, while covalent stapling maintains the functional state at the cost of permanent restriction of movement. Therefore, the first strategy permits temporary free interactions, whereas the latter imposes strict limitations.

The covalent staplings improve the structural integrity of proteins. This process enables stabilised helices to optimise their binding surfaces and increase target affinity. At the same time, the resulting secondary structure elements become less susceptible to proteolytic degradation. Therefore, cross-linking can convert protein fragments into rigid, small-molecule-like biologics with improved pharmacokinetics. The range of stabilisation techniques includes stapled helices, disulphide engineering and backbone cyclisation, which yield ultra-stable peptides to keep proteins' functionality for effective operation. Covalent stapling can improve protease resistance and binding affinity by pre-organising bioactive configurations. This 'molecular seatbelt approach', as we shall call it, imposes irreversible structural constraints; it was first developed in 2000, where hydrocarbon cross links were chemically introduced to maintain target peptides in their bioactive configurations (Schafmeister et al. 2000). Since then, numerous covalent stapling variations have been developed, including those summarised in several published reviews (Chen et al. 2025; Li et al. 2024; Esteban et al. 2024; Nielipinska et al. 2024; Li et al. 2020; Lau et al. 2015; Iegre et al. 2018; Walensky and Bird 2014; Ali et al. 2019). This approach often significantly reduces synthetic yields, thus complicating large-scale production, which does not apply to non-covalent stabilisation methods. Here, the lack of rigid cross links enables proteins to maintain their functionality. The non-covalent cross-linking techniques focus on reversibly constrained peptides with modified secondary structure elements, for example stabilised α-helices (Facchiano et al. 1998; Fornasari et al. 2008; Zhang et al. 2015), often resulting in only modest improvements in conformational stability and protease resistance, which may still be enough to design polypeptides with improved properties in biological contexts.

8.1 Non-covalent Stapling Methods

To maintain correct folding, various NCIs can be optimised. This is achieved through several non-covalent techniques such as helix capping, stabilising turns and forming inter-residue contacts. By strategically positioning H-bond donors and acceptors (placed several amino acid residues apart within the polypeptide chain, or close in the three-dimensional structures), folded polypeptides can become stabilised without additional covalent restraints. Flexible peptide chains can transform into stabilised helical structures by optimising existing NCIs or introducing various amino acid substitutions. Strategic amino acid substitutions and non-natural modifications can create networks that stabilise secondary structures and guide the polypeptide assembly into higher-order configurations (Marqusee et al. 1989).

8.1.1 Electrostatic Interactions

Electrostatic coordination is an effective method for stabilising protein and peptide structures. This can be achieved through direct ion-pairing or metal coordination, which helps stabilise the conformations of structured domains (Marqusee and Baldwin 1987). Introducing charged amino acid residues into the protein sequence can often stabilise or destabilise the helix. This strategic placement of residues (known as salt bridge engineering) further improves protein stability by strengthening electrostatic interactions between oppositely charged amino acids (e.g. within a single helical region), thereby creating molecular '*staples*' that favour the folded state over misfolded configurations (Marqusee and Baldwin 1987). This can be achieved by positioning charged residues (Arg/Lys and Asp/Glu) within secondary structures such as helices, beta-strands or domain interfaces. They support proteins through ionic interactions, limiting their conformational freedom and thereby improving solubility. Unlike covalent stapling methods, salt bridges maintain the protein's natural dynamics. These contacts are weak and reversible, yet they remain useful in preserving the biological activity of therapeutic proteins.

The geometric aspects of secondary structure elements must be considered in salt bridge engineering by ensuring optimal spacing of charged side chains at positions i to $i + 3$ and i to $i + 4$, while also taking into account the α-helical structure. Effective bridging between oppositely charged residues reinforces the functional folding and creates interaction networks that reduce the entropic cost of folding and offer protection against enzymatic breakdown. When placing salt bridges, some helpful guidelines are the following: buried hydrophobic environments really amplify interactions due to low dielectric effects; maintaining charge distances of around 3 Å helps maximise electrostatic potential; and cooperative clusters of bridges work together to provide greater stability than isolated pairs (Marqusee and Baldwin 1987; Marqusee et al. 1989). Although these engineered ionic locks are pH-sensitive and may weaken under high-salt conditions, their reversible and non-disruptive characteristics are important for biological applications, such as stabilising vaccine epitopes and reinforcing de novo helix bundles, where maintaining native-like dynamics is as important as ensuring conformational stability.

8.1.2 π-Interactions

Similar results can be achieved by tuning π-interactions in proteins. When in close proximity or with proper orientation, aromatic amino acid side chains can act as temporary non-covalent bridges, further stabilising protein structures through π-π

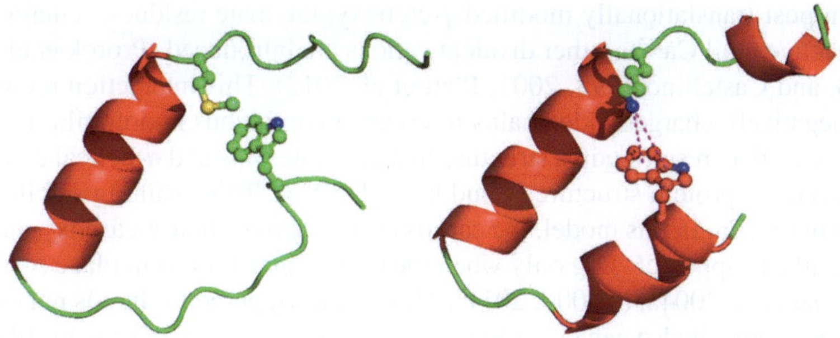

Fig. 6 A proposed structural optimisation of a helical peptide conformation via a single amino acid substitution. Met25 → Lys25. Cartoon representations of the native B-chain of H2 relaxin (left) and its modification containing the Met25 → Lys25 substitution. The predicted helical fragments are shown in red, and non-structured fragments are shown in green. The amino acid side chains at positions 3 and 25 are also shown. The figure was generated using AlphaFold3 (Abramson et al. 2024) and PyMOL

stacking (and also π-ion) interactions. The orientation of aromatic residues, including Phe, Tyr and Trp (and sometimes His, although it is also positively charged depending on pH), establishes π-π stacking networks (Bathgate et al. 2023). Positively and negatively charged side chains mediate π-ion bonds with electron-rich aromatic systems, resulting in enthalpic stabilisation due to marked quantum mechanical interactions within folded proteins (Infield et al. 2021; Yamada 2018; Mahadevi and Sastry 2013). Figure 6 illustrates an example of the π-cation stabilisation strategy, where a single strategic amino acid substitution can lead to conformational stabilisation of the B-chain in H2 relaxin, where substituting Met25 with Lys not only improves helix conformational stability but can also enhance peptide solubility and resistance to oxidation. Such non-covalent tethering strategies demonstrate structural adaptability, often forming temporary 'aromatic zippers' that facilitate stabilising conformations.

8.1.3 Metal Coordination

Metal coordination is another effective non-covalent method to stabilise proteins (Ma et al. 2009). Often, incorporating specific amino acid residues (e.g. His, Cys, Asp or Glu) into specific sequences, such as zinc fingers, Ca^{2+}-binding motifs or iron–sulphur clusters, promotes interactions with metal ions, thereby helping to structurally organise polypeptides. The interaction strength can range from very strong (almost 'covalent'), as seen in Zn^{2+} binding to Cys clusters, to highly labile, exemplified by bonds between Mg^{2+} and phosphorylated residues.

Metal chelation is clearly illustrated in the venoms of predatory sea snails (*Conus* sp.), where such metal-mediated peptide folding is evident in conantokins (Prorok et al. 1996b). In these peptides, non-functional, disordered chains become functional helices upon binding specific cations. These neuroactive peptides, which contain post-translationally modified γ-carboxyglutamate residues, remain flexible and inactive until Ca^{2+} or other divalent cations are introduced (Prorok et al. 1996b; Prorok and Castellino 1998, 2001; Platt et al. 2012). This interaction reconfigures their negatively charged side chains to create a continuous hydrophilic face along the helical edge, resulting in a structure that integrates 3_{10}- and α-helical elements of the secondary protein structure (Cnudde et al. 2007). This resulting stabilisation is shown in Fig. 7a. In this model, the carboxylate groups of four γ-carboxyglutamates behave like a zipper, closing only when the correct metal ion is in place (Chen et al. 2025; Dai et al. 2004a, b, 2009, 2011). This metal-triggered folding is not exclusive to marine snails; it also can occur in mammalian systems, where proteins like osteocalcin utilise γ-carboxyglutamate arrays for Ca^{2+}-dependent activation (Lacombe and Ferron 2015). This opens up the potential for designing stimulus-responsive peptides that remain inactive until they encounter their specific coordinator.

Zinc fingers also demonstrate how structure-stabilising metal chelation functions. These remarkable molecular structures coordinate Zn^{2+} cations to create flexible protein domains that excel at recognising DNA (Lee et al. 1989) (Fig. 7b). They typically feature several Cys and His residues, positioned in close proximity to coordinate a single Zn^{2+} cation. The resulting complex is known for its strong, nearly covalent bonds formed by the Zn^{2+}; zinc finger motifs can firmly wrap around DNA, almost like a molecular hand gripping the nucleic acid (Lee et al. 1989).

8.1.4 Hydrophobic Core Packing

Non-covalent forces explain why hydrophobic amino acid residues are generally hidden within the protein core, and the same forces also trigger the self-assembly process in proteins. When tightly organised side chains involving layers of branched

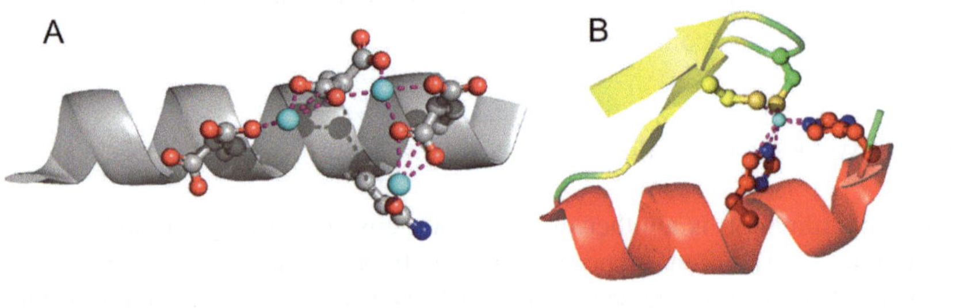

Fig. 7 (**a**) Metal-dependent helix stabilisation in conantokin-G by Ca^{2+}. The peptide backbone is shown as a grey ribbon, and the interacting residues are shown as sticks and spheres, featuring side-chain carbons (shown in grey), oxygens (red) and nitrogens (blue); Ca^{2+} atoms are shown as sky-blue spheres. H-bonds are shown as dashes (magenta). The figure was generated in PyMOL using the 2DPQ (Cnudde et al. 2007) in PDB model. (**b**) Zinc finger domain for DNA binding. The cartoon representation of the peptide backbone features the α-helical fragment (shown in red), an anti-parallel β-sheet (yellow) and connecting elements (green), with the interacting residues shown as sticks and spheres, featuring side-chain carbons (shown in their respective fragment's colours), nitrogens (blue) and sulphurs (gold); the Zn^{2+} atom is shown as a sky-blue sphere. H-bonds are shown as dashes (magenta). The figure was generated in PyMOL using the 1ZNF in PDB (Lee et al. 1989) model

aliphatic or aromatic residues come together, they create a fascinating three-dimensional ensemble. Similar to professional musicians in an orchestra, each methyl group and phenyl ring finds its ideal position to maximise coordination and packing density and minimise any potential voids. Incorporating α-methylated amino acids, such as α-aminoisobutyric acid (Aib), introduces a subtle yet significant helical stabilisation of the helix (Karle and Balaram 1990; Biondi et al. 2017; Aravinda et al. 2008). Adding an extra methyl group at the α-carbon imposes rigid conformational constraints, thus maintaining helicity by limiting backbone torsion angles to energetically favourable regions of Ramachandran space. Furthermore, it fortifies peptides against proteolytic breakdown. Two α-methyl groups of this achiral residue increase the hydrophobic moment (Islam et al. 2020), forcing the peptide chain to adopt its helical configuration as the methyl groups orient away from the solvent (Islam et al. 2020). However, this simple approach also has limitations: α,α-disubstituted residues may distort the classic α-helix into a 3_{10}-helix in polar environments (Pengo et al. 2003; Ceccacci et al. 2013). This can alter binding interfaces due to modified backbone geometry and greatly complicate chemical synthesis because of steric clashes, which reduce yields when residues are positioned consecutively.

9 Computational and Experimental Approaches

Modern software tools such as AlphaFold, Rosetta and other algorithms enable robust prediction of protein structures and their modifications (Leman et al. 2020; Abramson et al. 2024). These computational tools expand the growing universe of protein structures; they are highly effective and have already contributed to multiple recent breakthroughs (Varadi et al. 2024). Structural predictions and deep learning models are becoming increasingly reliable, allowing for virtual screening of numerous drug candidates before synthesis (Jumper et al. 2021). When combined with molecular dynamics (MD) simulations, structure prediction tools can verify whether designed non-covalent tethering methods maintain their connections during protein conformational changes (Cui et al. 2025). However, any *in silico* predictions should be verified experimentally, and reliable methods are available to determine the protein structure empirically. For example, circular dichroism (CD) spectra indicate whether synthesised peptides retain their helical structure after exposure to thermal stress; moreover, recent advances in data-driven machine learning approaches support structural determination (Zhao et al. 2021). X-ray crystallography provides atomic-level detail to confirm whether the structural changes predicted by simulations are accurate (Chruszcz et al. 2008). Additionally, mass spectrometry (MS) and proteolytic degradation assays can verify if strategically placed staples increase resistance to proteolysis, complemented by crystallography and nuclear magnetic resonance (NMR) spectroscopy to confirm spatial orientation. Recently, advances in high-resolution cryo-electron microscopy (cryo-EM) have revolutionised the determination of protein structures, particularly for large complexes and dynamic assemblies (Kuhlbrandt 2014; Cheng 2018). This has allowed researchers to capture alternative conformational states and gain insights into folding intermediates and misfolded species, supporting drug discovery and therapeutic design, while also providing experimental data for deep learning platforms. The productive dialogue between algorithms and experimental validation will make drug design more cost-effective, and optimised therapies will become more specific and less invasive. As we continue learning more about protein folding and structural stabilisation, we edge closer to designing protein-based therapeutics with the same accuracy and reliability that natural expression systems have perfected over billions of years of evolution.

References

Abramson J, Adler J, Dunger J, Evans R, Green T, Pritzel A, Ronneberger O, Willmore L, Ballard AJ, Bambrick J, Bodenstein SW, Evans DA, Hung CC, O'Neill M, Reiman D, Tunyasuvunakool K, Wu Z, Zemgulyte A, Arvaniti E, Beattie C, Bertolli O, Bridgland A, Cherepanov A, Congreve M, Cowen-Rivers AI, Cowie A, Figurnov M, Fuchs FB, Gladman H, Jain R, Khan YA, Low CMR, Perlin K, Potapenko A, Savy P, Singh S, Stecula A, Thillaisundaram A, Tong C, Yakneen S, Zhong ED, Zielinski M, Zidek A, Bapst V, Kohli P, Jaderberg M, Hassabis D, Jumper JM (2024) Accurate structure prediction of biomolecular interactions with AlphaFold 3. Nature 630(8016):493–500. https://doi.org/10.1038/s41586-024-07487-w

Adhav VA, Saikrishnan K (2023) The realm of unconventional noncovalent interactions in proteins: their significance in structure and function. ACS Omega 8(22):22268–22284. https://doi.org/10.1021/acsomega.3c01211

Alberts B (2017) Molecular biology of the cell, 6th edn. CRC Press, Boca Raton

Ali AM, Atmaj J, van Oosterwijk N, Groves MR, Dömling A (2019) Stapled peptide inhibitors: a new window for target drug discovery. Comput Struct Biotechnol J 17:263–281. https://doi.org/10.1016/j.csbj.2019.02.011

Anfinsen CB (1973) Principles that govern the folding of protein chains. Science 181(4096):223–230. https://doi.org/10.1126/science.181.4096.223

Aravinda S, Shamala N, Balaram P (2008) Aib residues in peptaibiotics and synthetic sequences: analysis of nonhelical conformations. Chem Biodivers 5(6):1238–1262. https://doi.org/10.1002/cbdv.200890114

Arrasate M, Finkbeiner S (2012) Protein aggregates in Huntington's disease. Exp Neurol 238(1):1–11. https://doi.org/10.1016/j.expneurol.2011.12.013

Bathgate RAD, Praveen P, Sethi A, Furuya WI, Dhingra RR, Kocan M, Ou Q, Valkovic AL, Gil-Miravet I, Navarro-Sanchez M, Olucha-Bordonau FE, Gundlach AL, Rosengren KJ, Gooley PR, Dutschmann M, Hossain MA (2023) Noncovalent peptide stapling using alpha-methyl-L-phenylalanine for alpha-helical peptidomimetics. J Am Chem Soc 145(30):20242–20247. https://doi.org/10.1021/jacs.3c05354

Behrendt R, White P, Offer J (2016) Advances in Fmoc solid-phase peptide synthesis. J Pept Sci 22(1):4–27. https://doi.org/10.1002/psc.2836

Bellais S, Arthur M, Dubost L, Hugonnet JE, Gutmann L, van Heijenoort J, Legrand R, Brouard JP, Rice L, Mainardi JL (2006) Aslfm, the D-aspartate ligase responsible for the addition of D-aspartic acid onto the peptidoglycan precursor of *Enterococcus faecium*. J Biol Chem 281(17):11586–11594. https://doi.org/10.1074/jbc.M511461200

Berisio R, Vitagliano L, Mazzarella L, Zagari A (2002) Crystal structure of the collagen triple helix model [(Pro-Pro-Gly)10]3. Protein Sci 11(2):262–270. https://doi.org/10.1110/ps.27902

Biondi B, Casciaro B, Di Grazia A, Cappiello F, Luca V, Crisma M, Mangoni ML (2017) Effects of Aib residues insertion on the structural-functional properties of the frog skin-derived peptide esculentin-1a(1-21)NH2. Amino Acids 49(1):139–150. https://doi.org/10.1007/s00726-016-2352-2

Cain JA, Solis N, Cordwell SJ (2014) Beyond gene expression: the impact of protein post-translational modifications in bacteria. J Proteome 97:265–286. https://doi.org/10.1016/j.jprot.2013.08.012

Calzolai L, Lysek DA, Güntert P, von Schroetter C, Riek R, Zahn R, Wüthrich K (2000) NMR structures of three single-residue variants of the human prion protein. Proc Natl Acad Sci USA 97(15):8340–8345. https://doi.org/10.1073/pnas.97.15.8340

Ceccacci F, Mancini G, Rossi P, Scrimin P, Sorrenti A, Tecilla P (2013) Deracemization and the first CD spectrum of a 310-helical peptide made of achiral alpha-amino-isobutyric acid residues in a chiral membrane mimetic environment. Chem Commun 49(94):10133–10135. https://doi.org/10.1039/c3cc45741e

Chandler D (2005) Interfaces and the driving force of hydrophobic assembly. Nature 437(7059):640–647. https://doi.org/10.1038/nature04162

Chen Y, Dai C, Han J, Xing Y, Yin F, Li Z (2025) Recent chemical biology insights towards reversible stapled peptides. Chembiochem 26(10):e202500052. https://doi.org/10.1002/cbic.202500052

Cheng Y (2018) Single-particle cryo-EM—how did it get here and where will it go. Science 361(6405):876–880. https://doi.org/10.1126/science.aat4346

Chin JW (2014) Expanding and reprogramming the genetic code of cells and animals. Annu Rev Biochem 83:379–408. https://doi.org/10.1146/annurev-biochem-060713-035737

Chruszcz M, Wlodawer A, Minor W (2008) Determination of protein structures—a series of fortunate events. Biophys J 95(1):1–9. https://doi.org/10.1529/biophysj.108.131508

Cnudde SE, Prorok M, Dai Q, Castellino FJ, Geiger JH (2007) The crystal structures of the calcium-bound con-G and con-T[K7γ] dimeric peptides demonstrate a metal-dependent helix-forming motif. J Am Chem Soc 129(6):1586–1593. https://doi.org/10.1021/ja0652671

Cooley RB, Arp DJ, Karplus PA (2010) Evolutionary origin of a secondary structure: pi-helices as cryptic but widespread insertional variations of alpha-helices that enhance protein functionality. J Mol Biol 404(2):232–246. https://doi.org/10.1016/j.jmb.2010.09.034

Crick F (1970) Central dogma of molecular biology. Nature 227(5258):561–563. https://doi.org/10.1038/227561a0

Cui X, Ge L, Chen X, Lv Z, Wang S, Zhou X, Zhang G (2025) Beyond static structures: protein dynamic conformations modeling in the post-AlphaFold era. Brief Bioinform 26(3):bbae164. https://doi.org/10.1093/bib/bbae164

Dai Q, Castellino FJ, Prorok M (2004a) A single amino acid replacement results in the Ca2+-induced self-assembly of a helical conantokin-based peptide. Biochemistry 43(42):13225–13232. https://doi.org/10.1021/bi048750j

Dai Q, Prorok M, Castellino FJ (2004b) A new mechanism for metal ion-assisted interchain helix assembly in a naturally occurring peptide mediated by optimally spaced gamma-carboxyglutamic acid residues. J Mol Biol 336(3):731–744. https://doi.org/10.1016/j.jmb.2003.12.071

Dai Q, Xiao C, Dong M, Liu Z, Sheng Z, Castellino FJ, Prorok M (2009) Non-strict strand orientation of the Ca2+-induced dimerization of a conantokin peptide variant with sequence-shifted gamma-carboxyglutamate residues. Peptides 30(5):866–872. https://doi.org/10.1016/j.peptides.2009.01.012

Dai Q, Dong M, Liu Z, Prorok M, Castellino FJ (2011) Ca2+-induced self-assembly in designed peptides with optimally spaced gamma-carboxyglutamic acid residues. J Inorg Biochem 105(1):52–57. https://doi.org/10.1016/j.jinorgbio.2010.09.002

Dhar J, Chakrabarti P (2015) Defining the loop structures in proteins based on composite beta-turn mimics. Protein Eng Des Sel 28(5):153–161. https://doi.org/10.1093/protein/gzv014

Dupai CD, Davies BW, Wilke CO (2021) A systematic analysis of the beta hairpin motif in the protein data Bank. Protein Sci 30(3):613–623. https://doi.org/10.1002/pro.4011

Eddinger GA, Gellman SH (2018) Differential effects of β3- versus β2-amino acid residues on the helicity and recognition properties of Bim BH3-derived α/β-peptides. Angew Chem Int Ed 57(44):13829–13832. https://doi.org/10.1002/anie.201808550

Ellis RJ (1990) The molecular chaperone concept. Semin Cell Biol 1(1):1–9. https://doi.org/10.1016/1043-4682(90)90018-Q

Englander SW, Mayne L (2014) The nature of protein folding pathways. Proc Natl Acad Sci USA 111(45):15873–15880. https://doi.org/10.1073/pnas.1411798111

Esteban JJ, Mason JR, Kaminski J, Ramachandran R, Luyt LG (2024) A survey of stapling methods to increase affinity, activity, and stability of ghrelin analogues. RSC Med Chem 15:254–266. https://doi.org/10.1039/D3MD00314E

Facchiano AM, Colonna G, Ragone R (1998) Helix stabilizing factors and stabilization of thermophilic proteins: an X-ray based study. Protein Eng 11:753–760. https://doi.org/10.1093/protein/11.9.753

Feng Z, Xu B (2016) Inspiration from the mirror: D-amino acid containing peptides in biomedical approaches. Biomol Concepts 7:179–187. https://doi.org/10.1515/bmc-2016-0013

Fetrow JS (1995) Omega loops: nonregular secondary structures significant in protein function and stability. FASEB J 9:708–717. https://doi.org/10.1096/fasebj.9.9.7601339

Fidler AL, Boudko SP, Rokas A, Hudson BG (2018) The triple helix of collagens—an ancient protein structure that enabled animal multicellularity and tissue evolution. J Cell Sci 131:jcs203950. https://doi.org/10.1242/jcs.203950

Fornasari MS, Parisi G, Echave J (2008) Teaching noncovalent interactions using protein molecular evolution. Biochem Mol Biol Educ 36:284–286. https://doi.org/10.1002/bmb.20102

Forood B, Feliciano EJ, Nambiar KP (1993) Stabilization of alpha-helical structures in short peptides via end capping. Proc Natl Acad Sci USA 90:838–842. https://doi.org/10.1073/pnas.90.3.838

Gertsman I, Fu CY, Huang R, Komives EA, Johnson JE (2010) Critical salt bridges guide capsid assembly, stability, and maturation behavior in bacteriophage HK97. Mol Cell Proteomics 9:1752–1763. https://doi.org/10.1074/mcp.M900578-MCP200

Gimenez-Andres M, Copic A, Antonny B (2018) The many faces of amphipathic helices. Biomolecules 8:45. https://doi.org/10.3390/biom8030045

Golmohammadi R, Valegard K, Fridborg K, Liljas L (1993) The refined structure of bacteriophage MS2 at 2.8 Å resolution. J Mol Biol 234:620–639. https://doi.org/10.1006/jmbi.1993.1616

Gulbis JM, Kelman Z, Hurwitz J, O'Donnell M, Kuriyan J (1996) Structure of the C-terminal region of p21(WAF1/CIP1) complexed with human PCNA. Cell 87:297–306. https://doi.org/10.1016/S0092-8674(00)81348-4

Hanukoglu I (2015) Proteopedia: Rossmann fold: a beta-alpha-beta fold at dinucleotide binding sites. Biochem Mol Biol Educ 43:206–209. https://doi.org/10.1002/bmb.20863

Hoover DM, Ludwig ML (1997) A flavodoxin that is required for enzyme activation: the structure of oxidized flavodoxin from Escherichia coli at 1.8 Å resolution. Protein Sci 6:2525–2537. https://doi.org/10.1002/pro.5560061202

Hu C, Yang J, Qi Z, Wu H, Wang B, Zou F, Mei H, Liu J, Wang W, Liu Q (2022) Heat shock proteins: biological functions, pathological roles, and therapeutic opportunities. MedComm 3:e161. https://doi.org/10.1002/mco2.161

Iakoucheva LM, Brown CJ, Lawson JD, Obradovic Z, Dunker AK (2002) Intrinsic disorder in cell-signaling and cancer-associated proteins. J Mol Biol 323:573–584. https://doi.org/10.1016/S0022-2836(02)00969-5

Iegre J, Ahmed NS, Gaynord JS, Wu Y, Herlihy KM, Tan YS, Lopes-Pires ME, Jha R, Lau YH, Sore HF, Verma C, O'Donoghue D, Pugh N, Spring DR (2018) Stapled peptides as a new technology to investigate protein-protein interactions in human platelets. Chem Sci 9:4638–4643. https://doi.org/10.1039/C8SC00721B

Imani Z, Mundlapati VR, Brenner V, Gloaguen E, Le Barbu-Debus K, Zehnacker-Rentien A, Robin S, Aitken DJ, Mons M (2023) Non-covalent interactions reveal the protein chain delta conformation in a flexible single-residue model. Chem Commun 59:1161–1164. https://doi.org/10.1039/D2CC05629B

Infield DT, Rasouli A, Galles GD, Chipot C, Tajkhorshid E, Ahern CA (2021) Cation-π interactions and their functional roles in membrane proteins. J Mol Biol 433:167035. https://doi.org/10.1016/j.jmb.2020.11.011

Irvine GB, El-Agnaf OM, Shankar GM, Walsh DM (2008) Protein aggregation in the brain: the molecular basis for Alzheimer's and Parkinson's diseases. Mol Med 14:451–464. https://doi.org/10.2119/2007-00100.Irvine

Islam R, Sviridov DO, Drake SK, Tunyi J, Abdoulayeva G, Freeman LA, Pastor RW, Remaley AT (2020) Incorporation of alpha-methylated amino acids into apolipoprotein A-I mimetic peptides improves their helicity and cholesterol efflux potential. Biochem Biophys Res Commun 526:349–354. https://doi.org/10.1016/j.bbrc.2020.03.090

Jena S, Dutta J, Tulsiyan KD, Sahu AK, Choudhury SS, Biswal HS (2022) Noncovalent interactions in proteins and nucleic acids: beyond hydrogen bonding and π-stacking. Chem Soc Rev 51:4261–4286. https://doi.org/10.1039/D1CS00902A

Jumper J, Evans R, Pritzel A, Green T, Figurnov M, Ronneberger O, Tunyasuvunakool K, Bates R, Zidek A, Potapenko A, Bridgland A, Meyer C, Kohl SAA, Ballard AJ, Cowie A, Romera-Paredes B, Nikolov S, Jain R, Adler J, Back T, Petersen S, Reiman D, Clancy E, Zielinski M, Steinegger M, Pacholska M, Berghammer T, Bodenstein S, Silver D, Vinyals O, Senior AW, Kavukcuoglu K, Kohli P, Hassabis D (2021) Highly accurate protein structure prediction with AlphaFold. Nature 596:583–589. https://doi.org/10.1038/s41586-021-03819-2

Karle IL, Balaram P (1990) Structural characteristics of alpha-helical peptide molecules containing Aib residues. Biochemistry 29:6747–6756. https://doi.org/10.1021/bi00479a015

Katen S, Zlotnick A (2009) The thermodynamics of virus capsid assembly. Methods Enzymol 455:395–417. https://doi.org/10.1016/S0076-6879(08)04215-0

Kim J, Yunn NO, Park M, Kim J, Park S, Kim Y, Noh J, Ryu SH, Cho Y (2022) Functional selectivity of insulin receptor revealed by aptamer-trapped receptor structures. Nat Commun 13:6500. https://doi.org/10.1038/s41467-022-34173-4

Koch NG, Baumann T, Budisa N (2021) Efficient unnatural protein production by pyrrolysyl-tRNA synthetase with genetically fused solubility tags. Front Bioeng Biotechnol 9:807438. https://doi.org/10.3389/fbioe.2021.807438

Kojasoy V, Tantillo DJ (2022) Impacts of noncovalent interactions involving sulfur atoms on protein stability, structure, folding, and bioactivity. Org Biomol Chem 21:11–23. https://doi.org/10.1039/D2OB01852C

Kuhlbrandt W (2014) Cryo-EM enters a new era. eLife 3:e03678. https://doi.org/10.7554/eLife.03678

Kumar P, Bansal M (2015) Dissecting pi-helices: sequence, structure and function. FEBS J 282:4415–4432. https://doi.org/10.1111/febs.13516

Kumar P, Paterson NG, Clayden J, Woolfson DN (2022) De novo design of discrete, stable 310-helix peptide assemblies. Nature 607:387–392. https://doi.org/10.1038/s41586-022-04885-2

Lacombe J, Ferron M (2015) Gamma-carboxylation regulates osteocalcin function. Oncotarget 6:19924–19925. https://doi.org/10.18632/oncotarget.4832

Lang K, Chin JW (2014) Cellular incorporation of unnatural amino acids and bioorthogonal labeling of proteins. Chem Rev 114:4764–4806. https://doi.org/10.1021/cr400355w

Lau YH, de Andrade P, Wu Y, Spring DR (2015) Peptide stapling techniques based on different macrocyclisation chemistries. Chem Soc Rev 44:91–102. https://doi.org/10.1039/C4CS00246F

Lee MS, Gippert GP, Soman KV, Case DA, Wright PE (1989) Three-dimensional solution structure of a single zinc finger DNA-binding domain. Science 245:635–637. https://doi.org/10.1126/science.2503871

Leman JK, Weitzner BD, Lewis SM, Adolf-Bryfogle J, Alam N, Alford RF, Aprahamian M, Baker D, Barlow KA, Barth P, Basanta B, Bender BJ, Blacklock K, Bonet J, Boyken SE, Bradley

P, Bystroff C, Conway P, Cooper S, Correia BE, Coventry B, Das R, de Jong RM, DiMaio F, Dsilva L, Dunbrack R, Ford AS, Frenz B, Fu DY, Geniesse C, Goldschmidt L, Gowthaman R, Gray JJ, Gront D, Guffy S, Horowitz S, Huang PS, Huber T, Jacobs TM, Jeliazkov JR, Johnson DK, Kappel K, Karanicolas J, Khakzad H, Khar KR, Khare SD, Khatib F, Khramushin A, King IC, Kleffner R, Koepnick B, Kortemme T, Kuenze G, Kuhlman B, Kuroda D, Labonte JW, Lai JK, Lapidoth G, Leaver-Fay A, Lindert S, Linsky T, London N, Lubin JH, Lyskov S, Maguire J, Malmström L, Marcos E, Marcu O, Marze NA, Meiler J, Moretti R, Mulligan VK, Nerli S, Norn C, O'Conchuir S, Ollikainen N, Ovchinnikov S, Pacella MS, Pan X, Park H, Pavlovicz RE, Pethe M, Pierce BG, Pilla KB, Raveh B, Renfrew PD, Burman SSR, Rubenstein A, Sauer MF, Scheck A, Schief W, Schueler-Furman O, Sedan Y, Sevy AM, Sgourakis NG, Shi L, Siegel JB, Silva DA, Smith S, Song Y et al (2020) Macromolecular modeling and design in Rosetta: recent methods and frameworks. Nat Methods 17:665–680. https://doi.org/10.1038/s41592-020-0848-2

Li SC, Deber CM (1993) Peptide environment specifies conformation. Helicity of hydrophobic segments compared in aqueous, organic, and membrane environments. J Biol Chem 268:22975–22978. https://doi.org/10.1016/j.jbc.1993.22975

Li H, Hu Y, Pu Q, He T, Zhang Q, Wu W, Xia X, Zhang J (2020) Novel stapling by lysine tethering provides stable and low hemolytic cationic antimicrobial peptides. J Med Chem 63:4081–4089. https://doi.org/10.1021/jm0000000

Li Y, Wu M, Fu Y, Xue J, Yuan F, Qu T, Rissanou AN, Wang Y, Li X, Hu H (2024) Therapeutic stapled peptides: efficacy and molecular targets. Pharmacol Res 203:107137. https://doi.org/10.1016/j.phrs.2024.107137

Ma MT, Hoang HN, Scully CC, Appleton TG, Fairlie DP (2009) Metal clips that induce unstructured pentapeptides to be alpha-helical in water. J Am Chem Soc 131:4505–4512. https://doi.org/10.1021/ja9000000

Mahadevi AS, Sastry GN (2013) Cation-pi interaction: its role and relevance in chemistry, biology, and material science. Chem Rev 113:2100–2138. https://doi.org/10.1021/cr3000000

Marqusee S, Baldwin RL (1987) Helix stabilization by Glu-…Lys+ salt bridges in short peptides of de novo design. Proc Natl Acad Sci USA 84:8898–8902. https://doi.org/10.1073/pnas.84.24.8898

Marqusee S, Robbins VH, Baldwin RL (1989) Unusually stable helix formation in short alanine-based peptides. Proc Natl Acad Sci USA 86:5286–5290. https://doi.org/10.1073/pnas.86.14.5286

Miller DM, Thomas SD, Islam A, Muench D, Sedoris K (2012) c-Myc and cancer metabolism. Clin Cancer Res 18:5546–5553. https://doi.org/10.1158/1078-0432.CCR-12-1123

Nielipinska D, Rubiak D, Pietrzyk-Brzezinska AJ, Malolepsza J, Blazewska KM, Gendaszewska-Darmach E (2024) Stapled peptides as potential therapeutics for diabetes and other metabolic diseases. Biomed Pharmacother 180:117496. https://doi.org/10.1016/j.biopha.2024.117496

Nowick JS (2008) Exploring beta-sheet structure and interactions with chemical model systems. Acc Chem Res 41:1319–1330. https://doi.org/10.1021/ar800045b

Ormo M, Cubitt AB, Kallio K, Gross LA, Tsien RY, Remington SJ (1996) Crystal structure of the Aequorea victoria green fluorescent protein. Science 273:1392–1395. https://doi.org/10.1126/science.273.5280.1392

Pengo P, Pasquato L, Moro S, Brigo A, Fogolari F, Broxterman QB, Kaptein B, Scrimin P (2003) Quantitative correlation of solvent polarity with the alpha–/3(10)-helix equilibrium: a heptapeptide behaves as a solvent-driven molecular spring. Angew Chem Int Ed 42:3388–3392. https://doi.org/10.1002/anie.200300756

Platt RJ, Han TS, Green BR, Smith MD, Skalicky J, Gruszczynski P, White HS, Olivera B, Bulaj G, Gajewiak J (2012) Stapling mimics noncovalent interactions of gamma-carboxyglutamates in conantokins, peptidic antagonists of N-methyl-D-aspartic acid receptors. J Biol Chem 287:20727–20736. https://doi.org/10.1074/jbc.M112.356148

Prorok M, Warder SE, Blandl T, Castellino FJ (1996a) Calcium binding properties of synthetic gamma-carboxyglutamic acid-containing marine cone snail "sleeper" peptides, conantokin-G and conantokin-T. Biochemistry 35:16528–16534. https://doi.org/10.1021/bi961336b
Prorok M, Castellino FJ (1998) Thermodynamics of binding of calcium, magnesium, and zinc to the N-methyl-D-aspartate receptor ion channel peptidic inhibitors, conantokin-G and conantokin-T. J Biol Chem 273:19573–19578. https://doi.org/10.1074/jbc.273.30.19573
Prorok M, Castellino FJ (2001) Structure-function relationships of the NMDA receptor antagonist conantokin peptides. Curr Drug Targets 2:313–322. https://doi.org/10.2174/1389450013348853
Prorok M, Geng JP, Warder SE, Castellino FJ (1996b) The entire gamma-carboxyglutamic acid- and helical stack-domains of human coagulation factor IX are required for optimal binding to its endothelial cell receptor. Int J Pept Protein Res 48:281–285. https://doi.org/10.1111/j.1399-3011.1996.tb00915.x
Ramachandran GN (1963) Stereochemistry of polypeptide chain configurations. J Mol Biol 7:95–99. https://doi.org/10.1016/S0022-2836(63)80023-6
Rosano GL, Ceccarelli EA (2014) Recombinant protein expression in Escherichia coli: advances and challenges. Front Microbiol 5:172. https://doi.org/10.3389/fmicb.2014.00172
Rufo CM, Moroz YS, Moroz OV, Stohr J, Smith TA, Hu X, DeGrado WF, Korendovych IV (2014) Short peptides self-assemble to produce catalytic amyloids. Nat Chem 6:303–309. https://doi.org/10.1038/nchem.1891
Schafmeister CE, Po J, Verdine GL (2000) An all-hydrocarbon cross-linking system for enhancing the helicity and metabolic stability of peptides. J Am Chem Soc 122:5891–5892. https://doi.org/10.1021/ja000558y
Shin SH, Chung S, Sanii B, Comolli LR, Bertozzi CR, De Yoreo JJ (2012) Direct observation of kinetic traps associated with structural transformations leading to multiple pathways of S-layer assembly. Proc Natl Acad Sci USA 109:12968–12973. https://doi.org/10.1073/pnas.1204611109
Sieradzki K, Tomasz A (1996) A highly vancomycin-resistant laboratory mutant of Staphylococcus aureus. FEMS Microbiol Lett 142:161–166. https://doi.org/10.1111/j.1574-6968.1996.tb08679.x
Soragni A, Janzen DM, Johnson LM, Lindgren AG, Thai-Quynh Nguyen A, Tiourin E, Soriaga AB, Lu J, Jiang L, Faull KF, Pellegrini M, Memarzadeh S, Eisenberg DS (2016) A designed inhibitor of p53 aggregation rescues p53 tumor suppression in ovarian carcinomas. Cancer Cell 29:90–103. https://doi.org/10.1016/j.ccell.2015.11.010
Soto C, Pritzkow S (2018) Protein misfolding, aggregation, and conformational strains in neurodegenerative diseases. Nat Neurosci 21:1332–1340. https://doi.org/10.1038/s41593-018-0235-9
Stefanis L (2012) Alpha-Synuclein in Parkinson's disease. Cold Spring Harb Perspect Med 2:a009399. https://doi.org/10.1101/cshperspect.a009399
Sun PD, Foster CE, Boyington JC (2004) Overview of protein structural and functional folds. Curr Protoc Protein Sci. Chapter 17:Unit 17.1. https://doi.org/10.1002/0471140864.ps1701s42
Tripathi NK, Shrivastava A (2019) Recent developments in bioprocessing of recombinant proteins: expression hosts and process development. Front Bioeng Biotechnol 7:420. https://doi.org/10.3389/fbioe.2019.00420
Van der Velden VH, Hulsmann AR (1999) Peptidases: structure, function and modulation of peptide-mediated effects in the human lung. Clin Exp Allergy 29:445–456. https://doi.org/10.1046/j.1365-2222.1999.00552.x
Varadi M, Bertoni D, Magana P, Paramval U, Pidruchna I, Radhakrishnan M, Tsenkov M, Nair S, Mirdita M, Yeo J, Kovalevskiy O, Tunyasuvunakool K, Laydon A, Zidek A, Tomlinson H, Hariharan D, Abrahamson J, Green T, Jumper J, Birney E, Steinegger M, Hassabis D, Velankar S (2024) AlphaFold protein structure database in 2024: providing structure coverage for over 214 million protein sequences. Nucleic Acids Res 52:D368–D375. https://doi.org/10.1093/nar/gkac1087
Vieira-Pires RS, Morais-Cabral JH (2010) 3(10) helices in channels and other membrane proteins. J Gen Physiol 136:585–592. https://doi.org/10.1085/jgp.201010430

Walensky LD, Bird GH (2014) Hydrocarbon-stapled peptides: principles, practice, and progress. J Med Chem 57:6275–6288. https://doi.org/10.1021/jm5004818

Wang K, McClure J, Tu A (1979) Titin: major myofibrillar components of striated muscle. Proc Natl Acad Sci USA 76:3698–3702. https://doi.org/10.1073/pnas.76.8.3698

Wang LQ, Zhao K, Yuan HY, Wang Q, Guan Z, Tao J, Li XN, Sun Y, Yi CW, Chen J, Li D, Zhang D, Yin P, Liu C, Liang Y (2020) Cryo-EM structure of an amyloid fibril formed by full-length human prion protein. Nat Struct Mol Biol 27:598–602. https://doi.org/10.1038/s41594-020-0421-2

Watson HC, Kendrew JC (1961) The amino-acid sequence of sperm whale myoglobin. Comparison between the amino-acid sequences of sperm whale myoglobin and of human hemoglobin. Nature 190:670–672. https://doi.org/10.1038/190670a0

Waudby CA, Dobson CM, Christodoulou J (2019) Nature and regulation of protein folding on the ribosome. Trends Biochem Sci 44:914–926. https://doi.org/10.1016/j.tibs.2019.07.009

Yamada S (2018) Cation-pi interactions in organic synthesis. Chem Rev 118:11353–11432. https://doi.org/10.1021/acs.chemrev.8b00264

Yamashita A, Singh SK, Kawate T, Jin Y, Gouaux E (2005) Crystal structure of a bacterial homologue of Na+/Cl—dependent neurotransmitter transporters. Nature 437:215–223. https://doi.org/10.1038/nature03978

Zhang C, Kim SH (2000) A comprehensive analysis of the Greek key motifs in protein beta-barrels and beta-sandwiches. Proteins 40:409–419. https://doi.org/10.1002/1097-0134(20001115)40:3<409::AID-PROT1024>3.0.CO;2-U

Zhang H, Lu H, Chingin K, Chen H (2015) Stabilization of proteins and noncovalent protein complexes during electrospray ionization by amino acid additives. Anal Chem 87:7433–7438. https://doi.org/10.1021/acs.analchem.5b01618

Zhao L, Zhang J, Zhang Y, Ye S, Zhang G, Chen X, Jiang B, Jiang J (2021) Accurate machine learning prediction of protein circular dichroism spectra with embedded density descriptors. JACS Au 1:2377–2384. https://doi.org/10.1021/jacsau.1c00305

Zwanzig R, Szabo A, Bagchi B (1992) Levinthal's paradox. Proc Natl Acad Sci USA 89:20–22. https://doi.org/10.1073/pnas.89.1.20

Applications of Nanotechnology in Healthcare

Md. Jahirul Islam, Mohammad Al Mamun, and Abu Hashem

Abstract Every facet of science and technology has seen the arrival and solidification of the field of nanoscience. The field of healthcare will witness great biomedical marvels because of nanoscale manufacturing technologies and instruments, including nanomachines, diagnostic tools, and nanomedicine, among many other applications. In the domains of diagnostics, regenerative medicine, disease treatment, dentistry, tissue engineering, oncology, gene therapy, drug delivery, and therapeutics, nanotechnology is demonstrating fruitful and advantageous applications. Scientists are working to harness the potential applications of nanotechnology in the diagnosis, treatment, and prevention of diseases, having recognized its benefits for the medical field. A thorough analysis has been carried out to integrate the most recent information regarding nanotechnology in healthcare in all of its facets, including how the technology is helping the healthcare sector, its long-term effects on health, and potential future remediation suggestions.

Keywords Nanotechnology · Healthcare · Application · Diagnosis · Treatment

Abbreviations

AFP	Alpha-fetoprotein
Ag-NPs	Silver nanoparticles
APCs	Antigen-presenting cells

M. J. Islam
Molecular Biotechnology Division, National Institute of Biotechnology, Dhaka, Bangladesh

M. Al Mamun
Department of Chemistry, Jagannath University, Dhaka, Bangladesh

A. Hashem (✉)
Microbial Biotechnology Division, National Institute of Biotechnology, Dhaka, Bangladesh
e-mail: hashem@nib.gov.bd

A. H. M. N. Nabi, H. Uddin Shekhar (eds.), *Biotechnological Advances in Healthomics*, Interdisciplinary Biotechnological Advances,
https://doi.org/10.1007/978-981-95-3612-2_14

Au-NPs	Gold nanoparticles
Bi-NPs	Bismuth nanoparticles
Cas9	CRISPR-associated protein 9
CRISPR	Clustered regularly interspaced short palindromic repeats
CEA	Carcinoembryonic antigen
CNTs	Carbon nanotubes
COVID-19	Coronavirus disease 2019
CRP	C-reactive protein
cTnI	Cardiac troponin I
CTCs	Circulating tumor cells
CT	Computed tomography
DDS	Drug delivery systems
EPR	Enhanced permeability and retention
ELISA	Enzyme-linked immunosorbent assay
EMA	European Medicines Agency
FDA	US Food and Drug Administration
Ga-NPs	Gadolinium nanoparticles
GalNAc	N-acetylgalactosamine
HPV	Human papillomavirus
LFAs	Lateral flow assays
LNP	Lipid nanoparticle
LOC	Lab-on-a-chip
LSPR	Localized surface plasmon resonance
Mn-NPs	Manganese nanoparticles
MRI	Magnetic resonance imaging
NDDS	Nanoparticle-based drug delivery systems
NHPs	Nanotechnology-enabled health products
NIR	Near-infrared
NPs	Nanoparticles
NSs	Nanosensors
PEG	Polyethylene glycol
PET	Positron emission tomography
PLGA	Poly (lactic-co-glycolic acid)
PSA	Prostate-specific antigen
Pt-NPs	Platinum nanoparticles
QDs	Quantum dots
SERS	Surface-enhanced Raman spectroscopy
SPR	Surface plasmon resonance
SPIONs	Superparamagnetic iron oxide nanoparticles
Ta-NPs	Tantalum nanoparticles
TLR	Toll-like receptor
UCNPs	Upconverting NPs
USPIONs	Ultra-small superparamagnetic iron oxide nanoparticles
Yb-NPs	Ytterbium nanoparticles

1 Introduction

Nanotechnology involves the manipulation of matter at the atomic and molecular scale, typically within the range of 1 to 100 nanometers. At this scale, materials exhibit unique physicochemical properties that differ significantly from their bulk counterparts. These properties enable new innovations across diverse fields, focusing on biological applications, such as food authentication (Khalil et al. 2021; Hashem et al. 2022c), pharmaceuticals (Shalauddin et al. 2023), pathogen detection (Al Mamun 2021), healthcare (Hashem et al. 2022d), biosensors (Al Mamun et al. 2023; Hashem et al. 2022a, b, c, d, e; Hossain et al. 2021), and environment monitoring (Hashem et al. 2021), where precision and efficiency are essential. The conceptual foundation of nanotechnology was laid by physicist Richard Feynman in his 1959 lecture, "There's Plenty of Room at the Bottom." In healthcare, early explorations involved liposomes for drug delivery and the use of colloidal gold (Barenholz 2012; Astruc 2016).

In the past few decades, nanotechnology has emerged as a transformative force, fundamentally changing advanced diagnostic and therapeutic strategies for various diseases (Farokhzad and Langer 2009; Howes et al. 2014; Hashem et al. 2022b). Its potential in healthcare is attributable to its ability to interface with biological systems at subcellular and molecular resolutions (Mitchell et al. 2021). Various nanomaterials, including nanoparticles (NPs) and nanocarriers, enable targeted drug delivery with enhanced bioavailability, reduced systemic toxicity, and controlled release kinetics. Such advancements were notably highlighted during the COVID-19 pandemic through the use of lipid nanoparticles (LNPs) in mRNA vaccine delivery (Tenchov et al. 2021; Hossain et al. 2022). Simultaneously, nanostructured biosensors achieve attomolar sensitivity in the detection of disease biomarkers, thus enabling earlier diagnosis (Nam et al. 2003; Krishnan et al. 2010). In the field of regenerative medicine, nanoengineered scaffolds mimic native extracellular matrices to guide tissue repair (Zhang and Webster 2009). Furthermore, the emergence of recent interdisciplinary interfaces, such as organ-on-chip technologies synergized with nanomaterials, exemplifies the cutting-edge convergence driving more predictive disease modeling and personalized therapeutic screening (Shi et al. 2024). Moreover, quantum dots and superparamagnetic nanoparticles are revolutionizing medical imaging modalities (Gao et al. 2004; Han et al. 2019).

This chapter aims to explore comprehensively the progress in nanotechnology applications in the healthcare sector over the past decade, analyzing their impacts on both diagnosis and treatment of disease. We begin by analyzing nanotechnology's contributions to medical diagnostics, where the integration of nanomaterials, including nanoparticles (NPs) and nanosensors (NSs), has revolutionized early and precise disease detection capabilities (Howes et al. 2014).

Building upon these diagnostic advances, we examine how similar nanoscale principles have enabled the development of sophisticated drug delivery systems (DDS). NPs and nanocarriers improve therapeutic effectiveness and concurrently reduce adverse side effects (Farokhzad and Langer 2009). Following this, we investigate the progress in regenerative medicine, emphasizing how nanomaterials and scaffolds facilitate tissue and organ repair with unprecedented precision (Zhang and Webster 2009). In the field of therapeutics, nanotechnology is enabling novel treatments, including cancer-targeting "smart" drugs (Davis et al. 2008) and antimicrobial agents designed to combat resistant infections (Huh and Kwon 2011). The integration of nanotechnology in biomedical devices, particularly in nano-enhanced implants and wearable sensors (Yao et al. 2018), extends diagnostic and therapeutic capabilities, creating a comprehensive monitoring and treatment ecosystem. The principal sectors of healthcare where nanotechnology-mediated nanosystems are contributing positively are illustrated in Fig. 1.

However, these rapid advancements raise specific concerns about nanoparticle accumulation in organs, potential cellular damage from long-term exposure, and ethical questions regarding human enhancement, requiring a balanced examination of therapeutic benefits and potential biological risks (Oberdörster et al. 2005). Beyond individual health, the global deployment of nanomedicine carries significant societal and environmental implications, including questions of equitable access across diverse populations, the lifecycle management and potential ecological impact of nanomaterials, and the broader societal readiness for these transformative technologies. Equally urgent are the ethical and social implications of nanotechnology, which encompass equitable access, privacy concerns, and the philosophical redefinition of human enhancement (Gupta et al. 2015).

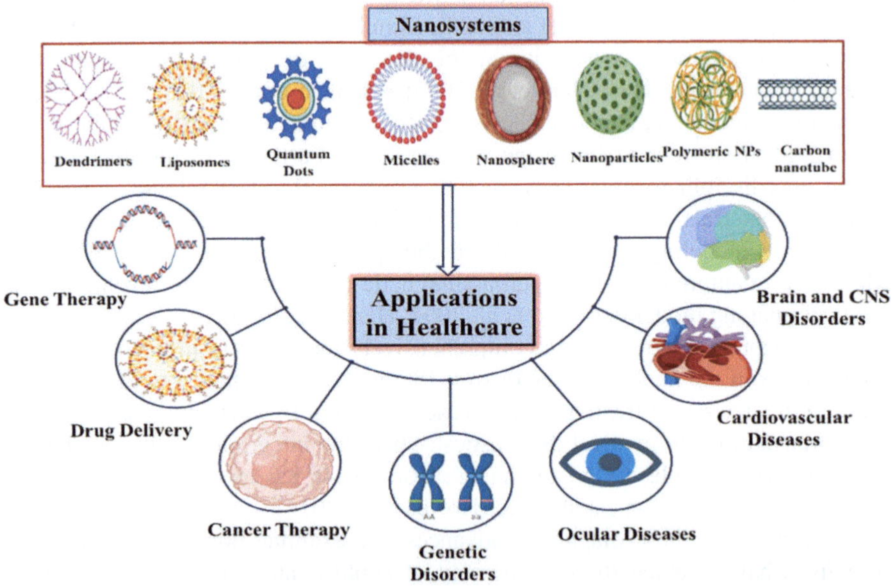

Fig. 1 Applications of various nanosystems in healthcare sectors. (Reprinted with permission under Creative Commons Attribution License from Anjum et al. (2021). Copyright 2021, MDPI)

Finally, we envision future directions, highlighting sophisticated medical diagnosis and improved treatment of disease, the nuances of personalized healthcare. This chapter discusses biomedical advancements through nanotechnology while encouraging readers to critically reflect on the associated challenges, ensuring alignment with global health equity and ethical responsibility.

2 Nanotechnology in Medical Diagnostics

Nanotechnology, the manipulation of matter on an atomic and molecular scale (usually 1–100 nanometers), represents a transformative frontier in medicine. By operating at the same dimensional scale as fundamental biological processes, nanotechnology offers extraordinary opportunities to interact with, visualize, and analyze biomolecules, cells, and tissues. This capability is particularly impactful in the field of medical diagnostics, where the quest for earlier, more sensitive, and more specific detection methods is paramount. The unique physical, chemical, and biological properties exhibited by materials at the nanoscale enable the development of novel diagnostic tools, highly effective imaging contrast agents, ultrasensitive biosensors, and rapid point-of-care devices. These collective developments have the potential to transform disease detection, moving toward identifying illnesses at their earliest molecular stages, often before clinical symptoms manifest (Malik et al. 2023; Hashem et al. 2022b). Nanotechnology-based diagnostic technologies are being developed as promising instruments for real-time, convenient, and cost-effective diagnosis and detection of various diseases. A schematic illustration of nanotechnology-based cancer diagnosis is depicted in Fig. 2.

2.1 Nanoparticle-Based Diagnostics Imaging Agents

Medical imaging is central to disease diagnosis, staging, and monitoring treatment response. Nanotechnology introduces an advanced class of imaging agents utilizing NPs, which present considerable advantages over traditional small-molecule contrast agents in modalities such as magnetic resonance imaging (MRI), computed tomography (CT), positron emission tomography (PET), and optical imaging. NPs enhance medical diagnostic imaging primarily through four key mechanisms. Firstly, NPs significantly improve contrast by delivering high payloads of tailored agents (e.g., superparamagnetic iron oxide for MRI, gold for CT, quantum dots for optical imaging, and radionuclides for PET/SPECT). Secondly, surface functionalization with targeting ligands (antibodies, peptides, and aptamers) enables selective accumulation at disease sites via specific biomarker binding, enhancing specificity. By engineering NPs with specific sizes, compositions, and surface properties, researchers can achieve improved pharmacokinetics, higher contrast payloads, reduced toxicity, and targeted delivery to disease sites (Sim and Wong 2021).

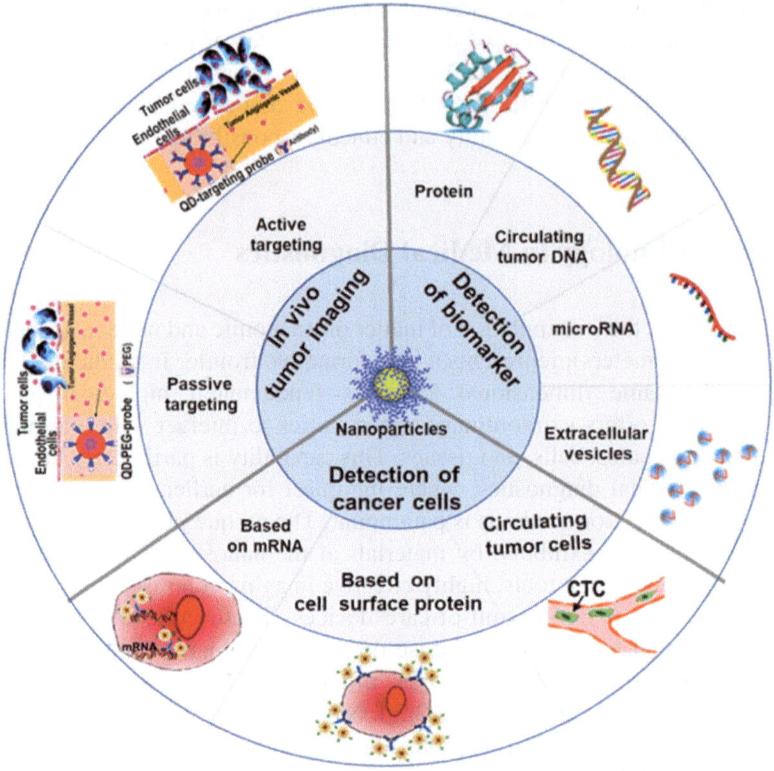

Fig. 2 Schematic representation of nanotechnology applications in cancer diagnosis. (Reprinted with permission under Creative Commons Attribution License from Zhang et al. (2019). Copyright 2019, Springer)

Thirdly, engineered multimodal nanoparticles integrate components for complementary techniques (e.g., CT/MRI/fluorescence) within a single agent, providing concurrent information. Crucially, their nanoscale size and high surface area-to-volume ratio facilitate substantial signal amplification by concentrating imaging probes at the target, improving sensitivity for early detection. Additionally, NP size critically influences bio-distribution, circulation time, tissue penetration, and targeting efficiency (Sim and Wong 2021). Collectively, these mechanisms enable precise, targeted, and highly sensitive imaging.

In MRI, NPs function by modifying the nuclear magnetic relaxation times of water protons in surrounding tissues. Generally, there are two main types of relaxation times: longitudinal relaxation time, T1 also known as spin–lattice relaxation and transverse relaxation time T2 also known as spin–spin relaxation. The T1 agents (positive contrast), like paramagnetic NPs containing gadolinium (Gd-NPs) and manganese (Mn-NPs) or ultra-small superparamagnetic iron oxide nanoparticles (USPIONs) (<5 nm), shorten T1 time, enhancing signals for applications like angiography (Wang et al. 2024b). T2 agents (negative contrast), primarily superparamagnetic iron oxide nanoparticles (SPIONs) (>5 nm), shorten T2 time, causing signal loss useful for imaging the liver, spleen, lymph nodes, and inflammation (Caspani et al. 2020). Dual-mode (T1-T2) agents are also being developed for increased versatility (Geraldes 2024).

In CT, contrast enhancement depends on differential X-ray attenuation. NPs composed of elements with high atomic numbers (high-Z) absorb X-rays strongly, providing substantial contrast against soft tissues. These high-Z element NPs (e.g., Au, Bi, Ta, and Yb) increase the X-ray attenuation coefficient in tissues where they accumulate. Targeting specific sites like tumors or atherosclerotic plaques is possible through surface functionalization (Hsu et al. 2020). The distinct X-ray absorption edges (K-edges) of these high-Z elements are particularly useful for spectral (multi-energy) CT, enabling simultaneous differentiation and quantification of multiple contrast agents (Mutreja et al. 2023).

In optical imaging, NPs offer unique properties for fluorescence- or scattering-based methods. Quantum dots (QDs), semiconductor nanocrystals like cadmium selenide zinc sulfide (CdSe/ZnS) or indium phosphide zinc sulfide (InP/ZnS), provide bright, size-tunable fluorescence with high photostability, used for cellular imaging, multiplexed biomarker detection, and sentinel lymph node mapping. Gold NPs (Au-NPs) exhibit strong scattering/absorption via localized surface plasmon resonance (LSPR), with gold nanoshells offering near-infrared (NIR) capabilities for deep-tissue imaging and therapy (Mashinchian et al. 2014). Upconverting NPs (UCNPs) absorb NIR light and emit visible/NIR light, enabling deep-tissue imaging with high signal-to-noise ratios (Kim et al. 2018). Additionally, in PET, NPs serve as carriers for radioisotopes (e.g., ^{64}Cu and ^{89}Zr), allowing highly sensitive quantification of NP biodistribution, often combined with MRI or CT (Thomas et al. 2019).

A major focus is on multimodal agents (e.g., PET/MRI and CT/optical) and theranostic platforms combining imaging with therapy (Hsu et al. 2020). These address limitations of single modalities and conventional agents (e.g., Gd toxicity concerns and rapid clearance of iodinated agents) by offering complementary information, longer circulation, and targeted delivery (Sim and Wong 2021). Theranostics, in particular, enables simultaneous diagnosis, targeted treatment, and response monitoring, advancing personalized medicine (Malik et al. 2023).

To provide a clearer overview of the diverse nanoparticle imaging agents discussed, Fig. 3 summarizes molecular imaging NPs and their applications in diagnosis.

Besides this, Table 1 summarizes key examples based on their composition, primary imaging modality, contrast mechanism, and diagnostic applications.

2.2 Nanosensors in Medical Diagnostics

Nanosensors (NSs) leverage the unique properties of nanomaterials to detect disease biomarkers with high sensitivity and specificity at the molecular level (Iyer et al. 2025). These devices typically combine a recognition element (e.g., antibody

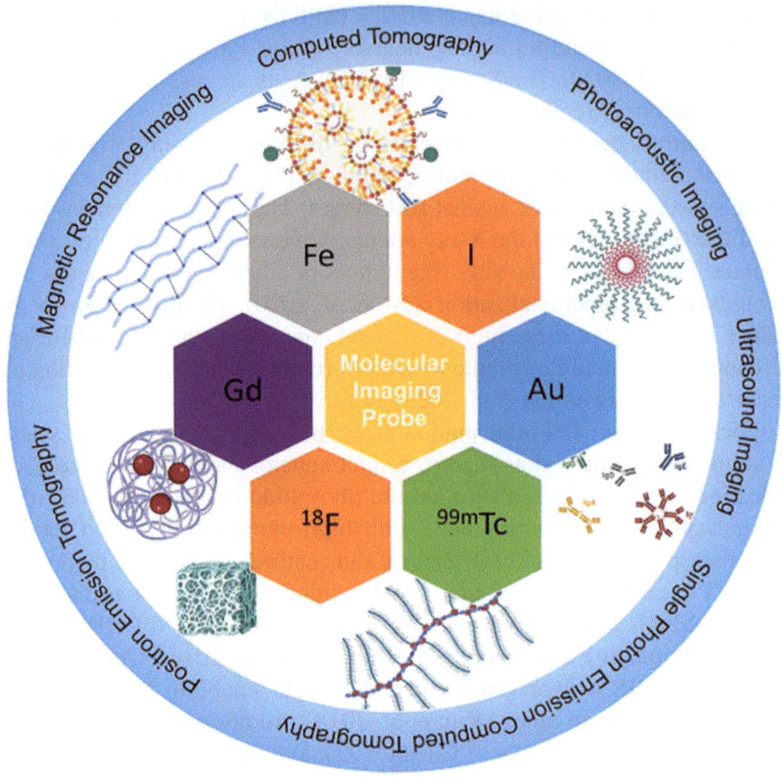

Fig. 3 Molecular imaging nanoprobes and their diagnostic applications. (Reprinted with permission under Creative Commons Attribution License from Kou and Yang (2024). Copyright 2024, Theranostics)

and aptamer) for selective analyte binding with a nanomaterial-based transducer that converts this binding into a measurable signal (Liu et al. 2021). The high surface area-to-volume ratio and distinct electronic or optical properties of nanomaterials like NPs, nanowires, and nanosheets enhance sensor performance, including sensitivity and response time (Paras et al. 2022). Surface functionalization ensures selectivity, while biocompatibility is key for biological applications (Chamorro-Garcia and Merkoçi 2016).

The preparation of these NSs for medical diagnostics is a multi-step process centered on functionalization. First, a nanomaterial is selected to act as the core signal transducer. Then, its surface is modified by attaching specific bioreceptors, such as antibodies or DNA strands. These bioreceptors are engineered to bind exclusively with a target analyte (e.g., a disease-specific protein). When the sensor is exposed to a biological sample, this specific binding event occurs, altering the readout properties (optical or electrical) of nanomaterial and generating a detectable signal that indicates the presence of the analyte.

Table 1 Selected nanoparticle-based diagnostic imaging agents

NP type	Example composition (core/shell/coating)	Primary imaging modality	Contrast mechanism	Key diagnostic applications	Reference
SPIONs	Fe_3O_4 or γ-Fe_2O_3/dextran, PEG, silica	MRI	T2 relaxation shortening (negative contrast)	Liver, spleen, lymph node imaging; inflammation/infection tracking; cell tracking; angiography (some); tumor imaging (EPR)	Lapusan et al. (2024)
USPIONs/ESIONs	Fe_3O_4 or γ-Fe_2O_3 (<5 nm)/coatings	MRI	T1 relaxation shortening (positive contrast)	Angiography; lymph node imaging; tumor imaging	Geraldes (2024)
Gd-NPs	Gd_2O_3, $NaGdF_4$/PEG	MRI	T1 relaxation shortening (positive contrast)	Angiography; tumor imaging; potential alternative to small molecule Gd agents	Geraldes (2024)
Mn-NPs	MnO, Mn_3O_4/coatings	MRI	T1 relaxation shortening (positive contrast)	Brain imaging; renal carcinoma imaging; liver imaging	Geraldes (2024)
Au-NPs	Au/PEG, targeting ligands	CT, optical	X-ray attenuation (high Z); SPR (scattering/abs.)	CT: Angiography, tumor imaging, atherosclerosis, cell tracking; optical: Cellular imaging, sensing	Sim and Wong (2021)
Bi-NPs	Bi_2S_3, BiF_3/coatings, targeting ligands	CT	X-ray attenuation (high Z, K-edge 90.5 keV)	Thrombus imaging; liver fibrosis diagnosis; GI tract imaging	Kim et al. (2018)
Ta-NPs	Ta_2O_5/PEG, silica	CT	X-ray attenuation (high Z, K-edge 67.3 keV)	Sentinel lymph node mapping; multimodal imaging (CT/fluorescence)	Hsu et al. (2020)
Yb-NPs	$NaYbF_4$, $BaYbF_5$/PEG, silica	CT, optical	X-ray attenuation (high Z, K-edge 61.3 keV); upconversion	Microthrombi imaging; hydrogel tracking; bone damage targeting; blood pool imaging; multimodal imaging (CT/fluorescence)	Mutreja et al. (2023)
QDs	CdSe/ZnS, InP/ZnS/coatings	Optical	Fluorescence (size-tunable emission)	Cellular imaging; biomarker detection; sentinel lymph node mapping; multiplexing	Mashinchian et al. (2014)
UCNPs	$NaYF_4$:Yb, Er/coatings	Optical	Upconversion (NIR excitation, vis/NIR emission)	Deep-tissue imaging; lymph node mapping; tumor imaging; high signal-to-noise imaging	Kim et al. (2018)
PET-labeled NPs	SPION-MANOTA-^{64}Cu	PET/MRI	Radioactivity (^{64}Cu); T2 relaxation shortening	High-sensitivity molecular imaging; biodistribution studies; liver imaging	Thomas et al. (2019)
Theranostic NPs	Au@Cu–Sb–S; ReS_2; drug-loaded SPIONs	CT/PA/therapy; CT/therapy; MRI/therapy	Combined imaging contrast and therapeutic action	Image-guided therapy; simultaneous diagnosis and treatment; monitoring therapeutic response (e.g., cancer and targeted drug delivery)	Wang et al. (2024b)

NSs are often classified by their transduction mechanism. Electrochemical NSs utilize materials like carbon nanotubes (CNTs), graphene, or metal NPs to detect electrical changes (current, impedance, and potential) upon analyte binding, targeting glucose, neurotransmitters, cancer markers, and pathogens. Optical NSs rely on changes in optical properties. Examples include colorimetric assays using AuNPs (LSPR shifts), fluorescence modulation using QDs, surface plasmon resonance (SPR) sensors detecting refractive index changes, and surface-enhanced Raman spectroscopy (SERS) using plasmonic NPs for molecular fingerprinting (Paras et al. 2022). These detect proteins, nucleic acids, ions, and pathogens (Iyer et al. 2025). Magnetic NSs employ magnetic NPs (typically SPIONs) as labels detected by magnetic readers or use magnetic relaxation changes for label-free detection of cardiac markers, cancer markers, and pathogens (Chamorro-Garcia and Merkoçi 2016). Mechanical NSs transduce binding into mechanical changes like cantilever bending, while nanochannel/nanopore sensors measure ionic current modulation as analytes pass through or bind within nanoscale pores, enabling single-molecule detection and DNA sequencing (Darwish et al. 2024; Malik et al. 2023).

The power of NSs lies in combining nanomaterial signal amplification with bioreceptor specificity. A significant trend is the move toward label-free detection methods (e.g., field-effect transistors, SPR, impedance, and SERS) to simplify assays, reduce costs, and enable real-time monitoring, which is advantageous for point-of-care testing (Chamorro-Garcia and Merkoçi 2016). Table 2 summarizes a comparative overview of the major types of NSs discussed.

Beyond conventional NSs, nanomaterials are now powering a new generation of omics technologies aimed at decoding cellular heterogeneity and tissue-level architecture in disease. Nanoparticle-enabled single-cell and spatial omics platforms integrate nanotechnology with transcriptomics, proteomics, and metabolomics to detect disease biomarkers at the single-cell level or within intact tissue microenvironments (Wang et al. 2024a). These systems use engineered nanostructures including magnetic nanoparticles for cell barcoding (Labib et al. 2018), Au-NPs for spatial signal amplification (Zhang et al. 2024a), and QDs for RNA detection and fluorescence multiplexing (Liu et al. 2018).

NPs also bridge different omics modalities as the spatial proteomics/matrix-assisted laser desorption/ionization (MALDI) workflow was combined with spatial transcriptomics. Zhang et al. integrated mass spectrometry imaging (metabolites and proteins) with microarray-based transcript mapping on the same tissue. This multi-omics imaging revealed co-localization of metabolic genes, lipids, and transcripts within tumor regions (Zhang et al. 2024a). Such integrated platforms, aided by nanomaterials, allow simultaneous biomarker readouts (genes, proteins, and metabolites) with spatial context.

Despite their significant potential, NSs face critical limitations that currently hinder their widespread clinical application. One of the primary hurdles is achieving a sufficient signal-to-noise ratio—distinguishing the true signal from a target biomarker against background "noise." The noise can originate from nonspecific bindings of other molecules in a complex biological sample or from the detection instrument itself. Since biomarkers are often present in tiny concentrations, the

Table 2 Overview of NS types for medical diagnostics

Sensor type	Transduction principle	Key nanomaterials employed	Example target analytes/biomarkers	Example diagnostic applications	Reference
Electrochemical	Amperometry, voltammetry, impedance, field-effect transistor (FET) conductance	CNTs, graphene, Au-NPs, Ag-NPs, Pt-NPs, metal oxides	Glucose, neurotransmitters, cancer markers (PSA, CEA, miRNA), pathogens, ions, and DNA	Diabetes monitoring, neurological studies, cancer detection, and infectious disease diagnosis	Darwish et al. (2024)
Optical	Colorimetry (LSPR), fluorescence, SERS, SPR, luminescence	Au-NPs, Ag-NPs, QDs, graphene oxide, UCNPs, nanorods	Cancer markers (AFP, CEA), cardiac markers (CRP), pathogens, DNA/miRNA, glucose, ions, and toxins	Cancer detection, cardiovascular risk assessment, infectious disease diagnosis, and environmental monitoring	Darwish et al. (2024)
Magnetic	Magnetoresistance, magnetic relaxation	SPIONs	Cardiac markers (CRP, cTnI), cancer markers (AFP), pathogens (salmonella), mRNA, and human papillomavirus (HPV) DNA	Cardiovascular diagnostics, cancer detection, and infectious disease detection	Chamorro-Garcia and Merkoçi (2016)
Mechanical	Cantilever bending, resonance frequency shift	Nanocantilevers (potentially with NP functionalization)	Proteins and DNA	Biomarker detection (less common than electrochemical/optical)	Darwish et al. (2024)
Nanochannel/pore	Ionic current modulation, optical changes	Fabricated nanopores/channels, protein pores (α-hemolysin)	DNA/RNA, proteins (thrombin), cancer markers (CTCs, miRNA), cells, ions, and enzyme activity	Single-molecule detection, DNA sequencing, cancer diagnostics, and monitoring enzyme activity	Malik et al. (2023)

resulting signal can be weak. A low signal-to-noise ratio can therefore lead to unreliable measurements, including false positives or false negatives, ultimately limiting the sensor's diagnostic accuracy.

In addition to signal-to-noise ratio, stability and reproducibility present major barriers to commercialization. The issue of stability arises because the physicochemical properties of nanomaterials are often sensitive to environmental factors like temperature and pH, which can lead to degradation and a limited shelf-life. Compounding this is the challenge of reproducibility. Manufacturing large batches of NSs with perfectly consistent size, shape, and surface characteristics is difficult. This batch-to-batch variation can cause inconsistent performance, making it difficult to standardize protocols and compare results across different devices, which is a critical step for regulatory approval.

2.3 Point-of-Care Nanodevices

Nanotechnology is essential in advancing point-of-care (POC) diagnostics, shifting testing from centralized laboratories closer to the patient for rapid results and improved management (Liu et al. 2021). This technological advancement is driven by the need for rapid results, improved disease management, and enhanced accessibility to healthcare diagnostics, particularly in resource-limited settings. The integration of nanotechnology addresses fundamental limitations of traditional POC devices by providing enhanced sensitivity, specificity, and miniaturization capabilities (Secchi et al. 2025; Gao et al. 2024). This involves enhancing performance and miniaturization in user-friendly formats like lateral flow assays (LFAs), lab-on-a-chip (LOC) systems, and wearable sensors.

LFAs, simple paper-based tests, benefit from nanotechnology primarily through enhanced sensitivity and specificity (Liu et al. 2021). This is achieved using superior nanoparticle labels (e.g., optimized NPs, QDs, and UCNPs) and signal amplification strategies (e.g., silver enhancement and catalytic amplification) to detect analytes like infectious agents, cardiac markers, and hormones at lower concentrations, bridging the gap with laboratory tests like enzyme-linked immunosorbent assay (ELISA) or polymerase chain reaction (PCR). Nanomaterial coatings also improve stability and reduce nonspecific binding (Liu et al. 2021).

LOC devices miniaturize laboratory functions onto chips using microfluidics (Arshavsky-Graham and Segal 2020). Nanotechnology integration includes using NPs as sensitive labels or sensors, creating nanostructured surfaces for efficient capture or reaction enhancement, incorporating nanozymes for on-chip assays, and using nanoporous membranes for separation (Gharib et al. 2022). LOC systems enable automated, multiplexed 'sample-in-answer-out' analysis with reduced sample volumes and faster times, beneficial for cancer, infectious disease, and cardiovascular diagnostics (Arshavsky-Graham and Segal 2020).

The convergence of nanotechnology with flexible electronics has enabled the development of sophisticated wearable diagnostic devices for continuous health monitoring (Jung et al. 2024). These systems integrate electrochemical or optical NSs with wearable form factors to enable noninvasive, real-time detection of biomarkers in accessible biofluids such as sweat, saliva, tears, and interstitial fluid (Jung et al. 2024). This allows noninvasive, real-time monitoring, particularly for chronic diseases like diabetes, providing personalized health data via connected devices like smartphones (Iyer et al. 2025).

A key driver for nanotechnology in POC is enhancing sensitivity to enable early diagnosis using rapid tests (Liu et al. 2021). Nanomaterials provide the necessary signal amplification to detect low-concentration biomarkers, approaching laboratory standards. This sensitivity also enables the shift toward noninvasive continuous monitoring using wearable devices (Jung et al. 2024). However, balancing enhanced performance with the essential POC requirements of simplicity, affordability, and accessibility remains an ongoing challenge.

2.4 Emerging Trends: Organoid and Organ-on-Chip Diagnostics

Nanomaterials are increasingly integral to the advancement of organoid and organ-on-chip platforms, enabling high-resolution disease modeling and personalized diagnostic applications (Shen et al. 2023; Shi et al. 2024). This integration of nanotechnology overcomes the limits of traditional assays by allowing continuous, real-time monitoring of biochemical and biophysical parameters (Zhang et al. 2024b). Recent advancements focus on integrating nanotechnology with organoids and microfluidic systems. High-throughput microfluidic chips integrated with nanopillar-based biosensors enable multiplexed detection of biomarkers, facilitating real-time monitoring of therapeutic response and disease progression (Rezaei et al. 2025). Carbon-based nanomaterials, including graphene and carbon nanotubes, have been embedded into liver- and brain-on-chip systems to enhance electrophysiological and metabolic sensing, thereby advancing disease modeling in neurodegeneration and hepatotoxicity (Rezaei et al. 2025). Patient-derived organoids functionalized with tailored nanomaterials serve as personalized screening tools, particularly in predicting tumor-specific responses to nanotherapeutics (Shen et al. 2023; Wensink et al. 2021). These innovations bridge gaps between in vitro models and clinical diagnostics, accelerating drug development and precision medicine (Labib et al. 2018).

3 Nanotechnology in Modern Medical Therapeutics

Owing to the unique features and physiobiological characteristics, nanostructures or nanocarriers have provided vast therapeutic techniques against microbial infections, cancers, tissue regeneration, tissue engineering, immunotherapies, and gene therapies through drug delivery systems. The therapeutic potentials and applications of nanobiotechnology in modern medical sciences are depicted in Fig. 4.

3.1 *Nanoparticle-Based Targeted Drug Delivery*

The impact of nanotechnology now spans virtually all aspects of healthcare. Its applications are not just improving diagnostics through more sensitive and rapid tools like nanoparticle-based imaging contrast agents, POC tests, and biosensors capable of detecting ultra-low concentrations of biomarkers. Its application is also revolutionizing drug delivery by creating carriers that enhance drug solubility, stability, bioavailability, and targeting, thereby increasing therapeutic efficacy while minimizing side effects (Chenxi et al. 2025). A primary focus lies in overcoming the limitations of conventional medicines through nanoparticle-based drug delivery. Another ambitious focus lies in enabling critical technologies for achieving precision medicine. Precision medicine identifies the need for highly specific, targeted interventions tailored to individual patient profiles, often based on molecular diagnostics revealing specific genetic mutations or biomarker expressions. However,

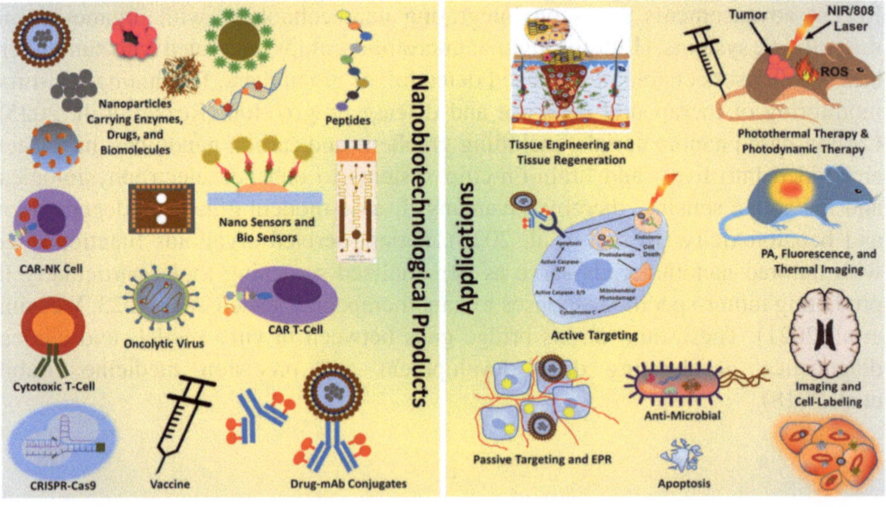

Fig. 4 Therapeutic applications of nanotechnology in modern medical sciences. (Reprinted with permission under Creative Commons Attribution License from Dutt et al. (2023). Copyright 2025, BioMed Central Ltd.)

conventional therapeutic and diagnostic modalities often lack the requisite specificity and sensitivity to effectively act upon this detailed molecular information. Nanotechnology provides the essential toolkit to bridge this gap. The ability to engineer nanomaterials, include NPs with precisely controlled size, shape, surface chemistry, and functionality, allows for the development of platforms capable of targeted drug delivery, highly sensitive diagnostic detection, and advanced imaging capabilities that align directly with the requirements of personalized care. This synergy is driving innovation across the healthcare landscape, offering unprecedented capabilities in developing integrated diagnostic and therapeutic (theranostic) systems. It underpins novel therapeutic strategies, including advanced cancer treatments, gene therapy vectors, and immunotherapies (Mitchell et al. 2021).

A diverse array of materials has been employed to construct NPs for drug delivery, each offering distinct properties and advantages. These can be broadly categorized into organic, inorganic, and carbon-based NPs. The organic NPs include liposomes, polymeric NPs, polymeric micelles, dendrimers, LNP, and solid lipid NPs (Heath 2015; Patra et al. 2018). The inorganic NPs involve metal, semiconductor, and silica NPs (Rodrigues et al. 2020). The carbon-based NPs encompass carbon nanotubes, graphene, and fullerenes (Devi et al. 2023).

To enhance the delivery specificity, NPs employ various targeting strategies. Passive targeting primarily relies on the enhanced permeability and retention (EPR) effect, first described in the 1980s (Hua et al. 2018). This phenomenon occurs in tissues with abnormal vasculature, such as solid tumors, where leaky blood vessels allow NPs (typically larger than the renal clearance threshold, ~10 nm, and often PEGylated for long circulation) to extravagate into the tissue, while poor lymphatic drainage leads to their retention (Heath 2015). While foundational for many early nanomedicines like Doxil®, the EPR effect is now understood to be highly heterogeneous, varying significantly between tumor types, patients, and even within the same tumor, limiting its universal effectiveness (Alvarez et al. 2017).

Active targeting seeks to improve specificity by functionalizing the nanoparticle surface with ligands—such as antibodies, peptides, aptamers, or small molecules like folate or sugars (e.g., N-acetylgalactosamine and GalNAc) that bind to specific receptors overexpressed on the surface of target cells. This binding aims to promote selective cellular uptake via receptor-mediated endocytosis (Patra et al. 2018). While conceptually attractive and successful in preclinical models, demonstrating a clear clinical advantage of active targeting over passive targeting for many nanoparticle systems has proven challenging (Hua et al. 2018). Nevertheless, specific examples like GalNAc-conjugated siRNA for highly efficient liver targeting via the asialoglycoprotein receptor highlight the potential of active targeting when a suitable high-affinity receptor system is available (Yan et al. 2022). This success underscores that the effectiveness of active targeting is highly dependent on the specific ligand–receptor interaction and target cell biology, representing a significant hurdle for broader application compared to the more general, albeit variable, EPR effect.

Once NPs reach the target site or cell, drug release is mediated by various mechanisms. Simple diffusion from the nanoparticle matrix, erosion or degradation of the carrier material (common for biodegradable polymers), or swelling of the nanoparticle can lead to drug release (Patra et al. 2018). More sophisticated designs utilize stimuli-responsive release, where the nanoparticle is engineered to release its payload in response to specific triggers. These triggers can be endogenous, such as the acidic pH within endosomes or the tumor microenvironment, specific enzymes overexpressed in diseased tissue, or the different redox potential inside cells compared to the bloodstream (Heath 2015).

Alternatively, external stimuli like light (for photothermal or photochemical release), temperature changes (using thermosensitive materials), ultrasound, or magnetic fields can be applied to trigger release with high spatial and temporal precision. The advantages offered by nanoparticle drug delivery are significant: improved drug solubility and stability (Patra et al. 2018), enhanced bioavailability and prolonged circulation times (Patra et al. 2018), the potential for targeted delivery leading to increased efficacy and reduced systemic toxicity (Gawne et al. 2023), controlled or sustained release profiles, protection of fragile drug payloads like nucleic acids or proteins (Patra et al. 2018), and the ability to overcome certain biological barriers (Alvarez et al. 2017). The major advantages of nanoparticle-based drug delivery system are illustrated in Fig. 5.

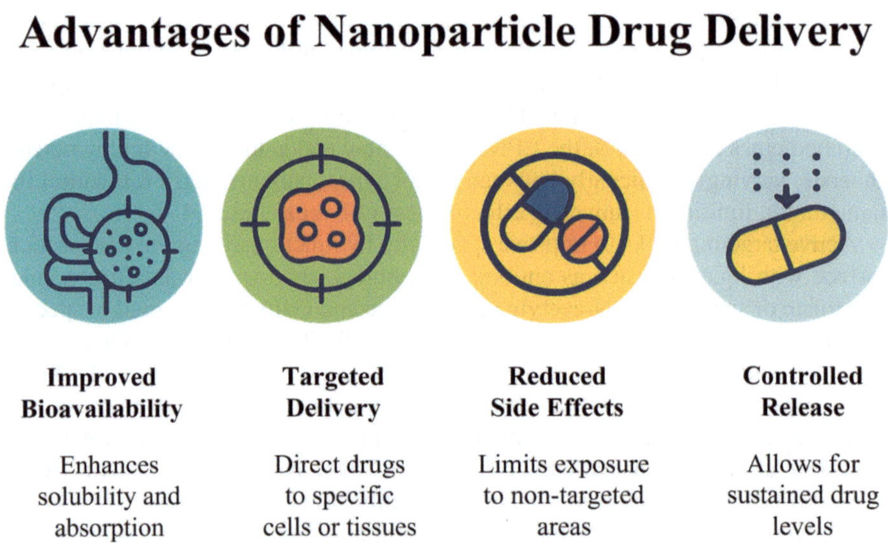

Fig. 5 Major advantages of nanoparticle-based drug delivery systems

3.2 *Nanomedicines and Nanoengineered Therapy*

3.2.1 Nanomedicines

Nanomedicines broadly refer to the application of nanotechnology for medical purposes. As with other medicinal products, nanomedicines necessitate rigorous regulatory approval prior to market introduction. Given the inherent complexity and novelty of these therapeutics, regulatory agencies such as the US Food and Drug Administration (FDA) and the European Medicines Agency (EMA) have adopted a cautious approach in defining and regulating nanomedicines. Despite these regulatory challenges, an increasing number of nanomedicines have successfully advanced to clinical application or are undergoing late-stage clinical trials, highlighting the significant promise of nanotechnology in medical interventions.

Table 3 provides a summary of nanomedicines currently in clinical stages, which span a diverse array of therapeutic areas and employ various nanoparticle platforms. Among these, liposomes and albumin NPs have been predominantly utilized to enhance pharmacokinetic and pharmacodynamic profiles, improve drug solubility, and mitigate toxicity-addressing issues such as cardiotoxicity associated with doxorubicin (e.g., encapsulated in Doxil®) and nephrotoxicity linked to amphotericin B. Additionally, these platforms have effectively tackled formulation challenges, as exemplified by the solvent-free formulation of paclitaxel in Abraxane® (Heath 2015).

The success of lipid nanoparticle (LNP) technology in delivering siRNA (as seen in Onpattro®) and mRNA (as utilized in COVID-19 vaccines) marks a pivotal advancement in the realm of novel therapeutic delivery systems, though the primary function of these LNPs remains focused on ensuring efficient delivery (Shan et al. 2022). Conversely, nanomedicines that are based on unique mechanisms derived from nanoscale properties, such as photothermal ablation utilizing gold nanoshells (e.g., AuroShell®) or advanced theranostic agents that integrate both imaging and therapeutic functionalities within a single particle, are predominantly still in the earlier phases of clinical development (Gawne et al. 2023).

3.2.2 Nanoengineered Therapy

Beyond improving the delivery of existing drugs, nanotechnology enables the development of entirely new therapeutic modalities or significantly enhances existing ones by engineering materials and devices at the nanoscale to interact with biological systems in specific ways. This "nanoengineered therapy" encompasses diverse strategies targeting major diseases like cancer, genetic disorders, and conditions requiring immune modulation.

For gene therapy, which aims to treat diseases by modifying gene expression, nanotechnology provides crucial nonviral vectors to deliver genetic material (DNA, mRNA, siRNA, and CRISPR components) safely and efficiently. This overcomes the limitations of naked nucleic acids (degradation, poor uptake, and

Table 3 Examples of clinical-stage nanomedicines

Product name/type	Nanoparticle platform	Therapeutic area	Key mechanism/advantage	Clinical status (approximately)	References
Doxil®/Caelyx®	PEGylated liposome	Cancer (Kaposi's sarcoma, ovarian, multiple myeloma)	Encapsulates doxorubicin; reduced cardiotoxicity, prolonged circulation (EPR effect)	Marketed (1995)	Shan et al. (2022)
Abraxane®	Albumin-bound nanoparticle	Cancer (breast, lung, and pancreatic)	Paclitaxel formulation: Improved solubility, avoids hypersensitivity reactions to solvent	Marketed (2005)	Heath (2015)
ONIVYDE®	Liposome	Cancer (pancreatic)	Encapsulates irinotecan; improved pharmacokinetics	Marketed (2015)	Shan et al. (2022)
VYXEOS®	Liposome	Cancer (acute myeloid leukemia [AML])	Co-delivers cytarabine and daunorubicin in synergistic ratio; improved survival	Marketed (2017)	Shan et al. (2022)
Hensify®	Hafnium oxide nanoparticle	Cancer (soft tissue sarcoma)	Radioenhancer for radiation therapy	Marketed (EU, 2019)	Shan et al. (2022)
FYARRO™	Albumin-bound nanoparticle	Cancer (malignant PEComa)	Sirolimus formulation; targeted delivery via albumin pathway	Marketed (2021)	Shan et al. (2022)
AmBisome®/ ABELCET®/ AMPHOTEC®	Liposomal/lipid complex	Fungal infections	Amphotericin B formulation; reduced nephrotoxicity	Marketed	Shan et al. (2022)
ARIKAYCE®	Liposomal inhalation suspension	Bacterial infection (Mycobacterium avium complex [MAC] lung disease)	Amikacin formulation; sustained release in lungs	Marketed (2018)	Shan et al. (2022)
Acticoat™	Silver nanocrystal dressing	Burn wound care	Antimicrobial effect of silver ions	Marketed	Shan et al. (2022)
Feraheme®/Rienso®	Iron oxide nanoparticle (SPION)	Iron deficiency anemia (chronic kidney disease)	Ferumoxytol; intravenous iron replacement	Marketed (2009)	Shan et al. (2022)

Onpattro®	LNP	Genetic disease (hereditary transthyretin [hATTR] amyloidosis)	Delivers patisiran siRNA to liver; silences transthyretin (TTR) gene	Marketed (2018)	Shan et al. (2022)
Comirnaty® (Pfizer/ BioNTech)	LNP	Infectious disease (COVID-19 prevention)	Delivers severe acute respiratory syndrome coronavirus 2 (SARS-CoV-2) spike protein mRNA; elicits immune response	Marketed (2020)	Younis et al. (2022)
Spikevax® (Moderna)	LNP	Infectious disease (COVID-19 prevention)	Delivers SARS-CoV-2 spike protein mRNA; elicits immune response	Marketed (2020)	Younis et al. (2022)
Lutathera®	Peptide receptor radionuclide therapy	Cancer (neuroendocrine tumors)	^{177}Lu-DOTATATE; targeted radiotherapy (theranostic principle)	Marketed (2018)	Gawne et al. (2023)
Pluvicto™	Radioligand therapy	Cancer (prostate)	^{177}Lu-PSMA-617; targeted radiotherapy (theranostic principle)	Marketed (2022)	Gawne et al. (2023)
AuroLase® therapy (using AuroShell® NPs)	Gold nanoshells	Cancer (prostate—Focal therapy)	Photothermal ablation (PTT) using NIR light	Clinical trials (pivotal)	Shan et al. (2022)
NTLA-2001	LNP	Genetic disease (hATTR amyloidosis)	Delivers clustered regularly interspaced short palindromic repeats (CRISPR)/CRISPR-associated protein 9 (Cas9) components for in vivo gene editing of TTR gene	Clinical trials (phase I/II)	Lee et al. (2025)
CALAA-01	Cyclodextrin-polymer nanoparticle	Cancer (solid tumors)	Delivers siRNA targeting ribonucleotide reductase M2 (RRM2); first targeted RNAi therapeutic in humans	Clinical trials (phase I)	Shan et al. (2022)

immunogenicity) and the safety concerns associated with viral vectors (immunogenicity, toxicity, and insertional mutagenesis) (Tong et al. 2023). Nonviral vectors, including LNPs, polymeric NPs (e.g., using PEI and PLGA), and inorganic NPs (e.g., gold), offer advantages like lower immunogenicity, potentially larger cargo capacity, and easier manufacturing (Tong et al. 2023). The delivery process involves protecting the nucleic acid payload via encapsulation or complexation, facilitating cellular uptake (often via endocytosis), enabling escape from the endosome to avoid degradation (a major hurdle), and, for DNA or nuclear-acting therapies, ensuring transport into the nucleus (Patra et al. 2018). The clinical success of LNP-based mRNA vaccines and siRNA therapeutics (patisiran) has validated nonviral platforms, driving further development for delivering diverse genetic therapies, including CRISPR/Cas9 gene-editing tools (Lee et al. 2025).

Nanotechnology also significantly enhances cancer immunotherapy, which harnesses the patient's immune system to fight cancer (Lu et al. 2024). NPs improve cancer vaccines by protecting antigens, enhancing their delivery to and uptake by antigen-presenting cells (APCs) like dendritic cells, and targeting delivery to lymph nodes (Saleh and Shojaosadati 2016; Koyande et al. 2022). A key advantage is the co-delivery of antigens and adjuvants (immune stimulants like TLR agonists) within the same NP, ensuring simultaneous signaling to APCs for robust T-cell activation (Saleh and Shojaosadati 2016). NPs can also facilitate antigen cross-presentation, crucial for activating cytotoxic T lymphocytes (Wen et al. 2019). Beyond vaccines, NPs modulate the tumor microenvironment by delivering agents to reprogram immunosuppressive cells (e.g., M2 tumor-associated macrophages to M1 macrophages), inhibit suppressive fibroblasts, or enhance T-cell activity within the tumor. They also enable targeted delivery of immune checkpoint inhibitors like anti-PD-1/PD-L1 agents (as antibodies or siRNA) to the tumor microenvironment, potentially increasing efficacy while reducing systemic toxicity and allowing synergistic combination therapies (Lu et al. 2024; Saleh and Shojaosadati 2016). The development of multifunctional NPs combining these strategies represents a powerful trend in nano-immunotherapy.

In addition to NPs, another new class of nanocarrier called programmable DNA nanostructures (e.g., DNA origami, tetrahedral cages, and DNA hydrogels) and metal-organic framework (MOFs) enable multifunctional cancer therapy and gene delivery. These scaffolds can be loaded with chemotherapeutics or siRNAs and engineered to respond to tumor-specific stimuli (pH, enzymes, and redox) for on-demand cargo release (Hong et al. 2025; Cai et al. 2022).

3.3 Nanotechnology in Regenerative Medicine

Regenerative medicine aims to repair, replace, or regenerate damaged tissues and organs, offering potential cures for conditions currently managed with limited success. Nanotechnology provides essential tools for this field by operating at the same dimensional scale as the fundamental components of tissues, cells, and the extracellular matrix (Davis et al. 2008). This allows for the creation of biomaterials and delivery systems that can interact with cells and guide tissue formation with unprecedented precision.

A central element of tissue engineering is the scaffold, which provides temporary structural support and guides new tissue growth. Nanotechnology revolutionizes scaffold design through several avenues. Techniques like electrospinning create nanofibers that mimic the nanoscale topography and architecture of the natural extracellular matrix (Mahara et al. 2023). Incorporating NPs (e.g., nano-hydroxyapatite, carbon nanotubes, and graphene) into scaffold materials improves functionalities like osteoconductivity, mechanical strength, and electrical conductivity, vital for tissues like bone, nerve, or cardiac muscle (Alvarez et al. 2017).

Furthermore, nanotechnology also significantly interfaces with stem cell therapies. NPs, like SPIONs, serve as agents for cell tracking and imaging, allowing researchers and clinicians to label stem cells before transplantation and monitor their location, migration, survival, and integration using noninvasive techniques such as MRI (Gawne et al. 2023). Nanostructured scaffolds or NPs can create specific microenvironments or "niches" that influence stem cell fate, promoting their survival, proliferation, or differentiation into desired cell types (Malik et al. 2023). Additionally, nanoscaffolds can function as carriers to deliver stem cells to the injury site, offering physical support and potentially releasing factors to enhance cell engraftment and function.

These principles find application across various tissues. In bone regeneration, nanoscale scaffolds incorporating hydroxyapatite or delivering bone morphogenetic proteins are commonly employed. For nerve regeneration, conductive scaffolds containing CNTs or graphene are designed to guide axonal outgrowth, often complemented by the delivery of neurotrophic factors (Rodrigues et al. 2020). Cardiac tissue repair explores the use of conductive patches and delivery systems for angiogenic factors like vascular endothelial growth factor (Alvarez et al. 2017). Similarly, nanofibrous scaffolds combined with nanoparticle delivery of antimicrobial agents or growth factors aid in skin regeneration (Mahara et al. 2023). Nanotechnology also presents solutions for challenging delivery routes, such as ocular drug delivery, where NPs are used to overcome physiological barriers in the eye (Rodrigues et al. 2020).

3.4 Nanotechnology in Precision Medicine

Precision medicine represents a paradigm shift from traditional "one-size-fits-all" approaches, aiming instead to tailor disease prevention, diagnosis, and treatment based on individual variability in genetics, environment, and lifestyle. In oncology, this involves leveraging specific molecular information about a patient's tumor to guide therapeutic strategies, predict responses, and determine prognosis. Nanotechnology provides indispensable tools to achieve the molecular-level specificity required by precision medicine (Mao et al. 2024). Its compatibility with the scale of biological processes enables direct interaction with cellular and molecular components, facilitating highly targeted interventions (Viborg et al. 2024). Nanotechnology underpins precision diagnostics through the development of ultra-sensitive NSs capable of detecting disease biomarkers—such as specific proteins, genetic mutations, or circulating tumor cells—at very low concentrations, enabling earlier diagnosis and detailed molecular profiling (Malik et al. 2023).

This diagnostic precision directly informs therapeutic strategies, particularly through nanoparticle-based drug delivery systems (NDDS) (Patra et al. 2018). Crucially for precision medicine, NDDS can be engineered for active targeting by functionalizing their surface with ligands (e.g., antibodies, peptides, and folate) that recognize specific receptors or biomarkers overexpressed on a patient's diseased cells, ensuring drug delivery tailored to individual molecular signatures (Hua et al. 2018). Additionally, stimuli-responsive NPs release their therapeutic payload in response to specific triggers within the individual's disease microenvironment (e.g., pH, redox potential, and enzymes) or external cues (e.g., light and heat), enabling controlled, on-demand therapy (Liu and Yuan 2025).

The integration of diagnostic and therapeutic capabilities within single nanoparticle platforms, known as nanotheranostics, further advances personalized treatment (Gawne et al. 2023). These systems allow for simultaneous imaging of target sites or biomarkers, delivery of targeted therapy (such as chemotherapy, photothermal therapy, or photodynamic therapy), and real-time monitoring of the therapeutic response within the individual patient, enabling dynamic adjustments to treatment plans. Nanotechnology also plays a pivotal role in personalized immunotherapy. Nanoparticle-based vaccines can efficiently co-deliver patient-specific tumor neoantigens, derived from individual tumor mutations, along with immunostimulatory adjuvants directly to APCs, often targeting lymph nodes to enhance antigen presentation and elicit potent, tailored antitumor T-cell responses (Hua et al. 2018). Beyond vaccines, NPs enhance other immunotherapies by facilitating the targeted delivery of immune checkpoint inhibitors (Huang and Zhang 2024), modulating the specific tumor microenvironment of a patient to overcome immunosuppression (Karimi et al. 2025), possessing intrinsic immunomodulatory properties (Feng et al. 2019), or improving the efficacy of adoptive cell therapies like chimeric antigen receptor (CAR)-T cell therapy (Baena et al. 2024). Collectively, nanotechnology provides a versatile and essential platform for realizing the goals of precision medicine, enabling the development of diagnostics and therapeutics that are specifically adapted to the unique molecular and physiological characteristics of each patient (Alghamdi et al. 2022).

4 Challenges

4.1 Regulatory and Safety Considerations in Nanotechnology

The incorporation of nanotechnology in healthcare has led to the emergence of nanotechnology-enabled health products (NHPs). The NHPs include nanomedicines and nano-enabled devices that hold the potential to significantly enhance medical treatments and diagnostic methodologies (Rodríguez-Gómez et al. 2025). Nevertheless, their unique properties also create significant regulatory and safety challenges.

4.1.1 Global Regulatory Oversight

Health product regulations encompass a complex array of legal, administrative, and technical measures instituted by governments to safeguard the safety, efficacy, and quality of medicinal products and medical devices. Additionally, they ensure that the information provided about these products is relevant and accurate. These regulations differ based on the specific legislative frameworks of various geographic regions. Currently, neither the FDA nor the EMA has a dedicated regulatory pathway for nanotechnology products; they are assessed under existing frameworks for conventional drugs and devices (Rodríguez-Gómez et al. 2025). A key challenge is the lack of a universally adopted definition of "nanomaterial," though the 1–100 nm range is a common reference point, with variations (e.g., the FDA considering up to 1000 nm in some cases and the European Union using a 50% particle threshold) (Rodríguez-Gómez et al. 2023). This ambiguity affects consistent classification and regulatory scrutiny. Table 4 outlines the key FDA and EMA guidelines and regulatory approaches for nanotechnology in healthcare (Rodríguez-Gómez et al. 2025): FDA guidelines on nanomaterials; EMA reports on nanotechnology-based medicinal products.

4.1.2 Nanotoxicology

Nanotoxicology studies the adverse effects of nanomaterials, whose unique physicochemical properties (small size, high surface area, shape, and surface chemistry) differ from bulk materials and govern their biological interactions. Their size allows penetration of biological barriers (Qamar et al. 2024).

Key toxicity mechanisms include oxidative stress, a primary pathway where NPs generate ROS, causing damage to lipids, proteins, DNA, and mitochondria (SCENIHR 2006). Inflammation is also common, as NPs often trigger inflammatory responses in lungs or systemically via accumulation in reticuloendothelial system organs like the liver, spleen, and kidneys (Zhao et al. 2019). Furthermore, some nanomaterials exhibit genotoxicity, damaging DNA and raising concerns about

Table 4 Overview of key FDA and EMA guidelines and regulatory approaches for nanotechnology in healthcare

Feature	FDA	EMA
Primary agencies	Food and Drug Administration	European medicines agency (centralized), national competent authorities (national/decentralized)
Governing framework	Existing statutory authorities based on product type (drug, device, biologic, etc.)	Existing frameworks (e.g., Directive 2001/83/EC for medicinal products and Medical Device Regulation (MDR) for devices)
Nano-specific pathway	No dedicated pathway; integrated into existing procedures	No dedicated pathway for nanomedicines; regulated case by case
Definition approach	No formal regulatory definition; considers size up to 1000 nm if engineered for properties	General 1–100 nm range referenced; EU Recommendation 2022/C 229/01 includes 50% number threshold; definitions not strictly binding for medicines
Key guidance/tools	"Drug products that contain nanomaterials" guidance; product-specific guidance; adopted International Organization for Standardization (ISO)/American Society for Testing and Materials (ASTM) standards; National Center for toxicological Research (NCTR)	Reflection papers on specific nano-categories, scientific advice (SA) procedure; MDR classification rules
Core approach	Product-focused, science-based assessment; risk identification and management; early consultation advised	Case-by-case assessment for nanomedicines (often as non-biological complex drugs [NBCDs]); centralized review for specific types; SA highly recommended
Key considerations	Engineered size/properties; critical quality attributes (CQAs); stability; characterization linked to safety/efficacy; current good manufacturing practice (cGMP)	Physicochemical complexity (NBCDs); potential for internal exposure (devices); suitability of development programs (via SA)

long-term effects like cancer (Xuan et al. 2023). Immunotoxicity is another concern, as interactions with the immune system can lead to hypersensitivity or immunosuppression; even materials considered biocompatible or coatings like PEG can become immunogenic (Zhao and Nalwa 2007). Other reported effects include membrane damage, cytoskeletal disruption, endoplasmic reticulum stress, blood clotting, and potential neurotoxicity (SCENIHR 2006).

Biodistribution, persistence, and accumulation are critical factors. NPs can distribute systemically, often accumulating in the liver, spleen, kidneys, and lungs

(Qamar et al. 2024). The formation of a "protein corona" alters their biological fate (Zhao et al. 2019). Many inorganic NPs are nonbiodegradable and can persist in tissues for long periods, raising concerns about chronic toxicity and long-term health effects, which remain largely unknown.

Toxicity is highly dependent on NP properties (size, shape, surface area, charge, composition, and coating) and extrinsic factors (dose, duration, and route) (Qamar et al. 2024). Assessing nanotoxicity requires new approaches beyond traditional toxicology, potentially focusing on particle number or surface area rather than just mass (SCENIHR 2006).

4.1.3 Regulatory Challenges in Evaluating Safety and Efficacy

The novel properties of nanomaterials pose significant challenges for regulatory assessment. Key issues include substantial knowledge gaps regarding nano–bio interactions, pharmacokinetics, biodistribution, and especially long-term toxicity (Paradise 2019). Characterization challenges arise from the lack of standardized, reliable methods to measure key physicochemical properties, particularly in biological matrices. This hinders the development of predictive models and validation of new testing strategies suitable for nano-specific effects (Kardani 2024). Consequently, regulators often employ a case-by-case assessment approach, which is resource-intensive, lacks predictability, and struggles with the volume of new nanoproducts (Rodríguez-Gómez et al. 2025). Designing informative clinical trials is also complex, requiring assessment of unique pharmacokinetics and biodistribution, alongside ethical challenges in communicating unknown risks (Wasti et al. 2023). Ensuring manufacturing consistency and addressing the lack of global harmonization add further complexity.

Regulators face the difficult task of managing potentially unknown risks stemming from novel nano–bio interactions. This creates a tension between the need for standardized guidelines and the inherent variability of nanomaterial behavior, leading to the pragmatic but challenging case-by-case approach (Rodríguez-Gómez et al. 2025).

4.2 Ethical, Social, and Environmental Implications of Nanotechnology

Beyond technical aspects, nanotechnology in healthcare raises complex ethical and social issues concerning privacy, consent, equity, public perception, and economic impacts.

4.2.1 Ethical Issues: Privacy, Consent, Equity, and Human Enhancement

Several key ethical issues, familiar within bioethics, are amplified by the unique capabilities of nanotechnology. Patient privacy emerges as a significant concern, particularly as advanced nanodevices, potentially integrated with artificial intelligence and machine learning, gain the capacity to collect vast quantities of sensitive physiological data. This capability raises issues about data security, potential misuse, and the need for robust legislative frameworks specifically addressing nano-related data (Wasti et al. 2023). The principle of informed consent also faces particular challenges. Due to the novelty of many nanomaterials, their full range of biological interactions and potential long-term effects are often unknown, making it difficult to fully disclose all potential risks to research participants or patients and thus complicating the process of obtaining truly informed consent (Wasti et al. 2023).

Perhaps one of the most prominent ethical concerns revolves around ensuring equitable access to the potential benefits of nanomedicine, an issue of distributive justice. High development and manufacturing costs are likely to translate into expensive treatments, potentially creating affordability barriers for individuals and healthcare systems. Furthermore, the concentration of nanotechnology research and development in wealthy, industrialized nations risks creating a "nano-divide," leaving developing countries unable to access or afford these advanced technologies and potentially exacerbating global health inequities (Wasti et al. 2023). Proactive measures, including international cooperation, frameworks like responsible research and innovation (RRI), and initiatives promoting equitable access, are widely called for to mitigate these disparities.

Additionally, nanotechnology's potential for precise intervention fuels the ongoing debate about its application beyond therapy into the realm of human enhancement, raising questions about fairness, social stratification, and the definition of human normalcy. Concerns about the potential for misuse, such as for surveillance purposes, also contribute to the ethical landscape (Cobb and Macoubrie 2004). While these ethical categories are familiar, nanotechnology infuses them with new dimensions due to its potential for unprecedented biological interaction, data generation, high costs, and fine-grained manipulation, demanding specific ethical scrutiny.

4.2.2 Social Issues: Public Perception, Economic Landscape, and Access Disparities

The societal integration of nanotechnology in healthcare is significantly shaped by public attitudes, economic forces, and the resulting patterns of access. Public perception studies indicate generally low awareness and detailed understanding of nanotechnology among the general public, although initial outlooks have often been cautiously positive, with benefits expected to outweigh risks (Cobb and Macoubrie 2004). However, considerable uncertainty often remains (Feindt and Poortvliet 2020). Importantly, trust in institutions, particularly scientists and regulatory bodies, emerges as a key factor influencing risk and benefit perceptions, often acting as a heuristic in the absence of detailed knowledge. Attitudes are also molded by broader views on science, personal relevance, values, and how information is presented.

Economically, the nanomedicine market demonstrates strong projected growth, with forecasts suggesting expansion to potentially over USD 500 billion by the early 2030s, driven by factors like therapeutic potential, rising chronic disease prevalence, demand for targeted delivery, personalized medicine trends, R&D investment, and regulatory improvements (2025). Despite this optimism, significant economic hurdles persist, including high development and manufacturing costs and complex regulatory pathways (Rodríguez-Gómez et al. 2025).

These economic factors directly contribute to accessibility disparities. As highlighted under equity concerns, the high cost of nanomedicines impacts affordability and, combined with the concentration of R&D in wealthier nations, creates substantial access barriers for low- and middle-income countries (Wasti et al. 2023). Broader societal concerns also encompass potential environmental consequences from nanomaterial release and occupational health risks for workers involved in their production (Tabrizi and Li 2025). A potential disconnect exists between strong market forecasts and these unresolved scientific, regulatory, ethical, and social challenges, suggesting that failure to address issues like long-term safety, regulatory ambiguity, cost, and equity could impede translation, erode trust, or lead to inequitable benefit distribution. Given the flexibility of public perception and its reliance on trust, ongoing, transparent public engagement is essential for responsible innovation (Cobb and Macoubrie 2004).

4.2.3 Environmental Concerns

The proliferation of nanomaterials in healthcare applications has generated significant environmental concerns that warrant rigorous academic scrutiny. While nanomedicine offers unprecedented therapeutic possibilities, the environmental implications of these technologies extend far beyond their immediate medical benefits, presenting complex challenges for sustainable development and ecological preservation (Mahapatra et al. 2013). The environmental concerns associated with healthcare nanomaterials encompass multiple interconnected domains, including exposure pathways, bioaccumulation potential, ecological toxicity, and inadequate regulatory frameworks for post-use environmental management.

The primary environmental concern stems from the unique physicochemical properties of nanomaterials that confer both therapeutic efficacy and potential ecological hazards (Hauser et al. 2019; Ray et al. 2009). When nanomaterials released into environmental compartments through excretion, disposal, or manufacturing processes, nanomaterials can undergo various transformations, including aggregation, dissolution, and organic matter binding, which may alter their toxicological profiles and environmental fate (Lead et al. 2018; Spurgeon et al. 2020). The bioaccumulation potential of these materials in aquatic and terrestrial ecosystems presents particular concern, as studies have demonstrated differential uptake patterns across trophic levels, with zooplankton showing significantly higher bioconcentration factors compared to phytoplankton and fish (Zheng and Nowack 2022; Mintis et al. 2024). Furthermore, the inadequacy of current environmental risk assessment frameworks to address nano-specific properties, combined with limited data on long-term environmental concentrations and effects, creates substantial uncertainty regarding the cumulative ecological impact of healthcare nanomaterials (Baun and Hansen 2008; Scott-Fordsmand et al. 2017). The intersection of nanomedicine advancement and environmental stewardship thus necessitates comprehensive lifecycle assessments and the development of robust regulatory mechanisms to ensure that the benefits of these innovative technologies do not compromise ecological integrity for future generations (Mahapatra et al. 2018; Nizam et al. 2021).

5 Future Perspective

The evolution of nanotechnology in healthcare demonstrates a clear progression driven by addressing the limitations of conventional approaches. Several emerging trends are shaping the future of nanotechnology in healthcare. A prominent trend is the integration with personalized medicine, where nanotechnology enables highly sensitive diagnostics for biomarker identification and tailored therapies based on individual patient profiles (Alghamdi et al. 2022). The development of theranostics, platforms combining diagnostic and therapeutic functions within a single nanoparticle system, represents a significant advance (Mao et al. 2024). These systems allow for simultaneous imaging, treatment delivery, and monitoring of therapeutic response, facilitating patient stratification and optimizing treatment strategies (Gawne et al. 2023). The recent clinical success of radiopharmaceutical theranostics like Lutathera® and Pluvicto™ underscores the potential of this approach (Gawne et al. 2023).

Another major trend involves stimuli-responsive NPs designed to release their therapeutic payload in response to specific triggers within the body (e.g., pH changes in tumors, specific enzymes, and redox potential) or external stimuli (e.g., light, heat, ultrasound, and magnetic fields) (Hua et al. 2018). This approach offers enhanced spatiotemporal control over drug release, improving targeting and reducing off-target effects. Biomimetic and bioinspired approaches are also gaining traction, involving the design of NPs that mimic natural biological structures like viruses, exosomes, or cell membranes to improve biocompatibility, evade immune clearance, and enhance cellular uptake (Yan et al. 2022).

Additionally, the advances in nonviral vectors, spurred by successes like mRNA vaccines, will continue to expand the possibilities for safe and effective gene and RNA therapies (Lee et al. 2025). Nanotechnology will be increasingly employed to modulate the immune system and the tumor microenvironment, enhancing the power of immunotherapy (Lu et al. 2024). The push toward green nanotechnology promises more sustainable and potentially safer manufacturing processes (Banihashemi Jozdani et al. 2024).

Furthermore, nanotechnology is increasingly converging with other cutting-edge technologies, such as artificial intelligence (AI) for rational drug design and discovery, 3D printing for creating personalized implants and tissue scaffolds, and integration with wearable sensors and telemedicine platforms (Mahara et al. 2023; Malik et al. 2023).

With advances in early diagnosis, smart drug delivery, and precision medicine, nanotechnology has the potential to significantly transform the healthcare industry. Future developments are anticipated in areas such as tissue engineering and regenerative medicine, real-time health monitoring through wearable technology and nanosensors, and precision medicines with minimal side effects. As nanotechnology, artificial intelligence, and digital health converge, clinical outcomes are expected to improve, leading to more personalized treatments.

6 Conclusion

This chapter scrutinizes how nanotechnology is being used in healthcare, with a particular emphasis on the usage of NPs in different medical specialties. Disease prevention, diagnosis, and therapy could all be altered by nanotechnology, which manipulates matter at the atomic or molecular level. Tissue engineering, imaging, medication delivery, and diagnostics are some of its uses. By delivering medications straight to specific cells, NPs can increase therapeutic efficacy and reduce adverse effects. They can also be employed as contrast agents to enhance the clarity and detail of images in medical imaging procedures. Nanotechnology in diagnostics can result in quicker and more precise testing that identifies diseases early. By fostering improved growth and integration, NPs can be used in tissue engineering to produce artificial organs or tissues. Nanotechnology in healthcare does, however, have many drawbacks and restrictions, including safety issues and the difficulty of manufacturing and regulating it.

In summary, revolutionary developments in materials science, environmental sustainability, and medicine have resulted from the quick development of nanotechnology. Stimuli-responsive nanomaterials, nanoscale diagnostics, and targeted medication delivery systems are important innovations. But there are still a lot of unanswered questions, especially about long-term toxicity, environmental effects, and regulatory issues related to widespread use. A new era of innovation is being ushered in by the merging of nanotechnology with other cutting-edge technologies including synthetic biology, organoid modeling, and artificial intelligence (AI). While organoid systems provide human-relevant models for evaluating nanoscale interventions, artificial intelligence (AI) speeds up the design of nanomaterials and prediction toxicity. Unprecedented possibilities are promised by this interdisciplinary integration, but responsible and equitable implementation also requires strong ethical, legal, and safety frameworks. Future studies should concentrate on guaranteeing the safety of nanomaterials and investigating novel medical uses in order to fully realize the potential of nanotechnology in healthcare.

References

Al Mamun M, Atique M, Hossain MS, Rahman M, Sarker S, Hasan M, Khan M, Alam M, Rahman M, Chowdhury S (2021) Electrochemical biosensors with Aptamer recognition layer for the diagnosis of pathogenic bacteria: barriers to commercialization and remediation. TrAC—Trends in Analytical Chemistry 145:116458. https://doi.org/10.1016/j.trac.2021.116458

Al Mamun M, Atique M, Hossain MS, Rahman M, Sarker S, Hasan M, Khan M, Alam M, Rahman M, Chowdhury S (2023) Determination of the Aptamer Probe Density by double layer and redox capacitance of CNT-based electrochemical DNA-Aptasensors. In: 2023 IEEE regional symposium on micro and Nanoelectronics (RSM). IEEE, pp 74–77. https://doi.org/10.1109/RSM59033.2023.10327172

Alghamdi MA, Ahmed T, Khan M, Alqahtani F, Alharbi S (2022) The promise of nanotechnology in personalized medicine. J Personal Med 12(5):673. https://doi.org/10.3390/jpm12050673

Alvarez MM, Singh P, Kim J, Johnson R, Patel A (2017) Emerging trends in micro-and nanoscale technologies in medicine: from basic discoveries to translation. ACS Nano 11:5195–5214. https://doi.org/10.1021/acsnano.7b01493

Anjum S, Khan M, Ali F, Rahman T, Siddiqui M (2021) Emerging applications of nanotechnology in healthcare systems: grand challenges and perspectives. Pharmaceuticals 14(8):707. https://doi.org/10.3390/ph14080707

Arshavsky-Graham S, Segal E (2020) Lab-on-a-chip devices for point-of-care medical diagnostics. In: Microfluidics in biotechnology. Springer Nature, Switzerland AG, pp 247–265. https://doi.org/10.1007/10_2020_127

Astruc D (2016) Introduction to nanomedicine. Molecules 21(1):4. https://doi.org/10.3390/molecules21010004

Baena JC, Lopez R, Martinez F, Gonzalez E, Torres A (2024) CAR T cell nanosymbionts: revealing the boundless potential of a new dyad. Int J Mol Sci 25(23):13157. https://doi.org/10.3390/ijms252313157

Banihashemi Jozdani SM, Wang H, Zhang L, Kim S, Lee J (2024) Emerging trends in the biomedical application of carbon-based nanomaterials. Nano Biomed Engineering 16:357–369. https://doi.org/10.26599/NBE.2024.9290091

Barenholz YC (2012) Doxil®—the first FDA-approved nano-drug: lessons learned. J Control Release 160(2):117–134. https://doi.org/10.1016/j.jconrel.2012.03.020

Baun A, Hansen SF (2008) Environmental challenges for nanomedicine. Nanomedicine 3(5):605–608. https://doi.org/10.2217/17435889.3.5.605

Cai X, Bao X, Wu Y (2022) Metal–organic frameworks as intelligent drug nanocarriers for cancer therapy. Pharmaceutics 14(12):2641. https://doi.org/10.3390/pharmaceutics14122641

Caspani S, Rossi F, Lombardi G, Bianchi M (2020) Magnetic nanomaterials as contrast agents for MRI. Materials 13(11):2586. https://doi.org/10.3390/ma13112586

Chamorro-Garcia A, Merkoçi A (2016) Nanobiosensors in diagnostics. Nanobiomedicine 3:2016. https://doi.org/10.1177/1849543516663574

Chenxi Z, Li H, Wang Y, Zhou J, Tang K (2025) Nanoparticle-enhanced drug delivery systems: an up-to-date review. J Mol Liq 424:126999. https://doi.org/10.1016/j.molliq.2025.126999

Cobb MD, Macoubrie J (2004) Public perceptions about nanotechnology: risks, benefits and trust. J Nanopart Res 6:395–405. https://doi.org/10.1007/s11051-004-3394-4

Darwish MA, Alqahtani F, Khan M, Zhang L, Patel S (2024) Advancements in nanomaterials for nanosensors: a comprehensive review. Nanoscale Adv 6:4015–4046. https://doi.org/10.1039/D4NA00214H

Davis ME, Chen Z, Shin DM (2008) Nanoparticle therapeutics: an emerging treatment modality for cancer. Nat Rev Drug Discov 7:771–782. https://doi.org/10.1038/nrd2614

Devi N, Kumar P, Singh A, Sharma R, Gupta S (2023) Carbon-based nanomaterials: carbon nanotube, fullerene, and carbon dots. In: Nanomaterials: advances and applications. Springer, Singapore, pp 27–57. https://doi.org/10.1007/978-981-19-7963-7_2

Dutt Y, Singh P, Kaur M, Sharma R, Patel A (2023) Therapeutic applications of nanobiotechnology. J Nanobiotechnol 21:148. https://doi.org/10.1186/s12951-023-01909-z
Farokhzad OC, Langer R (2009) Impact of nanotechnology on drug delivery. ACS Nano 3(1):16–20. https://doi.org/10.1021/nn900002m
Feindt PH, Poortvliet PM (2020) Consumer reactions to unfamiliar technologies: mental and social formation of perceptions and attitudes toward nano and GM products. J Risk Res 23(4):475–489. https://doi.org/10.1080/13669877.2019.1591487
Feng X, Liu Y, Zhang H, Chen Y, Wu Z (2019) Immunomodulatory nanosystems. Adv Sci 6(17):1900101. https://doi.org/10.1002/advs.201900101
Gao X, Chen H, Li Y, Zhang L (2004) In vivo cancer targeting and imaging with semiconductor quantum dots. Nat Biotechnol 22:969–976. https://doi.org/10.1038/nbt994
Gao F, Li H, Zhang Y, Wang Z, Liu J (2024) A nanoparticle-assisted signal-enhancement technique for lateral flow immunoassays. J Mater Chem B 12:6735–6756. https://doi.org/10.1039/D4TB00865K
Gawne PJ, Smith R, Li H, Zhang Y, Patel S (2023) New opportunities and old challenges in the clinical translation of nanotheranostics. Nat Rev Mater 8:783–798. https://doi.org/10.1038/s41578-023-00581-x
Geraldes CF (2024) Rational design of magnetic nanoparticles as T1–T2 dual-mode MRI contrast agents. Molecules 29(6):1352. https://doi.org/10.3390/molecules29061352
Gharib G, Khan M, Ali F, Zhang Y, Patel S (2022) Biomedical applications of microfluidic devices: a review. Biosensors 12(11):1023. https://doi.org/10.3390/bios12111023
Gupta N, Fischer AR, Frewer LJ (2015) Ethics, risk and benefits associated with different applications of nanotechnology: a comparison of expert and consumer perceptions of drivers of societal acceptance. Nanoethics 9:93–108. https://doi.org/10.1007/s11569-015-0222-5
Han X, Li Y, Zhang H, Wang J, Liu Z (2019) Applications of nanoparticles in biomedical imaging. Nanoscale 11:799–819. https://doi.org/10.1039/C8NR07769J
Hashem A, Ali F, Khan M, Patel S, Zhang Y (2021) Nanomaterials based electrochemical nucleic acid biosensors for environmental monitoring: a review. Appl Surf Sci Adv 4:100064. https://doi.org/10.1016/j.apsadv.2021.100064
Hashem A, Ali F, Khan M, Patel S, Zhang Y (2022a) A unique oligonucleotide probe hybrid on graphene decorated gold nanoparticles modified screen-printed carbon electrode for pork meat adulteration. Electrocatalysis 14:179–194. https://doi.org/10.1007/s12678-022-00779-7
Hashem A, Ali F, Khan M, Patel S, Zhang Y (2022b) Nucleic acid-based electrochemical biosensors for rapid clinical diagnosis: advances, challenges, and opportunities. Crit Rev Clin Lab Sci 59(3):156–177. https://doi.org/10.1080/10408363.2021.1997898
Hashem A, Ali F, Khan M, Patel S, Zhang Y (2022c) Rapid and sensitive detection of box turtles using an electrochemical DNA biosensor based on a gold/graphene nanocomposite. Beilstein J Nanotechnol 13:1458–1472. https://doi.org/10.3762/bjnano.13.120
Hashem A, Ali F, Khan M, Patel S, Zhang Y (2022d) Applications of nanocellulose as biosensing platforms for the detection of functional biomacromolecules: a review. Malay Nano Int J 2(1):15–45. https://doi.org/10.22452/mnij.vol2no1.2
Hashem A, Ali F, Khan M, Patel S, Zhang Y (2022e) Application of microbial fuel cells as biosensors. In: Ahmad A et al (eds) Microbial fuel cells for environmental remediation. Sustainable materials and technology. Springer, Singapore, pp. 349–387. https://doi.org/10.1007/978-981-19-2681-5_17
Hauser M, Li G, Nowack B (2019) Environmental hazard assessment for polymeric and inorganic nanobiomaterials used in drug delivery. J Nanobiotechnol 17:56. https://doi.org/10.1186/s12951-019-0489-8
Heath JR (2015) Nanotechnologies for biomedical science and translational medicine. Proc Natl Acad Sci 112(47):14436–14443. https://doi.org/10.1073/pnas.1515202112
Hong Y, Li H, Zhang X, Wang Y, Chen J, Zhao L (2025) Advances in programmable DNA nanostructures enabling stimuli-responsive drug delivery and multimodal biosensing. RSC Chem Biol 2025. https://doi.org/10.1039/D5CB00057B

Hossain MAM, Rahman M, Khan M, Alam M, Chowdhury S, Hasan M, Sarker S, Atique M, Al Mamun M (2021) DNA-based methods for species identification in food forensic science. In: Galanakis CM (ed) Food toxicology and forensics. Academic Press, Amsterdam, pp 181–211. https://doi.org/10.1016/B978-0-12-822360-4.00007-8

Hossain MAM, Rahman M, Khan M, Alam M, Chowdhury S, Hasan M, Sarker S, Atique M, Al Mamun M (2022) Advancements in detection approaches of severe acute respiratory syndrome coronavirus 2. Malays J Med Sci 29(6):15–33. https://doi.org/10.21315/mjms2022.29.6.3

Howes PD, Rana S, Stevens MM (2014) Plasmonic nanomaterials for biodiagnostics. Chem Soc Rev 43:3835–3853. https://doi.org/10.1039/C3CS60346F

Hsu JC, Li H, Zhang Y, Wang X, Chen J (2020) Nanoparticle contrast agents for X-ray imaging applications. WIREs Nanomed Nanobiotechnol 12(6):e1642. https://doi.org/10.1002/wnan.1642

Hua S, Li H, Zhang Y, Chen X, Wang J (2018) Current trends and challenges in the clinical translation of nanoparticulate nanomedicines: pathways for translational development and commercialization. Front Pharmacol 9:790. https://doi.org/10.3389/fphar.2018.00790

Huang X, Zhang W (2024) Overcoming T cell exhaustion in tumor microenvironment via immune checkpoint modulation with nano-delivery systems for enhanced immunotherapy. Small Methods 8(8):2301326. https://doi.org/10.1002/smtd.202301326

Huh AJ, Kwon YJ (2011) "Nanoantibiotics": a new paradigm for treating infectious diseases using nanomaterials in the antibiotics resistant era. J Control Release 156:128–145. https://doi.org/10.1016/j.jconrel.2011.07.002

Iyer KA, Li H, Zhang Y, Wang J, Chen X (2025) Nanoscale materials in biomedical applications of sensors: insights from a comprehensive landscape analysis. ACS Appl Nano Material 8(3):1305–1331. https://doi.org/10.1021/acsanm.4c04463

Jung HH, Kim S, Lee J, Park H, Choi M (2024) Wearable electrochemical sensors for real-time monitoring in diabetes mellitus and associated complications. Soft Sci 4:15. https://doi.org/10.20517/ss.2024.02

Kardani SL (2024) Nanocarrier-based formulations: regulatory challenges, ethical and safety considerations in pharmaceuticals. Asian J Pharm 18(2):246–351. https://doi.org/10.22377/ajp.v18i02.5444

Karimi S, Li H, Zhang Y, Wang J, Chen X (2025) For and against tumor microenvironment: nanoparticle-based strategies for active cancer therapy. Mater Today Bio 31:101626. https://doi.org/10.1016/j.mtbio.2025.101626

Khalil I, Li H, Zhang Y, Chen X, Wang J (2021) DNA/Nano based advanced genetic detection tools for authentication of species: strategies, prospects and limitations. Mol Cell Probes 59:101758. https://doi.org/10.1016/j.mcp.2021.101758

Kim D, Li H, Zhang Y, Wang J, Chen X (2018) Recent development of inorganic nanoparticles for biomedical imaging. ACS Central Sci 4(3):324–336. https://doi.org/10.1021/acscentsci.7b00574

Kou H, Yang H (2024) Molecular imaging nanoprobes and their applications in atherosclerosis diagnosis. Theranostics 14(12):4747–4772. https://doi.org/10.7150/thno.96037

Koyande NP, Li H, Zhang Y, Wang J, Chen X (2022) Advances in nanotechnology for cancer immunoprevention and immunotherapy: a review. Vaccine 10(10):1727. https://doi.org/10.3390/vaccines10101727

Krishnan S, Li H, Zhang Y, Wang J, Chen X (2010) Attomolar detection of a cancer biomarker protein in serum by surface plasmon resonance using superparamagnetic particle labels. Angew Chem Int Ed 50(5):1175–1178. https://doi.org/10.1002/anie.201005607

Labib M, Li H, Zhang Y, Wang J, Chen X (2018) Single-cell mRNA cytometry via sequence-specific nanoparticle clustering and trapping. Nat Chem 10:489–495. https://doi.org/10.1038/s41557-018-0025-8

Lapusan R, Borlan R, Focsan M (2024) Advancing MRI with magnetic nanoparticles: a comprehensive review of translational research and clinical trials. Nanoscale Adv 6:2234–2259. https://doi.org/10.1039/D3NA01064C

Lead JR, Li H, Zhang Y, Wang J, Chen X (2018) Nanomaterials in the environment: behavior, fate, bioavailability, and effects—an updated review. Environ Toxicol Chem 37(8):2029–2063. https://doi.org/10.1002/etc.4147

Lee H, Li H, Zhang Y, Wang J, Chen X (2025) CRISPR-Cas9 gene therapy: non-viral delivery and stimuli-responsive nanoformulations. Molecules 30(3):542. https://doi.org/10.3390/molecules30030542

Liu X, Yuan H (2025) Responsive nanomaterials in biomedicine, patent path and prospect analysis. Front Bioeng Biotechnol 13:1539991. https://doi.org/10.3389/fbioe.2025.1539991

Liu Y, Li H, Zhang Y, Wang J, Chen X (2018) Enhanced mRNA FISH with compact quantum dots. Nat Commun 9:4461. https://doi.org/10.1038/s41467-018-06740-x

Liu Y, Li H, Zhang Y, Wang J, Chen X (2021) Ultrasensitive and highly specific lateral flow assays for point-of-care diagnosis. ACS Nano 15(3):3593–3611. https://doi.org/10.1021/acsnano.0c10035

Lu Q, Li H, Zhang Y, Wang J, Chen X (2024) Nanoparticles in tumor microenvironment remodeling and cancer immunotherapy. J Hematol Oncol 17:16. https://doi.org/10.1186/s13045-024-01535-8

Mahapatra I, Li H, Zhang Y, Wang J, Chen X (2013) Potential environmental implications of nano-enabled medical applications: critical review. Environ Sci: Processes Impacts 15:123–144. https://doi.org/10.1039/C2EM30640A

Mahapatra I, Li H, Zhang Y, Wang J, Chen X (2018) Expert perspectives on potential environmental risks from nanomedicines and adequacy of the current guideline on environmental risk assessment. Environ Sci Nano 5:1873–1889. https://doi.org/10.1039/C8EN00053K

Mahara G, Li H, Zhang Y, Wang J, Chen X (2023) Revolutionising health care: exploring the latest advances in medical sciences. J Glob Health 13:03042. https://doi.org/10.7189/jogh.13.03042

Malik S, Muhammad K, Waheed Y (2023) Emerging applications of nanotechnology in healthcare and medicine. Molecules 28(18):6624. https://doi.org/10.3390/molecules28186624

Mao Y, Li H, Zhang Y, Wang J, Chen X (2024) Advances and prospects of precision nanomedicine in personalized tumor theranostics. Front Cell Develop Biol 12:1514399. https://doi.org/10.3389/fcell.2024.1514399

Mashinchian O, Li H, Zhang Y, Wang J, Chen X (2014) Impacts of quantum dots in molecular detection and bioimaging of cancer. Bioimpacts 4(3):149–166. https://doi.org/10.15171/bi.2014.008

Mintis DG, Li H, Zhang Y, Wang J, Chen X (2024) NanoBioAccumulate: modelling the uptake and bioaccumulation of nanomaterials in soil and aquatic invertebrates via the Enalos DIAGONAL Cloud Platform. Comput Struct Biotechnol J 25:243–255. https://doi.org/10.1016/j.csbj.2024.09.028

Mitchell MJ, Li H, Zhang Y, Wang J, Chen X (2021) Engineering precision nanoparticles for drug delivery. Nat Rev Drug Discov 20:101–124. https://doi.org/10.1038/s41573-020-0090-8

Mutreja I, Li H, Zhang Y, Wang J, Chen X (2023) High atomic number nanoparticles to enhance spectral CT imaging aspects. Materials Adv 4:3967–3988. https://doi.org/10.1039/D3MA00231D

Nam JM, Thaxton CS, Mirkin CA (2003) Nanoparticle-based bio-bar codes for the ultrasensitive detection of proteins. Science 301(5641):1884–1886. https://doi.org/10.1126/science.108875

Nizam NUM, Hanafiah MM, Woon KS (2021) A content review of life cycle assessment of nanomaterials: current practices, challenges, and future prospects. Nano 11(12):3324. https://doi.org/10.3390/nano11123324

Oberdörster G, Maynard A, Donaldson K, Castranova V, Fitzpatrick J, Ausman K et al (2005) Principles for characterizing the potential human health effects from exposure to nanomaterials: elements of a screening strategy. Part Fibre Toxicol 2:8. https://doi.org/10.1186/1743-8977-2-8

Paradise J (2019) Regulating nanomedicine at the food and drug administration. AMA J Ethics 21(4):E347–E355. https://doi.org/10.1001/amajethics.2019.347

Paras A, Li H, Zhang Y, Wang J, Chen X (2022) A review on low-dimensional nanomaterials: nanofabrication, characterization and applications. Nano 13(1):160. https://doi.org/10.3390/nano13010160

Patra JK, Das G, Fraceto LF, Campos EVR, Rodriguez-Torres MP, Acosta-Torres LS, Diaz-Torres LA, Grillo R, Swamy MK, Sharma S et al (2018) Nano based drug delivery systems: recent developments and future prospects. J Nanobiotechnol 16:71. https://doi.org/10.1186/s12951-018-0392-8

Qamar W, Li H, Zhang Y, Wang J, Chen X (2024) An insight into impact of nanomaterials toxicity on human health. PeerJ 12:e17807. https://doi.org/10.7717/peerj.17807

Ray PC, Yu H, Fu PP (2009) Toxicity and environmental risks of nanomaterials: challenges and future needs. J Environ Sci Health C 27(1):1–35. https://doi.org/10.1080/10590500802708267

Research C (2025) Nanomedicine market size, growth, share and forecast 2032, Credence Research last Updated July 14, 2025. Available from: https://www.fortunebusinessinsights.com/nanomedicine-market-110983

Rezaei Z, Li H, Zhang Y, Wang J, Chen X (2025) Enhancing organoid technology with carbon-based nanomaterial biosensors: advancements, challenges, and future directions. Adv Drug Deliv Rev 222:115592. https://doi.org/10.1016/j.addr.2025.115592

Rodrigues FS, Li H, Zhang Y, Wang J, Chen X (2020) Emerging trends in nanomedicine for improving ocular drug delivery: light-responsive nanoparticles, mesoporous silica nanoparticles, and contact lenses. ACS Biomater Sci Eng 6(12):6587–6597. https://doi.org/10.1021/acsbiomaterials.0c01347

Rodríguez-Gómez FD, Li H, Zhang Y, Wang J, Chen X (2023) Classification system for nanotechnology-enabled health products with both scientific and regulatory application. Front Med 10:1212949. https://doi.org/10.3389/fmed.2023.1212949

Rodríguez-Gómez FD, Li H, Zhang Y, Wang J, Chen X (2025) Regulatory pathways and guidelines for nanotechnology-enabled health products: a comparative review of EU and US frameworks. Front Med 12:1544393. https://doi.org/10.3389/fmed.2025.1544393

Saleh T, Shojaosadati SA (2016) Multifunctional nanoparticles for cancer immunotherapy. Hum Vaccin Immunother 12(7):1863–1875. https://doi.org/10.1080/21645515.2016.1147635

SCENIHR (2006) The appropriateness of existing methodologies to assess the potential risks associated with engineered and adventitious products of nanotechnologies. European Commission, C7:1–79. Available from: https://ec.europa.eu/health/ph_risk/committees/04_scenihr/docs/scenihr_o_003b.pdf

Scott-Fordsmand JJ, Li H, Zhang Y, Wang J, Chen X (2017) Environmental risk assessment strategy for nanomaterials. Int J Environ Res Public Health 14:1251. https://doi.org/10.3390/ijerph14101251

Secchi V, Li H, Zhang Y, Wang J, Chen X (2025) Advanced techniques and nanotechnologies for point-of-care testing. Front Nanotechnol 6:1465429. https://doi.org/10.3389/fnano.2024.1465429

Shalauddin M, Li H, Zhang Y, Wang J, Chen X (2023) Nanocellulose-based carbon nanocomposite for the electrochemical sensing application for pharmaceuticals: a review. Malay NANO Int J 3(1):1–16. https://doi.org/10.22452/mnij.vol3no1.1

Shan X, Li H, Zhang Y, Wang J, Chen X (2022) Current approaches of nanomedicines in the market and various stage of clinical translation. Acta Pharm Sin B 12(7):3028–3048. https://doi.org/10.1016/j.apsb.2022.02.025

Shen C, Li H, Zhang Y, Wang J, Chen X (2023) Intersection of nanomaterials and organoids technology in biomedicine. Front Immunol 14:1172262. https://doi.org/10.3389/fimmu.2023.1172262

Shi Y, Li H, Zhang Y, Wang J, Chen X (2024) Nanomaterials in organoids: from interactions to personalized medicine. ACS Nano 18(49):33276–33292. https://doi.org/10.1021/acsnano.4c13330

Sim S, Wong NK (2021) Nanotechnology and its use in imaging and drug delivery. Biomed Rep 14:42. https://doi.org/10.3892/br.2021.1418

Spurgeon DJ, Lahive E, Schultz CL (2020) Nanomaterial transformations in the environment: effects of changing exposure forms on bioaccumulation and toxicity. Small 16(36):2000618. https://doi.org/10.1002/smll.202000618

Tabrizi E, Li B (2025) From innovation to application: safety concerns in nanomaterial implant coatings. Nanomedicine 20(12):1373–1376. https://doi.org/10.1080/17435889.2025.2480045
Tenchov R, Li H, Zhang Y, Wang J, Chen X (2021) Lipid nanoparticles—from liposomes to mRNA vaccine delivery, a landscape of research diversity and advancement. ACS Nano 15(11):16982–17015. https://doi.org/10.1021/acsnano.1c04996
Thomas G, Li H, Zhang Y, Wang J, Chen X (2019) Innovative magnetic nanoparticles for PET/MRI bimodal imaging. ACS Omega 4(2):2637–2648. https://doi.org/10.1021/acsomega.8b03283
Tong L, Li H, Zhang Y, Wang J, Chen X (2023) Research status and prospect of non-viral vectors based on siRNA: a review. Int J Mol Sci 24(4):3375. https://doi.org/10.3390/ijms24043375
Viborg N, Kleine-Kohlbrecher D, Rønø B (2024) Personalized neoantigen DNA cancer vaccines: current status and future perspectives. J Cell Immunol 6(1):15–24. https://doi.org/10.33696/immunology.6.188
Wang R, Li H, Zhang Y, Wang J, Chen X (2024a) Applications of nanotechnology for spatial omics: biological structures and functions at nanoscale resolution. ACS Nano 19(1):73–100. https://doi.org/10.1021/acsnano.4c11505
Wang W, Li H, Zhang Y, Wang J, Chen X (2024b) Utilization of nanomaterials in MRI contrast agents and their role in therapy guided by imaging. Front Bioeng Biotechnol 12:1484577. https://doi.org/10.3389/fbioe.2024.1484577
Wasti S, Li H, Zhang Y, Wang J, Chen X (2023) Ethical and legal challenges in nanomedical innovations: a scoping review. Front Genet 14:1163392. https://doi.org/10.3389/fgene.2023.1163392
Wen R, Li H, Zhang Y, Wang J, Chen X (2019) Nanoparticle systems for cancer vaccine. Nanomedicine 14(5):627–648. https://doi.org/10.2217/nnm-2018-0147
Wensink GE, Li H, Zhang Y, Wang J, Chen X (2021) Patient-derived organoids as a predictive biomarker for treatment response in cancer patients. NPJ Precision Oncol 5:30. https://doi.org/10.1038/s41698-021-00168-1
Xuan L, Li H, Zhang Y, Wang J, Chen X (2023) Nanoparticles-induced potential toxicity on human health: applications, toxicity mechanisms, and evaluation models. MedComm 4(4):e327. https://doi.org/10.1002/mco2.327
Yan Y, Li H, Zhang Y, Wang J, Chen X (2022) Non-viral vectors for RNA delivery. J Control Release 342:241–279. https://doi.org/10.1016/j.jconrel.2022.01.008
Yao S, Swetha P, Zhu Y (2018) Nanomaterial-enabled wearable sensors for healthcare. Adv Healthc Mater 7(1):1700889. https://doi.org/10.1002/adhm.201700889
Younis MA, Li H, Zhang Y, Wang J, Chen X (2022) Clinical translation of nanomedicines: challenges, opportunities, and keys. Adv Drug Deliv Rev 181:114083. https://doi.org/10.1016/j.addr.2021.114083
Zhang L, Webster TJ (2009) Nanotechnology and nanomaterials: promises for improved tissue regeneration. Nano Today 4:66–80. https://doi.org/10.1016/j.nantod.2008.10.014
Zhang Y, Li H, Zhang X, Wang J, Chen X (2019) Nanotechnology in cancer diagnosis: progress, challenges and opportunities. J Hematol Oncol 12:137. https://doi.org/10.1186/s13045-019-0833-3
Zhang H, Li H, Zhang Y, Wang J, Chen X (2024a) Mass spectrometry imaging for spatially resolved multi-omics molecular mapping. NPJ Imaging 2:20. https://doi.org/10.1038/s44303-024-00025-3
Zhang J, Li H, Zhang Y, Wang J, Chen X (2024b) Integration of nanobiosensors into organ-on-chip systems for monitoring viral infections. Nano Convergence 11:47. https://doi.org/10.1186/s40580-024-00455-0
Zhao Y, Nalwa HS (2007) Nanotoxicology: interactions of nanomaterials with biological systems. Vol. 19, American Scientific Publishers
Zhao Y, Sultan D, Liu Y (2019) Biodistribution, excretion, and toxicity of nanoparticles. In: Theranostic Bionanomaterials. Elsevier, pp 27–53. https://doi.org/10.1016/B978-0-12-815341-3.00002-X
Zheng Y, Nowack B (2022) Meta analysis of bioaccumulation data for nondissolvable engineered nanomaterials in freshwater aquatic organisms. Environ Toxicol Chem 41(5):1202–1214. https://doi.org/10.1002/etc.5312

Marine-Derived Anticancer Agents: Synergizing Omics and Biotechnology for Breakthroughs

M. R. Washikur and Mohammad Nazir Hossain

Abstract Marine biodiversity covers >70% of Earth's surface and harbors a vast array of species (>300,000 described across 34 of 36 phyla) that produce unique secondary metabolites. The marine environment has yielded a growing pipeline of anticancer leads: For example, trabectedin (from the tunicate *Ecteinascidia*) is an alkylating agent that binds DNA, salinosporamide A (from a marine *Streptomyces*) is a potent 20S proteasome inhibitor, and bryostatin-1 (from the bryozoan *Bugula neritina*) is a macrocyclic lactone that modulates protein kinase C. Modern omics technologies (genomics, transcriptomics, proteomics, metabolomics, and their integration) are revealing biosynthetic gene clusters in uncultured marine microbes and elucidating mechanisms of these compounds. Complementary biotechnological advances—including synthetic biology, genome editing (e.g., clustered regularly interspaced short palindromic repeats [CRISPR]), metabolic engineering, and optimized microbial fermentation—are enabling sustainable production of complex marine molecules. Similarly, nanotechnology and novel delivery systems (nanoparticles and antibodies–drug conjugates) are improving the stability and targeting of marine-derived drugs. Case studies of trabectedin, salinosporamide A (marizomib), and bryostatin-1 exemplify these themes. Key challenges remain in scalable supply, structural complexity, and clinical translation of marine compounds. Future perspectives include integrating marine pharmacology with healthomics—leveraging patient multi-omics and systems biology to guide discovery and precision use of marine-derived anticancer therapies.

Keywords Marine natural products · Anticancer agents · Omics technologies · Synthetic biology · Drug delivery systems

M. R. Washikur · M. N. Hossain (✉)
Department of Genetic Engineering and Marine Biotechnology, Bangladesh Maritime University, Dhaka, Bangladesh
e-mail: nazir.geb@bmu.edu.bd

A. H. M. N. Nabi, H. Uddin Shekhar (eds.), *Biotechnological Advances in Healthomics*, Interdisciplinary Biotechnological Advances,
https://doi.org/10.1007/978-981-95-3612-2_15

Abbreviations

ABS	Access and benefit-sharing
ADC	Antibody–drug conjugate
BBNJ	Biodiversity beyond national jurisdiction
BGCs	Biosynthetic gene clusters
CNS	Central nervous system
CNV	Copy number variation
CRISPR	Clustered regularly interspaced short palindromic repeats
DEGs	Differentially expressed genes
DSI	Digital sequence information
FDA	Food and Drug Administration
HDAC	Histone deacetylase
HPLC	High-performance liquid chromatography
IC_{50}	Half Maximal inhibitory concentration
LC-MS	Liquid chromatography–mass spectrometry
MNPs	Marine natural products
NMR	Nuclear magnetic resonance
OSMAC	One strain–many compounds
PKC	Protein kinase C
TCM	Traditional Chinese medicine
TC-NER	Transcription-coupled nucleotide excision repair
WHO	World Health Organization

1 Introduction

Cancer remains one of the leading causes of morbidity and mortality worldwide, accounting for nearly ten million deaths in 2020 alone (Sung et al. 2021), and despite significant advances in conventional chemotherapy, radiotherapy, and immunotherapy, the overall five-year survival rates for many malignancies have plateaued, underscoring the urgent need for novel therapeutic agents with unique mechanisms of action (Bray et al. 2018; Siegel et al. 2020). In this context, natural products have historically been a prolific source of anticancer medications, with over 60% of approved oncology therapies originating from or inspired by natural compounds (Newman and Cragg 2020). While terrestrial plants and microorganisms have been extensively mined for bioactive metabolites, the marine environment represents a largely unexplored source of chemical diversity, given its immense biodiversity, extreme ecological niches, and unique evolutionary pressures (Blunt et al. 2015). A wide range of marine organisms, including sponges, tunicates, mollusks, and cyanobacteria, as well as fungi and deep-sea crustaceans, are able to produce secondary metabolites (Fig. 1), with potent cytotoxic, antiproliferative, and pro-apoptotic activities, many of which exhibit unique chemical scaffolds not seen

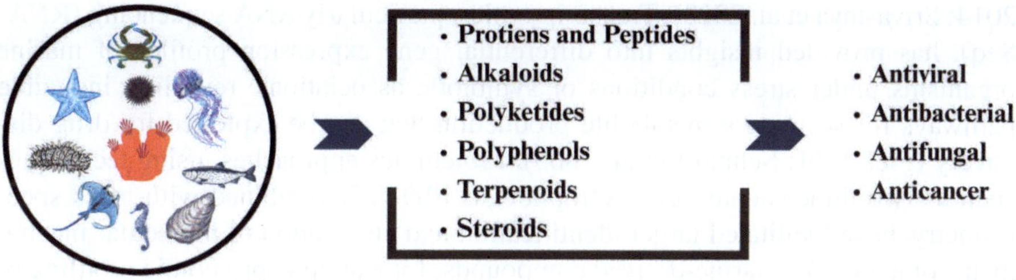

Fig. 1 Certain biological activity displayed by bioactive substances generated from marine resources

in terrestrial sources (Haefner 2003; Molinski et al. 2008). Landmark examples of marine-derived anticancer agents include cytarabine (Ara-C), originally isolated from the sponge *Cryptotethya crypta* and approved for leukemia treatment in 1969 (Barreca et al. 2020). In 2007, the European Medicines Agency approved trabectedin (ecteinascidin-743), a tetrahydroisoquinoline alkaloid derived from the tunicate Ecteinascidia turbinata, for soft tissue sarcoma and ovarian cancer (Ferrandina et al. 2013; Jordan et al. 2015). Eribulin mesylate, a structurally simplified synthetic counterpart of halichondrin B produced from the sponge *Halichondria okadai*, has been approved for metastatic breast cancer and liposarcoma (Dybdal-Hargreaves et al. 2015). Salinosporamide A (marizomib), a potent proteasome inhibitor from the marine actinomycete *Salinispora tropica*, currently undergoing phase III clinical trials for multiple myeloma (Feling et al. 2003; Potts et al. 2011) and bryostatin-1, a macrolactone from the bryozoan *Bugula neritina* that modulates protein kinase C and has been evaluated in multiple phase II trials for lymphoma, leukemia, and solid tumors (Stanwell et al. 1994; König et al. 2006). Despite these successes, only a small fraction of marine biodiversity has been explored for drug discovery, in part due to challenges associated with sustainable harvesting, low natural abundance of bioactives, the difficulty of culturing marine microorganisms, and the structural complexity of many marine metabolites that hampers large-scale synthesis (Faulkner 2002; Fenical and Jensen 2006). To overcome these obstacles, the integration of advanced omics technologies—including genomics, transcriptomics, proteomics, and metabolomics—has revolutionized the field by enabling the systematic exploration of biosynthetic gene clusters (BGCs), regulatory networks, and metabolic pathways in marine organisms and their symbionts (Li and Qin 2005; ul Hassan and Jin 2025). For instance, genome sequencing and mining of Salinispora species have revealed a wealth of cryptic BGCs encoding diverse polyketide synthases (PKSs) and nonribosomal peptide synthetases (NRPSs), facilitating the heterologous expression of these clusters in tractable hosts such as Streptomyces and *Escherichia coli* to yield novel metabolites and increase production yields (Jensen et al. 2015; Baltz 2017). Similarly, metagenomics approaches applied to marine sponge microbiomes have uncovered previously inaccessible gene clusters from uncultivated symbiotic bacteria, leading to the discovery of novel anticancer compounds like the cytotoxic polyketide discodermolide and the macrolide laulimalide (Wilson et al.

2014; Srivastava et al. 2023). Transcriptomics, particularly RNA sequencing (RNA-Seq), has provided insights into differential gene expression profiles of marine organisms under stress conditions or symbiotic associations, revealing inducible pathways for secondary metabolite production that can be exploited for drug discovery (Piel 2011; Schmidt et al. 2005). Proteomics approaches, using techniques such as two-dimensional gel electrophoresis (2D-GE) combined with mass spectrometry, have facilitated target identification and elucidation of molecular mechanisms of action for marine-derived compounds; for example, proteomic profiling of cancer cells treated with salinosporamide A has pinpointed perturbations in proteasome subunit expression, confirming its mechanism as a potent proteasome inhibitor (Potts et al. 2011; Groll et al. 2006). Metabolomics, employing high-resolution liquid chromatography-mass spectrometry (LC-MS) and nuclear magnetic resonance (NMR) spectroscopy, has enabled the comprehensive profiling of marine extracts, aiding dereplication to avoid rediscovery of known compounds and guiding fractionation efforts toward novel bioactive metabolites (Bayona et al. 2022; van der Hooft et al. 2020). Integrating multi-omics datasets within a systems biology framework allows researchers to construct metabolic networks, predict biosynthetic pathways, and prioritize candidate compounds using machine learning algorithms, thereby accelerating lead identification and optimization (Medema and Fischbach 2015; Crüsemann et al. 2017). Concurrent with omics-driven discovery, biotechnological innovations have addressed the supply and scalability challenges that historically limited the clinical translation of marine natural products (MNPs). Synthetic biology techniques, such as refactoring of BGCs and pathway engineering, have enabled the heterologous production of complex marine metabolites in genetically tractable chassis, exemplified by the successful expression of the trabectedin precursor in *Pichia pastoris* (Peña et al. 2018) and the reconstruction of the bryostatin pathway in *Saccharomyces cerevisiae*, resulting in sustainable production of bryostatin analogs with improved pharmacological properties (Keasling 2012; Qin et al. 2015). CRISPR/CRISPR-associated protein 9 (Cas9) genome-editing tools have further enhanced yield optimization by enabling precise modifications of regulatory elements and elimination of competing pathways in microbial hosts (Jiang et al. 2013; Zhao et al. 2021). In parallel, advances in fermentation technology and downstream processing, including the development of seawater-based bioreactors and scalable purification protocols, have improved the feasibility of large-scale production of marine-derived agents (Thomas et al. 2010; Finore et al. 2014). Nanotechnology-based drug delivery systems, such as liposomal encapsulation and polymeric nanoparticles (NPs), have been employed to enhance the solubility, stability, and tumor-targeting of marine compounds, thereby reducing off-target toxicity and improving therapeutic indices; for example, liposomal formulations of eribulin have demonstrated superior efficacy and reduced cardiotoxicity in preclinical models (Khafoor et al. 2023; Chen et al. 2024). Moreover, aquaculture and mariculture efforts have enabled sustainable biomass production of marine organisms, such as sponge cell culture and controlled cultivation of tunicates, mitigating ecological impact and ensuring a steady supply of raw material (Sankarapandian et al. 2023; Piel et al. 2004). Despite these advances, several challenges persist, including the ethical and legal complexities of marine bioprospecting governed by international agreements such as the Nagoya Protocol, the conservation of fragile marine ecosystems, and the need for standardized protocols to ensure reproducibility and comparability of omics data across laboratories (Bekiari 2023; Flemsæter 2020). Additionally, the structural complexity of many marine natural products poses synthetic and analytical challenges that demand continuous innovation in organic synthesis, fermentation, and analytical chemistry (Walsh and Fischbach 2010; Jayaraman et al. 2024). Nevertheless, the convergence of omics and biotechnology holds immense promise for the future of marine-derived anticancer drug discovery, offering a holistic approach like AI that leverages genomic insights, functional assays, and engineering strategies to expedite the translation of novel bioactive compounds into clinical applications (Hong 2011; Gangwal and Lavecchia 2025). As high-throughput sequencing technologies continue to decrease in cost and computational pipelines for multi-omics integration become more sophisticated, researchers are poised to further unlock the immense chemical diversity harbored in marine environments, with the potential to identify targeted therapies for resistant tumor subtypes, discover synergistic drug combinations, and contribute to the burgeoning field of precision oncology (Bertrand et al. 2014; Harvey et al. 2015). Ultimately, a multidisciplinary paradigm that synergizes marine biology, omics-driven systems biology, synthetic biology, and advanced bioprocessing will be essential to realize the full therapeutic potential of marine-derived anticancer agents, paving the way for breakthrough treatments that address unmet clinical needs and improve patient outcomes (Paterson and Anderson 2005).

Illustration of the diverse biological activities is exhibited by bioactive compounds derived from marine organisms, highlighting their potential as anticancer agents. This figure showcases the range of biological activities from marine sources like sponges, sea cucumbers, and crabs. It underscores the chemical diversity of marine natural products and their relevance to anticancer drug discovery by illustrating how these compounds target critical cellular pathways, providing a foundation for their therapeutic potential as discussed in Sect. 2.

2 Marine Biodiversity as a Resource for Anticancer Drug Discovery

The vast expanse of the marine environment, covering approximately 70% of the Earth's surface, harbors an unparalleled biodiversity encompassing up to 36 distinct phyla—significantly surpassing the 17 phyla found in terrestrial ecosystems. This immense biological diversity translates into a rich chemical repertoire, offering a plethora of unique secondary metabolites with potential therapeutic applications, particularly in oncology. Marine organisms, ranging from microorganisms like bacteria and fungi to macroorganisms such as sponges, tunicates, and mollusks, have evolved to produce a myriad of bioactive compounds as adaptive responses to the challenging and competitive marine environment (Ruiz-Torres et al. 2017; Chakraborty et al. 2009; Barreca et al. 2020). The exploration of marine natural products (MNPs) has led to the discovery of over 36,000 compounds, with hundreds more identified annually. Notably, several MNPs have progressed to clinical use as anticancer agents. For instance, cytarabine (Ara-C), derived from the Caribbean sponge *Cryptotethya crypta*, has been a cornerstone in the treatment of acute myelogenous leukemia for decades. Similarly, trabectedin, isolated from the tunicate *Ecteinascidia turbinata*, is approved for soft tissue sarcomas and ovarian cancer. Eribulin mesylate, a synthetic analog of halichondrin B from the sponge *Halichondria okadai*, is utilized in treating metastatic breast cancer. These examples underscore the significant contributions of marine biodiversity to anticancer drug discovery (Barreca et al. 2020; Wang et al. 2020; Jaspars et al. 2016; Khalifa et al. 2019). Beyond these approved drugs, numerous marine-derived compounds exhibit potent anticancer properties. Discodermolide, sourced from the deep-sea sponge *Discodermia dissoluta*, is recognized for its microtubule-stabilizing activity and efficacy against multidrug-resistant cancer cell lines. Curacin A, produced by the cyanobacterium *Lyngbya majuscula*, demonstrates antimitotic activity by binding to the colchicine site on tubulin. Additionally, bryostatins from the bryozoan Bugula neritina and eleutherobin from the coral *Erythropodium caribaeorum* are under investigation for their anticancer potential (Gunasekera et al. 1990; Nobili et al. 2009; Nagle et al. 1995; Britton et al. 2001). The marine environment's unique conditions have driven the evolution of novel biosynthetic pathways, resulting in structurally diverse and biologically active compounds. Advancements in omics technologies, such as genomics and metabolomics, have facilitated the identification and characterization of these bioactive molecules, accelerating the drug discovery process. However, challenges persist, including sustainable sourcing, structural complexity, and ecological considerations. Innovative approaches, including synthetic biology and aquaculture, are being explored to overcome these hurdles and harness the full potential of marine biodiversity in anticancer drug development (Barreca et al. 2020; Jaspars et al. 2016; Francesch et al. 2024; Deshmukh et al. 2018; Ruiz-Torres et al. 2017; Khalifa et al. 2019).

In summary, the marine ecosystem stands as a vast and largely untapped reservoir of anticancer agents. Continued interdisciplinary research integrating marine biology, biotechnology, and omics sciences holds promise for the discovery (Fig. 2) (Santhiravel et al. 2025) and development of novel therapeutics to combat cancer.

This line plot depicts the total number of new marine natural products discovered each year from 2008 to 2022. This figure quantifies the steady increase in the discovery of marine natural products over a 15-year period, reflecting the growing interest in marine biodiversity for drug development. It highlights the accelerating pace of identifying novel compounds with potential anticancer properties, as enabled by advancements in omics and screening technologies, aligning with the discussion in Sect. 2 on the untapped potential of marine ecosystems.

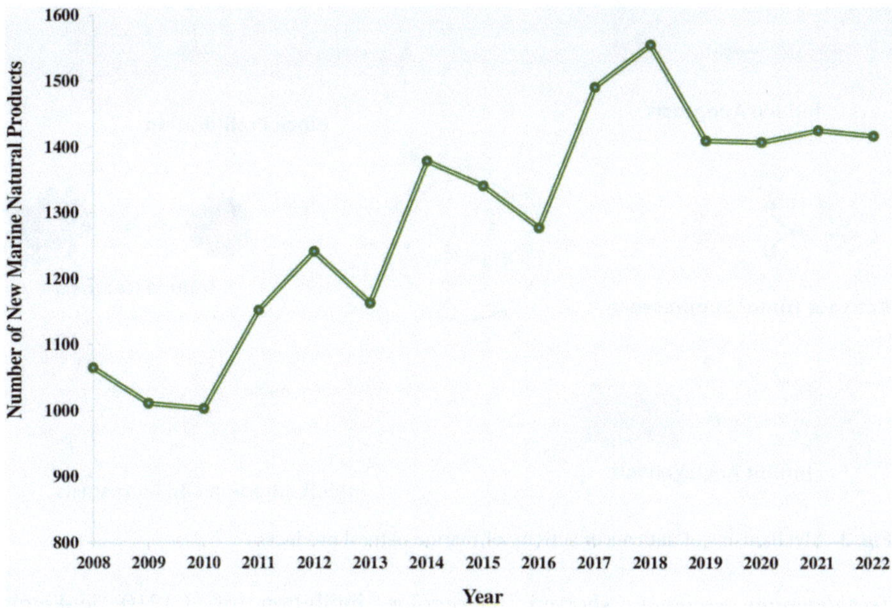

Fig. 2 Annual discovery of new marine natural products (2008–2022)

3 Marine-Derived Anticancer Compounds: Mechanisms and Examples

Marine ecosystems, encompassing a vast array of biodiversity, have yielded numerous bioactive compounds with potent anticancer properties. These marine-derived compounds exhibit diverse mechanisms of action (Fig. 3) (Santhiravel et al. 2025), targeting various cellular pathways to exert their cytotoxic effects.

Diagram illustrates the molecular mechanisms by which marine-derived compounds exert anticancer effects, including cytotoxicity, antiproliferative effects, and apoptosis induction. This figure provides a visual summary of the diverse mechanisms of action of marine-derived anticancer agents, such as trabectedin, salinosporamide A, and bryostatin-1, as detailed in Sect. 3. It emphasizes how these compounds target key cellular processes like DNA repair, protein degradation, and signaling pathways, offering unique therapeutic strategies to overcome drug resistance in cancer treatment.

Marine organisms continue to yield structurally unique alkaloids, peptides, and steroids that exhibit potent anticancer activities through diverse mechanisms. For example, the cyclic peptide scleritodermin A, isolated from the sponge *Scleritoderma nodosum*, displayed submicromolar cytotoxicity against A2780 (ovarian), HCT116 (colon), and SKBR3 (breast) carcinoma cell lines (IC_{50} = 0.94 μM, 1.92 μM, and 0.97 μM, respectively) by disrupting microtubule dynamics and inducing G_2/M arrest. Likewise, the cyclodepsipeptide didemnin B, sourced from the tunicate

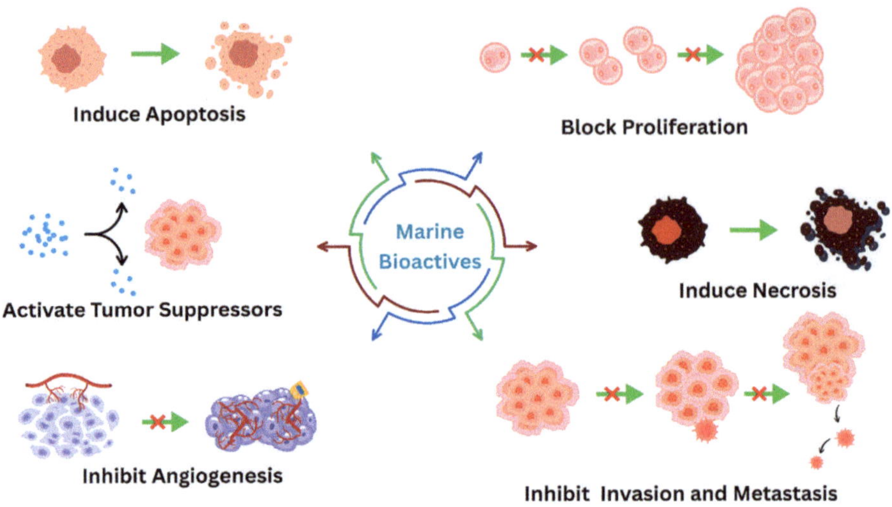

Fig. 3 Mechanisms of anticancer activity of marine natural products

Tridemnum solidum, showed nanomolar inhibition of L1210 leukemia (IC_{50} = 2 nM) and significant antiproliferative effects against B16 melanoma and P388 leukemia cells (IC_{50} = < 15 ng/mL). Sponge-derived arenastatin A and geodiamolide H likewise target the mitotic spindle, whereas the macrocyclic lactone bryostatin-1 modulates protein kinase C (PKC), triggering both pro-apoptotic signaling and growth arrest (PKC isoform-specific). Furthermore, marine cyanobacteria have yielded apratoxin D, which binds elongation factor eEF1A and induces apoptosis in MDA-MB-231 breast cancer cells ($IC_{50} \approx 2.6$ nM). These examples illustrate that marine alkaloids and peptides can either (1) disrupt mitotic microtubule dynamics (e.g., discodermolide and dolastatin 10), (2) inhibit proteasome activity (e.g., salinosporamide A), or (3) modulate kinase-mediated survival pathways (e.g., bryostatin-1), thereby overcoming drug resistance typically encountered with terrestrial chemotypes. Some other examples of marine bioactives possessing anticancer activities are presented in Table 1 (Santhiravel et al. 2025).

One notable example is trabectedin (ET-743), originally isolated from the tunicate *Ecteinascidia turbinata*. Trabectedin binds to the minor groove of DNA, causing DNA backbone cleavage and disrupting the cell cycle, leading to apoptosis. It also interferes with the nucleotide excision repair system, enhancing its cytotoxicity in actively transcribing cancer cells. Furthermore, trabectedin inhibits the expression of the MDR1 gene, which encodes a membrane pump responsible for multidrug resistance, thereby restoring the efficacy of other chemotherapeutic agents (Dalisay et al. 2024).

Curacin A, derived from the cyanobacterium *Lyngbya majuscula*, is another marine compound with significant anticancer activity. It binds to the colchicine site on tubulin, inhibiting microtubule polymerization, which is essential for cell division. This disruption leads to cell cycle arrest and apoptosis in various cancer cell lines, including renal, colon, and breast cancers (Nagle et al. 1995).

Table 1 Representative marine-derived bioactives with documented anticancer activities across major compound classes

Peptide	Source	Anticancer activity
Theopapuamides A–C	Marine sponge *Siliquariaspongia mirabilis*	Cytotoxicity against HCT-116 colon carcinoma cells (IC_{50} = 2.1–4.0 μg/mL)
Samoamide A	*Symploca* sp., American Samoa	Potent cytotoxicity against H460 non–small cell lung cancer cells (IC_{50} = 1.1 μM)
Sansalvamide A	*Fusarium* sp., marine plant *H. wrightii*	Cytotoxic to colon, pancreatic, prostate, breast sarcoma, and melanoma cell lines
Azoanazide	Marine sediment-derived fungus, *Aspergillus insulicola*	Cytotoxic to HCT-116 cells (IC_{50} < 15 ng/mL)
Microsclerodermin A	Sponge *Amphibleptula* spp.	Inhibits NF-κB transcriptional activity
Urukthapelstatin A	*Mechercharomyces* spp.	Cytotoxic to A549 lung cancer cells (IC_{50} = 12 nM)
Lagunamide A and D	Cyanobacterium *Lyngbya majuscula*	Antiproliferative against A549 lung adenocarcinoma
Scopularides A and B	Fungus *Scopulariopsis brevicaulis*	Cytotoxic to colon tumor lines HT29, Panc89, and Colo357
Desmethoxymajusculamide C	Fijian *Lyngbya majuscula*	Cytotoxic to colon carcinoma (IC_{50} = 0.4 μM)
Rollamide A and B	Marine sponge *Eurypon laughlini*	Cytotoxic against SKBR3 and A2780 (IC_{50} = 0.4–5.8 μM)
Diazonamide A	Ascidian *Diazona angulata*	Active against C46, MCF7, PC3, and A549 cells
Kuloekahikiliide-2	Cephalaspidean mollusk (*Philinopsis speciosa*)	Cytotoxic to P388, SK-MD-MB-435, and A-10 (IC_{50} = 4.2–59.1 nM)
Peptide YALPAH	*Setipinna taty* (half-fin anchovy)	Inhibits PC-3 prostate cancer cells (IC_{50} = 111.4 μM)
Pardaxin	Red Sea Moses sole *Pardachirus marmoratus*	Induces apoptosis in HT1080 fibrosarcoma and HeLa cells
Alkaloid	**Source**	**Anticancer activity**
Discorhabdin and discodermolide	Sponge *Latrunculia apicalis* and *Latrunculia biformis*	Cytotoxic activity toward HCT116 cell line
Ascomylactam A	Mangrove endophytic fungus *Ascomycota* sp.	Suppressed A549 and NCI-H460 cancer cell lines (6 mg/kg per day)
(−)-Agelamide D	Marine sponge *Agelas* sp.	Improves efficacy of radiation therapy in xenograft Hep3B cells

(continued)

Table 1 (continued)

Neoamphimedine	Marine sponge *Xestospongia* sp.	Reduces the growth of KB and HCT116 tumors
Crambescidine-816	Marine sponge *Crambe crambe*	Reduces tumor development in colorectal carcinoma (CRC) HTC-116 cells in vivo in zebrafish embryos
Isofistularin-3	Marine sponge *Aplysina aerophoba*	Suppresses development of neuroblastoma (15–25 μM) and prostate (20–25 μM) cancers
Monanchoxymycalin C	Marine sponge *M. pulchra*	Cytotoxic to human prostate cancer via JNK1/2 activation and non-apoptotic cell death
Phenolics	Source	Anticancer activity
Penimethavone A (flavone)	*P. chrysogenum* (gorgonian coral, South China Sea)	Active against rhabdomyosarcoma and cervical cancer (HeLa); IC_{50} = 8.18 and 8.41 μM
Phloroglucinol	Brown seaweeds	Enhances 5-fluorouracil activity in HT29 colorectal cancer cells
Alternariol and (coumarin)	*Alternata* (from soft coral *L. arboreum*)	Antiproliferative versus L1210 and CCRF-CEM leukemia cells
Alternariol methyl ether	*Alternata* (soft coral *L. arboreum*)	Active against H-125 and Colon-38 leukemia lines
Moromycin B and saquayamycins B_1 and B	*Streptomyces* sp. (marine-derived)	Cytotoxicity against MCF-7, MDA-MB-231, and BT-474 breast cancer cells (IC_{50} = 0.16–0.67 μM)
Bis(2,3-dibromo-4,5-dihydroxy-phenyl)-methane	Marine algae *OC1610.4*	Active against RKO, HeLa, U87, HT-29, BEL7402, and HCT116 (IC_{50} = 11.37–105.58 μg/mL)
Laurebiphenyl (terpenophenolics)	Red macroalga *L. tristicia*	Inhibits BCC-823, A549, HCT-8, BEL7402, and HeLa; IC_{50} = 1.22–1.91 μg/mL
Fradimycin B	*Streptomyces fradiae* PIZ0025 (marine)	Active versus HCT-15, SW620, and C6 cells (IC_{50} = 0.13–0.47 μM)
Fucoidphloroeothol G	*Ecklonia cava* (marine brown macroalga)	Inhibits HeLa, HT1080, and HT29 (IC_{50} = 226.5–298.2 μM)
Galvaquinone B and lupinacidin A (anthraquinones)	Sea anemone *Gyractis sesere* (Easter Island)	IC_{50} = 5.0 μM (galvaquinone B); 8.8 and 3.1 μM (lupinacidin A) against Calu-3 and H2887 cells

(continued)

Table 1 (continued)

Northeriridin B (hydroquinone)	*Cyanobium* sp. (marine picocyanobacterium)	Effective on MDA-MB-231, SKBR3, MDA-MB-468, A375, HT116 (IC_{50} = 0.061–3.2 μM)
Dicoronol L	*Didymella* sp. SCISO F46 (marine sponge-derived fungus)	Cytotoxic to HeLa, HT-29, DU145, HL60 (IC_{50} = 5.7–9.6 μM)
Plastoquinone	*Sargassum micracanthum* (brown macroalga)	IC_{50} = 1.69 μg/mL against PC-12 and HeLa
Phlorofucofuroeckol	*E. cava* (brown macroalga)	Suppresses MDA-MB-231 migration via NF-κB and TLR2 pathway
Terpenoids	Source	Target cell lines
Paclitaxel	Natural terpenoid (originally from *Taxus brevifolia*). Also found in hazelnut by-products	Various human tumor lines IC_{50} = 2.5–7.5 nM
Cryptosphaerolide	Ascidian-derived *Ascomycete* strain CNL-52	HCT-116 IC_{50} = 4.5 μM
Epoxyphomalin D	*Paraconiothyrium* cf. *sporulosum* from sponge *Ectyplasia perox*	PC3M (prostate) and BXF1218L (bladder) IC_{50} = 0.72 μM, 1.43 μM
Saccharoquinoline	*Saccharomonospora* sp. CNQ-490	HCT-116 IC_{50} = potent (exact value not stated)
Guanahanolide A	*Streptomyces* sp. (fermentation extract)	HCT-116, HTB-26, and MCF-7 IC_{50} = 11.9, 10.1, 7.8 μM
Pestalachloride B, E and pestalactone isomers	*Pestalotiopsis heterocornis* from sponge *Phakellia fusca*	BGC-823, H460, PC-3, and SMMC-7721 IC_{50} = 6.8–87.8 μM
Steroid type	Source	Target cell lines
19-oxygenated steroids	Soft coral *Lobophytum mollis*	MCF-7 and NCI-1299 IC_{50} = 15.7, 15.1 μM
4α-methylated steroids	Soft coral *L. mollis*	MCF-7, NCI-1299, and HepG2 IC_{50} = 30.8–46.3 μM

Source: Santhiravel et al. (2025)

Discodermolide, isolated from the deep-sea sponge *Discodermia dissoluta*, stabilizes microtubules, similar to the mechanism of paclitaxel. It binds to tubulin with high affinity, promoting microtubule assembly and inducing G2/M phase cell cycle arrest. Discodermolide has shown potent activity against multidrug-resistant cancer cell lines and exhibits immunosuppressive properties (Gunasekera et al. 1990).

Salinosporamide A (NPI-0052), produced by the marine actinomycete *Salinispora tropica*, is a potent proteasome inhibitor. It irreversibly inhibits the chymotrypsin-like activity of the 20S proteasome, leading to the accumulation of misfolded proteins and induction of apoptosis. Salinosporamide A has demonstrated cytotoxic effects in tumor cells resistant to bortezomib, another proteasome inhibitor (von Schwarzenberg and Vollmar 2013).

Bryostatin 1, a macrocyclic lactone isolated from the bryozoan *Bugula neritina*, modulates protein kinase C (PKC) activity. By binding to the regulatory domain of PKC, bryostatin 1 influences cell differentiation, proliferation, and apoptosis. It has been investigated for its potential to enhance the efficacy of other anticancer agents and to overcome drug resistance (von Schwarzenberg and Vollmar 2013).

Dolastatin 10, obtained from the sea hare *Dolabella auricularia*, is a linear peptide that inhibits microtubule assembly by binding to tubulin. This action disrupts the mitotic spindle formation, leading to cell cycle arrest and apoptosis. Due to its potent cytotoxicity, dolastatin 10 has served as a lead compound for the development of synthetic analogs used in antibody–drug conjugates (von Schwarzenberg and Vollmar 2013).

Cryptophycins, produced by cyanobacteria, are another class of microtubule inhibitors. They bind to tubulin, preventing microtubule polymerization and inducing apoptosis. Cryptophycin-52, in particular, has shown high-affinity binding to microtubule ends, leading to the stabilization of microtubule dynamics and potent antiproliferative effects (Schwartz et al. 1990).

Coibamide A, isolated from the cyanobacterium Leptolyngbya, is a cyclic depsipeptide that induces apoptosis through activation of caspase-3 and caspase-7. It also causes cell cycle arrest in the G1 phase and has shown efficacy against apoptosis-resistant glioblastoma cell lines (Seyed et al. 2021).

Plocabulin (PM060184), derived from the sponge *Lithoplocamia lithistoides*, is a tubulin-binding agent that disrupts microtubule dynamics, leading to mitotic arrest and apoptosis. It has demonstrated potent antitumor activity in various xenograft models, including breast, colon, lung, prostate, gastric, and renal cancers. Notably, plocabulin retains its efficacy in P-glycoprotein overexpressing cell lines, which are often resistant to other antitubulin agents (Jimenez et al. 2019).

Astaxanthin, a xanthophyll carotenoid found in marine organisms like microalgae and sea urchins, exhibits anticancer properties by modulating multiple signaling pathways. It downregulates STAT3, JAK2, BCL-2, and NF-κB, while upregulating pro-apoptotic proteins such as BAX and caspases. These actions lead to the inhibition of proliferation, induction of apoptosis, and suppression of invasion and migration in cancer cells (Alves and Diederich 2021).

Sulfated polysaccharides and bioactive compounds from Caulerpa racemosa, a seaweed native to the Bay of Bengal, exhibit anticancer, antimicrobial, and immunomodulatory properties—potentially inducing apoptosis in cancer cells by modulating Bcl-2, BAX, and caspase-3 expression, and showing promise for therapies targeting cervical cancer and human melanoma (Chowdhury et al. 2023). Collectively, these marine-derived compounds highlight the vast potential of oceanic biodiversity in providing novel anticancer agents with unique mechanisms of action. Their diverse targets and pathways offer promising avenues for the development of effective cancer therapies.

4 Omics Technologies Enabling Marine Drug Discovery

The development of omics technology has greatly aided the study of marine biodiversity for drug discovery. These high-throughput approaches—genomics, transcriptomics, proteomics, metabolomics, and integrated multi-omics—provide comprehensive insights into the molecular makeup of marine organisms, making it easier to identify and characterize novel bioactive compounds with potential anticancer properties.

4.1 Genomics

Genomics is the process of sequencing and analyzing an organism's whole genetic material. In marine drug discovery, genomics has helped to identify biosynthetic gene clusters (BGCs) that produce secondary metabolites. Advancements in next-generation sequencing (NGS) technologies have enabled genome assembly from uncultured marine microbes, revealing a tremendous store of genetic variety and metabolic potential (Albarano et al. 2020). For instance, the identification of novel BGCs in marine actinomycetes has led to the discovery of new polyketides and nonribosomal peptides with anticancer activities (Ziemert et al. 2016).

4.2 Transcriptomics

Transcriptomics is the study of RNA transcripts produced by the genome under various situations. In the context of marine drug development, transcriptome investigations assist clarify gene expression patterns related to the manufacture of bioactive compounds. By comparing transcriptomes under different environmental or developmental conditions, researchers can identify genes upregulated during the production of secondary metabolites. This approach has been applied to marine dinoflagellates, leading to the identification of genes involved in the biosynthesis of potent cytotoxins like saxitoxin (Hackett et al. 2013; Srivastava et al. 2023).

4.3 Proteomics

Proteomics entails the large-scale study of proteins, including their expression, structure, and functions. In marine organisms, proteomic analyses have been utilized to identify enzymes involved in the biosynthesis of secondary metabolites and to understand the molecular mechanisms underlying their bioactivities. For example, proteomic profiling of marine sponges has revealed the presence of unique enzymes that catalyze the formation of complex bioactive compounds with anticancer properties (Slattery et al. 2012). Additionally, proteomics aids in the identification of protein targets of marine-derived compounds, facilitating the elucidation of their modes of action (Gamage et al. 2022).

4.4 Metabolomics

Metabolomics is the comprehensive study of metabolites in a biological system. In marine drug discovery, metabolomic approaches are used to characterize the chemical diversity of marine organisms to uncover novel chemicals with therapeutic potential. Advanced analytical techniques such as mass spectrometry (MS) and nuclear magnetic resonance (NMR) spectroscopy are used to detect and characterize metabolites (Patti et al. 2012). Metabolomics has proven essential in the discovery of new anticancer agents from marine sources, such as the identification of novel cytotoxic compounds in marine cyanobacteria (Tan 2007).

4.5 Integrated Multi-omics

The integration of multiple omics approaches—known as multi-omics—enables a comprehensive understanding of the complex biological systems in marine organisms. By combining genomics, transcriptomics, proteomics, and metabolomics data, researchers can comprehensively map biosynthetic pathways and regulatory networks involved in bioactive chemical formation (Misra et al. 2019). This integrative strategy enhances the efficiency of marine drug discovery by enabling the identification of novel compounds and their biosynthetic origins, as well as understanding their mechanisms of action (Johnson et al. 2016). For instance, multi-omics analyses have facilitated the discovery of new anticancer compounds in marine-derived fungi by correlating gene expression profiles with metabolite production (Ong et al. 2019).

Li and Zhou (2022) reported that recent multi-omics approaches—combining genomics, transcriptomics, proteomics, metabolomics, and clinical survival data—have become essential for systematically identifying biomarkers and targets of anticancer compounds, including both approved drugs and complementary therapies (e.g., traditional Chinese medicine). For instance, Li and Zhou (2022) performed an integrated pan-cancer analysis on 326 molecular targets of TCM-derived compounds—first collected from public databases (Traditional Chinese Medicine Integrated Database [TCMID], SymMap, and HIT) and filtered via ChEMBL and GDSC for bioactivity data—then evaluated them across four omics layers:

1. *Differential gene expression (DEG) analysis* (fixed dose rate [FDR] < 0.01, $|\log_2$ fold change$| \geq 1$) across 23 The Cancer Genome Atlas (TCGA) tumor cohorts (e.g., lung adenocarcinoma [LUAD], breast carcinoma [BRCA], head and neck squamous cell carcinoma [HNSC]), revealing that many TCM targets were moderately (but consistently) upregulated or downregulated in multiple cancers, suggesting "small multiple effects" rather than singular, high-fold changes seen with canonical oncogenes.

2. *Copy number variation (CNV) assessment*, which showed that TCM targets (e.g., ERBB2, MYC, EGFR, FLT4, TEK, GLI1, TERT, and TOP2A) exhibited amplifications in ~40–60% of samples in cancers such as BRCA, ESCA, bladder carcinoma (BLCA), and LUAD—comparable to, or overlapping with, known oncogenes and western drug targets.

3. *Somatic mutation frequency analysis*, identifying that several TCM targets (e.g., EGFR, MYC, LRP1B, and TTN) ranked among the top 5% most frequently mutated genes in pancreatic (PAAD), skin (SKCM), head and neck (HNSC), and breast (BRCA) cancers, similar to targets of FDA-approved anticancer agents.

4. *Survival correlation (Kaplan–Meier) analysis*, which linked high expression of TCM targets (e.g., EGFR in bladder carcinoma [BLCA] and head and neck squamous cell carcinoma [HNSC], TERT in HNSC and mesothelioma) to significantly worse overall survival ($p < 0.01$), demonstrating prognostic value.

Through this multilayered pipeline, 13 TCM-related molecules (ERBB2, MYC, FLT4, TEK, GLI1, TOP2A, PDE10A, SLC6A3, GPR55, TERT, EGFR, KCNA3, and HDAC4) emerged as high-confidence therapeutic biomarkers—being differentially expressed (Fig. 4) (Li and Zhou 2022), highly mutated, CNV-amplified, and prognostically significant across diverse cancers. This approach underscores how multi-omics integration can nominate novel or underappreciated targets, guide combination strategies (e.g., integrating TCM compounds with checkpoint inhibitors), and prioritize candidates for functional validation.

Flowchart outlines the multi omics pipeline for identifying therapeutic biomarkers of anticancer compounds, integrating genomics, transcriptomics, proteomics, and metabolomics data (Li and Zhou 2022). This figure illustrates the integrated multi-omics approach used to identify biomarkers for marine-derived and other natural anticancer compounds, as discussed in Sect. 4.5. It highlights the workflow's steps, including differential gene expression, copy number variation, mutation frequency, and survival correlation analyses, demonstrating how such strategies prioritize high-confidence therapeutic targets for precision oncology.

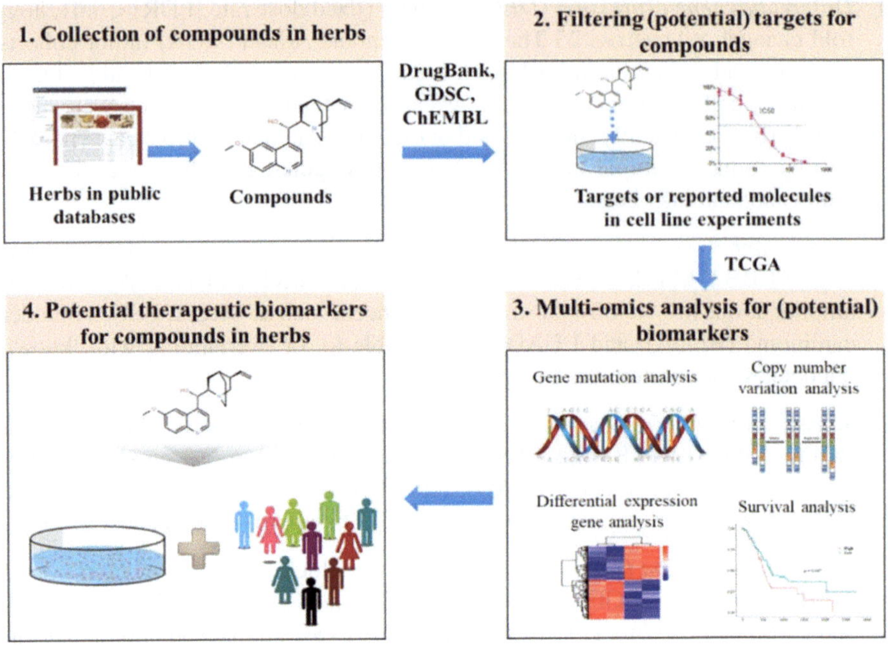

Fig. 4 Workflow for identifying potential therapeutic biomarkers for anticancer compounds in TCM via multi-omics analysis

5 Biotechnological Advances Facilitating Marine Drug Development

The progress of biotechnology has greatly sped the discovery and development of marine-derived medications. Synthetic biology, gene editing, fermentation, bioprocessing, and drug delivery methods have all made marine drug development more efficient and scalable.

5.1 Synthetic Biology

Synthetic biology allows for the design and development of novel biological pathways, which aid in the manufacture of complex marine natural products (MNPs) in heterologous hosts. By reconstructing biosynthetic gene clusters (BGCs) in model organisms like *Escherichia coli* or *Streptomyces*, researchers can produce bioactive compounds without relying on typically restricted natural sources. This approach not only ensures a sustainable supply of MNPs but also allows for the generation of novel analogs with improved pharmacological properties (Kiran et al. 2018).

5.2 CRISPR and Genome Editing

The CRISPR/Cas9 system has revolutionized genome editing, offering precise and efficient manipulation of genetic material. In marine biotechnology, CRISPR has been employed to activate silent BGCs in marine-derived microorganisms, leading to the discovery of previously unexpressed bioactive compounds. For instance, CRISPR-mediated gene editing in marine fungi has resulted in the production of novel secondary metabolites with potential anticancer activities (Jiang et al. 2013; Zhao et al. 2021; Wang et al. 2020).

5.3 Fermentation and Aquaculture

Fermentation processes are pivotal for the large-scale production of MNPs. Optimizing fermentation conditions, such as pH, temperature, and nutrient availability, can significantly enhance the yield of desired compounds (Contreras-Angulo et al. 2025). Additionally, aquaculture techniques have been developed to cultivate marine organisms like sponges and algae, providing a renewable source of bioactive compounds (Benkendorff 2009).

Alim et al. (2021) reported that the sulfated polysaccharide H3-a1 from the edible brown seaweed Rosenvingea, a species suitable for aquaculture, has demonstrated neuroprotective, antifungal (including activity against fluconazole-resistant strains), and anticancer effects against HL-60, MCF-7, and hepatocellular carcinoma cell lines, with low cytotoxicity—highlighting its potential as a sustainable candidate for antiproliferative drug development.

These methods address the challenges of limited natural availability and environmental concerns associated with wild harvesting.

Moreover, aquaculture not only provides a scalable and controlled means of producing marine bioresources but also plays a vital role in replenishing depleted populations through cultured restocking strategies, thereby contributing to the conservation and long-term sustainability of marine ecosystems.

5.4 Bioprocessing and Downstream Technologies

Efficient downstream processing is essential for the purification and formulation of marine-derived drugs. Advanced techniques, including membrane filtration, chromatography, and solvent extraction, have been optimized for marine bioproducts (Muffler and Ulber 2005). These technologies ensure the isolation of high-purity compounds, critical for pharmaceutical applications. Moreover, integrating process analytical technologies (PATs) allows for real-time monitoring and control, enhancing process efficiency and product consistency (Gerzon et al. 2022).

5.5 *Drug Delivery Innovations*

Innovative drug delivery systems have been developed to increase the bioavailability and targeted distribution of marine-derived medicines. MNPs have been encapsulated using nanoparticle-based carriers, liposomes, and hydrogels, which protect them from degradation and allow for regulated release (Chen et al. 2024). For example, chitosan-based nanoparticles have been used to deliver marine-derived anticancer agents, enhancing their therapeutic efficacy while minimizing side effects (Sachdeva et al. 2023).

Chitosan-based medications derived from marine organisms may offer a lucrative anticancer drug delivery industry. Despite its importance, the cGAS–stimulator of interferon genes (STING)-dependent immune signaling pathway has received little attention in the realm of anticancer treatment, despite the fact that it deserves more investigation. Additionally, Janus kinase/signal transducer and activator of transcription (JAK/STAT) signaling is part of several signaling transduction pathways that are mediated by various anticancer regulators, including immunity, cell death, and tumor formation (Fig. 5) (Saeed et al. 2021; Ahmad et al. 2025).

This figure shows the schematic representation of how FDA-approved marine-derived anticancer drugs modulate the JAK/STAT signaling pathway to exert therapeutic effects. This figure depicts the molecular interactions between marine-derived drugs and the JAK/STAT signaling pathway, a key regulator of cell proliferation and immune response, as discussed in Sect. 5.5. It illustrates how compounds like

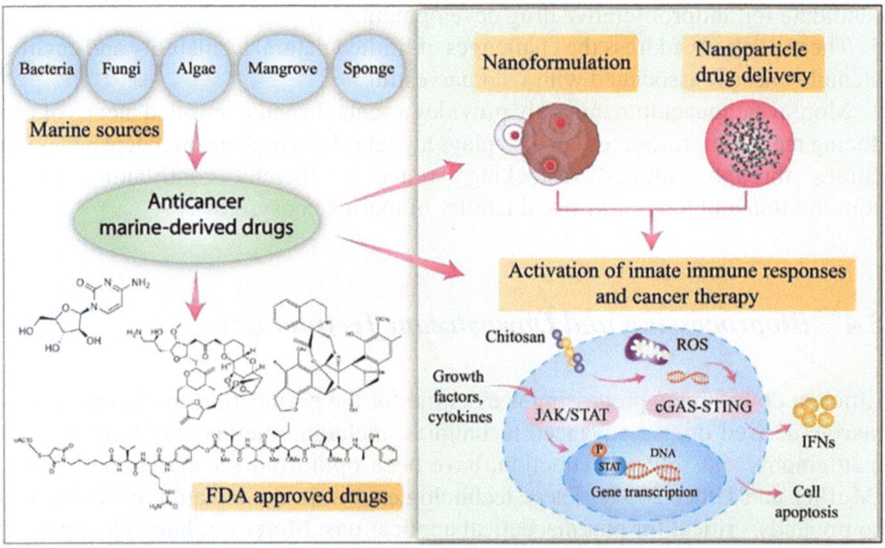

Fig. 5 Regulation of JAK/STAT signaling pathways by marine-based FDA-approved anticancer medicines

astaxanthin influence this pathway to inhibit cancer progression, providing insights into their potential as immunomodulatory and anticancer agents.

6 Case Studies

Case 1: Trabectedin (ET-743, YONDELIS)

Trabectedin is a tetrahydroisoquinoline alkaloid found in the Caribbean tunicate *Ecteinascidia turbinata* and was discovered in 1980s. The tunicate extract was found to be extremely cytotoxic after being screened by the US National Cancer Institute, and the structure of the active molecule (formerly known as ecteinascidin 743) was solved in 1984. Trabectedin binds covalently in the minor groove of DNA, forming adducts at the N^2 position of guanine and bending the DNA helix. This distortion interferes with transcription factors and DNA-binding proteins and poisons transcription-coupled nucleotide excision repair (TC-NER), ultimately generating lethal DNA strand breaks in highly transcribed genes. The C–carbinolamine subunit of trabectedin protrudes from DNA, altering interactions with nuclear proteins and even modulating tumor microenvironment cytokines, which may contribute to its antitumor effects.

After promising preclinical activity, trabectedin underwent extensive clinical trials. It showed single-agent activity in soft tissue sarcoma (STS) and ovarian cancer, leading to EU approval in 2007 for advanced STS. In a pivotal phase III study, trabectedin significantly improved progression-free survival in liposarcoma and leiomyosarcoma after anthracycline therapy. In October 2015, the FDA approved trabectedin (YONDELIS) for the treatment of unresectable or metastatic liposarcoma and leiomyosarcoma after previous chemotherapy. Trabectedin has also been studied in combination regimens; notably, a combination with pegylated liposomal doxorubicin was approved in Europe for recurrent platinum-sensitive ovarian cancer.

Production of trabectedin posed major challenges because the natural source yields only ~0.0001–0.00001% w/w of the compound. Total syntheses required dozens of steps at early stage, but a landmark semisynthetic route using the bacterial product saframycin B (cyanosafracin B) was developed by Cuevas and colleagues. This biosynthetic precursor (from *Pseudomonas fluorescens*) enabled gram-scale supply via fermentation and semi-synthesis, effectively solving the supply bottleneck. In practice, PharmaMar (Johnson & Johnson) maintains fermenters producing safracin B, which is then elaborated into trabectedin.

Today, trabectedin remains an approved drug for advanced STS in Europe, the USA, and other regions. It is used mainly after anthracycline failure in leiomyosarcoma and liposarcoma and as second-line therapy in relapsed ovarian cancer. Ongoing studies explore new indications and combinations. Its development exemplifies the path from a rare marine natural product to a marketed anticancer agent through innovative chemistry and clinical research (Cancer Network Staff 2015; D'Incalci and Galmarini 2010; Rath et al. 2011; Son et al. 2024; Cuevas and Francesch 2009).

Case 2: Salinosporamide A (Marizomib, NPI-0052)

Salinosporamide A is a γ-lactam–β-lactone natural product isolated in 2003 from *Salinispora tropica*, a seawater-inhabiting actinomycete discovered in ocean sediments. Fenical and Jensen (Scripps Institution) first reported its structure and potent activity. Salinosporamide A (also called marizomib) is a potent, irreversible inhibitor of the 20S proteasome. It covalently binds to the N-terminal Thr^1 of the β-subunits of the proteasome via opening its β-lactone ring, blocking all three proteolytic activities (chymotrypsin-like, trypsin-like, and caspase-like). Unique among proteasome inhibitors, marizomib's small non-peptidic structure allows broad-spectrum inhibition and good blood–brain barrier penetration, making it attractive for CNS tumors. In cell culture, salinosporamide causes proteasome inhibition with low nanomolar potency, leading to accumulation of ubiquitinated proteins, apoptosis, and sensitization to chemotherapies.

Salinosporamide entered clinical trials rapidly due to its potency. By 2006, it was in phase I trials for multiple myeloma and solid tumors. It advanced into phase II and III trials, especially in glioblastoma. A recent randomized phase III study in newly diagnosed glioblastoma (adding marizomib to temozolomide-based chemoradiation) showed no significant benefit in overall or progression-free survival. Other trials have explored marizomib in multiple myeloma and mantle cell lymphoma. To date, marizomib has not gained regulatory approval and remains experimental.

Large-scale production of salinosporamide is achieved by fermentation of *S. tropica*. The wild-type strain is grown in a high-salt (seawater) medium under cGMP conditions. Nereus Pharmaceuticals developed protocols to overcome corrosion in stainless steel fermenters, achieving high titers (250–300 mg/L) in 500- to 1500-L tanks. Resin adsorption is used in situ to capture the antibiotic, followed by extraction and chromatographic purification. This saline fermentation process was notable as the first manufacture of a marine microbe–derived drug by seawater fermentation. The production scale and yields are now sufficient to supply large clinical trials (multigram quantities).

Currently, salinosporamide A remains in clinical development. It is being tested in late-stage trials for certain cancers (e.g., brain tumors) and in combination regimens. Its main challenges are demonstrating clear clinical benefit and managing its potent myelosuppressive/toxic side effects (e.g., central nervous system toxicity). Interest in salinosporamide has also expanded to neurological diseases (e.g., Alzheimer's) due to its blood–brain barrier permeability and proteasome activity. Thus, marizomib is a prominent example of a marine bacterial metabolite advanced through modern fermentation and clinical development (Potts and Lam 2010; Sülzen et al. 2025; Gulder and Moore 2010; Lam 2009).

Case 3: Bryostatin-1

Bryostatin-1 is a large macrocyclic lactone first isolated in the early 1980s from the bryozoan *Bugula neritina*. In a historic harvest, Pettit's team used ~500 kg of *Bugula neritina* to isolate only 120 mg of bryostatin-1. Structure elucidation was reported in 1982 by Pettit and coworkers. Bryostatin-1 modulates protein kinase C

(PKC) signaling: It binds the regulatory C1 domain of PKC isoforms and acts as a potent agonist on short-term exposure but leads to PKC downregulation (desensitization) with prolonged exposure. This dual effect differentiates it from the tumor-promoting phorbol esters that also bind PKC. In cancer cell lines, bryostatin-1 inhibits proliferation, induces differentiation and apoptosis, and can markedly potentiate the effects of chemotherapy (in a sequence-dependent fashion). Its complex mechanism—including causing PKC isoform-specific responses—remains an active research area.

Bryostatin-1 entered clinical trials in the 1990s and early 2000s for various cancers (leukemias, melanoma, ovarian, breast, etc.) based on its preclinical promise. However, phase II trials generally showed minimal single-agent activity. In total, dozens of trials in hematologic and solid tumors showed at best modest effects and some biological activity, often accompanied by significant myalgia and other toxicities. No definitive cancer indication was established. Many investigators concluded that bryostatin-1 would need to be used only in combination with other therapies (e.g., with paclitaxel or gemcitabine) to achieve benefit.

The natural supply of bryostatin-1 was a major problem: massive quantities of bryozoan yielded only grams of compound (e.g., 14 tons of *B. neritina* yielded ~18 g of bryostatin-1). This scarcity spurred intense synthetic chemistry efforts. First syntheses (circa 1998–2008) took >50 steps, but elegant advances by Wender and others have since reduced this to ~30 steps with multigram output. Although current synthetic routes can provide gram-scale material, production remains complex. More recently, simplified bryostatin analogs ("bryologues") have been designed for easier synthesis and similar biological activity.

Today, bryostatin-1 is not approved for any cancer therapy. Its anticancer development has largely ceased due to marginal efficacy and supply difficulties. Interest in bryostatin-1 has revived in other fields: Notably, it is under investigation as a neuroregenerative agent (for Alzheimer's disease and stroke recovery) and in human immunodeficiency virus (HIV) "shock-and-kill" latency studies, capitalizing on its PKC-modulating and memory-enhancing effects. In oncology, it remains an experimental agent, exemplifying the promise and challenges of translating a potent marine natural product into a therapeutic drug (Halford 2011; Kortmansky and Schwartz 2003).

7 Challenges and Limitations in Marine-Derived Anticancer Drug Development

Marine-derived anticancer drug discovery faces significant scientific and technical hurdles. For one, marine organisms typically produce target compounds in extremely low concentrations and often in complex mixtures. Isolating a single active molecule from a marine extract is difficult, since "intricate chemistry" and "low abundance" plague these sources. Drugs from natural sources can be enhanced by

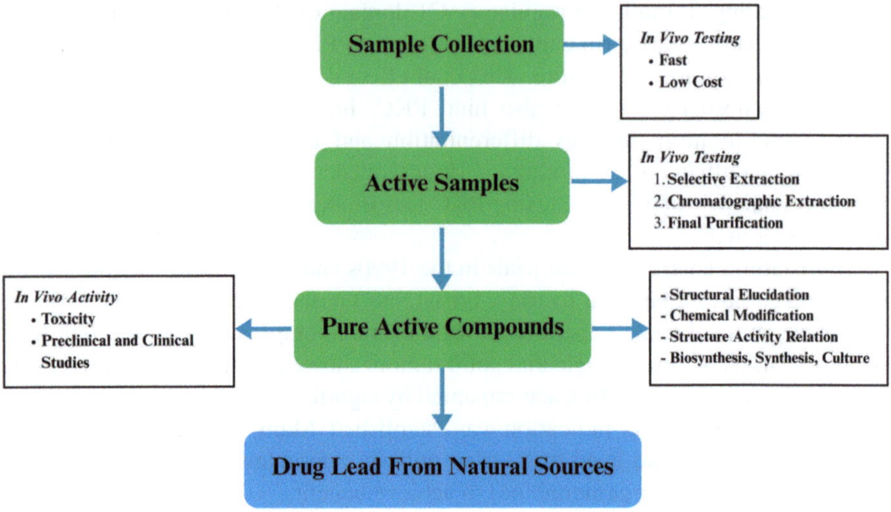

Fig. 6 Discovery and sequential extraction process for bioactive compounds from natural resources

synthetic means and have also been utilized as drug leads. Figure 6 depicts the full sequential extraction process for the source and buildup of bioactive chemicals (Ahmad et al. 2025).

The diagram illustrates the sequential extraction and isolation process for obtaining bioactive compounds from marine natural resources for drug development. This figure outlines the step-by-step process of extracting and isolating bioactive molecules from marine organisms, addressing challenges like low abundance, as discussed in Sect. 7. It highlights techniques such as solvent extraction and chromatography, emphasizing their role in overcoming supply constraints and enabling the development of marine-derived anticancer drugs.

For example, the approved drug trabectedin originally came from the tunicate *Ecteinascidia* but was present at ~0.0001% of the biomass—roughly 1 g of trabectedin required about 1 tonne of sea squirt tissue (Barreca et al. 2020). Likewise, many sponge-derived compounds have extraordinarily complex structures (numerous stereocenters, macrocycles, and halogenations) that make chemical synthesis challenging (Barreca et al. 2020; Liang et al. 2019). Liang et al. (2019) noted that such "evolutionarily optimized" metabolites are "usually quite complex and can be difficult to recreate in the laboratory."

These structural complexities, combined with only microgram to milligram sample sizes, impede both structure elucidation and biological testing. In practice, only a few potent marine drugs have made it to market by overcoming these limits: For example, trabectedin itself is now produced by semi-synthesis from a cultured bacterium, and halichondrin B (from sponges) led to the synthetic analog eribulin. But in general, the supply problem remains severe, slowing lead development and forcing reliance on laborious total syntheses or biosynthetic engineering (Barreca et al. 2020; Liang et al. 2019).

Culturing and scale-up pose another technical barrier. Most marine invertebrates (sponges, corals, and tunicates) cannot be maintained or farmed easily, so initial drug leads come from wild collection. This is not only costly but often ecologically unsustainable. One promising strategy is to use marine microbes (free-living or symbiotic) as production hosts. For instance, many sponge-associated bacteria can be grown in the laboratory and have yielded new bioactive compounds (Mehbub et al. 2024). However, these microbes themselves often require special media or growth conditions that mimic the ocean environment. As Mehbub et al. (2024) point out, researchers still grapple with "accessing harsh marine environments and maintaining laboratory conditions" for marine cultures. Thus, developing robust microbial fermentation processes or mariculture systems remains an ongoing challenge. Without such advances, even discovery of a promising compound may stall if only minute amounts can be obtained from nature or difficult laboratory cultures.

Beyond the science, economic and regulatory factors significantly constrain marine drug development. Bringing any new cancer drug to market is a multidecade, high-cost endeavor. Recent analyses estimate the R&D cost for an average new anticancer agent at roughly USD 0.9–4.5 billion (Schlander et al. 2021).

Most candidates fail: Historically, only a few percent of oncology drugs entering phase I trials ever receive approval. Clinical development itself is slow; total time from discovery to launch often spans a decade or more. For example, analysis of FDA-approved oncology drugs shows an average of ~10.5 years from phase I trial start to approval. In addition to terrestrial creatures, NPs are also being investigated in marine habitats. As a result, natural medications derived from marine sources are extremely powerful and specific in treating a number of human ailments. The US Food and Drug Administration and the European Medicines Agency (EMEA) have authorized at least eight related medications, demonstrating the substantial advancements in natural marine drug development in recent decades (Fig. 7) (Ahmad et al. 2025).

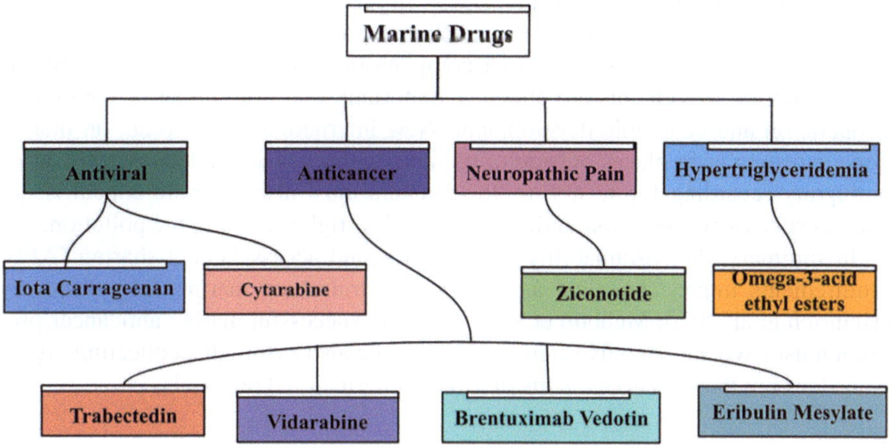

Fig. 7 FDA- or EMEA-approved marine drugs

The chart lists the marine-derived drugs approved by the FDA or EMEA for clinical use, highlighting their therapeutic applications in oncology. This figure presents a summary of marine-derived drugs, such as trabectedin and eribulin that have received regulatory approval for cancer treatment, as noted in Sect. 7. It underscores the successful translation of marine natural products into clinical therapies, illustrating their impact on addressing unmet needs in oncology and the regulatory milestones achieved.

These lengthy timelines and high attrition mean that sponsors must commit enormous resources before any return. Intellectual property considerations add further complexity: companies often patent synthetic routes, derivatives, or production methods for marine compounds (to secure exclusivity), which can impede open research collaboration. At the market stage, approved marine drugs often target rare cancers (e.g., trabectedin for soft tissue sarcoma) or are used in combination therapies, limiting patient population and sales. Price pressures and reimbursement rules can also limit uptake: Some observers have noted that high launch prices for novel anticancer drugs (including marine-derived ones) raise concerns about access and cost-effectiveness. In sum, the high cost of R&D, protracted regulatory paths, patent strategies, and narrow markets all create formidable barriers that many marine leads cannot overcome.

Finally, environmental and ethical issues constrain marine bioprospecting. Many source organisms are rare, slow-growing, or part of fragile ecosystems. Removing large quantities of sponge, coral or other invertebrates from their habitats can damage reefs and reduce biodiversity. Thus, sustainable supply is a critical concern. Experts emphasize that exploitation of marine resources must go hand-in-hand with conservation: "it is imperative that we prioritize the conservation of [sponge] habitats and protect the delicate balance" of ocean ecosystems when developing these drugs (Mehbub et al. 2024).

Legal frameworks add another layer of complexity. The Nagoya Protocol (2014) obligates researchers and companies to share benefits fairly with countries providing genetic resources (Heinrich et al. 2020). In practice, this means that any discovery of a bioactive molecule from, say, a Pacific sponge must involve agreements on royalties, technology transfer or other compensations to the source country. While benefit-sharing agreements are ethically important, they can hinder or complicate cooperation and economic development. New international talks (e.g., on marine genetic resources in places outside of national authority) are underway, generating ambiguity regarding future norms. Researchers must also examine ethical issues such as species conservation, indigenous peoples' rights, and marine pollution.

In summary, the sustainability, biodiversity, and access–benefit-sharing (ABS) requirements impose practical and moral constraints on marine drug discovery (Heinrich et al. 2020; Mehbub et al. 2024). Any successful marine anticancer program must navigate not only scientific hurdles but also ensure that collecting organisms does not threaten ecosystems or violate international protocols.

8 Future Directions and Perspectives in Marine-Derived Anticancer Research

Technological Advancements Cutting-edge technologies are transforming marine drug discovery. Machine learning and AI are being integrated with genomics and metabolomics ("multi-omics") to accelerate identification of novel compounds (Manochkumar et al. 2023).

For example, multi-platform "one strain–many compounds" (OSMAC) workflows now combine high-resolution mass spectrometry (GNPS spectral networking) with genome mining tools (AntiSMASH) to discover new analogues of bioactive peptides from marine microbes (Saide et al. 2021). Figure 8 shows the drug discovery using genome mining (Ahmad et al. 2025).

The workflow depicts the application of genome mining techniques to identify novel marine natural products with anticancer potential. This figure illustrates the genome mining process used to uncover biosynthetic gene clusters in marine microbes, as discussed in Sect. 8. It highlights how tools like AntiSMASH and GNPS spectral networking facilitate the discovery of new bioactive compounds, emphasizing the role of omics-driven approaches in accelerating marine drug discovery.

In one study, such integrated omics approaches revealed two new surugamide-A analogs from a Streptomyces strain by correlating combining HPLC-MS/MS data with biosynthetic gene clusters. Advances in synthetic biology and bioengineering further enable sustainable production of marine metabolites. Genetically engineered microbial hosts can now express complex marine biosynthetic pathways, overcoming supply constraints. Marine microorganisms (e.g., bacteria and cyanobacteria) are increasingly seen as eco-friendly "cell factories" because they grow in

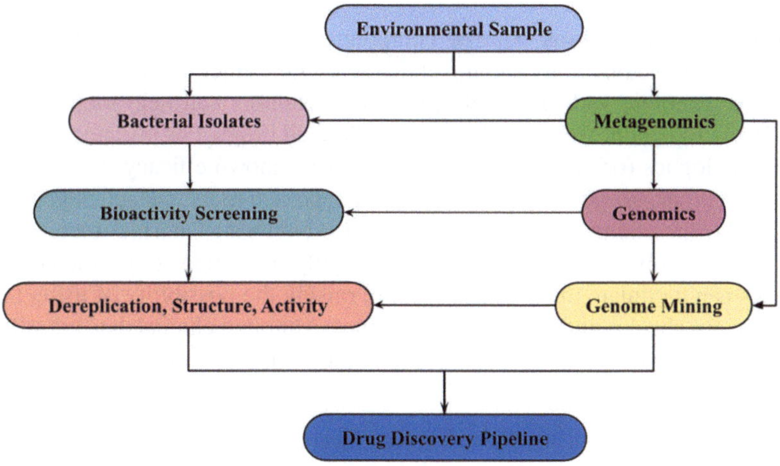

Fig. 8 Drug discovery using genome mining

controlled bioreactors, allowing scalable biomass for drug production (Saide et al. 2021).

Automation and high-throughput screening (robotic assays and microfluidics) are also being used with marine extracts to improve the speed and efficiency of lead discovery. AI-driven analytics, big-data bioinformatics, and automated processes promise to greatly expand the chemical space of marine natural products and match compounds to cancer targets more quickly (Manochkumar et al. 2023).

Policy and Regulatory Improvements International policy frameworks are evolving to balance innovation with conservation. The Nagoya Protocol (2014) on access and benefit-sharing (ABS) established compliance rules for genetic resources; today it has been ratified by ~140+ countries, with over 130 national ABS laws enacted (Covington and Burling LLP 2024).

Recent policy updates address gaps in the original treaty. In late 2024, the Convention on Biological Diversity established an agreement on digital sequence information (DSI), mandating benefit-sharing for genomic data obtained from genetic resources. Meanwhile, the 2023 UN High Seas Treaty (BBNJ) expands ABS obligations to marine genetic resources (MGRs) from areas outside state authority (Van Vooren et al. 2023; Casolari et al. 2025). This new "marine biodiversity" treaty creates the first global mechanism to share benefits (financial and non-monetary) from collecting and using high-seas MGRs (Van Vooren et al. 2023).

Compliance with these rules will require clearer guidance for researchers; as Casolari et al. (2025) noted, improved communication between policy changes and scientific practice is urgently needed. Sustainable bioprospecting is also a major focus: Both governments and industry are encouraging ethical collection (e.g., permits, quotas, and minimal-impact sampling) and development of synthetic or aquaculture-based alternatives to wild harvest. Taken together, these regulatory trends aim to ensure marine biodiversity is accessed fairly and preserved, even as new anticancer agents are sought.

Therapeutic Innovations Marine compounds are being positioned in the vanguard of precision oncology and novel therapies. One emerging area is personalized medicine: genomic and proteomic profiling of tumors can help identify patients most likely to benefit from specific marine-derived drugs. For instance, trabectedin and its analogues (originally from tunicates) have shown efficacy in translocation-driven sarcomas, suggesting a biomarker-led approach (personalized to genetic subtypes). At the same time, researchers are exploring marine metabolites in combination with immunotherapies. Marine polysaccharides (e.g., fucoidans) and peptides can modulate the immune system, raising the possibility of using them as adjuvants for checkpoint inhibitors or vaccines. Although clinical data are still early, these "immune–marine" combinations are an active research trend (Ying and Hao 2023).

Notably, several marine-derived drugs have unique mechanisms that address hard-to-treat cancers. For example, plitidepsin (from a tunicate) targets elongation factor eEF1A2 and is approved for multiple myeloma, including relapsed cases.

Antibody–drug conjugates (ADCs) built with marine cytotoxins are also yielding breakthroughs: POLIVY (CD79b–MMAE) and belantamab mafodotin (BCMA–MMAF) were FDA-approved in 2019–2020 for refractory lymphomas and myeloma. Other examples include lurbinectedin (a DNA-binding marine agent approved for resistant small-cell lung cancer) and marizomib (also known as Salinosporamide A, a proteasome inhibitor used in trials for glioma and myeloma) (Saide et al. 2021). These illustrate how marine drugs can target rare or drug-resistant tumors that evade standard therapies. Future clinical strategies will likely leverage these novel mechanisms—for instance, matching a patient's mutational profile to a marine drug's target—and combining marine agents with genomically guided regimens to improve outcomes.

Collaborative Frameworks Strong partnerships and data-sharing initiatives are fueling progress. Academia–industry collaborations and public–private partnerships are on the rise, pooling expertise in oceanography, chemistry, and pharmacology. Large consortia (often international) are forming around marine biodiscovery projects. For example, open-access databases and analytical platforms have been established to share resources: the Comprehensive Marine Natural Products Database (CMNPD) now catalogues over 31,000 marine-derived compounds with chemical, bioactivity, and source data (Lyu et al. 2021).

Other platforms like the GNPS spectral network and the Natural Products Atlas enable researchers worldwide to deposit and annotate mass spectra and structures of marine molecules. These shared infrastructures accelerate dereplication and lead prioritization. In parallel, standardized biorepositories and DNA sequence archives (e.g., SeqCode and BOLD for barcode data) are being expanded for marine specimens, facilitating genome mining and synthetic biology efforts. Such open-access frameworks democratize marine NP research: a team in one country can build on the discoveries of another (reducing duplication of effort). Sequence-based strategies, bioinformatics, and heterogeneous expression may also allow for the development of new NPs from bacterial strains that have not yet been cultivated (Fig. 9) (Ahmad et al. 2025).

The diagram shows the process of discovering natural products from uncultivated marine bacterial strains through sequence-based genome mining. This figure details the use of sequence-based strategies and bioinformatics to identify novel natural products from uncultivated marine bacteria, as described in Sect. 8. It emphasizes the potential of metagenomics and synthetic biology to access previously inaccessible biosynthetic pathways, advancing the discovery of marine-derived anticancer agents.

On the policy side, international consortia (e.g., the Global Ocean Alliance) and networks like the Marine Biodiscovery Centre (United Kingdom) or The European Marine Biological Resource Infrastructure Cluster (EMBRIC) consortium encourage coordinated research and compliance with ABS. As a result, the field is evolving into a more collaborative, global endeavor. Through cross-sector partnerships and shared data infrastructure, the marine biodiscovery community is better equipped to translate ocean-derived anticancer leads into therapies (Lyu et al. 2021; Saide et al. 2021).

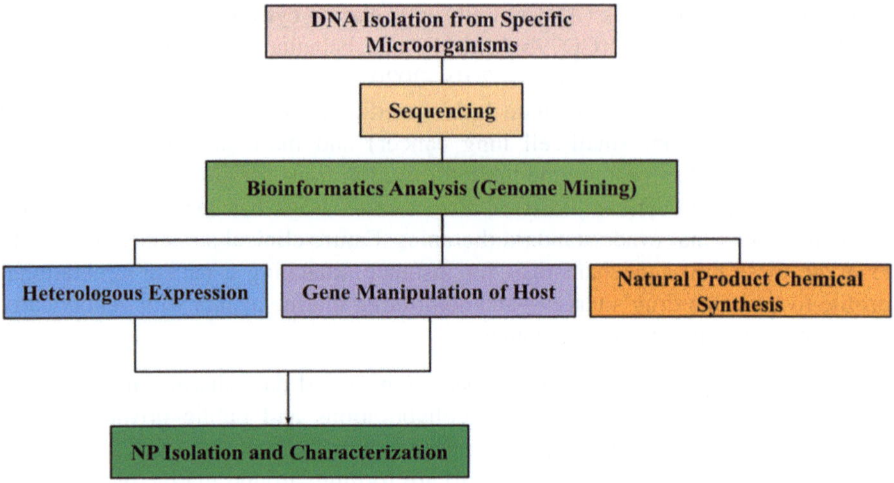

Fig. 9 Natural products discovery by genome mining of uncultivated bacterial strains

9 Conclusion

Marine ecosystems, with their unparalleled biodiversity, have proven to be rich reservoirs of structurally novel bioactive molecules that can inspire new cancer therapies (Khalifa et al. 2019; Lindequist 2016). The uniqueness of marine life—from sponges and tunicates to microalgae and deep-sea bacteria—has yielded dozens of compounds with potent cytotoxicity and novel mechanisms of action. Modern omics tools (genomics, transcriptomics, proteomics, and metabolomics) have greatly accelerated the discovery and characterization of these compounds. For example, de novo genome mining routinely uncovers cryptic biosynthetic gene clusters in marine microbes, and high-throughput metabolomics can now profile thousands of compounds from a single extract (Zuo and Kwok 2021; El-Seedi et al. 2025; Saeed et al. 2021; Barreca et al. 2020). At the same time, advances in biotechnology—such as heterologous expression of marine biosynthetic pathways and chemoenzymatic synthesis—help to overcome longstanding supply challenges. Several case studies illustrate this synergy: trabectedin, an alkaloid from the tunicate *Ecteinascidia*, became the first marine-derived anticancer agent for soft-tissue sarcoma (Europe, 2007) and is now produced semi-synthetically (Zuo and Kwok 2021; El-Seedi et al. 2025; Saeed et al. 2021); eribulin (HALAVEN), a synthetic analogue of a sponge-derived polyether, is approved for breast cancer; and antibody–drug conjugates use marine peptides (dolastatins) to target hematologic malignancies. These examples underscore that marine natural products provide new chemical scaffolds that can evade common drug resistance mechanisms. Importantly, only a handful of marine-derived drugs have reached the clinic to date—a limitation often attributed to the difficulty of accessing and sustainably sourcing marine organisms (Lindequist 2016; Zuo and Kwok 2021; El-Seedi et al. 2025; Saeed et al. 2021). Challenges such as low abundance of active metabolites and complex synthesis have constrained throughput, but they also motivate innovation in cultivation, synthetic biology, and precision chemistry (Barreca et al. 2020; Agarwal et al. 2024).

The broader implications of marine-derived oncology agents are profound. In the era of precision medicine, these novel natural compounds expand the armamentarium against cancer subtypes that have become resistant to conventional therapies. By providing unique modes of action, marine drugs can fill gaps left by terrestrial libraries. Moreover, marine biotechnology is now explicitly linked to sustainable development goals: Policymakers view marine bioprospecting as a pillar of the emerging bioeconomy (Zuo and Kwok 2021; El-Seedi et al. 2025; Saeed et al. 2021) and an avenue for sustainable drug discovery. For instance, recent studies highlight the vast patenting of deep-sea marine genes by major companies, signaling both economic interest and the need for stewardship of marine resources (Albarano et al. 2020). These efforts are conceptually aligned with "health-omics" initiatives that integrate genomic and metabolomic data with clinical insights. In practical terms, multi-omics analyses of marine organisms can reveal bioactive pathways that are relevant to human health and may even guide the personalization of therapies (Barreca et al. 2020). Thus, by bridging ocean science and human health, marine-derived drug research exemplifies a systems-level approach to oncology in which the environment, the organism, and the patient are considered together.

Looking forward, sustained progress will depend on interdisciplinary and sustainable strategies. First, continued exploration of marine biodiversity is essential: Untapped niches such as deep-sea vents, polar waters, and microbial consortia likely harbor many more lead compounds. These efforts should be guided by responsible bioprospecting frameworks—for example, global patent analyses suggest prioritizing microbial and deep-sea genomes while ensuring benefit-sharing and ecosystem protection (Ziemert et al. 2016). Second, integration of omics technologies with artificial intelligence and data science will be transformative. Machine learning models applied to large marine metabolomic and genomic datasets can predict bioactivity and streamline candidate selection, reducing costs and accelerating discovery (Agarwal et al. 2024). Third, biotechnology must advance sustainable production methods: Engineering microbial cell factories or using chemoenzymatic synthesis can decouple drug supply from wild harvesting, addressing the "supply problem" noted by Lindequist (2016). In fact, combining biosynthetic enzymes with modern organic chemistry has already simplified the synthesis of complex marine molecules (Barreca et al. 2020; Agarwal et al. 2024).

Finally, the chapter's themes point to a vision of health that embraces ocean resources in a sustainable, patient-centric manner. Marine-derived drugs not only contribute new treatments but also inspire the concept of healthomics, where holistic data from the environment and omics layers inform cancer prevention and therapy. To achieve this vision, future research should foster collaborations among marine biologists, chemists, bioinformaticians, and clinicians. Cross-sector partnerships (academic, industry, and government) can help translate marine bioactives into approved therapeutics while adhering to conservation and ethical standards. By continuing to leverage marine biodiversity through cutting-edge omics and biotechnology, the field is poised to deliver innovative, sustainable anticancer agents and to chart new frontiers at the intersection of oceanography and human health.

References

Agarwal N, Ahmad S, Singh V (2024) Artificial intelligence and machine learning for analysis of multi-omics. In: Mani I, Singh V (eds) Multi-omics analysis of the human microbiome. Springer, Singapore, pp 1–20. https://doi.org/10.1007/978-981-97-1844-3_16

Ahmad M, Li H, Zhang Y, Wang J, Chen X (2025) Plant and marine-derived natural products: sustainable pathways for future drug discovery and therapeutic development. Front Pharmacol 15:1497668. https://doi.org/10.3389/fphar.2024.1497668

Albarano L, Li H, Zhang Y, Wang J, Chen X (2020) Genome mining as a new challenge in natural products discovery. Mar Drugs 18(4):199. https://doi.org/10.3390/md18040199

Alim MA, Li H, Zhang Y, Wang J, Chen X (2021) Phytochemical profiling, antioxidant potentiality and antibacterial activity of the ethanolic extracts of Rosenvingea sp. of Bay of Bengal. J Bangladesh Acad Sci 45(1):59–71. https://doi.org/10.3329/jbas.v45i1.54260

Alves C, Diederich M (2021) Marine natural products as anticancer agents. Mar Drugs 19(8):447. https://doi.org/10.3390/md19080447

Baltz RH (2017) Gifted microbes for genome mining and natural product discovery. J Ind Microbiol Biotechnol 44(4–5):573–588. https://doi.org/10.1007/s10295-016-1815-x

Barreca M, Li H, Zhang Y, Wang J, Chen X (2020) Marine anticancer agents: an overview with a particular focus on their chemical classes. Mar Drugs 18(12):619. https://doi.org/10.3390/md18120619

Bayona LM, de Voogd NJ, Choi YH (2022) Metabolomics on the study of marine organisms. Metabolomics 18(3):17. https://doi.org/10.1007/s11306-022-01874-y

Bekiari M (2023) In: Garcia MDG, Cortês A (eds) Marine bioprospecting: understanding the activity and some challenges related to environmental protection, scientific research, ethics, and the law. Blue Planet Law. Springer, Cham, pp 237–252. https://doi.org/10.1007/978-3-031-24888-7_19

Benkendorff K (2009) Aquaculture and the production of pharmaceuticals and nutraceuticals. In: Burnell G, Allan G (eds) New technologies in aquaculture: improving production efficiency, quality and environmental management. Woodhead Publishing, Cambridge, pp 866–891. https://doi.org/10.1533/9781845696474.5.866

Bertrand S, Li H, Zhang Y, Wang J, Chen X (2014) Metabolite induction via microorganism co-culture: a potential way to enhance chemical diversity for drug discovery. Biotechnol Adv 32(6):1180–1204. https://doi.org/10.1016/j.biotechadv.2014.03.001

Blunt JW, Li H, Zhang Y, Wang J, Chen X (2015) Marine natural products. Nat Prod Rep 32(2):160–258. https://doi.org/10.1039/C4NP00144C

Bray F, Ferlay J, Soerjomataram I, Siegel RL, Torre LA, Jemal A (2018) Global cancer statistics 2018: GLOBOCAN estimates of incidence and mortality worldwide for 36 cancers in 185 countries. CA Cancer J Clin 68(6):394–424. https://doi.org/10.3322/caac.21492

Britton R, Li H, Zhang Y, Wang J, Chen X (2001) Antimitotic diterpenoids from Erythropodium caribaeorum: isolation artifacts and putative biosynthetic intermediates. Tetrahedron Lett 42(16):2953–2956. https://doi.org/10.1016/S0040-4039(01)00347-1

Cancer Network Staff (2015) FDA approves trabectedin (Yondelis) for advanced soft-tissue sarcoma. Cancer Network. Available from: https://www.cancernetwork.com/view/fda-approves-trabectedin-yondelis-advanced-soft-tissue-sarcoma

Casolari F, Li H, Zhang Y, Wang J, Chen X (2025) Unpacking policy developments in marine natural product research: a scientist's guide to DSI and BBNJ. Nat Prod Rep. https://doi.org/10.1039/d4np00070f

Chakraborty C, Li H, Zhang Y, Wang J, Chen X (2009) Anticancer drugs discovery and development from marine organisms. Curr Top Med Chem 9(16):1536–1545. https://doi.org/10.2174/156802609789909803

Chen Q, Li H, Zhang Y, Wang J, Chen X (2024) Novel drug delivery systems: an important direction for drug innovation research and development. Pharmaceutics 16(5):674. https://doi.org/10.3390/pharmaceutics16050674

Chowdhury KR, Li H, Zhang Y, Wang J, Chen X (2023) Screening of anti-inflammatory and analgesic activities of Caulerpa racemosa from the Bay of Bengal, Bangladesh. Biores Commun 9(2):1–10. https://doi.org/10.3329/brc.v9i2.67090

Contreras-Angulo LA, Li H, Zhang Y, Wang J, Chen X (2025) Bioactive compounds extraction from marine microorganisms by fermentation. In: Heredia JB, Gutiérrez-Grijalva EP, Cabanillas-Bojórquez LA (eds) Bioactive compounds extraction from marine resources and wastes. Springer, Singapore, pp 151–174. https://doi.org/10.1007/978-981-96-1253-6_7

Covington & Burling LLP (2024) The Nagoya protocol at its 10th anniversary: lessons learned and new challenges from 'access and benefit-sharing'. Covington & Burling LLP. Available from: https://www.cov.com/en/news-and-insights/insights/2024/10/the-nagoya-protocol-at-its-10th-anniversary-lessons-learned-and-new-challenges-from-access-and-benefit-sharing

Crüsemann M, Li H, Zhang Y, Wang J, Chen X (2017) Prioritizing natural product diversity in a collection of 146 bacterial strains based on growth and extraction protocols. J Nat Prod 80(3):588–597. https://doi.org/10.1021/acs.jnatprod.6b00722

Cuevas C, Francesch A (2009) Development of Yondelis (trabectedin, ET-743). A semisynthetic process solves the supply problem. Nat Prod Rep 26(3):322–337. https://doi.org/10.1039/b808331m

D'Incalci M, Galmarini CM (2010) A review of trabectedin (ET-743): a unique mechanism of action. Mol Cancer Ther 9(8):2157–2163. https://doi.org/10.1158/1535-7163.mct-10-0263

Dalisay DS, Li H, Zhang Y, Wang J, Chen X (2024) Marine-derived anticancer agents targeting apoptotic pathways: exploring the depths for novel cancer therapies. Mar Drugs 22(3):114. https://doi.org/10.3390/md22030114

Deshmukh SK, Prakash V, Ranjan N (2018) Marine fungi: a source of potential anticancer compounds. Front Microbiol 8:2536. https://doi.org/10.3389/fmicb.2017.02536

Dybdal-Hargreaves NF, Risinger AL, Mooberry SL (2015) Eribulin mesylate: mechanism of action of a unique microtubule-targeting agent. Clin Cancer Res 21(11):2445–2452. https://doi.org/10.1158/1078-0432.CCR-14-3252

El-Seedi HR, Li H, Zhang Y, Wang J, Chen X (2025) Marine natural products as a source of novel anticancer drugs: an updated review (2019–2023). Nat Prod Bioprospect 15:13. https://doi.org/10.1007/s13659-024-00493-5

Faulkner DJ (2002) Marine natural products. Nat Prod Rep 19(1):1–49. https://doi.org/10.1039/B009029H

Feling RH, Li H, Zhang Y, Wang J, Chen X (2003) Salinosporamide A: a highly cytotoxic proteasome inhibitor from a novel microbial source, a marine bacterium of the new genus Salinospora. Angew Chem Int Ed 42(3):355–357. https://doi.org/10.1002/anie.200390115

Fenical W, Jensen PR (2006) Developing a new resource for drug discovery: marine actinomycete bacteria. Nat Chem Biol 2(12):666–673. https://doi.org/10.1038/nchembio841

Ferrandina G, Li H, Zhang Y, Wang J, Chen X (2013) Trabectedin as single agent in the salvage treatment of heavily treated ovarian cancer patients: a retrospective, multicenter study. Gynecol Oncol 130(3):505–510. https://doi.org/10.1016/j.ygyno.2013.06.025

Finore I, Poli A, Rai M et al (2014) Fermentation technologies for the optimization of marine microbial exopolysaccharide production. Mar Drugs 12(5):3005–3024. https://doi.org/10.3390/md12053005

Flemsæter F (2020) Regulating marine bioprospecting: exploring the establishment of new regulatory regimes in the blue bioeconomy. Ocean Coastal Manag 194:105207. https://doi.org/10.1016/j.ocecoaman.2020.105207

Francesch AM, Li H, Zhang Y, Wang J, Chen X (2024) Pharmaceuticals from marine sources: past, present and future. Pharm J Available from: https://pharmaceutical-journal.com/article/research/pharmaceuticals-from-marine-sources-past-present-and-future

Gamage NTG, Li H, Zhang Y, Wang J, Chen X (2022) Proteomic applications in aquatic environment studies. Proteomes 10(3):32. https://doi.org/10.3390/proteomes10030032

Gangwal A, Lavecchia A (2025) Artificial intelligence in natural product drug discovery: current applications and future perspectives. J Med Chem 68(4). https://doi.org/10.1021/acs.jmedchem.4c01257

Gerzon G, Sheng Y, Kirkitadze M (2022) Process analytical technologies—advances in bioprocess integration and future perspectives. J Pharm Biomed Anal 207:114379. https://doi.org/10.1016/j.jpba.2021.114379
Groll M, Berkers CR, Ploegh HL, Ovaa H (2006) Crystal structure of the boronic acid-based proteasome inhibitor bortezomib in complex with the yeast 20S proteasome. Structure 14(3):451–456. https://doi.org/10.1016/j.str.2005.11.019
Gulder TAM, Moore BS (2010) Salinosporamide natural products: potent 20S proteasome inhibitors as promising cancer chemotherapeutics. Angew Chem Int Ed 49(49):9346–9367. https://doi.org/10.1002/anie.201000728
Gunasekera SP, Longley RE, Pomponi SA, Wright AE (1990) Discodermolide: a polyketide natural product that stabilizes microtubules. J Org Chem 55(19):4912–4915. https://doi.org/10.1021/jo00303a029
Hackett JD, Anderson DM, Erdner D, Bhattacharya D (2013) Evolution of saxitoxin synthesis in cyanobacteria and dinoflagellates. Mol Biol Evol 30(1):70–78. https://doi.org/10.1093/molbev/mss142
Haefner B (2003) Drugs from the deep: marine natural products as drug candidates. Drug Discov Today 8(12):536–544. https://doi.org/10.1016/s1359-6446(03)02713-2
Halford B (2011) The Bryostatins' tale: with the promise of treating cancer, Alzheimer's, and HIV, this family of marine natural products continues to intrigue scientists more than four decades after its discovery. Chem Eng News Available from: https://cen.acs.org/articles/89/i43/BryostatinsTale.html
Harvey AL, Edrada-Ebel R, Quinn RJ (2015) The re-emergence of natural products for drug discovery in the genomics era. Nat Rev Drug Discov 14(2):111–129. https://doi.org/10.1038/nrd4510
Heinrich M, Molnár A, Gertsch J et al (2020) Access and benefit sharing under the Nagoya protocol—quo vadis? Six Latin American case studies assessing opportunities and risk. Front Pharmacol 11:765. https://doi.org/10.3389/fphar.2020.00765
Hong J (2011) Natural product diversity and its role in chemical biology and drug discovery. Curr Opin Chem Biol 15(3):350–354. https://doi.org/10.1016/j.cbpa.2011.03.004
Jaspars M, Edvardsen K, Andersen JH et al (2016) The marine biodiscovery pipeline and ocean medicines of tomorrow. J Mar Biol Assoc U K 96(1):151–158. https://doi.org/10.1017/S0025315415002106
Jayaraman M, Li H, Zhang Y, Wang J, Chen X (2024) Computational insights into potential marine natural products as selective inhibitors of Mycobacterium tuberculosis InhA: a structure-based virtual screening study. Comput Biol Chem 108:107991. https://doi.org/10.1016/j.compbiolchem.2023.107991
Jensen PR, Moore BS, Fenical W (2015) The marine actinomycete genus Salinispora: a model organism for secondary metabolite discovery. Nat Prod Rep 32(5):738–751. https://doi.org/10.1039/c4np00167b
Jiang W, Bikard D, Cox D, Zhang F, Marraffini LA (2013) RNA-guided editing of bacterial genomes using CRISPR-Cas systems. Nat Biotechnol 31(3):233–239. https://doi.org/10.1038/nbt.2508
Jimenez PC, Li H, Zhang Y, Wang J, Chen X (2019) Enriching cancer pharmacology with drugs of marine origin. Br J Pharmacol 177(1):3–27. https://doi.org/10.1111/bph.14876
Johnson CH, Ivanisevic J, Siuzdak G (2016) Metabolomics: beyond biomarkers and towards mechanisms. Nat Rev Mol Cell Biol 17(7):451–459. https://doi.org/10.1038/nrm.2016.25
Jordan K, Forster M, Markman B et al (2015) Trabectedin: supportive care strategies and safety profile. Crit Rev Oncol Hematol 94(3):279–290. https://doi.org/10.1016/j.critrevonc.2015.02.012
Keasling JD (2012) Synthetic biology and the development of tools for metabolic engineering. Metab Eng 14(3):189–195. https://doi.org/10.1016/j.ymben.2012.01.004
Khafoor AA, Karim AS, Sajadi SM (2023) Recent progress in synthesis of nano-based liposomal drug delivery systems: a glance at their medicinal applications. Results Surf Interfaces 11:100124. https://doi.org/10.1016/j.rsurfi.2023.100124
Khalifa SAM, Li H, Zhang Y, Wang J, Chen X (2019) Marine natural products: a source of novel anticancer drugs. Mar Drugs 17(9):491. https://doi.org/10.3390/md17090491

Kiran GS, Manilal A, Selvin J et al (2018) Synthetic biology approaches: towards sustainable exploitation of marine bioactive molecules. Int J Biol Macromol 112:1278–1288. https://doi.org/10.1016/j.ijbiomac.2018.01.149

König GM, Wright AD, Lindel T et al (2006) Natural products from marine organisms and their associated microbes. Chembiochem 7(2):229–238. https://doi.org/10.1002/cbic.200500087

Kortmansky J, Schwartz GK (2003) Bryostatin-1: a novel PKC inhibitor in clinical development. Cancer Investig 21(6):924–936. https://doi.org/10.1081/cnv-120025095

Lam R (2009) Large scale saline fermentation process for providing clinical supply of proteasome inhibitor NPI-0052. RAFT VIII Conference Proceedings. Available from: https://sim.confex.com/sim/raft8/techprogram/P12955.HTM

Li X, Qin L (2005) Metagenomics-based drug discovery and marine microbial diversity. Trends Biotechnol 23(11):539–543. https://doi.org/10.1016/j.tibtech.2005.08.006

Li R, Zhou W (2022) Multi-omics analysis to screen potential therapeutic biomarkers for anti-cancer compounds. Heliyon 8:e09616. https://doi.org/10.1016/j.heliyon.2022.e09616

Liang X, Luo D, Luesch H (2019) Advances in exploring the therapeutic potential of marine natural products. Pharmacol Res 147:104373. https://doi.org/10.1016/j.phrs.2019.104373

Lindequist U (2016) Marine-derived pharmaceuticals—challenges and opportunities. Biomol Ther 24(6):561–571. https://doi.org/10.4062/biomolther.2016.181

Lyu C, Li H, Zhang Y et al (2021) CMNPD: a comprehensive marine natural products database towards facilitating drug discovery from the ocean. Nucleic Acids Res 49(D1):D509–D515. https://doi.org/10.1093/nar/gkaa763

Manochkumar J, Li H, Zhang Y et al (2023) A critical review of machine-learning for "multi-omics" marine metabolite datasets. Comput Biol Med 165:107425. https://doi.org/10.1016/j.compbiomed.2023.107425

Medema MH, Fischbach MA (2015) Computational approaches to natural product discovery. Nat Chem Biol 11(9):639–648. https://doi.org/10.1038/nchembio.1884

Mehbub MF, Lei J, Franco C, Zhang W (2024) Marine sponge-derived natural products: trends and opportunities for the decade of 2011–2020. Front Mar Sci 11:1462825. https://doi.org/10.3389/fmars.2024.1462825

Misra BB, Langefeld CD, Olivier M, Cox LA (2019) Integrated omics: tools, advances, and future approaches. J Mol Endocrinol 62(1):R21–R45. https://doi.org/10.1530/JME-18-0055

Molinski TF, Dalisay DS, Lievens SL, Saludes JP (2008) Drug development from marine natural products. Nat Rev Drug Discov 8(1):69–85. https://doi.org/10.1038/nrd2487

Muffler K, Ulber R (2005) Downstream processing in marine biotechnology. In: Ulber R, Le Gal Y (eds) Marine biotechnology II. Springer, Berlin, pp 63–103. https://doi.org/10.1007/b135823

Nagle DG, Paul VJ, Mooberry SL et al (1995) Absolute configuration of curacin a, a novel anti-mitotic agent from the tropical marine cyanobacterium Lyngbya majuscula. Tetrahedron Lett 36(8):1189–1192. https://doi.org/10.1016/0040-4039(95)00030-G

Newman DJ, Cragg GM (2020) Natural products as sources of new drugs over the nearly four decades from 01/1981 to 09/2019. J Nat Prod 83(3):770–803. https://doi.org/10.1021/acs.jnatprod.9b01285

Nobili S, Lippi D, Witort E et al (2009) Natural compounds for cancer treatment and prevention. Pharmacol Res 59(6):365–378. https://doi.org/10.1016/j.phrs.2009.01.017

Ong JFM, Chua KH, Yap JKY et al (2019) Integrated genomic and metabolomic approach to the discovery of potential anti-quorum sensing natural products from microbes associated with marine samples from Singapore. Mar Drugs 17(1):72. https://doi.org/10.3390/md17010072

Paterson I, Anderson EA (2005) The renaissance of natural products as drug candidates. Science 310(5747):451–453. https://doi.org/10.1126/science.1116364

Patti GJ, Yanes O, Siuzdak G (2012) Metabolomics: the apogee of the omics trilogy. Nat Rev Mol Cell Biol 13(4):263–269. https://doi.org/10.1038/nrm3314

Peña DA, Lin-Cereghino J, Lin-Cereghino GP (2018) Metabolic engineering of Pichia pastoris. Metab Eng 50:2–15. https://doi.org/10.1016/j.ymben.2018.04.017

Piel J (2011) Approaches to capturing and designing biologically active small molecules produced by uncultured microbes. Ann Rev Microbiol 65:431–453. https://doi.org/10.1146/annurev-micro-090110-102805

Piel J, Hui DQ, Wen G et al (2004) Targeting modular polyketide synthases with iteratively acting acyltransferases from metagenomes of uncultured bacterial consortia. Environ Microbiol 6(9):921–927. https://doi.org/10.1111/j.1462-2920.2004.00531.x

Potts BC, Lam KS (2010) Generating a generation of proteasome inhibitors: from microbial fermentation to total synthesis of salinosporamide a (marizomib) and other salinosporamides. Mar Drugs 8(4):835–880. https://doi.org/10.3390/md8040835

Potts BC, Ding J, Gustafson KR et al (2011) Marizomib, a proteasome inhibitor for all seasons: preclinical profile and a framework for clinical trials. Curr Cancer Drug Targets 11(3):254–284. https://doi.org/10.2174/156800911794519716

Qin J, Li J, Wang Q et al (2015) Modular pathway rewiring of Saccharomyces cerevisiae enables high-level production of L-ornithine. Nat Commun 6:8224. https://doi.org/10.1038/ncomms9224

Rath CM, Akee R, Reynolds KA (2011) Meta-omic characterization of the marine invertebrate microbial consortium that produces the chemotherapeutic natural product ET-743. ACS Chem Biol 6(11):1244–1256. https://doi.org/10.1021/cb200244t

Ruiz-Torres V, Gamboa-Suasnavart L, Morin E et al (2017) An updated review on marine anticancer compounds: the use of virtual screening for the discovery of small-molecule cancer drugs. Molecules 22(7):1037. https://doi.org/10.3390/molecules22071037

Sachdeva B, Chouhan D, Gupta A et al (2023) Chitosan nanoparticles-based cancer drug delivery: application and challenges. Mar Drugs 21(4):211. https://doi.org/10.3390/md21040211

Saeed AFUH, Su J, Ouyang S (2021) Marine-derived drugs: recent advances in cancer therapy and immune signaling. Biomed Pharmacother 134:111091. https://doi.org/10.1016/j.biopha.2020.111091

Saide A, Molinski TF, Dalisay DS (2021) Promising activities of marine natural products against hematopoietic malignancies. Biomedicine 9(6):645. https://doi.org/10.3390/biomedicines9060645

Sankarapandian V, Ransangan J, Venmathi Maran BA (2023) Marine biotechnology and its applications in drug discovery. In: Shah MD, Ransangan J, Venmathi Maran BA (eds) Marine biotechnology: applications in food, drugs and energy. Springer, Singapore, pp 1–20. https://doi.org/10.1007/978-981-99-0624-6_9

Santhiravel S, Dave D, Shahidi F (2025) Bioactives from marine resources as natural health products: a review. Pharmacol Rev 77(1):100006. https://doi.org/10.1124/pharmrev.123.001227

Schlander M et al (2021) How much does it cost to research and develop a new drug? A systematic review and assessment. PharmacoEconomics 39(8):1243–1269. https://doi.org/10.1007/s40273-021-01065-y

Schmidt EW, Nelson JT, Rasko DA et al (2005) Patellamide A and C biosynthesis by a microcin-like pathway in Prochloron didemni, the cyanobacterial symbiont of Lissoclinum patella. Proc Natl Acad Sci 102(20):7315–7320. https://doi.org/10.1073/pnas.0501424102

Schwartz RE, Reyes GR, Rossi DJ (1990) Pharmaceuticals from cultured algae. J Ind Microbiol 5(2–3):113–123. https://doi.org/10.1007/BF01573860

Seyed MA, Ayesha S (2021) Marine-derived pipeline anticancer natural products: a review of their pharmacotherapeutic potential and molecular mechanisms. Future J Pharm Sci 7:203. https://doi.org/10.1186/s43094-021-00350-z

Siegel RL, Miller KD, Jemal A (2020) Cancer statistics, 2020. CA Cancer J Clin 70(1):7–30. https://doi.org/10.3322/caac.21590

Slattery M, Bock T, Clarke C et al (2012) Marine proteomics: a critical assessment of an emerging technology. J Nat Prod 75(10):1833–1877. https://doi.org/10.1021/np300366a

Son K, Li H, Zhang Y et al (2024) Trabectedin derails transcription-coupled nucleotide excision repair to induce DNA breaks in highly transcribed genes. Nat Commun 15:1388. https://doi.org/10.1038/s41467-024-45664-7

Srivastava P, Sahgal M, Dasila H (2023) Microbial symbiosis in marine ecosystem. In: Soni R, Suyal DC, Morales-Oyervides L, Fouillaud M (eds) Current status of marine water microbiology. Springer, Singapore, pp 1–20. https://doi.org/10.1007/978-981-99-5022-5_3

Stanwell C, Woodburn JR, Powis G et al (1994) The role of protein kinase C isoenzymes in the growth inhibition caused by bryostatin 1 in human A549 lung and MCF-7 breast carcinoma cells. Int J Cancer 56(4):585–592. https://doi.org/10.1002/ijc.2910560420

Sülzen H, Li H, Zhang Y et al (2025) Structural insights into Salinosporamide A mediated inhibition of the human 20S proteasome. bioRxiv. https://doi.org/10.1101/2025.01.28.635221

Sung H, Ferlay J, Siegel RL et al (2021) Global cancer statistics 2020: GLOBOCAN estimates of incidence and mortality worldwide for 36 cancers in 185 countries. CA Cancer J Clin 71(3):209–249. https://doi.org/10.3322/caac.21660

Tan LT (2007) Bioactive natural products from marine cyanobacteria for drug discovery. Phytochemistry 68(7):954–979. https://doi.org/10.1016/j.phytochem.2007.01.012

Thomas TRA, Kaulekar DP, Lokabarathi PA (2010) Marine drugs from sponge-microbe association: a review. Mar Drugs 8(4):1417–1468. https://doi.org/10.3390/md8041417

ul Hassan SS, Jin H (2025) Harnessing marine bacteria for next-gen antibiotics: potent inhibition of S. aureus and Riemerella anatipestifer through in vitro, omics, and chemoinformatics approach with enhanced production of secondary metabolites through zinc sulphate. Front Mar Sci 11:1461607. https://doi.org/10.3389/fmars.2024.1461607

van der Hooft JJJ, Ridder L, de Vos RC et al (2020) Linking genomics and metabolomics to chart specialized metabolic diversity. Chem Soc Rev 49(11):3296–3314. https://doi.org/10.1039/D0CS00162G

Van Vooren B, Gevrenova Y, Finan S (2023) Historic marine biodiversity treaty creates new access and benefit-sharing obligations for life sciences companies. Inside EU Life Sciences. Available from: https://www.insideeulifesciences.com/2023/06/21/historic-marine-biodiversity-treaty-creates-new-access-and-benefit-sharing-obligations-for-life-sciences-companies/

von Schwarzenberg K, Vollmar AM (2013) Targeting apoptosis pathways by natural compounds in cancer: Marine compounds as lead structures and chemical tools for cancer therapy. Cancer Lett 332(2):295–303. https://doi.org/10.1016/j.canlet.2010.07.004

Walsh CT, Fischbach MA (2010) Natural products version 2.0: connecting genes to molecules. J Am Chem Soc 132(8):350–354. https://doi.org/10.1021/ja909118a

Wang E, Wu X, Li Y et al (2020) From seabed to bedside: a review on promising marine anticancer compounds. Biomolecules 10(2):248. https://doi.org/10.3390/biom10020248

Wilson MC, Mori T, Rückert C et al (2014) An environmental bacterial taxon with a large and distinct metabolic repertoire. Nature 506(7486):58–62. https://doi.org/10.1038/nature12959

Ying Y, Hao W (2023) Immunomodulatory function and anti-tumor mechanism of natural polysaccharides: a review. Front Immunol 14:1147641. https://doi.org/10.3389/fimmu.2023.1147641

Zhao D, van der Lee R, Li S et al (2021) CRISPR-based metabolic pathway engineering. Metab Eng 63:148–159. https://doi.org/10.1016/j.ymben.2020.10.004

Ziemert N, Alanjary M, Weber T (2016) The evolution of genome mining in microbes—a review. Nat Prod Rep 33(8):988–1005. https://doi.org/10.1039/c6np00025h

Zuo W, Kwok HF (2021) Development of marine-derived compounds for cancer therapy. Mar Drugs 19(6):342. https://doi.org/10.3390/md19060342

MIX
Papier aus verantwortungsvollen Quellen
Paper from responsible sources
FSC® C105338

FSC
www.fsc.org

If you have any concerns about our products,
you can contact us on
ProductSafety@springernature.com

In case Publisher is established outside the EU,
the EU authorized representative is:
Springer Nature Customer Service Center GmbH
Europaplatz 3, 69115 Heidelberg, Germany

Printed by Libri Plureos GmbH
in Hamburg, Germany